T0190477

Lecture Notes in Computer Science 12693

More information about this subseries at http://www.springer.com/series/7407

Juan Romero · Tiago Martins ·
Nereida Rodríguez-Fernández (Eds.)

Artificial Intelligence in Music, Sound, Art and Design

10th International Conference, EvoMUSART 2021
Held as Part of EvoStar 2021
Virtual Event, April 7–9, 2021
Proceedings

Editors
Juan Romero
University of A Coruña
A Coruña, Spain

Tiago Martins
University of Coimbra
Coimbra, Portugal

Nereida Rodríguez-Fernández
University of A Coruña
A Coruña, Spain

ISSN 0302-9743 ISSN 1611-3349 (electronic)
Lecture Notes in Computer Science
ISBN 978-3-030-72913-4 ISBN 978-3-030-72914-1 (eBook)
https://doi.org/10.1007/978-3-030-72914-1

LNCS Sublibrary: SL1 – Theoretical Computer Science and General Issues

This Springer imprint is published by the registered company Springer Nature Switzerland AG
The registered company address is: Gewerbestrasse 11, 6330 Cham, Switzerland

Preface

EvoMUSART 2021—the 10th International Conference on Artificial Intelligence in Music, Sound, Art and Design—took place from 7 to 9 April 2021, in Seville, Spain, as part of Evo*, the leading European Event on Bio-Inspired Computation.

Following the success of previous events and the importance of the field of Artificial Intelligence, specifically, evolutionary and biologically inspired (artificial neural network, swarm, alife) music, sound, art and design, evoMUSART has become an Evo* conference with independent proceedings since 2012.

Although the use of Artificial Intelligence for artistic purposes can be traced back to the 1970s, the use of Artificial Intelligence for the development of artistic systems is a recent, exciting and significant area of research. There is a growing interest in the application of these techniques in fields such as: visual art and music generation, analysis and interpretation; sound synthesis; architecture; video; poetry; design; and other creative tasks.

The main goal of evoMUSART 2021 was to bring together researchers who are using Artificial Intelligence techniques for artistic tasks, providing the opportunity to promote, present and discuss ongoing work in the area. As always, the atmosphere was fun, friendly and constructive.

EvoMUSART has grown steadily since its first edition in 2003 in Essex, UK, when it was one of the Applications of Evolutionary Computing workshops. Since 2012 it has been a full conference as part of the Evo* co-located events.

EvoMUSART 2021 received 66 submissions. The peer-review process was rigorous and double-blind. The international Programme Committee, listed below, was composed of 79 members from 19 countries. EvoMUSART continued to provide useful feedback to authors: among the papers sent for full review, there were on average 3 reviews per paper. The number of accepted papers was 24 long talks (36.4% acceptance rate) and 7 posters accompanied by short talks, meaning an overall acceptance rate of 47%.

As always, the EvoMUSART proceedings cover a wide range of topics and application areas, including generative approaches to music and visual art, deep learning and architecture. This volume of proceedings collects the accepted papers.

As in previous years, the standard of submissions was high, and good-quality papers had to be rejected. We thank all the authors for submitting their work, including those whose work was not accepted for presentation on this occasion.

The work of reviewing is done voluntarily and generally with little official recognition from the institutions where reviewers are employed. Nevertheless, professional reviewing is essential to a healthy conference. Therefore we particularly thank the members of the Programme Committee for their hard work and professionalism in providing constructive and fair reviews.

EvoMUSART 2021 was part of the Evo* 2021 event, which included three additional conferences: EuroGP 2021, EvoCOP 2021 and EvoApplications 2021. Many people helped to make this event a success.

We thank SPECIES, the Society for the Promotion of Evolutionary Computation in Europe and its Surroundings, for its sponsorship and HTWK Leipzig University of Applied Sciences for their patronage of the event.

We thank the local organising team lead by Francisco Fernández de Vega (University of Extremadura, Spain) and Federico Divina (University Pablo de Olavide, Spain) and also University Pablo de Olavide in Sevilla, Spain for supporting the local organisation.

We thank João Correia (University of Coimbra, Portugal) for Evo* publicity, and website and social media service, and Nuno Lourenço for coordinating the Submission System.

Finally, and above all, we would like to express our most heartfelt thanks to Anna Esparcia-Alcázar (SPECIES, Europe), for her dedicated work and coordination of the event. Without her work, and the work of Jennifer Willies in the past years, Evo* would not enjoy its current level of success as the leading European event on Bio-Inspired Computation.

April 2021

Juan Romero
Tiago Martins
Nereida Rodríguez-Fernández

Organization

EvoMUSART 2021 was part of Evo* 2021, Europe's premier co-located events in the field of evolutionary computing, which also included the conferences EuroGP 2021, EvoCOP 2021 and EvoApplications 2021.

Organizing Committee

Conference Chairs

Juan Romero	University of A Coruña, Spain. PhotoILike
Tiago Martins	University of Coimbra, Portugal

Publication Chair

Nereida Rodríguez-Fernández	University of A Coruña, Spain. PhotoILike

Programme Committee

Mauro Annunziato	ENEA, Italy
Aurélien Antoine	McGill University, Canada
Daniel Ashlock	University of Guelph, Canada
Peter Bentley	University College London, UK
Gilberto Bernardes	University of Porto, Portugal
Eleonora Bilotta	University of Calabria, Italy
Daniel Bisig	Zurich University of the Arts, Switzerland
Tim Blackwell	Goldsmiths, University of London, UK
Jean-Pierre Briot	Sorbonne University, France & Pontifical Catholic University of Rio de Janeiro, Brasil
Andrew Brown	Griffith University, Australia
Marcelo Caetano	McGill University, Canada
Amilcar Cardoso	University of Coimbra, Portugal
Luz Castro Pena	University of A Coruña, Spain
Carmine Emanuele Cella	University of Bologna, Italy
Bing Yen Chang	The Hong Kong University of Science and Technology (HKUST), China
Vic Ciesielski	RMIT University, Australia
João Correia	University of Coimbra, Portugal
Pedro Cruz	Northeastern University, USA
Camilo Cruz Gambardella	Monash University, Australia
João Miguel Cunha	University of Coimbra, Portugal
David De Roure	University of Oxford, UK
Hans Dehlinger	University of Kassel, Germany

Gustavo Reis	Polytechnic Institute of Leiria (IPL), Portugal
Nereida Rodríguez-Fernández	University of A Coruña, Spain
Juan Romero	University of A Coruña, Spain
Brian Ross	Brock University, Canada
Jonathan E. Rowe	University of Birmingham, UK
Iria Santos	University of A Coruña, Spain
Rob Saunders	Leiden University, The Netherlands
Marco Scirea	University of Southern Denmark, Denmark
Daniel Silva	University of Porto, Portugal
Kivanç Tatar	Simon Fraser University, Canada
Luis Teixeira	Fraunhofer AICOS, Portugal
Stephen Todd	Goldsmiths, University of London, UK
Julian Togelius	New York University, USA
Hannu Toivonen	University of Helsinki, Finland
Álvaro Torrente Patiño	University of A Coruña, Spain
Paulo Urbano	University of Lisbon, Portugal
Anna Ursyn	University of Northern Colorado, USA

Contents

Long Talks

Short Talks

Short Talks

Long Talks

Sculpture Inspired Musical Composition

One Possible Approach

Francisco Braga[1,2(✉)] and H. Sofia Pinto[1,2(✉)]

[1] INESC ID, Lisbon, Portugal
sofia@inesc-id.pt
[2] Instituto Superior Técnico, Lisbon, Portugal
francisco.braga@tecnico.ulisboa.pt

Abstract. In this paper, we present an inspirational system that takes a 3D model of a sculpture as starting point to compose music. It is considered that cross-domain mapping can be an approach to model inspiration. Our approach does not consider the interpretation of the sculpture but rather looks at it abstractly. The results were promising: the majority of the participants gave a classification of 4 out of 5 to the preferred interpretations of the compositions and related them to the respective sculpture. This is a step to a possible model for inspiration.

Keywords: Computational creativity · Inspiration · Genetic algorithm · Sculpture · Musical composition

1 Introduction and Motivation

Whenever we, humans, attempt to create something new, regardless of the field, we search for inspiration either from inside or from the world around us. Trash and Elliot consider two processes [19]: being inspired *by* and being inspired *to*. While being inspired *by* can be seen as the stimulus, being inspired *to* involves motivation. These two components are related with two important objects or ideas involved in the act of inspiration: the trigger - the stimulus object, associated with the "inspired *by*" process - and the target - the object to which the resulting motivation is targeted, associated with the "inspired *to*" process [19].

Existing ideas or objects inspire new ones in all sort of areas, from engineering, where solutions to new problems may be influenced by existing solutions to similar problems, to product design, where inspiration may come from shapes in nature [6], or art, where existing artworks can help one to create original art pieces [15]. Even though in the above cases inspiration comes from the same domain as the initial problem, it can also come from different domains. In the latter case, the challenge is usually harder because there is no direct connection.

Modeling inspiration computationally is far from being a reality. However, some systems address cross-domain inspiration as the mapping between two artifacts from different domains [10,17]. While Horn *et al.* mapped between two visual domains, images into 3D-printable ceramic vases [10], Teixeira and Pinto

J. Romero et al. (Eds.): EvoMUSART 2021, LNCS 12693, pp. 3–19, 2021.
https://doi.org/10.1007/978-3-030-72914-1_1

mapped images into sounds [17], involving two different domains. The basis for this mapping has as starting point trying to get the essence of an already existing artifact and one of the main challenges is how to relate two different domains.

This paper describes our effort in creating an inspirational system that composes music inspired by sculptures. We decided to draw inspiration from existing human-made artworks to maintain the human hand in the equation. The features used to characterize both domains and their respective mapping, between the sculpture and music domains, were guided by the aesthetics of the authors. Other features and possible mappings could have been explored. Although the approach described here is only one of an infinite number of possibilities, based on our evaluation results, it was successful. Three goals have driven our research:

- Build a system that acts both creatively and generates a product that is considered creative.
- The music pieces composed by the system are inspired by the sculpture, therefore, in a certain way, the music is associated with the sculpture.
- Build a system that composes music deemed aesthetically pleasing.

In the next section, we describe related works. Then we describe our approach to sculpture inspired musical composition, starting by describing the representation, shape, and texture features we used. We also describe the mapping&composer approach to music composition that was implemented. Evaluation describes how our listeners perceived the composed music. We conclude with a discussion on our results and some future directions.

2 Related Work

Our work focuses on two main areas: the generation of musical compositions and the act of inspiration. Regarding the generation of musical compositions, there is relevant work both in the algorithmic composition and in the computational creativity fields. As to inspiration, it is still an under-explored topic [10].

2.1 Generation of Musical Compositions

Algorithmic composition can be described as *"the process of using some formal process to make music with minimal human intervention"* [1]. The simplest form of algorithmic composition lies in randomness. Many different examples can be inserted here, such as the technique attributed to composer Wolfgang Amadeus Mozart, called *Musikalisches Würfelspiel* ("musical dice game") [1], which consists of using a dice to combine already existing musical fragments. There are also some examples of more complex rule-based algorithmic composition, such as the Twelve-tone technique [18], invented by the Austrian composer Arnold Schoenberg. In the computational realm, some systems combine randomness with rules. Lejaren Hiller and Leonard Isaacson made a system responsible for the first computer-generated composition [8]. It uses a paradigm of generator/modifier/selector. It starts by generating random integers that represent musical elements, then the obtained values are validated according to rules where the values that fail are re-generated.

In the field of computational creativity, several artificial intelligence methods have been used to address this goal, from Mathematical Models to Learning Algorithms [16]. The two methods that are more relevant to our work are Knowledge-Based systems and evolutionary methods. Regarding Knowledge-Based systems, in the 1980s, composer David Cope developed a system that would later be called EMI (Experiments in Music Intelligence) [4] that aims at creating new pieces of music based on a specific style or author. Concerning evolutionary methods, GenJam [3] is a system that generates jazz solos given a certain chord progression. The system proposed by Biles solves the jazz solo search problem using a genetic algorithm. Variations [11] is a system that applies a genetic algorithm to the composition process with as few restrictions as possible.

It is relevant to see that, although the complex rule-based approach found on EMI was very successful in creating pieces in a specific style [4], systems that use genetic algorithms are the ones most associated with novelty. In the latter, the level of constraints imposed by these rules will change the final result drastically. In the system Variations [11], the author imposes as few rules as possible, leading to the composition of a not so well defined piece of music, with very few structural elements. On the other hand, considering GenJam [3], Biles imposes many rules and constraints, leading to more structured results, as intended, but lacking more in novelty.

2.2 Inspirational Systems

Although inspiration is a complex topic, there are some examples of research on inspirational systems in the field of computational creativity. Horn et al. [10] proposed a cross-domain inspiration framework with the system Visual Information Vases (VIV). In this system, the authors consider cross-domain inspiration to be a cross-domain analogy mapping. Using this model for inspiration, the system produces 3D printable vases inspired by 2D images provided by the user. The image is processed, and using the obtained color palette from dominant and salient colors, the mapping to 3D printable vases is made considering four aesthetic measures: activity, warmth, weight, and hardness. VIV then uses a genetic algorithm to produce the vase with an aesthetic profile close to the one from the image's color palette.

Teixeira and Pinto used a similar cross-domain inspiration framework in their system Cross-Domain Analogy: From Image to Music [17]. This system generates a piece of music inspired by a given image. By considering a similar inspiration model as the one previously described, this system processes the image, and from the extracted features tries to map the emotional mood present in the image to the musical artifact. Two artifacts are produced from the mapping, one with very few rules and one with more constraints. The authors then use a Genetic Algorithm to evolve these two artifacts, thus obtaining a balance between them. The algorithm does not impose more rules but favors the most musically pleasing artifacts as guided by the fitness function.

3 Sculpture Inspired Musical Composition

In our system, we used a model for inspiration similar to the inspirational systems previously described. Regarding musical composition, we tried to achieve a balance between randomness and rules regarding the mapping from sculpture features to musical elements. A genetic algorithm was also used with this in mind. A diagram of our system's architecture can be seen in Fig. 1. Below we go through each module describing our approach and implementation.

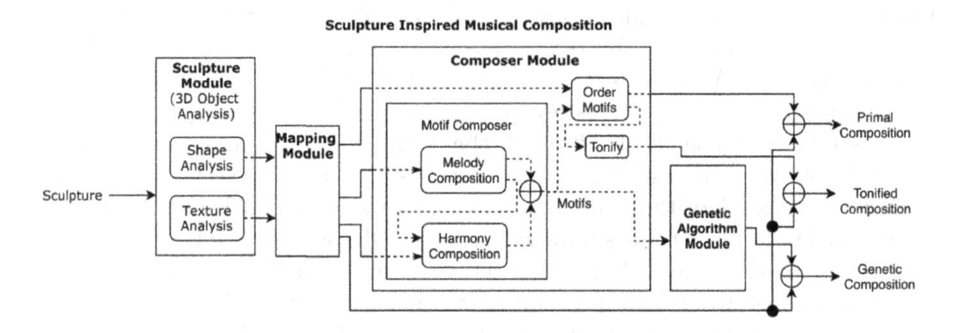

Fig. 1. System architecture

3.1 Sculpture Module

When first looking at a sculpture, several characteristics come to our attention. These can be divided into two groups: shape features, regarding the shape of the sculpture, and texture features, regarding the color or texture of the sculpture. It was decided not to address the semantic meaning of sculptures because this would be a complex task both computationally and in terms of interpretation, since each individual may have a different perception of a sculpture. So, a more general approach that does not address semantics was chosen, leading to a more abstract interpretation.

Representation. There are many possibilities to obtain a computational representation of a sculpture, such as retrieving it from images, but the one we followed was using 3D objects since all the information regarding both shape and texture can be more easily obtained.

There are several different possible representations of the 3D objects. We decided to use Polygonal Meshes [5] since it is a simple representation, yet it has all the information needed. In this representation, the object is represented by its boundary surface, composed of several planar shapes, which are defined by a series of vertices connected by edges. In our work, we used the most commonly used shape, the triangle (triangular meshes).

Although we are using 3D objects, when it comes to the texture, it refers to a 2D surface. A simple analogy is a gift wrap. The gift itself is a 3D object, but the wrapping alone is a 2D surface. To do the mapping of the texture in the 3D object to 2D, a mapping is used, that takes the coordinates of a 3D object, (x, y, z), and converts them to 2D coordinates, (U, V). Having specified our representation choice, we can move to the extracted feature.

Shape Features. First considering shape features, three main features were extracted: curvature, angles and segments.

There are many possibilities to quantify the curvature of a given surface. From all the available options, we chose the mean curvature [13], where each vertex of the mesh has its mean curvature value.

Regarding the angles, these are measured using the normal vector of the faces. To obtain each vertex's angle, the angles between all its faces are measured, and only the maximum is considered. One important aspect to note is that, since the angle is obtained from the normal vector, if the faces form a plane surface, the angle is 0°, while two overlapping faces have an angle of 180°. Using the angles, we obtained another measure, which we called zero angle predominance. This measure verifies if the vertices with 0° angles in an object are inserted in a plane surface, or if they occur separately. If they are inserted in a plane surface, then our focus goes towards what arises from it, or to the limits of that surface. To obtain this measure, first, we get the number of vertices that have an angle of 0°. Then, we obtain how many also have at least half of their neighbors with an angle of 0°. Then we calculate the ratio of the second to the first.

Finally, the segmentation is performed using the spectral clustering algorithm [20]. This algorithm fits our purpose since its roots are on graph theory, and a mesh can be seen as a graph. It receives the adjacency matrix of the object's vertices and the number of clusters, returning a label for each vertex. Using this, we divide the mesh into segments. We decided to use a fixed number of eight clusters, however, this number could be changed from sculpture to sculpture if desired. An example of segmentation of a 3D Mesh using Spectral Clustering can be seen in Fig. 2.

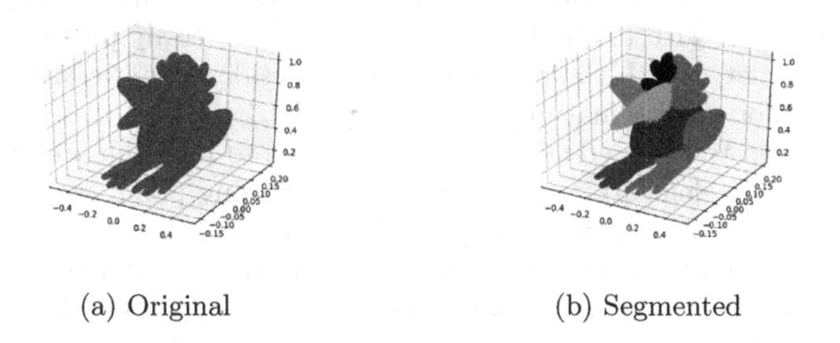

(a) Original (b) Segmented

Fig. 2. 3D mesh segmentation example using spectral clustering

Texture Features. Since we are dealing with a 2D image regarding the texture, the feature extraction process is performed using the same feature extraction algorithms as in a 2D image.

Considering color representation, the ones more relevant to us are RGB and HSV. On the one hand, in RGB, any color is described as a combination of values of the three principal additive colors: Red, Green, and Blue. On the other hand, HSV refers to a combination of Hue, Saturation, and Value. Hue refers to the color itself, saturation to the intensity (purity) of a particular hue, and value to the lightness or darkness of that color. Both models were used for different purposes.

There are several different color features. The two main categories histogram-based methods and color statistics [21]. Color histogram-based methods mainly represent the distribution of colors in an image. Although there are several ways to do this, the most common and useful way is the histogram of the distribution over the color model (3D in the case of RGB and HSV) or each channel of the color model. Color Statistics are statistical measures, such as the value, standard deviation, median, percentiles, among others. Aside from these two categories, several features can be obtained by processing the image. From those, the two extracted and used in our work are:

- Most Common Colors - To obtain these colors, a simple counting approach was used. First, nine equally spaced values were defined for each channel of the RGB model (thus a total of 729 distinct colors). Having the image's pixels represented in the same model, for each pixel, the closest value for each channel is chosen, and that color's count is incremented.
- Perceived Brightness - This can be seen as a conversion from the original colors to a greyscale, based on the brightness of each color. In our work, we used a formula where different brightness weights are attributed to each of the RGB channels [7]: $PB = \sqrt{0.299 * R^2 + 0.587 * G^2 + 0.114 * B^2}$

3.2 Mapping and Composer Module

Since we aim at composing sculpture inspired music, an analogy from one domain to the other was made. However, since music is such a vast field, it becomes quite impossible not to restrict. As such, we decided to focus our approach on modal music since, with modal harmony, one can more easily pass on a specific sensation, associated with the respective mode. Besides this, we also only used the most common time signature in western music, $\frac{4}{4}$.

The first step towards creating the analogy was to map the sculpture's features into musical elements and concepts. As a starting point, we did not aim at a too precise approach, for example, where each point would be converted into a note. We intended to obtain the musical elements through the analysis of the sculpture's features in a more general manner, by considering either the sculpture or each of the sculpture's segments as a whole. With this in mind, let us start with a high-level association between the two groups of sculpture's features, shape and texture, and the considered elements of a musical piece: melody, harmony, and rhythm.

If we listen to a melody without any harmony (chords), we may perceive a particular sensation or emotion. Once we listen to it with harmony, this sensation or emotion may vary drastically from the initial one. Moving to the sculpture domain, if we think about a sculpture's shape, we obtain mainly sensations associated with its characteristics, such as smooth or rough. However, once we add the texture/color, we get, once again, a much better-defined sensation, or even emotion. One can say that the same way texture/color gives context to the shape, harmony gives context to the melody. As such, our approach's foundations lie in the association of the sculpture's shape with melody, and the texture/color with harmony. It is important to note that this is not strict, and some associations were made from texture to melody. There is still one part of the music piece left to map, the rhythm.

As already stated, the rhythm is associated with the time signature and the tempo. Concerning time signature, as already stated, we decided to only use the most common time signature in western music, $\frac{4}{4}$. Although we initially wanted both shape and texture to be considered for the tempo, only the texture was used. This decision was made when confronted with the final results, as such, it shall be later on explained. Regarding the composition, we decided to consider motifs, i.e., short recurrent musical ideas. As seen before, a sculpture can be segmented. As such, we decided to compose each motif using features obtained from each segment. To obtain the size of the motif, we used each segment's size in relation to the whole sculpture.

Motif Composition - Melody. Starting with the melody, this element consists of a linear sequence of notes, while each note consists of pitch and duration. Accordingly, we tried to associate these two characteristics with shape features.

The pitch of the notes was obtained using the sculpture's lines or curves. If the lines of the sculpture are smooth, the pitches of the melody would have to give this same sensation, as well as the contrary. To obtain the line's description, we used the angles. Using the angles histogram, we related the musical interval for the next note concerning the current note with each *bin* of the histogram. Since in one octave there are twelve possible intervals, the histogram was calculated using twelve *bins*. With the histogram, we created a probability distribution to obtain a musical interval, thus having the probability for the next note's interval concerning the current note. The mapping made from angles to musical intervals can be seen in Table 1.

Table 1. Mapping from angles to musical intervals

Angles (degrees)	0°-15°	15°-30°	30°-45°	45°-60°	60°-75°	75°-90°	90°-105°	105°-120°	120°-135°	135°-150°	150°-165°	165°-180°
Interval	M2	M3	m3	P1	M6	m6	P5	P4	m7	m2	M7	TT

At this point, we decided to make use of the Zero Angle Predominance feature. As explained before, this feature identifies if a sculpture's angles of 0° are an important element, or if they correspond to a plane. If they correspond to a plane, the protrusions or objects that arise from that plane or the angle between this and another plane should matter the most. As such, in this case, the probability for the interval associated with the angles near 0° should be reduced in favor of the others. To do so, using a maximum cut for the probability of 80%, from the zero angle predominance ratio we obtain how much the cut is. What has been cut is then equally distributed by the remaining non-zero probabilities.

One issue with this approach is that if one particular *bin* does not contain any occurrences, that interval will never be used. We acknowledged that an interval of angles could not only be mapped to the respective musical interval but also the neighbors' musical intervals considering the order of Table 1. With this in mind, after obtaining the musical interval, we also apply a normal distribution to give the neighboring musical intervals a chance to be chosen. We used what is know as the standard normal distribution ($\mu = 0$ and $\sigma = 1$). Using a value obtained from the normal distribution, there are three possible cases: if the value is between -1 and 1 (probability of 68.3%), the original musical interval is chosen; If the value is below -1 (probability of 15.85%), the left neighbor musical interval is chosen, unless the original value is the first position, in this case, the original musical interval is chosen; If the value is above 1 (probability of 15.85%), the right neighbor musical interval is chosen, unless the original value is the last position, in this case, the original musical interval is chosen.

The only aspect left to decide is the direction of the interval, being the only musical element in the melody mapped from the texture. Here, we decided to use of the perceived brightness. Colors with a higher perceived brightness value would be related to upwards intervals, and colors with lower values with downwards intervals. From the histogram of perceived brightness values, the value that occurred more often was considered. It ranges from 0 to 1 and sets the probability of the interval direction being upwards between 20% and 80%.

The duration of a note will significantly influence the way we perceive a music piece. If we consider a sculpture with an irregular surface curvature value, one can easily associate it with a dense music piece, where the notes keep changing in a small fragment of time. The opposite can also be related. As such, we aimed at relating the variation of the sculpture's curvature value with the note's duration. To obtain the curvature value variation, we used autocorrelation. We gathered the curvature value of each point, grouped them by neighbors, and used autocorrelation in these groups. We decided to use the mean of all results, and map this value with what would become the most probable rhythmic value. The mapping can be seen in Table 2.

Table 2. Mean autocorrelation and most probable rhythmic value mapping

Mean Autocorrelation	0-0.05	0.05-0.1	0.1-0.4	0.4-0.7	0.7-1
Most Probable Rhythmic Value	𝅘𝅥𝅯	𝅘𝅥𝅮	♩	𝅗𝅥	𝅝

We did not want only one type of rhythmic value. We were looking for a general way to relate the mean autocorrelation with the rhythmic value without making a direct relation. To achieve this, the Standard Normal Distribution was used once again. Using this tool, we can set the most probable rhythmic value while maintaining the possibility of using other rhythmic values. For each note, a rhythmic value is chosen based on the most probable one and a value obtained using the normal distribution.

Motif Composition - Harmony. Having the various motifs that will form the melody, we can now move to harmony. It was decided to relate the harmony with the sculpture's texture/color. As such, we want to harmonize each motif according to the respective segment's color. For each segment, we obtained the most common colors, as explained before. Having the HSV color model in mind, for the same Hue, and the highest Saturation (100%), the Value channel will give us how dark or how pure that Hue is, ranging from black (Value close to 0%) to the pure Hue (Value close to 100%). Similarly, the musical modes can also be ordered from darker (Locrian) to brighter (Lydian) [14]. In Table 3, a summary of the modes of the major scale is shown, with the degrees of the major scale, the respective mode, intervals, and chord.

Table 3. Summary of the major scale modes (characteristics notes in bold)

Degree of major scale	Mode	Intervals	Chord (Tetrad)
I	Ionian	P1 M2 M3 **P4** P5 M6 **M7**	IΔ
II	Dorian	P1 M2 m3 P4 P5 **M6** m7	IIm7
III	Phrygian	P1 **m2** m3 P4 P5 m6 m7	IIIm7
IV	Lydian	P1 M2 M3 **A4** P5 M6 M7	IVΔ
V	Mixolydian	P1 M2 M3 P4 P5 M6 **m7**	V7
VI	Aeolian	P1 M2 m3 P4 P5 **m6** m7	VIm7
VII	Locrian	P1 m2 m3 P4 **d5** m6 m7	VIIm7♭5

As such, having the most common colors, we related the Value channel from the HSV color model to musical modes. A color with a lower value in the Value channel would be associated with a darker mode, while one with a higher value with a brighter mode. However, not all Hues should be matched with all modes, since, for the same Value, some colors are perceived as darker than others. As such, the Hues were divided into three groups. For each group, the Value channel maps to a range of modes. The division in groups can be seen in Fig. 3.

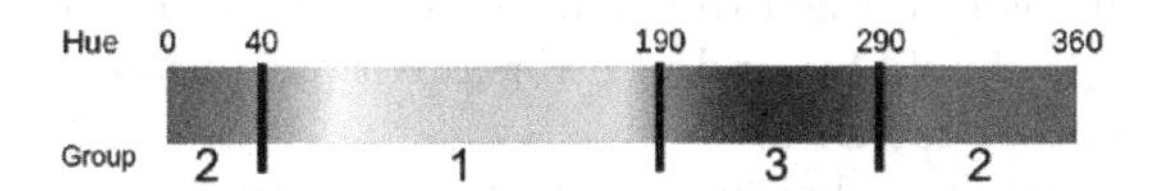

Fig. 3. Hue value based groups

Besides these three groups, two special cases are not accounted for by the Hue values: If the color is black, or very close to black (Value channel below 20 %), the mode is set to Locrian (darkest); if the color is white, or very close to white (Saturation channel below 10%), the mode is set to Lydian (the brightest). It is important to note that even if the Saturation is near 0% if the Value channel is close to 0%, the color will be close to black. As such, the first verification to be made is if the color is close to black (Value $< 20\%$). The range of modes per color group association and the special cases can be seen in Fig. 4.

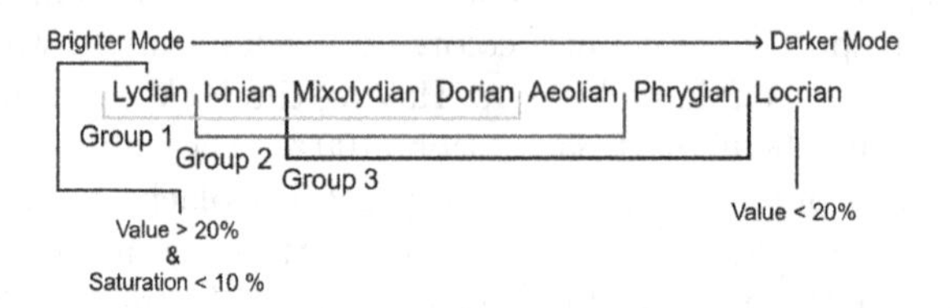

Fig. 4. Range of modes per color group and special cases

To obtain the chord, the closest tone of each motif is found, and then the mode's chord in that tonality is obtained. For each segment, we get the two most common colors, from there the two respective modes, and finally, the two chords. The chord obtained from the most common color is placed first in the motif and the one from the second most common chord second. The number of bars for the first chord is obtained by dividing the motif's number of bars by two and rounding up, while the remaining bars are left for the second chord.

3.3 Order Motifs and Tempo

To join the motifs in a logical order, we decided to get the probability distribution of intervals from the histogram of angles, although this time from the whole sculpture. From there, we calculate which of all possible permutations of the motifs' order best corresponds to the probability distribution of the intervals, being this the chosen order.

The only aspect considered left to map is the tempo. For this element, we used the variation of color among segments. To get this variation, first, we obtained the most common colors of the whole sculpture. Then, for each segment, we verified how many of the most common colors of that segment matched the ones from the whole sculpture. In the end, we obtained a ratio of the colors that matched the total colors verified that ranges from 0 to 1. The higher the ratio, the lower the tempo should be, and vice versa. For our purposes, we considered a tempo range of 80 to 160 BPM.

Having the melody and harmony from the ordered motifs, and respective tempo, we obtain the first version of the music piece, designated as the raw composition. The second version was obtained by correcting notes in the motifs that did not belong to the respective scale, changing them to the closest note found on the scale. This way, we obtain a version that respects each motif's scale. We called it a "tonified" composition. It was also decided to use a Genetic Algorithm since it provides a way to search for better results while retaining some randomness that may lead to exciting results.

3.4 Genetic Algorithm

A Genetic Algorithm is a search algorithm inspired by the process of natural evolution first introduced by Holland et al. [9]. Regarding representation, we defined each individual (candidate solution) as a music piece. As already stated, we considered a music piece to be the gathering of several motifs (melody and harmony), their ordering, and tempo. Therefore, each individual is a music piece, and its genes are motifs. A crucial property in each individual is the ordering of the motifs (genes). The evolution process is divided into several phases. These phases are explained below, while a visual summary of the genetic algorithm can be seen in Fig. 5.

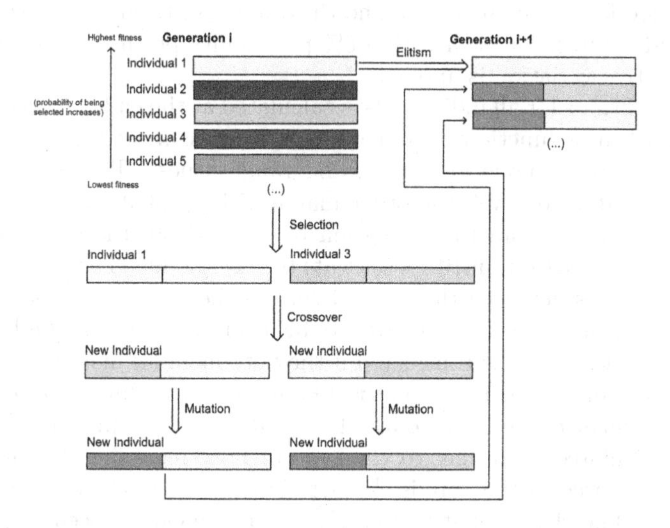

Fig. 5. Genetic algorithm summary

Initialization. This phase is the initial stage of the algorithm. We decided to have a population of 100 individuals. The number of motifs generated per segment is the chosen number of individuals. Each individual is created by randomly choosing a motif from each segment. The initial order of the motifs is randomly chosen.

Elitism. In each generation, 10% of the individuals with the highest fitness are directly passed to the new generation. This guarantees some level of quality.

Selection. In this stage, a pair of individuals is selected. The probability of each individual being chosen is based on the individual's fitness. The higher the fitness, the higher the probability of being chosen. As such, one individual may be selected several times or no times at all.

Crossover. Having a pair of selected individuals, there is an 80% probability of suffering crossover. We defined two types of crossover, both equally probable: motif and note crossover. In the first, half of the individual's motifs are randomly selected and crossed over. This crossover occurs between motifs generated from the same segment. In the second, the pitch of the individuals is swapped. Since one motif may be longer than the other, among the proper starting points, one is randomly chosen.

Mutation. The individuals obtained have a probability of being mutated. This involves altering the individual's genes. Four types of mutation were used, each with an independent probability. The first is order mutation, with a 10% probability. In this case, the order of the motifs is randomly altered. The second is the "tonify" mutation, with a 10% probability for each motif. In this case, notes that do not belong to the motif's scale are corrected to the closest note found on the scale. The third and fourth are the Inversion and Retrograde mutations, each with a 5% probability per motif. In these cases, the respective variation is applied.

Fitness. The fitness for all individuals is calculated at the end of each generation through a fitness function. In our case, three measures were used to obtain the fitness: order, mode definition, and range fitness. The order definition fitness evaluates how well the order matches the probability distribution of intervals obtained from the histogram of angles from the whole sculpture. This measure ranges from 10 (perfect fit) to -10 (no fit). The mode definition fitness evaluates how well the motif describes the respective mode. For this, each note has a weight according to its beat and its duration. Only the beat with the highest value is considered if the note includes more than one beat. If it is in the first beat, the weight is $4 * duration$, the third $3 * duration$, the second $2 * duration$, and the fourth $1 * duration$. The value for this measure is then calculated according to the type of note regarding the mode. Two concepts are used: if the note is characteristic of the mode and if the note is consonant. The characteristic intervals of a mode can be seen in Table 3. A note is considered consonant if the chord already includes that note, or if the note does not create the interval m2 with any note from the chord. With this in mind, this fitness measure is then calculated as described in Table 4. The range measure penalizes -4 for pitches that are not in an acceptable range. We defined that acceptable range between 55 and 90 in MIDI.

Termination. To terminate the algorithm, we decided to have a fixed number of generations. Regarding the final music piece, our interest moved towards an approach that considers the *evolutionary process as artwork* [12]. To do this, we decided to obtain the fittest individual in each fixed number of generations. In the end, the individuals are joined and form the final music piece. We decided to terminate the algorithm once it reached 300 generations, and join the fittest individual from every 100 generations.

Table 4. Mode definition Fitness measure for different types of notes

			weight **<= 1**	***weight*** **> 1**
Note in Mode	Characteristic	Consonant	2*weight	
		Dissonant	2*weight	-weight
	Not Characteristic	Consonant	weight	
		Dissonant	weight	-weight
Note not in Mode			weight	-weight

3.5 Implementation

This project was implemented using the programming language Python (https://www.python.org/), more specifically Python 3.6.8. Some packges were used for different purposes: PyMesh (https://github.com/PyMesh/PyMesh) to process meshes, an unofficial pre-built OpenCV package (https://pypi.org/project/opencv-python/) to process textures, MIDIUtil (https://pypi.org/project/MIDIUtil/) to export the music in MIDI format, Numpy (https://numpy.org/) to handle arrays and Scikit-learn (https://scikit-learn.org) for clustering.

4 Evaluation

Six 3D Models of sculptures were used as our dataset. Two 3D models were found online, *Sungod* and *Tir (Shooting Altar)*, by the artist Niki de Saint Phalle, and four were captured and processed by us, *Unkown Sculpture*, *Sr. Estúpido* by Robert Panda, *Galo de Barcelos* and *Les Baigneuses* by Niki de Saint Phalle. These were the selected sculptures due to the variety in shape and texture, and thus better evaluate our system's capability of generalization. We were limited both by the availability of 3D models online, and sculptures to be scanned. The end result were three distinct sculptures by the same author, and three other sculptures with different authors and periods. The dataset can be seen in Fig. 6. For the six sculptures, the genetic composition was generated. This was the only composition considered for the evaluation since we believe this is the one that represents our work as intended. Having the six compositions, three possible interpretations of each were generated using virtual instruments. All the 3D models of the sculptures and respective interpretations of the composed music are available at http://web.tecnico.ulisboa.pt/ist181444/.

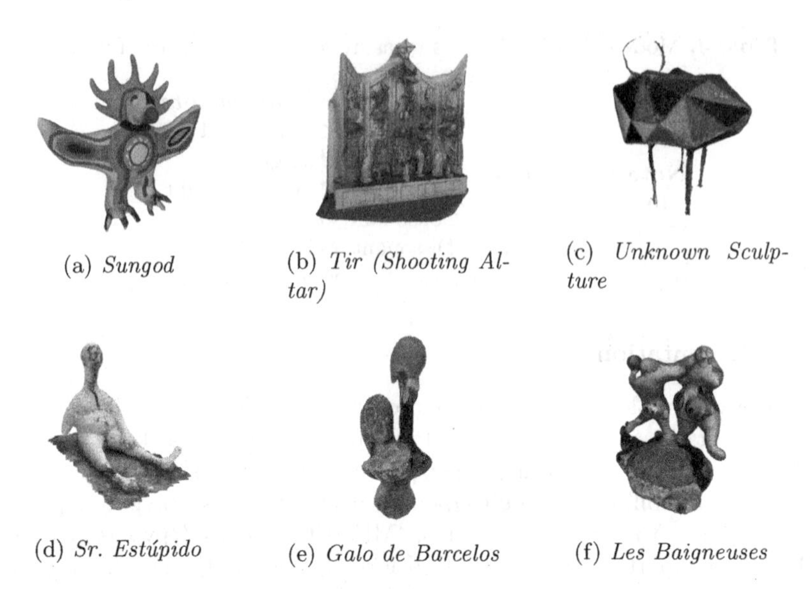

(a) *Sungod* (b) *Tir (Shooting Altar)* (c) *Unknown Sculpture*

(d) *Sr. Estúpido* (e) *Galo de Barcelos* (f) *Les Baigneuses*

Fig. 6. Dataset

We decided to use online surveys to perform the evaluation. Two surveys were made. From one survey, answers from 62 participants were collected, from the other, 36. Each survey contained three compositions. Having three interpretations for each composition, most of the survey concerned the preferred interpretation of the three presented. For each, we evaluated the quality of the music piece and the association with the sculpture. The least preferred interpretation was also evaluated, to establish a ground floor for interpretation.

Regarding the quality of the music, the statistics for all compositions can be seen in Table 5. In this question, the participants rated the music for the preferred and the least preferred interpretations. For the preferred version, on a scale of 1 to 5, the obtained median was 4, as such at least 50% of the participants rated the pieces as 4 or higher. These results attest the musical quality of the composed music pieces. The least preferred interpretation had lower results, as expected, but it still obtained a median and mode of 3.

Table 5. How would you rate this music? - Statistics

How would you rate this music?	Preferred	Least Preferred
Mean	3.55	2.79
Median	4	3
Mode	4	3
Std. Deviation	0.92	1.03

Considering the sculpture-music association, the participants were asked to select descriptors among a given set that better describe the preferred interpretation and the sculpture (in separate). The descriptors used were: Happy, Exciting, Smooth, Mellow, Sad; Rough, Bitter, Harsh, Aggressive, Boring, and Other. To verify the association, it was checked whether the descriptors assigned to the sculpture corresponded with those of the music. The histogram of the descriptors was made, and for each sculpture-music pair, the Bhattacharyya Coefficient [2] was calculated (Table 6) as a measure of the similarity.

Table 6. Bhattacharyya Coefficient for the music/sculpture descriptions

Sculpture	*Tir (Shooting Altar)*	*Unknown Sculpture*	*Sungod*	*Les Baigneuses*	*Galo de Barcelos*	*Sr. Estúpido*
Bhattacharyya Coefficient	0.4981	0.8963	0.8539	0.8848	0.9426	0.8491

Two special cases were found. The first is the *Tir (Shooting Altar)* sculpture, where its semantics leads to a drastically different perspective from the one obtained through our approach. In this case, the description given by the participants for the sculpture was distant from the composition, as we can see by the Bhattacharyya Coefficient in Table 6. The composition was not considered related to the sculpture since, for both the preferred and the least preferred interpretations, about 70% of the participants rated the relation as either 1 or 2 on a scale of 1 to 5. The second is the *Unknown Sculpture*, where there is much variation on how humans perceived the sculpture. The Bhattacharyya Coefficient is high because both the music and sculpture descriptions were ambiguous, without any characteristic descriptor. Despite this variation, for the preferred composition, at least 50% of the participants rated the relation as 3 or higher. For the least preferred interpretation, 51.6% of the participants rated the relation either as 1 or 2.

For the typical case, which was found in the remaining four of the six music/sculpture pairs, most people related the music to the sculpture. Regarding the description, there was a clear connection between the sculpture and the music as seen by the Bhattacharyya Coefficient in Table 6. For the sculpture *Sr. Estúpido*, the median for the relation rating for the preferred interpretation was 5, while for *Les Baigneuses* and *Sungod* 4, and for *Galo de Barcelos* as 3. For the least preferred, the median for the rating was 3 for the four cases.

5 Conclusions and Future Work

We have presented a system that aims at composing music inspired by 3D models of sculptures. To build it we have used an existing model for inspiration [17], involving different human senses, and applied it to a pair of domains that, to the best of our knowledge, have never been explored before. One of the unique aspects of our system is the chosen mapping between the two different domains, which is the base for the applied inspirational model. Only the *Genetic* Composition was considered for evaluation. This evaluation was made based on answers from online surveys, which focused on assessing two of our goals: the quality of the musical composition, and the sculpture-music association. The musical compositions were well rated, thus providing evidence of their quality. Regarding the sculpture-music association, despite two exceptional cases, we believe the results also provide evidence of its association.

It is important to stress that this paper describes one possible approach lead by the aesthetics of the authors, among an infinite number of other possible approaches. Some different possibilities were tested during the process, however, these are a small fraction compared to what could be explored. Our work was also limited regarding the musical concepts and elements used, therefore other approaches could be explored. For example, in the music realm a more general approach could be implemented, not only considering modal harmony, but also tonal or atonal. As for the association other possibilities could be tested. Although the shape-harmony and texture-melody association were used in our work, other possibilities could be explored, and even available for a user to choose. Other approaches to the genetic algorithm could also be investigated. One possibility would be to give more freedom to the algorithm by not limiting the crossover between motifs generated from the same segment of the sculpture. This could be compensated by using a fitness measure that would favor individuals where all the segments are represented. Another aspect that could be worked on is evaluation since only one of the three output compositions was evaluated. The three could be assessed to verify, for example, which one is preferred and which one is considered to be more related to the sculpture. Besides this, having human interpretations of the compositions could significantly improve evaluation. The quality of music played by real musicians is different from interpretations generated using a computer.

Acknowledgments. This work was supported by Fundação para a Ciência e a Tecnologia, under project UIDB/50021/2020 We thank all comments and suggestions from Nuno Correia and Rui Pereira Jorge.

References

1. Alpern, A.: Techniques for Algorithmic Composition of Music (1995)
2. Bhattacharyya, A.: On a measure of divergence between two statistical populations defined by their probability distributions. Bull. Calcutta Math. Soc. **35**, 99–109 (1943)
3. Biles, J.A.: GenJam: a genetic algorithm for generating jazz solos. In: International Computer Music Conference, vol. 94, pp. 131–137 (1994)
4. Cope, D., Mayer, M.J.: Experiments in Musical Intelligence, vol. 12. A-R Editions (1996)
5. Dugelay, J.L., Baskurt, A., Daoudi, M.: 3D Object Processing: Compression, Indexing and Watermarking. Wiley, Hoboken (2008)

6. Eckert, C., Stacey, M.: Fortune favours only the prepared mind: why sources of inspiration are essential for continuing creativity. Creativity Innov. Manag. **7**(1), 9–16 (1998)
7. Finley, D.R.: HSP Color Model - Alternative to HSV (HSB) and HSL (2006). http://alienryderflex.com/hsp.html
8. Hiller, L.A., Isaacson, L.M.: Experimental Music. Composition with an Electronic Computer. Greenwood Publishing Group Inc. (1979)
9. Holland, J.H.: Adaptation in Natural and Artificial Systems. University of Michigan Press, Ann Arbor (1975)
10. Horn, B., Smith, G., Masri, R., Stone, J.: Visual information vases: towards a framework for transmedia creative inspiration. In: International Conference on Computational Creativity, pp. 182–188 (2015)
11. Jacob, B.: Composing with genetic algorithms. In: Proceedings of the International Computer Music Conference, pp. 452–455 (1995)
12. Johnson, C.G.: Fitness in evolutionary art and music: what has been used and what could be used? In: Machado, P., Romero, J., Carballal, A. (eds.) EvoMUSART 2012. LNCS, vol. 7247, pp. 129–140. Springer, Heidelberg (2012). https://doi.org/10.1007/978-3-642-29142-5_12
13. Koenderink, J.J., Van Doorn, A.J.: Surface shape and curvature scales. Image Vis. Comput. **10**(8), 557–564 (1992)
14. Miller, R.: Modal Jazz Composition & Harmony. Advance Music, Rottenburg (1996)
15. Okada, T., Ishibashi, K.: Imitation, inspiration, and creation: cognitive process of creative drawing by copying others' artworks. Cogn. Sci. **41**(7), 1804–1837 (2017)
16. Papadopoulos, G., Wiggins, G.: AI methods for algorithmic composition: a survey, a critical view and future prospects. In: AISB Symposium on Musical Creativity, Edinburgh, UK, vol. 124, pp. 110–117 (1999)
17. Teixeira, J., Pinto, H.S.: Cross-domain analogy: from image to music. In: Proceedings of the 5th International Workshop on Musical Metacreation (2017)
18. The Editors of Encyclopaedia Britannica: 12-tone Music. https://www.britannica.com/art/12-tone-music. Accessed 04 Mar 2020
19. Trash, T., Elliot, A.: Inspiration: core characteristics, component process, antecedent, and function. J. Pers. Soc. Psychol. **87**, 957–973 (2004)
20. Xu, D., Tian, Y.: A comprehensive survey of clustering algorithms. Ann. Data Sci. **2**(2), 165–193 (2015)
21. Álvarez, M.J., González, E., Bianconi, F., Armesto, J., Fernández, A.: Colour and texture features for image retrieval in granite industry. Dyna **77**, 121–130 (2010)

Network Bending: Expressive Manipulation of Deep Generative Models

Terence Broad[1,2](✉), Frederic Fol Leymarie[1], and Mick Grierson[2]

[1] Goldsmiths, University of London, London, UK
{t.broad,ffl}@gold.ac.uk
[2] Creative Computing Institute, University of the Arts London, London, UK
m.grierson@arts.ac.uk

Abstract. We introduce a new framework for manipulating and interacting with deep generative models that we call *network bending*. We present a comprehensive set of deterministic transformations that can be inserted as distinct layers into the computational graph of a trained generative neural network and applied during inference. In addition, we present a novel algorithm for analysing the deep generative model and clustering features based on their spatial activation maps. This allows features to be grouped together based on spatial similarity in an unsupervised fashion. This results in the meaningful manipulation of sets of features that correspond to the generation of a broad array of semantically significant features of the generated images. We outline this framework, demonstrating our results on state-of-the-art deep generative models trained on several image datasets. We show how it allows for the direct manipulation of semantically meaningful aspects of the generative process as well as allowing for a broad range of expressive outcomes.

Keywords: Neural networks · Generative models · Expressive manipulation

1 Introduction

We introduce a new framework for the direct manipulation of deep generative models that we call *network bending*. This framework allows for *active divergence* [7] from the original training distribution in a flexible way that provides a broad range of expressive outcomes. We have implemented a wide array of image filters that can be inserted into the network and applied to any assortment of features, in any layer, in any order. We use a plug-in architecture to dynamically insert these filters as individual layers inside the computational graph of the pre-trained generative neural network, ensuring efficiency and minimal dependencies. As this process is altering the computation graph of the model, changes get applied to the entire distribution of generated results. We also present a novel approach to grouping together features in each layer. This is based on the spatial similarity of the activation map of the features and is done to reduce the dimensionality of

J. Romero et al. (Eds.): EvoMUSART 2021, LNCS 12693, pp. 20–36, 2021.
https://doi.org/10.1007/978-3-030-72914-1_2

the parameters that need to be configured by the user. It gives insight into how *groups* of features combine to produced different aspects of the image. We show results from these processes performed on generative models for images (using StyleGAN2, the current state-of-the-art for unconditional image generation [26]) trained on several different datasets, and map out a pipeline to harness the generative capacity of deep generative models in producing novel and expressive outcomes.

2 Related Work

2.1 Deep Generative Models

A generative model consists in the application of machine learning to learn a configuration of parameters that can approximately model a given data distribution. This was historically a very difficult problem, especially for domains of high data dimensionality such as for audio and images. With the advent of deep learning and large training datasets, great advances were made in the last decade. Deep neural networks are now capable of generating realistic audio [14,31] and images [11,25,26]. In the case of images, Variational Autoencoders [28,35] and Generative Adversarial Networks (GANs) [19] have been major breakthroughs that provide powerful training methods. Over the past few years there has been major improvements to their fidelity and training stability, with application of convolutional architecture [34], progressively growing architecture [24], leading to the current state of the art in producing unconditional photo-realistic samples in StyleGAN [25] and then StyleGAN2 [26]. One class of conditional generative models that take inputs in the form of semantic segmentation maps can be used to perform semantic image synthesis, where an input mask is used to generate an image of photographic quality [22,32].

Understanding and manipulating the *latent space* of generative models has subsequently been a growing area of research. Semantic latent manipulation consists in making informed alterations to the latent code that correspond to the manipulation of different semantic properties present in the data. This can be done by operating directly on the latent codes [36] or by analysing the activation space of latent codes to discover interpretable directions of manipulation in latent space [21]. Evolutionary methods have been applied to search and map the latent space [8,17] and interactive evolutionary interfaces have also been built to operate on the latent codes [37] for human users to explore and generate samples from generative models.

2.2 Analysis of Deep Neural Networks

Developing methods for understanding the purpose of the internal features (aka hidden units) of deep neural networks has been an on-going area of research. In computer vision and image processing applications, there have been a number of approaches, such as through visualisation, either by sampling patches that maximise the activation of hidden units [41,44], or by using variations of backpropagation to generate salient image features [38,41]. A more sophisticated approach is *network dissection* [5] where hidden units responsible for the detection of semantic properties are identified by analysing their responses to semantic concepts and quantifying their alignment. Network dissection was later adapted and applied to generative models [5], by removing individual units, while using in combination a bounding box detector trained on the ADE20K Scene dataset [43]. This led to the ability to identify a number of units associated with the generating of certain aspects of the scene. This approach has since been adapted for music generation [10].

2.3 Manipulation of Deep Generative Models

The manipulation of deep generative models is itself a nascent area of research. An interactive interface built upon the GAN Dissection approach [5] was presented with the GANPaint framework in 2019 [3]. This allows users to 'paint' onto an input image in order to edit and control the spatial formation of hand-picked features generated by the GAN.

An approach that alters the computational graph of the model such that a change alters the entire distribution of results, is presented as an algorithm for "rewriting the rules of a generative model" [2]. In this approach, the weights from a single convolutional layer are used as an associative memory. Using a copy-paste interface, a user can then map a new element onto a generated output. The algorithm uses a process of constrained optimisation to edit values in the weight matrix to find the closest match to the copy-paste target. Once the rules of the weight matrix have been altered, all results from the generator have also been altered.

3 Clustering Features

As most of the layers in current state of the art GANs, such as StyleGAN2, have very large numbers of convolutional features, controlling each one individually would be far too complicated to build a user interface around and to control these in a meaningful way. In addition, because of the redundancy existing in these models, manipulating individual features does not normally produce any kind of meaningful outcome. Therefore, it is necessary to find some way of grouping them together into more manageable ensembles of sets of features. Ideally such sets of features would correspond to the generation of distinct, semantically meaningful aspects of the image, and manipulating each set would correspond to the manipulation of specific semantic properties in the resulting generated sample. In order to achieve this, we present a novel approach, combining metric learning and a clustering algorithm to group sets of features in each layer based on the spatial similarity of their activation maps. We train a separate convolutional neural network (CNN) for each layer of the StyleGAN2 model with a bottleneck architecture (first introduced by Grézl et al. [20]) to learn a highly compressed feature representation; the later is then used in a metric learning approach in combination with the k-means clustering algorithm [13,29] to group sets of features in an unsupervised fashion. In our experiments we have performed this feature clustering process on models trained on three different datasets: the FFHQ [25], LSUN churches and LSUN cats datasets [40].

3.1 Architecture

For each layer of the StyleGAN2 model, we train a separate CNN on the activation maps of all the convolutional features. As the resolution of the activation maps and number of features varies for the different layers of the model (a breakdown of which can be seen in Table 1) we employ an architecture that can dynamically be changed, by increasing the number of convolutional blocks, depending on what depth is required.

Table 1. Table showing resolution, number of features of each layer, the number of ShuffleNet [42] convolutional blocks for each CNN model used for metric learning, the number of clusters calculated for each layer using k-means and the batch size used for training the CNN classifiers. Note: LSUN church and cat models have only 12 layers.

Layer	Resolution	#features	CNN depth	#clusters	Batch size
1	8 × 8	512	1	5	500
2	8 × 8	512	1	5	500
3	16 × 16	512	2	5	500
4	16 × 16	512	2	5	500
5	32 × 32	512	3	5	500
6	32 × 32	512	3	5	500
7	64 × 64	512	4	5	200
8	64 × 64	512	4	5	200
9	128 × 128	256	5	4	80
10	128 × 128	256	5	4	80
11	256 × 256	128	6	4	50
12	256 × 256	128	6	4	50
13	512 × 512	64	7	3	20
14	512 × 512	64	7	3	20
15	1024 × 1024	32	8	3	10
16	1024 × 1024	32	8	3	10

We employ the ShuffleNet architecture [42] for the convolutional blocks in the network, which is one of the state-of-the-art architectures for efficient inference (in terms of memory and speed) in computer vision applications. For each convolutional block we utilise a feature depth of 50 and have one residual block per layer. The motivating factor in many of the decisions made for the architecture design was not focused on achieving the best accuracy per se. Instead, we want a network that can learn a sufficiently good metric while also being reasonably quick to train (with 12–16 separate classifiers per GAN model). We also want a lightweight enough network, such that it could be used in a real-time setting where clusters can quickly be calculated for an individual latent encoding, or it could be used efficiently when processing large batches of samples.

After the convolutional blocks, we flatten the final layer ($4 \times 4 \times 50$) and learn from it a mapping into a narrow bottleneck ($v \in \mathbb{R}^{10}$), before re-expanding the dimensionality of the final layer to the number of convolutional features present in the GAN layer. The goal of this bottleneck is to force the network to learn a highly compressed representation of the different convolutional features in the GAN. While this invariably looses some information, most likely negatively affecting classification performance during training, this is in-fact the desired result. We want to force the CNN to combine features of the activation maps with similar spatial characteristics so that they can easily be grouped together by the clustering algorithm. Another motivating factor is that the clustering algorithm we have chosen (k-means) does not scale well for feature spaces with high dimensionality.

3.2 Training

We generated a training set of the activations of every feature for every layer of 1000 randomly sampled images, and a test set of 100 samples for the models trained on all of the datasets used in our experiments. We trained each CNN using the softmax feature learning approach [16], a reliable method for distance metric learning. This method employs the standard softmax training regime [9] for CNN classifiers. Each classifier has been initialised with random weights and then trained for 100 epochs using the Adam optimiser [27] with a learning rate of 0.0001 and with $\beta_1 = 0.9$ and $\beta_2 = 0.999$. All experiments were carried out on a single NVIDIA GTX 1080ti. The batch size used for training the classifiers for the various layers can be seen in Table 1.

After training, the softmax layer is discarded and the embedding of the final layer is used as the discriminative feature vector where the distances between points in feature space permit to gauge the degree of similarity of two samples. The one difference in our approach to standard softmax feature learning is that we use the second to last layer, the feature vector from the bottleneck, giving a more compressed feature representation than what standard softmax feature learning would offer.

3.3 Clustering Algorithm

Once the CNNs for every layers have been trained, they can then be used to extract feature representations of the activation maps of the different convolutional features corresponding to each individual layer of the GAN. There are two approaches to this. The first is to perform the clustering on-the-fly for a specific latent for one sample. A user would want to do this to get customised control of a specific sample, such as a latent that has been found to produce the closest possible reproduction of a specific person from the StyleGAN2 model trained on the FFHQ dataset [1,26]. The second approach is to perform clustering based on an average of features' embedding drawn from many random samples, which can be used to find a general purpose set of clusters.

The clustering algorithm for a single example is activated by a forward pass of the GAN performed without any additional transformation layers being inserted, this to obtain the unmodified activation maps. The activation map X_{df} for each layer d and feature f is fed into the CNN metric learning model for that layer C_d to get the feature vector v_{df}. The feature vectors for each layer are then aggregated and fed to the k-means clustering algorithm—using Lloyd's method [29] with Forgy initialization [13,18]. This results in a pre-defined number of clusters for each layer. Sets of features for each layer can then be manipulated in tandem by the user.

Alternatively, to find a general purpose set of clusters, we first calculate the mean feature vector $\bar{v}_{df}$ that describes the spatial activation map for each convolutional feature in each layer of StyleGAN2 from a set of N randomly generated samples—the results in the paper are from processing 1000 samples. Then we perform the same clustering algorithm as previously for individual samples on the mean feature vectors.

4 Transformation Layers

We have implemented a broad variety of deterministically controlled transformation layers that can be dynamically inserted into the computational graph of the generative model. The transformation layers are implemented natively in PyTorch [33] for speed and efficiency. We treat the activation maps of each feature of the generative model as 1-channel images in the range −1 to 1. Each transformation is applied to the activation maps individually before they are passed to the next layer of the network. The transformation layers can be applied to all the features in a layer, or a random selection, or by using pre-defined groups automatically determined based on spatial similarity of the activation maps (Sect. 3). Figure 1 shows a comparison of a selection of these transformations applied to all the features layer-wide in various layers.

4.1 Numerical Transformations

We begin with simple numerical transformations $f(x)$ that are applied to individual activation units x. We have implemented four distinct numerical transformations: the first is *ablation*, which can be interpreted as $f(x) = x \cdot 0$. The second is *inversion*, which is implemented as $f(x) = 1 - x$. The third is *multiplication by a scalar* p implemented as $f(x) = x \cdot p$. The final transformation is *binary thresholding* (often referred to as posterisation) with threshold t, such that:

$$f(x) = \begin{cases} 1, & \text{if } x \geq t \\ 0, & \text{otherwise} \end{cases} \tag{1}$$

4.2 Affine Transformations

For this set of transformations we treat each activation map X for feature f as an individual matrix, that simple affine transformations can be applied too. The first two are horizonal and vertical *reflections* that are defined as:

$$X \begin{bmatrix} -1 & 0 & 0 \\ 0 & 1 & 0 \\ 0 & 0 & 1 \end{bmatrix}, \quad X \begin{bmatrix} 1 & 0 & 0 \\ 0 & -1 & 0 \\ 0 & 0 & 1 \end{bmatrix} \tag{2}$$

The second is *translations* by parameters p_x and p_y such that:

$$X \begin{bmatrix} 1 & 0 & p_x \\ 0 & 1 & p_y \\ 0 & 0 & 1 \end{bmatrix} \tag{3}$$

The third is *scaling* by parameters k_x and k_y such that:

$$X \begin{bmatrix} k_x & 0 & 0 \\ 0 & k_y & 0 \\ 0 & 0 & 1 \end{bmatrix} \tag{4}$$

Note that in this paper we only report on using uniform scalings, such that $k_x = k_y$. Finally, fourth is *rotation* by an angle θ such that:

$$X \begin{bmatrix} cos(\theta) & -sin(\theta) & 0 \\ sin(\theta) & cos(\theta) & 0 \\ 0 & 0 & 1 \end{bmatrix} \tag{5}$$

Other affine transformations can easily be implemented by designing the matrices accordingly.

4.3 Morphological Transformations

We have implemented two of the possible basic mathematical morphological transformation layers, performing *erosion* and *dilation* [39] when applied to the activation maps, which can be interpreted as 1-channel images. These can be configured with the parameter r which is the radius for a circular kernel (aka structural element) used in the morphological transformations.

5 Manipulation Pipeline

In our current implementation,[1] transforms are specified in `YAML` configuration files [6], such that each transform is specified with 5 items: (i) the layer, (ii) the transform itself, (iii) the transform parameters, (iv) the layer type (i.e. how the features are selected in the layer: across all features in a layer, to pre-defined clusters, or to a random selection of features), and (v) the parameter associated with the layer type (either the cluster index, or the percentage of features the filter will randomly be applied to). There can be any number of transforms defined in such a configuration file.

After loading the configuration, we either lookup which features are in the cluster index, or randomly apply indices based on the random threshold parameter. Then the latent is loaded, which can either be randomly generated, or be predefined in latent space z, or be calculated using a projection in latent space w [1,26]. The latent code is provided to the generator network and inference is performed. As our implementation is using PyTorch [33], a dynamic neural network library, these transformation layers can therefore be inserted dynami-

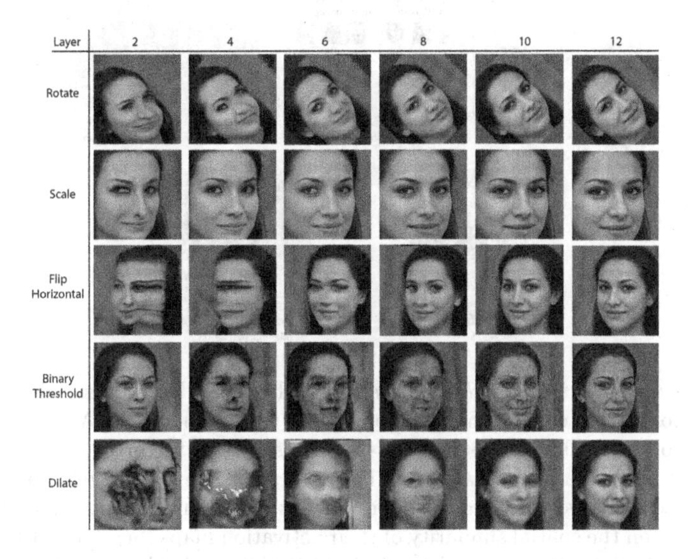

Fig. 1. A comparison of various transformation layers inserted and applied to all of the features in different layers in the StyleGAN2 network trained on the FFHQ dataset, showing how applying the same filters in different layers can make wide-ranging changes the generated output. The rotation transformation is applied by an angle $\theta = 45$. The scale transformation is applied by a factor of $k_x = k_y = 0.6$. The binary threshold transformation is applied with a threshold of $t = 0.5$. The dilation transformation is applied with a structuring element with radius $r = 2$ pixels.

[1] Our implementation and the datasets we have used for training the clustering models are publicly available and can be found at: https://github.com/terrybroad/network-bending.

cally during inference as and when they are required, and applied only to the specified features as defined by the configuration. Once inference in unrolled, the generated output is returned. Figure 2 provides a visual overview of the pipeline, as well as a comparison between a modified and unmodified generated sample.

5.1 Chaining Transformations

From the perspective of building tools that impact the generation of expressive and novel samples, performing one transformation at a time can be quite restricting. With our approach, we are not limited in this manner, and a user can explore more complicated effects by chaining multiple transformations. In Fig. 3 a few examples of combining multiple transformations, when applied to different sets of features in different layers, illustrate how our proposed architecture can generate very unusual and highly distinctive results. This significantly broadens the space of possible outcomes to explore, allowing for surprising results when different transformations interact with each other.

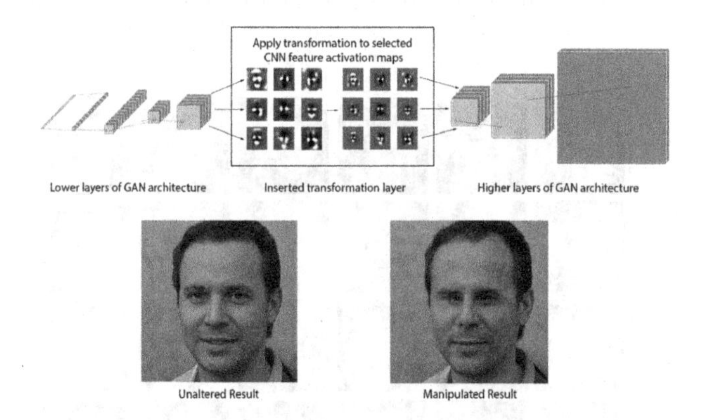

Fig. 2. Overview of our *network bending* approach where deterministically controlled transformation layers can be inserted into a pre-trained network. As an example, a transformation layer that scales the activation maps by a factor of $k_x = k_y = 0.6$ is applied (Sect. 4.2) to a set of features in layer 5 responsible for the generation of eyes, which has been discovered in an unsupervised fashion using our algorithm to cluster features based on the spatial similarity of their activation maps (Sect. 3). At the bottom left we show the sample generated by StyleGAN2 [26] trained on the FFHQ dataset without modification, while to its right we show the same sample generated with the scaling transform applied to the selected features. NB: the GAN network architecture diagram shown on the top row is for illustrative purpose only.

5.2 Stochastic Layers

Utilising stochastic layers, where the filters are applied to a random selections of features, can provide an alternative workflow than simply the straightforward

Fig. 3. A broad range of styles and novel outcomes can be achieved by chaining transformations. The 4 images show some samples of different configurations of transformations applied to different sets of features on different layers. All results are produced using the StyleGAN2 model trained on the FFHQ dataset.

direct manipulation of parameters for producing a single output. For instance, to take a real world example, these network bending techniques were used in the production of a series of EP (extended play record) artworks for the music band *0171*, which provides an illustration of some of the affordances of the stochastic layers. A series of artworks were commissioned for 5 singles in an EP, such that the works had to be variations on the same theme, and while each artwork had to be unique, they shared a common visual aesthetic. A workflow was specifically developed to fit the brief and make use of possibilities afforded by the many variations on a theme made possible by using stochastic transformation layers.

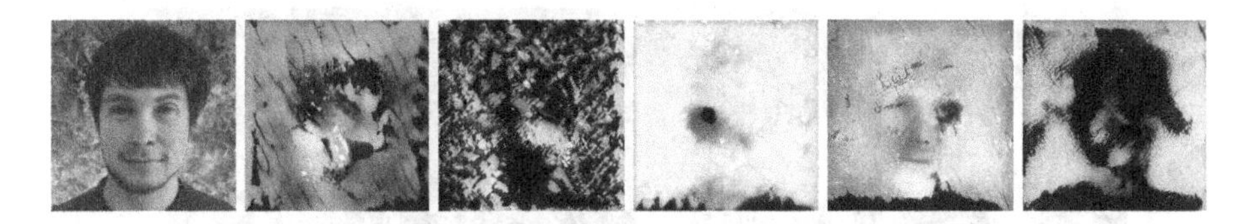

Fig. 4. Left: the projected image into the FFHQ StyleGAN2 latent space of one of the band members of *0171*. Other images: The 5 EP selected artworks that were made using the same latent code, but with multiple stochastic network bending transformations.© *0171*, 2020.

A photograph of one of the band members was taken, and then the digital image was projected into the StyleGAN2 latent space of the model trained on the FFHQ dataset [1,26]. Different configurations of random layers were applied, and large batches of results were generated. When a configuration produced stylistically interesting and varied results, a hand-picked selection was saved, to later be shown to the band. Through a process of iteration from different input photographs, and different transform configurations, a final configuration was found that matched the aesthetic that the band wanted to convey. After generating a large number of samples from this configuration, the best ones were highlighted and the band finally picked their favourite 5 samples, which were then used as the artworks for the singles in the EP (Fig. 4).

6 Discussion

The main motivation of the clustering algorithm presented in this paper was to simplify the parameter space in a way that allows for more meaningful and controllable manipulations whilst also enhancing the expressive possibilities afforded by interacting with the system. Our results show that the clustering algorithm is capable of discovering groups of features that correspond to the generation of different semantic aspects of the results, which can then be manipulated in tandem. These semantic properties are discovered in an unsupervised fashion, and are discovered across the entire hierarchy of features present in the generative model. For example, Fig. 5 shows the manipulation of groups of features across a broad range of layers that control the generation of: the entire face, the spatial formation of facial features, the eyes, the nose, textures, facial highlights and overall image contrast.

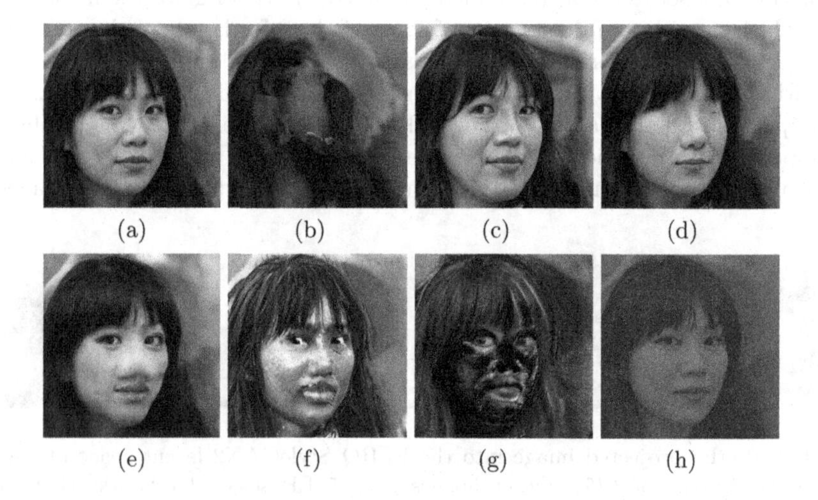

Fig. 5. Clusters of features in different layers of the model are responsible for the formation of different image attributes. (a) The unmanipulated result. (b) A cluster in layer 1 has been multiplied by a factor of -1 to completely remove the facial features. (c) A cluster in layer 3 has been multiplied by a factor of 5 to deform the spatial formation of the face. (d) A cluster in layer 6 has been ablated to remove the eyes. (e) A cluster in layer 6 has been dilated to enlarge the nose. (f) A cluster in layer 9 has been multiplied by a factor of 5 to distort the formation of textures and edges. (g) A cluster of features in layer 10 have been multiplied by a factor of -1 to invert the highlights on facial regions. (h) A cluster of features in layer 15 has been multiplied by a factor of 0.1 to desaturate the image. All transformations have been applied to sets of features discovered using our feature clustering algorithm (Sect. 3) in the StyleGAN2 model trained on the FFHQ dataset.

Grouping and manipulating features in a semantically meaningful fashion is an important component for allowing expressive manipulation. However, artists are often also ready to consider surprising, unexpected results, to allow for the creation of new aesthetic styles, which can become uniquely associated to an individual or group of creators. Therefore the tool needs to allow for unpredictable as well as predictable possibilities, which can be used in an exploratory fashion and can be mastered through dedicated and prolonged use [15]. There is usually a balance between utility and expressiveness of a system [23]. While it will be required to build an interface and perform user studies to more conclusively state that our approach has struck such a balance, our current results do show that both predictable semantic manipulation and more unpredictable, expressive outcomes are possible. This is a good indication that our approach represents a good initial step, and with further refinements it can become an innovative powerful tool for producing expressive outcomes, when using deep generative models.

6.1 Active Divergence

One of the key motivations of our network bending approach, was to allow for the direct manipulation of generative models, such that the results were novel and divergent from the original training data, a goal that has been referred to as *active divergence* [7]. One common criticism of using deep generative models in an artistic and creative context, is that they can only re-produce samples that *fit* the distribution of samples in the training set. However, by introducing deterministic controlled filters into the computation graph during inference, these models can be used to produce a large array of novel results. Figure 1 shows how the results vary drastically by applying the same transformation with the same parameters to different layers. Because our method alters the computational graph of the model, these changes to the results take effect across the entire distribution of possible results that can be generated. The results we have obtained markedly lie outside the distribution of training images, and allow for a very large range of possible outcomes, of which a small set of examples is seen in Fig. 3, which shows the broad range of outcomes possible when various transformation applied to different sets of features in different layers are combined. We emphasise that such outcomes, to the best of our knowledge, could not reasonably be produced using any other existing method of image manipulation or generation.

6.2 Comparison with Other Methods

With respect to the semantic analysis and manipulation of a generative model, our approach of clustering features and using a broad array of transformation layers is a significant advance over previous works [3–5, 10]. This recent thread of techniques only interrogate the function of individual features, and as such are unlikely to be capable of capturing a full account of how a deep network generates results, since such networks tend to be robust to the transformation of individual features.

We also show that sets of features, which may not be particularly responsive to certain transformations, are very responsive to others. Figure 6 shows that in the model trained on the LSUN church dataset, a cluster of features, that when ablated has little noticeable effect on the result, can produce significant changes when using another transformation on the same cluster, here removing the trees and revealing the church building that was obscured by the foliage in the original result. This, we argue, shows that the functionality of features, or sets of features, cannot be understood only through ablation (which is the approach used in GAN dissection [5]), because of the high levels of redundancy present in the learned network parameters. We show that their functionality can be better understood by applying a wide range of deterministic transformations, of which different transformations are better suited to revealing the utility of different sets of features (Figs. 5 and 6).

Fig. 6. Groups of features that are not particularly sensitive to ablation may be more sensitive to other kinds of transformation. Left: original unmodified input. Middle: a cluster of features in layer 3 that has been ablated. Right: the same cluster of features that has been multiplied by a scalar of 5. As can be seen ablation had a negligible effect, only removing a small roof structure which was behind the foliage. On the other hand, multiplying by a factor of 5 removes the trees whilst altering the building structure to have gable roof sections on both the left and right sides of the church - which are now more prominent and take precedence in the generative process. Samples are taken from the StyleGAN2 model trained on the LSUN church dataset.

Our method of analysis is completely *unsupervised*, and does not rely on auxiliary models trained on large labelled datasets (such as in [5,22,32]) or other kinds of domain specific knowledge. This approach therefore can be applied to any CNN based GAN architecture used for image generation which has been trained on any dataset. This is of particular relevance to artist who create their own datasets and would want to apply these techniques to models they have trained on their own data. Labelled datasets, especially the pixel labelled datasets used in semantic image synthesis, are prohibitively time consuming (or expensive) to produce for all but a few artists or organisations. Having a method of analysis that is completely unsupervised and can be applied to unconditional generative models is important in opening up the possibility that such techniques become adopted more broadly.

The framework we have presented is the first approach to manipulating generative models that focuses on allowing for a large array of novel expressive outcomes. In contrast to other methods that manipulate deep generative models [2,3], our approach allows the manipulation of any feature or set of features in any layer, with a much broader array of potential transformations. By allowing for the combination of many different transformations, it is evident that the outcomes can diverge significantly from the original training data, allowing for a much broader range of expressive outcomes and new aesthetic styles than would be possible with methods derived from semantic image synthesis [22,32] or semantic latent manipulation [21,36].

7 Conclusion and Future Work

In this paper we have introduced a novel approach for the interaction with and manipulation of deep generative models that we call *network bending* which we have demonstrated on generative models trained on several datasets. By inserting deterministic filters inside the network, we present a framework for performing manipulation inside the networks' black-box and utilise it to generate samples that have no resemblance to the training data, or anything that could be created easily using conventional media editing software. We also present a novel clustering algorithm that is able to group sets of features, in an unsupervised fashion, based on spatial similarity of their activation maps. We demonstrated that this method is capable of finding sets of features that correspond to the generation of a broad array of semantically significant aspects of the generated images. This provides a more manageable number of sets of features that a user could interact with. We propose that using our approach, possibly in conjunction with other methods, for a better understanding and navigating of the latent space of a model, can provide a very powerful set of tools for the production of novel and expressive images.

In future work we plan to build an interface around this framework, and to perform user studies to understand how artists would want to and how they end up using this framework, and to refine the parameter space to allow for a balance between utility and expressiveness of possible outcomes. At the time of publication, this network bending framework—inserting deterministically controlled transformation layers and applying them to clustered sets of features—has been adapted and applied in the domains of audio synthesis [30] and audio-reactive visual synthesis [12]. In future work we look to further extend this framework to generative models of other domains such as those that produce text, video or 3D images and meshes.

References

1. Abdal, R., Qin, Y., Wonka, P.: Image2StyleGAN: how to embed images into the StyleGAN latent space? In: Proceedings of the IEEE International Conference on Computer Vision, pp. 4432–4441 (2019)

2. Bau, D., Liu, S., Wang, T., Zhu, J.-Y., Torralba, A.: Rewriting a deep generative model. In: Vedaldi, A., Bischof, H., Brox, T., Frahm, J.-M. (eds.) ECCV 2020. LNCS, vol. 12346, pp. 351–369. Springer, Cham (2020). https://doi.org/10.1007/978-3-030-58452-8_21
3. Bau, D., et al.: Semantic photo manipulation with a generative image prior. ACM Trans. Graph. (TOG) **38**(4), 1–11 (2019)
4. Bau, D., Zhou, B., Khosla, A., Oliva, A., Torralba, A.: Network dissection: quantifying interpretability of deep visual representations. In: Proceedings of the IEEE Conference on Computer Vsion and Pattern Recognition, pp. 6541–6549. openaccess.thecvf.com (2017)
5. Bau, D., et al.: GAN dissection: visualizing and understanding generative adversarial networks. In: International Conference on Learning Representations (November 2018)
6. Ben-Kiki, O., Evans, C., Ingerson, B.: YAML ain't markup language (YAML^{TM}) version 1.1. Working Draft 2008–05 11 (2009)
7. Berns, S., Colton, S.: Bridging generative deep learning and computational creativity. In: Proceedings of the 11th International Conference on Computational Creativity (2020)
8. Bontrager, P., Roy, A., Togelius, J., Memon, N., Ross, A.: DeepMasterPrints: generating MasterPrints for dictionary attacks via latent variable evolution. In: 2018 IEEE 9th International Conference on Biometrics Theory, Applications and Systems (BTAS), pp. 1–9. IEEE (2018)
9. Bridle, J.S.: Probabilistic interpretation of feedforward classification network outputs, with relationships to statistical pattern recognition. In: Soulié, F.F., Hérault, J. (eds.) Neurocomputing. NATO ASI Series (Series F: Computer and Systems Sciences), vol. 68. Springer, Heidelberg (1990). https://doi.org/10.1007/978-3-642-76153-9_28
10. Brink, P.: Dissection of a generative network for music composition. Master's thesis (2019)
11. Brock, A., Donahue, J., Simonyan, K.: Large scale GAN training for high fidelity natural image synthesis. In: International Conference on Learning Representations (2019)
12. Brouwer, H.: Audio-reactive latent interpolations with StyleGAN. In: NeurIPS 2020 Workshop on Machine Learning for Creativity and Design (2020)
13. Celebi, M.E., Kingravi, H.A., Vela, P.A.: A comparative study of efficient initialization methods for the k-means clustering algorithm. Expert Syst. Appl. **40**(1), 200–210 (2013)
14. Dhariwal, P., Jun, H., Payne, C., Kim, J.W., Radford, A., Sutskever, I.: Jukebox: a generative model for music. arXiv preprint arXiv:2005.00341 (2020)
15. Dobrian, C., Koppelman, D.: The 'E' in NIME: musical expression with new computer interfaces. In: NIME, vol. 6, pp. 277–282 (2006)
16. Dosovitskiy, A., Springenberg, J.T., Riedmiller, M., Brox, T.: Discriminative unsupervised feature learning with convolutional neural networks. In: Advances in Neural Information Processing Systems, pp. 766–774 (2014)
17. Fernandes, P., Correia, J., Machado, P.: Evolutionary latent space exploration of generative adversarial networks. In: Castillo, P.A., Jiménez Laredo, J.L., Fernández de Vega, F. (eds.) EvoApplications 2020. LNCS, vol. 12104, pp. 595–609. Springer, Cham (2020). https://doi.org/10.1007/978-3-030-43722-0_38
18. Forgy, E.W.: Cluster analysis of multivariate data: efficiency versus interpretability of classifications. Biometrics **21**, 768–769 (1965)

19. Goodfellow, I., et al.: Generative adversarial nets. In: Advances in Neural Information Processing Systems, pp. 2672–2680 (2014)
20. Grézl, F., Karafiát, M., Kontár, S., Cernocky, J.: Probabilistic and bottle-neck features for LVCSR of meetings. In: 2007 IEEE International Conference on Acoustics, Speech and Signal Processing-ICASSP 2007, vol. 4, pp. IV-757. IEEE (2007)
21. Härkönen, E., Hertzmann, A., Lehtinen, J., Paris, S.: GANSpace: discovering interpretable GAN controls. arXiv preprint arXiv:2004.02546 (2020)
22. Isola, P., Zhu, J.Y., Zhou, T., Efros, A.A.: Image-to-image translation with conditional adversarial networks. In: Proceedings of the IEEE Conference on Computer Vision and Pattern Recognition, pp. 1125–1134 (2017)
23. Jacobs, J., Gogia, S., Měch, R., Brandt, J.R.: Supporting expressive procedural art creation through direct manipulation. In: Proceedings of the 2017 CHI Conference on Human Factors in Computing Systems, pp. 6330–6341 (2017)
24. Karras, T., Aila, T., Laine, S., Lehtinen, J.: Progressive growing of GANs for improved quality, stability, and variation. In: International Conference on Learning Representations (2017)
25. Karras, T., Laine, S., Aila, T.: A style-based generator architecture for generative adversarial networks. In: Proceedings of the IEEE Conference on Computer Vision and Pattern Recognition, pp. 4401–4410 (2019)
26. Karras, T., Laine, S., Aittala, M., Hellsten, J., Lehtinen, J., Aila, T.: Analyzing and improving the image quality of StyleGAN. arXiv preprint arXiv:1912.04958 (2019)
27. Kingma, D.P., Ba, J.: Adam: a method for stochastic optimization. arXiv preprint arXiv:1412.6980 (2014)
28. Kingma, D.P., Welling, M.: Auto-encoding variational Bayes. In: International Conference on Learning Representations (2013)
29. Lloyd, S.: Least squares quantization in PCM. IEEE Trans. Inf. Theory **28**(2), 129–137 (1982)
30. McCallum, L., Yee-King, M.: Network bending neural vocoders. In: NeurIPS 2020 Workshop on Machine Learning for Creativity and Design (2020)
31. Oord, A.V.D., et al.: WaveNet: a generative model for raw audio. arXiv preprint arXiv:1609.03499 (2016)
32. Park, T., Liu, M.Y., Wang, T.C., Zhu, J.Y.: Semantic image synthesis with spatially-adaptive normalization. In: Proceedings of the IEEE Conference on Computer Vision and Pattern Recognition, pp. 2337–2346 (2019)
33. Paszke, A., et al.: PyTorch: an imperative style, high-performance deep learning library. In: Advances in Neural Information Processing Systems, pp. 8024–8035 (2019)
34. Radford, A., Metz, L., Chintala, S.: Unsupervised representation learning with deep convolutional generative adversarial networks. In: International Conference on Learning Representations (2016)
35. Rezende, D.J., Mohamed, S., Wierstra, D.: Stochastic backpropagation and approximate inference in deep generative models. In: Proceedings of the 31st International Conference on Machine Learning (2014)
36. Shen, Y., Gu, J., Tang, X., Zhou, B.: Interpreting the latent space of GANs for semantic face editing. In: Proceedings of the IEEE/CVF Conference on Computer Vision and Pattern Recognition, pp. 9243–9252 (2020)
37. Simon, J.: GANBreeder app (November 2018). https://www.joelsimon.net/ganbreeder.html. Accessed 1 Mar 2020

38. Simonyan, K., Vedaldi, A., Zisserman, A.: Deep inside convolutional networks: visualising image classification models and saliency maps. arXiv preprint arXiv:1312.6034 (2013)
39. Soille, P.: Erosion and dilation. In: Morphological Image Analysis. Springer, Heidelberg (1999). https://doi.org/10.1007/978-3-662-03939-7_3
40. Yu, F., Seff, A., Zhang, Y., Song, S., Funkhouser, T., Xiao, J.: LSUN: construction of a large-scale image dataset using deep learning with humans in the loop. arXiv preprint arXiv:1506.03365 (2015)
41. Zeiler, M.D., Fergus, R.: Visualizing and understanding convolutional networks. In: Fleet, D., Pajdla, T., Schiele, B., Tuytelaars, T. (eds.) ECCV 2014. LNCS, vol. 8689, pp. 818–833. Springer, Cham (2014). https://doi.org/10.1007/978-3-319-10590-1_53
42. Zhang, X., Zhou, X., Lin, M., Sun, J.: ShuffleNet: an extremely efficient convolutional neural network for mobile devices. In: Proceedings of the IEEE Conference on Computer Vision and Pattern Recognition, pp. 6848–6856 (2018)
43. Zhou, B., Zhao, H., Puig, X., Fidler, S., Barriuso, A., Torralba, A.: Scene parsing through ADE20K dataset. In: Proceedings of the IEEE Conference on Computer Vision and Pattern Recognition (CVPR), pp. 5122–5130 (2017)
44. Zhou, B., Khosla, A., Lapedriza, A., Oliva, A., Torralba, A.: Object detectors emerge in deep scene CNNs. In: International Conference on Learning Representations (2015)

SyVMO: Synchronous Variable Markov Oracle for Modeling and Predicting Multi-part Musical Structures

Nádia Carvalho[1(✉)] and Gilberto Bernardes[1,2(✉)]

[1] Faculty of Engineering, University of Porto, Porto, Portugal
{n.svc,gba}@fe.up.pt
[2] INESC TEC, Porto, Portugal

Abstract. We present *SyVMO*, an algorithmic extension of the Variable Markov Oracle algorithm, to model and predict multi-part dependencies from symbolic music manifestations. Our model has been implemented as a software application named *INCITe* for computer-assisted algorithmic composition. It learns variable amounts of musical data from style-agnostic music represented as multiple viewpoints. To evaluate the SyVMO model within *INCITe*, we adopted the Creative Support Index survey and semi-structured interviews. Four expert composers participated in the evaluation using both personal and exogenous music corpus of variable size. The results suggest that *INCITe* shows great potential to support creative music tasks, namely in assisting the composition process. The use of *SyVMO* allowed the creation of polyphonic music suggestions from style-agnostic sources while maintaining a coherent melodic structure.

Keywords: Automatic music generation · Pattern analysis · Variable Markov Oracle · Context-dependent · Variable-length · Musical composition

1 Introduction

Stemming from information theory principles and the postulate of music as a low entropy phenomenon [11], musical informatics has been historically exploring the algorithmic modeling and prediction of musical structure.

To address the temporal and hierarchical nature of the musical structure, algorithmic methods are typically informed by a sequence of past events, i.e., a *context*, to model existing structures and predict or generate new structures [10]. These models aim to capture different degrees of inter-dependency across the component elements of the musical structure.

Existing models focus mostly on the temporal dimension of music. The operational property of such methods is to slice the musical time into component elements (e.g., notes or chords) and provide the conditional distribution or paths over an *alphabet* – i.e., a comprehensive collection of musical events – given a

J. Romero et al. (Eds.): EvoMUSART 2021, LNCS 12693, pp. 37–51, 2021.
https://doi.org/10.1007/978-3-030-72914-1_3

context.[1] To encode harmony, pitch-class sets or chord notations are typically adopted to address the vertical structure. However, both misrepresent the inter-part dependencies towards a balanced information content [7]. The former is quite sensitive to chord notes instances because of its octave invariance. The latter is quite abstract, as it discards the multiple possibilities for instantiating the chord structure.

The balance between familiarity to known compositional traits and novelty introduced by unfamiliar and unpredictable structures is of utter importance in designing generative music systems [6]. This notion of balance is captured by the Wundt curve, a hedonic function that relates the levels of novelty and expectation to the 'pleasantness' of creative works [5]. Representations with a low level of abstraction eliminate almost all redundancy, which is vital for achieving the novelty-familiarity balance. Representations with a high level of abstraction poorly capture the stylistic idiosyncrasies of the musical corpus [7].

In our work, we extend the Variable Markov Oracle (VMO) algorithm by proposing a synchronous VMO (SyVMO), which addresses the aforementioned balance. It models multi-part dependencies from symbolic music manifestations as synchronous VMOs. In this context, the SyVMO enhances the potential for greater novelty in the generation of multi-part musical structures, and models both vertical and horizontal (melodic) traits of musical structure.

The remainder of this paper is structured as follows. Section 2 details the state-of-art in temporal music modeling techniques, particularly those related to statistical inference, such as Markov Models. Section 3 explores the construction and properties of the VMO algorithm. Section 4 describes SyVMO, our algorithmic extension to the VMO. Section 5 expands on the generation of new musical sequences from the proposed SyVMO model. Section 6 explores the application of SyVMO in the context of a software prototype for computer-assisted musical creation, named *INCITe*. Section 7 discusses the preliminary evaluation of *INCITe*. Finally, Sect. 8, presents the conclusions and avenues for future work.

2 State of the Art

Markov models are the earliest and most adopted temporal-modeling techniques in generative music [9]. They model the statistical occurrences of an alphabet across time, from which one can predict and generate musical sequences. Different Markov Models algorithms have been proposed in algorithmic composition: Markov Chains, Hidden Markov Models, Skip-grams, and Variable Markov Models. These algorithms are not complete and incremental. String-matching automaton algorithms promote this completeness and incremental fashion [20]. The automaton algorithms that have been proposed in algorithmic composition are the Factor Oracle (FO) and its extensions: the Audio Oracle (AO) and the VMO, which inherits characteristics from Markov Models. Both Markov

[1] A notable exception to this case is counterpoint modeling and generation [13], where alphabet events typically encode vertical pitch aggregates when processing polyphonic music structures.

and Oracle-based modeling algorithms can process different functional aspects of music, such as melody, harmony, rhythm, timbre, and interaction. However, they are mostly used individually, except for melody and rhythm that are usually linked [15]. The group of symbols modeled by these algorithms is called the alphabet and all past events used to construct the model is referred to as context.

Markov Chains are the simplest and most popular technique. Modeling and prediction consider fixed and finite contexts referred to as the model's order [17,18]. Hidden Markov Models [14] are adopted to model musical events where not all the information is known to predict future events, which can, nonetheless, be derived from the existing information. *Skip*-grams [21] were proposed to parse non-contiguous elements from the musical structure. Finally, the Variable Markov Models [4] allow the adoption of a non-fixed range from previous events. The *Continuator* [19] is a representative system that uses Markov Models to learn and interactively play in a particular style.

The FO algorithm was introduced by [1] as a time- and memory-efficient online string-matching acyclic automaton algorithm.[2] It has recently been proposed for modeling musical structures in an online, incremental fashion and is commonly presented in the literature as tackling current limitations of Markov Models [3]. FO is particularly useful in satisfying the incremental and fast online learning, time-bounded generation of musical sequences, and implementation of multi-attribute models accounting for the multi-dimensionality of music [22]. AO [12] was presented as an extension of the FO to process continuous time-series, such as audio.

The VMO [24] extends the two last models for clustering multivariate time series without *a priori* assumptions on the number of clusters by introducing a threshold value. It adopts an entropy measure to calculate the events' similarity as a measure of information rate (IR) [23].

3 VMO

The FO family algorithms are acyclic word graphs, capable of recognizing at least the factors of a sequence p, having fewer states as possible (minimum is the length of p plus 1) with linear complexity in time and space. The model's repeated patterns are denoted by two types of links between states: *factor links* and *suffix links*. Factor links indicate paths across states that produce similar patterns by continuing forward. Symbols are associated with these transitions as they denote the construction of the sequence. Each state has a factor link to the next contiguous state and can have one or more factor links to other, non-contiguous states. Suffix links indicate paths across states that share the largest similar subsequence from the input sequence. These transitions do not have symbols associated [1,2].

[2] Efficient data structure for representing substring index of a given string which allows for storage, processing, and retrieval of compressed information about all its substrings.

The VMO explores FO and AO construction by explicitly identifying event clusters while maintaining the algorithm's online nature. It both treats the cluster labels as FO's symbolic sequence and stores the links across events and their associated cluster. The threshold defining the degree of similarity is computed as a measure of IR, i.e., the amount of entropy of the musical structure. Low threshold values result in a large number of clusters. In the extreme case, each data point is considered as a separate cluster. In this scenario, no real pattern can be abstracted from the structure. A higher threshold value reduces the number of data clusters up to the extreme case of a single cluster containing all data points. In this case, the VMO structure does not retain any original time series characteristics [24]. The online construction of a VMO part is shown in Algorithms 1.1 and 1.2.

A VMO, corresponding to an independent musical part, can model the relations between musical events by considering musical features per event and comparing their distance. Clustering similar events and identifying the longest repeated sequences allows modeling complex, highly-structured musical sequences and identifying repeated patterns.

To model multi-part musical structures, we could only use the VMOs by collapsing all vertical information from polyphonic music as symbols. Therefore, melodic patterns may not be recognized whenever changes occur in the remaining parts. As the vertical events are sliced at every new event, we lose the duration attributes of the individual musical events. One approach that attempts to preserve this information consists of connecting related inter-part events using ties. However, it depends on multiple ties for a single inter-part event, related to the multiple notes that it contains, leading to pattern matching difficulties. This inter-part VMO is particularly efficient in modeling chorale textures, where there is a chord for every (or nearly every) melodic note, making the individual parts inter-dependent. On the other hand, it does not capture every part's contrapuntal singularities [16], as it fails to recognize any variation, repetitions, or similar melodic patterns within and across parts.

Algorithm 1.1. Online Construction of a part VMO

Require: Time Series as $O = O_1 O_2 ... O_T$

1: Create an Oracle P with initial state p_0
2: $sfx_P[0] \leftarrow -1$, $B \leftarrow \emptyset$, $N \leftarrow 1$
3: **for** $t = 1 \rightarrow T$ **do**
4: $\quad Oracle(P = p_1 ... p_t) \leftarrow AddSymbol(Oracle(P = p_1 ... p_{t-1}), O_t)$
5: **return** $Oracle(P = p_1 ... p_T)$

Algorithm 1.2. Adding Symbol to VMO

Require: Oracle $P = p_1...p_t$, time series instance O_{t+1}
1: Create a new state $t+1$
2: $q_{t+1} \leftarrow 0$, $sfx_P[t+1] \leftarrow 0$
3: Create a new transition from t to $t+1$, $\delta(t, q_{t+1}) = t+1$
4: $k \leftarrow sfx_P[t]$
5: **while** $k > -1$ **do**
6: $D \leftarrow$ distances between O_{t+1} and $O[\delta(k, :)]$
7: **if** all distances in D is greater than θ **then**
8: $\delta(k, q_{t+1}) \leftarrow t+1$
9: $k \leftarrow sfx_p[k]$
10: **else**
11: Find the forward link from k that minimizes D, $k' \leftarrow \delta(k, :)[argmin(D)]$
12: $sfx_p[k] \leftarrow k'$
13: **break**
14: **if** $k = -1$ **then**
15: $sfx_P[t+1] = 0$
16: Initialize a new cluster with current frame index, $b_{N+1} \leftarrow t+1$
17: $B \leftarrow [B; b_{N+1}]$
18: Assign a label to the new cluster, $q_{t+1} \leftarrow N+1$
19: Update number of clusters, $N \leftarrow N+1$
20: **else**
21: Assign cluster label based on assigned suffix link, $q_{t+1} \leftarrow q_{k'}$
22: $b_{q_{k'}} \leftarrow [b_{q_{k'}}; t+1]$
23: **return** $Oracle(P = p_1...p_{t+1})$

4 SyVMO

We propose SyVMO, an algorithmic extension of the VMO to tackle the limitation of modeling and predicting inter-part dependencies in polyphonic music structures. SyVMO can theoretically model any number of parts, each represented by a single VMO component.

For each VMO component in the SyVMO model, we store each symbol's onset times in the original score. The onset times allow the (vertical or temporal) synchronization of all VMO components. This information is instrumental in preserving the temporal dimension of the modeled music (across parts) while allowing melodic patterns from each part to be recognized.

The VMO component featuring the largest number of states in a *SyVMO* model is identified as the leading VMO. This VMO maximizes the reach for temporal events at finer degrees by 'defining' the Tatum of the texture, or, in other words, the smallest time interval between successive notes in a musical texture. The leading VMO is an additional part of the modeled music, that includes all modeled VMOs vertically segmented. The states of the leading VMO include symbols for all vertical elements. A new state is created for each new onset across all parts. A SyVMO with four melodic parts and a leading VMO is shown in Fig. 1.

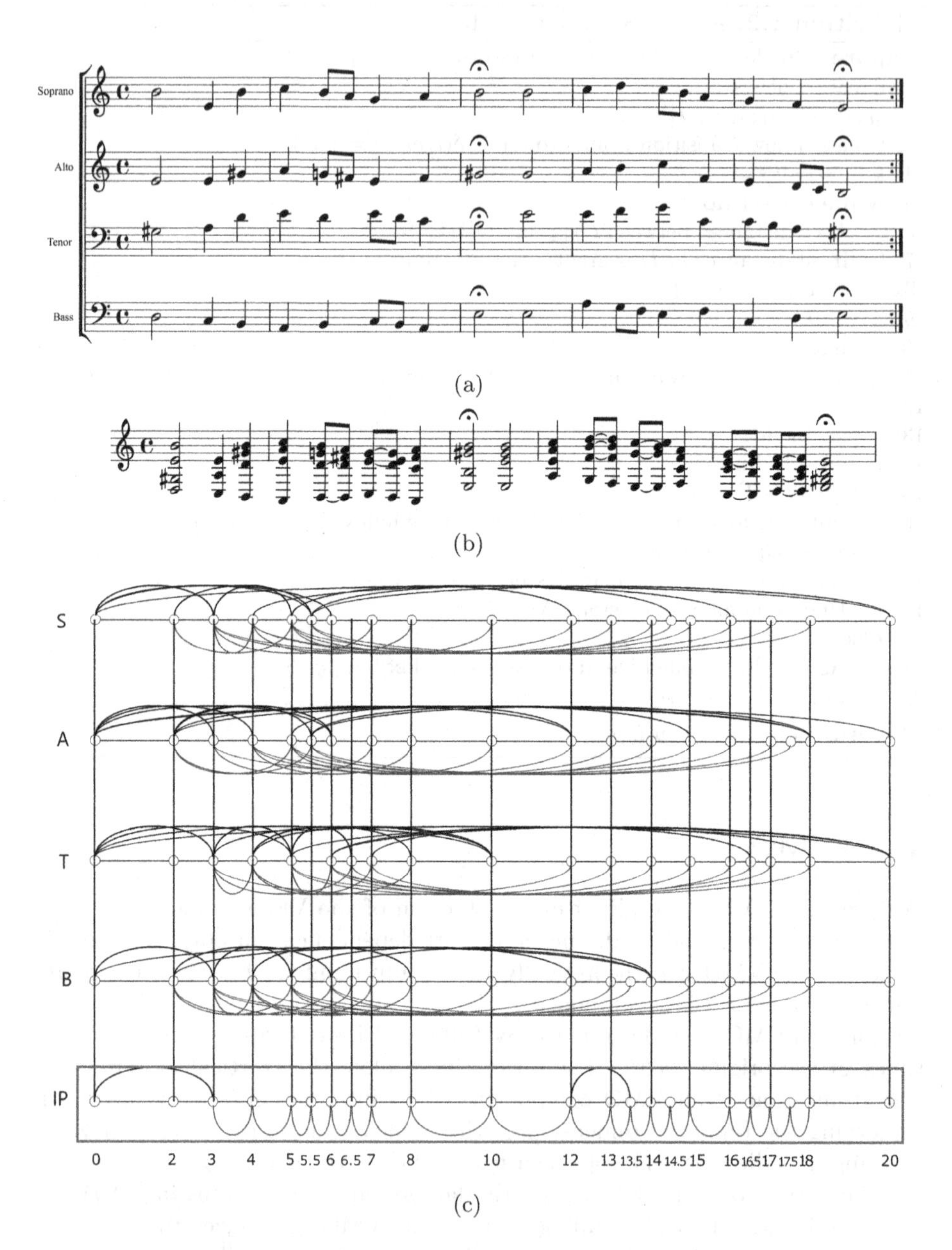

Fig. 1. A SyVMO model (1), depicting a multi-part musical structure (a), in which the last VMO, surrounded by a green box, matches the artificial line extracted from inter-part information (b). The vertical lines symbolize the states aligned at a time offset. The first state of the SyVMO is a non-musical state that serves as the starting point of the model. (Color figure online)

The musical properties of the SyVMO extend those of the original VMO. It maintains their capacity of abstracting the linear patterns as it represents every independent part of a musical structure. Therefore, both the individual qualities of each musical part and the global structure of the vertical harmonic texture will be modeled. This is reinforced by the existence of the leading VMO, representing the inter-part relationships as groupings of simultaneous musical events. By modeling the several parts as individual VMOs, synchronized (or aligned in time) according to the leading VMO, the SyVMO expands the modeling of homophonic to contrapuntal polyphonic textures. Patterns can thus be easily extracted from these forms either in a vertical or linear sense.

5 Generation of Musical Structures from SyVMO

We can generate new musical sequences that retain the original music's characteristics at both linear and vertical levels using a SyVMO model. We can extract a sequence of events by traversing the model links across parts (each component VMO). The algorithm will add new events to the generated musical sequence in an iterative fashion. At each step, we evaluate the existing transitions across all part oracles connected by forward and suffix links and decide the next symbol in the generation for every part of the sequence. This process consists of three steps. The first step retrieves the current state for each component VMO. If a component VMO does not have a state at a particular onset, the dictionary will have a no-state flag for that component VMO. The second step determines all possible transitions from each symbol and decides which transition to retain in the generated sequence. The third consists of calculating and retrieving the block of symbols that compose the next selected event in the sequence.

While computing all possible transitions from the current state, we take into account three conditions. 1) If no component VMO is currently in its last state and no suffix links are found, the chosen transition will be the direct forward link for every component VMO. 2) If any component VMO is currently in its last state, a suffix link will be chosen, preferably those that lead to a state with synchronous events in all component VMOs. In the latter case, the algorithm retrieves the symbols associated with the direct forward links of the selected states across all component VMOs. 3) If all component VMOs have suffix links departing from the current states (which not the last one), a randomly decision defines the probability of transitioning by a forward or suffix link. In both cases, the preference will be given to transitions that lead to states with synchronized onset events in all component VMOs. The choice across the retrieved suffix links during generation maximize the selection of the longest repeated suffix.

To calculate the symbols to be included in the generated sequence, we assume the vertical structure of the various oracles by connecting blocks of synchronized states (i.e., blocks of states that belong to the same time frame), as shown in Fig. 2. Therefore, instead of retrieving a single state for every step of the algorithm, we consider a block of states that fall within the longest duration found across all parts, as described in Algorithm 1.3. To this end, retrieve all

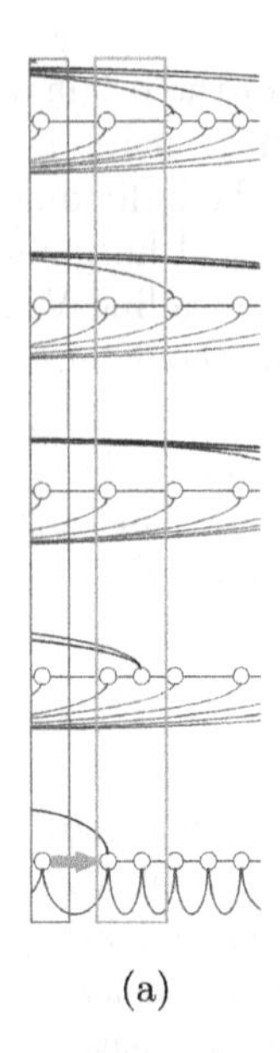

(a)

(b)

Fig. 2. In (a), we can observe a block identification for a forward transition, where the current state is the one in the green box and the transition chosen was the one made by the blue arrow in the artificial line. The block stored is the one in the blue box. In (b), we can see the respective musical score. (Color figure online)

symbols at the selected time frame across all component VMOs. If a component VMO does not have an event onset at the selected time frame, an artificial (rest) symbol is created until a new event is found. Similarly, we also evaluate the block's end for every component VMO, creating artificial symbol if its last events do not match the maximal duration of the block. The last step guarantees a synchronous block whose inter-part relations from the original modeled material are retained in the generation.

The SyVMO presents a clear advantage to a single inter-part VMO in generating polyphonic musical sequences. The larger number of connections from which to choose the next event to be generated is a clear evidence. This is particularly valid because the number of suffix links in an inter-part VMO (and the number of patterns encountered) is usually lower than a VMO capturing from individual parts. By using every individual part VMO, in addition to the inter-part VMO, we can access all the forward and suffix links at every slice, thus increasing the number of possible transitions for the next generated event.

Algorithm 1.3. Block Recognition and Storing

Require: $Oracles = P_1P_2...P_N$, states at resulting 'symbol' for every oracle $next\_states$, $offsets$ at every state of each oracle.

1: $min\_offset \leftarrow min([offsets[i][next\_states[i]]|\ i \in Oracles])$
2: $max\_offset \leftarrow max([offsets[i][next\_states[i]+1]|\ i \in Oracles])$
3: **for** $i = 1 \rightarrow N$ **do**
4: **if** $offsets_i[next\_states[i]] \neq min\_offset$ **then** # If there is no state starting at that time for the oracle, add an artificial state
5: $Sequences[i] \leftarrow [Sequences[i];\ 'None\_' + string(offsets_i[next\_states[i]] - min\_offset])$
6: **for** $j = next\_states[i] \rightarrow length(offsets_i)$ **do**
7: **if** $offsets_i[j+1] \leq max\_offset$ **then**
8: $Sequences[i] \leftarrow [Sequences[i];\ j]$
9: $ktraces[i] \leftarrow [ktraces[i];\ j]$
10: **else**# Add an artificial state to avoid a possible missing block at the end
11: $Sequences[i] \leftarrow [Sequences[i];\ 'None\_' + string(max\_offset - offsets_i[j])]$
12: **Continue**

6 *INCITe*: Application of SyVMO to Computer-Assisted Algorithmic Composition

We developed *INCITe*,[3] a Python application to assist in the music composition process. It integrates with the notation software Musescore[4] as a plugin. The core of the application relies on the SyVMO algorithm. The software models stylistic traits from variable-size symbolic music, e.g., a single musical composition or a corpus of works. Novel musical content is suggested to the user (i.e., composer) based on the SyVMO output. Multiple *ranked* suggestions of musical sequences foster and provoke the composer at creating variations or continuations of a given musical context.

Figure 3 shows the architecture of INCITe, which features two main component modules: the logic module and the interface. The former addresses the logic behind *INCITe* and communicates with the latter, which displays the graphical user interface, where the parameterization of the system can be defined.

The Logic module is divided into two sub-modules: representation and generation. The representation sub-module, shown in Fig. 4a, addresses the encoding of symbolic musical manifestations as Multiple Viewpoint Systems [11]. It consists of multiple parsers for extracting and storing symbolic musical information from MusicXML files. The generation sub-module, shown in Fig. 4b, addresses the modeling and generation of stylistic-driven music structures using the SyVMO. The ranking order of the generated musical sequences adopts a proximity-based metric based on the number non-contiguous (suffix and for-

[3] The implementation for both SyVMO and *INCITe* can be found online at https://github.com/NadiaCarvalho/INCITe.git, last access on February 2021.
[4] https://musescore.org/en, last access on February 2021.

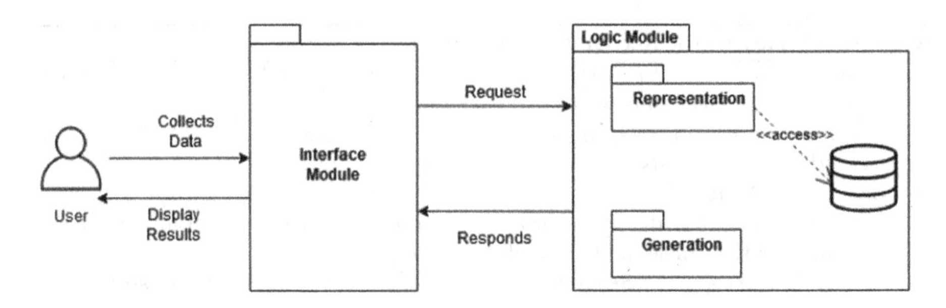

Fig. 3. General architecture of *INCITe*.

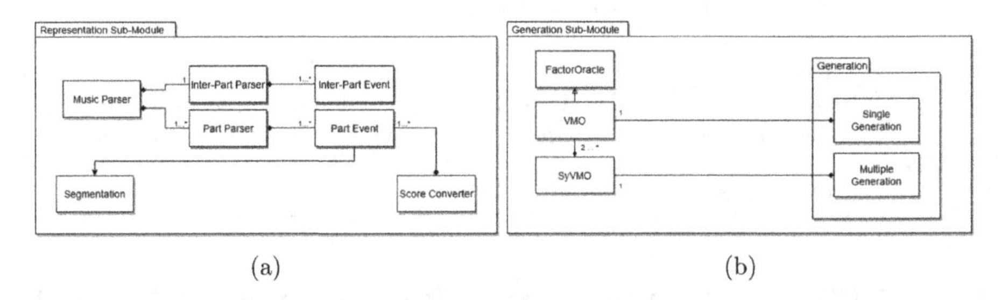

Fig. 4. Structure of *INCITe*'s representation (a) and generation (b) sub-modules.

ward link transitions). The smaller the resulting number, the greater the degree of similarity to the modeled source.

The *INCITe*'s interface (Fig. 5) allows the composer to interact with the processes incorporated in the first module, either in the form of a single cross-platform application or as a plugin for the music notation editor Musescore.

7 Evaluation and Results

The preliminary evaluation of *INCITe* consisted of two experiments, aiming to evaluate complementary aspects of the application, including the use of the SyVMO and its creative potential for an expert-based community of composers. In this context, participants were recruited with a purposive sampling, namely from a pool of post-graduate composition students. We collected data from four participants with an average age of 25.

The twofold evaluation was conducted online. The first was a task-oriented test, conducted in an informal setting and designed to assess the relevance of the *INCITe* features and the usability of its interface. The second phase adopted the Creativity Support Index [8] questionnaire to evaluate the creative potential of *INCITe*.

In the task-oriented test, participants were asked to conduct three activities, which assessed the modeling and generation modes of *INCITe*. Once the tasks were finalized, an interview with open-questions has followed.

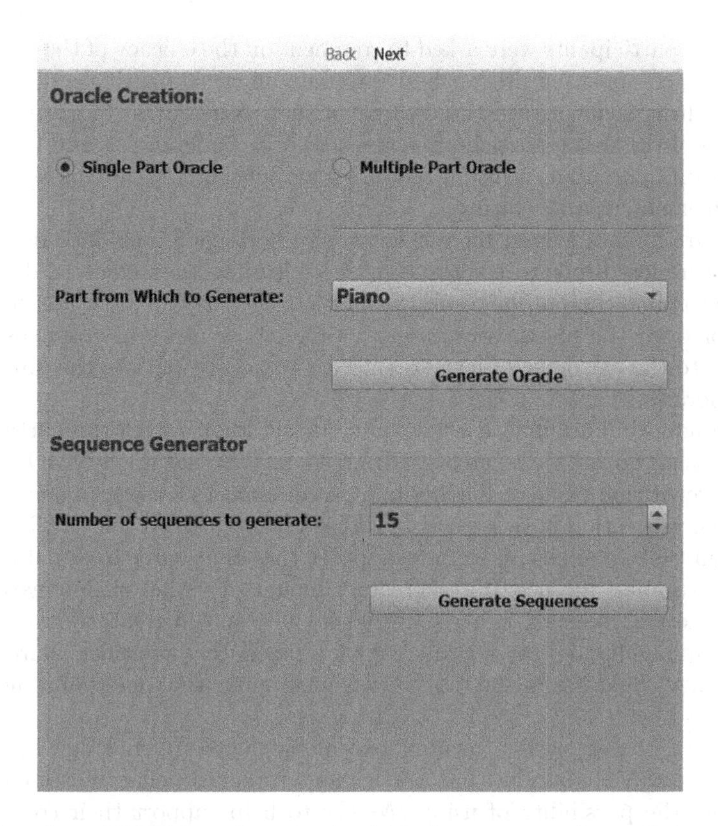

Fig. 5. Interface of *INCITe* at the point in which the user must choose if the generation should refer to a single VMO (and if so, the correspondent part) or a SyVMO and the number of sequences that should be generated.

To cover a wide range of features, we developed three different use case scenarios, allowing the participants to learn the tool's essential functions and experiment with their own compositions. The use case scenarios consisted of 1) using pre-defined musical style using multiple viewpoints, 2) one-part, melodic generation from a user-defined corpus, and 3) multi-part polyphonic generation of a user-defined corpus.

The first task aimed at acquainting the participants with *INCITe* and evaluate the application's capacity to learn the musical style from Bach chorales. A corpus of six Bach's chorales (BWVs 8.6, 9.7, 45.7, 86.6, 124.6, and 139.6) were adopted to create the SyVMO model. The second and third tasks were used to evaluate the application's ability to process the participants' compositions. The second task adopted single part modeling from a participant's original music, and the third task adopted compositions with multiple parts. The second and third tasks adopted a SyVMO from a single work so that the participants can better identify the links between the resulting generation and the model. The participants freely selected the multiple viewpoints adopted in the SyVMO. In the end, the participants were asked to comment on the efficacy of the tool in the flow of their composition process and elaborate on the similarity and usefulness of the resulting sequences in the context of their own works. In greater detail, at the end of the tasks, participants were asked to 1) detail the interface's suitability and its potential utility in real-life situations and 2) provide suggestions to be implemented in the future.

The participants agreed on the harmonic motion's plausibility in the first task yet were less prone to the resulting voice leading. Nevertheless, depending on the viewpoints chosen, the result can vary significantly in capturing the source style. Moreover, the phrase structures, notably those resulting from suspended time (due to the existence of fermatas), seem to poorly capture the phrasing in Bach's chorales.

When adopting original, contemporary music idioms from the participants' works in tasks two and three, an overall agreement was that the SyVMO well captured the repetition of ideas. Furthermore, no differences between monophonic or polyphonic generation have been noted. Participants agreed that *INCITe* could be a useful tool, in exposing to the composer that the source material is rather static, questioning the need for a greater amount of variation. Furthermore, a greater variety of modeled source material would result in greater variation. The participants highlighted these traits, which immediately identified source material that they could use in the tool. It was unanimous that the tool achieved its purpose.

All the tasks were easily executed, except for the relatively long time needed to create the SyVMO model. The participants reported being impressed by the results and the possibility of using *INCITe* to help support their composition process.

Based on concepts and theories from creativity research, the Creativity Support Index (CSI) was proposed by Carroll and Latulipe [8] as a protocol for evaluating creative tools. The test is divided into two parts. In the first part, the participant must fill a short survey with six statements, one for each factor (exploration, collaboration, engagement, effort/reward trade-off, transparency, and expressiveness). A level of agreement for each statement on a twenty-point scale, where zero stands for total disagreement and 20 for total agreement, was requested. Then, the participant has to rate each factor's relative importance against all the others. The global value of the CSI is computed by multiplying each factor level by its importance. A global creativity index is computed as the sum of these values normalized to the [0, 100].

The results of the Creative Support Index are shown in Fig. 6b. In the first part of the test, participants had a general agreement on each factor's level, showing the positive impact of *INCITe*. Simultaneously, all the factors have an average value in the [15.5, 19] range. *Exploration* and *Expressiveness* were the highest-rated factors of interest to the participants and highly ranked in the context of the application, as it was possible to generate countless sequences without making a lot of repetitive steps. Moreover, the potential for using the composer's musical pieces and not only style-limited music, as in most tools,

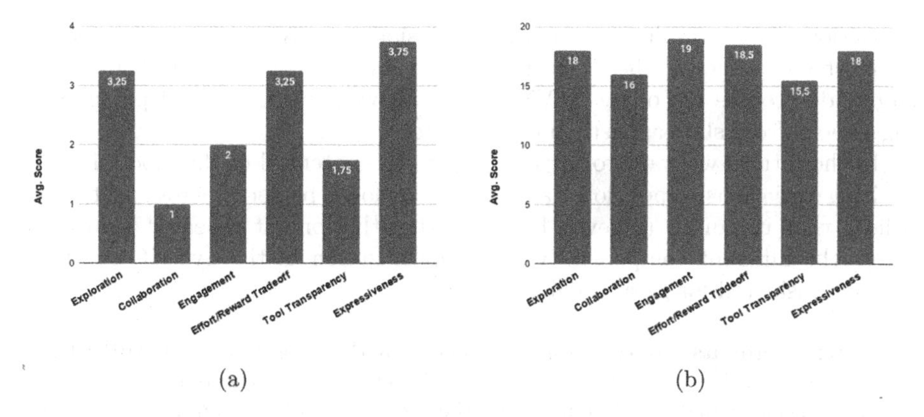

Fig. 6. Average CSI factor interest (a), measured from 0–5 and average score by factor (b), measured from 0–20.

was a relevant feature of the application. *Effort/Reward Tradeoff* and *Engagement* had the highest factor scores. All participants agreed that the resulting sequences, especially those generated from models from their works, were worth the time they took to generate. The participants all concurred that they would use *INCITe* with frequency in their composition process, particularly in early composition phases. They highlighted the potential of *INCITe* in provoking new ideas for musical compositions even if they were discarded in the final composition. The CSI test's final score was 89.1 out of 100, with the participants scores ranging from 81 to 98. The final score suggests that *INCITe* shows great potential to support creative music tasks, namely in assisting the composition process.

8 Conclusions and Future Work

This paper presented an original contribution to the field of algorithmic modeling and generation of symbolic musical manifestations, consisting of an extension to the VMO algorithm to model and predict multi-part dependencies. SyVMO can theoretically model any number of parts. A VMO represents each part, and a leading VMO is used as a guiding oracle. In musical terms, the SyVMO expands the modeling of homophonic to contrapuntal polyphony, as it can capture stylistic traits from both melodic and harmonic musical content.

We made a preliminary evaluation of our SyVMO model by implementing a software prototype, *INCITe*, that assists in the music composition process. It models stylistic traits from variable-size symbolic music and suggests new musical content that shares stylistic traits with the modeled source. Multiple *ranked* suggestions foster and provoke the composer at creating variations or continuations of a given musical context. The information captured by the models is always dependent on the user, which has total freedom to choose the viewpoints adopted to represent the musical material and their relative weight. The evaluation process results support the conclusion that *INCITe* has the potential to assist composers in their creative activity by unlocking creative processes, largely due to the use of the SyVMO, that allowed the creation of polyphonic suggestions from style-agnostic musical sources.

In the future, we want to further explore the potential of the model in recognizing variations, repetitions, and similar melodic patterns across parts. We believe that it can be achieved by comparing the longest repeated sequences across the different component VMOs. Optimization of the SyVMO creation will be considered for greater efficiency.

Acknowledgements. Research partially funded by the project "Co-POEM:Platform for the Collaborative Generation of European Popular Music" (ES01-KA201-064933) which has been funded with support from the European Commission and the project "Experimentation in music in Portuguese culture: History, contexts and practices in the 20th and 21st centuries" (POCI-01-0145-FEDER-031380) co-funded by the European Union through the Operational Program Competitiveness and Internationalization, in its ERDF component, and by national funds, through the Portuguese Foundation for Science and Technology; and by the European Union's Horizon 2020 research. This publication reflects the views only of the author, and the Commission cannot be held responsible for any use which may be made of the information contained therein.

References

1. Allauzen, C., Crochemore, M., Raffinot, M.: Factor oracle: a new structure for pattern matching. In: Pavelka, J., Tel, G., Bartošek, M. (eds.) SOFSEM 1999. LNCS, vol. 1725, pp. 295–310. Springer, Heidelberg (1999). https://doi.org/10.1007/3-540-47849-3_18
2. Assayag, G., Bloch, G., Chemillier, M., Cont, A., Dubnov, S.: OMax brothers: a dynamic topology of agents for improvisation learning. In: Proceedings of the ACM International Multimedia Conference and Exhibition, pp. 125–132 (2006)
3. Assayag, G., Dubnov, S.: Using factor oracles for machine improvisation. Soft. Comput. **8**(9), 604–610 (2004)
4. Begleiter, R., El-Yaniv, R., Yona, G.: On prediction using variable order Markov models. J. Artif. Intell. Res. **22**(1), 385–421 (2004)
5. Berlyne, D.E.: Novelty, complexity, and hedonic value. Percept. Psychophys. **8**(5), 279–286 (1970)
6. Bevington, J., Knox, D.: Cognitive factors in generative music systems. In: Proceedings of the 9th Audio Mostly: A Conference on Interaction With Sound. AM 2014. Association for Computing Machinery, New York (2014)
7. Carvalho, N., Bernardes, G.: Towards balanced tunes: a review of symbolic music representations and their hierarchical modeling. In: Proceedings of the International Conference on Computational Creativity (ICCC), pp. 236–242 (2020)
8. Cherry, E., Latulipe, C.: The creativity support index. In: Proceedings of the 27th International Conference on Human Factors in Computing Systems, pp. 4009–4014 (2009)
9. Conklin, D.: Music generation from statistical models. In: Proceedings of the AISB 2003 Symposium on Artificial Intelligence and Creativity in the Arts and Sciences, pp. 30–35 (2003)

10. Conklin, D., Anagnostopoulou, C.: Representation and discovery of multiple viewpoint patterns. In: Proceedings of the International Computer Music Conference (ICMC), pp. 479–485 (2001)
11. Conklin, D., Witten, I.H.: Multiple viewpoint systems for music prediction. J. New Music Res. **24**(1), 51–73 (1995)
12. Dubnov, S., Assayag, G., Cont, A.: Audio oracle: a new algorithm for fast learning of audio structures. In: Proceedings of the International Computer Music Conference (ICMC), pp. 224–227 (2007)
13. Farbood, M., Schoner, B.: Analysis and synthesis of Palestrina-style counterpoint using Markov chains. In: Proceedings of the International Computer Music Conference (ICMC), vol. 2, pp. 471–474 (2001)
14. Frankel-Goldwater, L.: Computers composing music: an artistic utilization of hidden markov models for music composition. J. Undergrad. Res. **5**(1 and 2), 17–20 (2007)
15. Herremans, D., Chuan, C.H., Chew, E.: A functional taxonomy of music generation systems. ACM Comput. Surv. **50**(5), 1–30 (2017)
16. Laitz, S.G.: The Complete Musician, 2nd edn. Oxford University Press Inc., New York (2008)
17. Manaris, B., Johnson, D., Vassilandonakis, Y.: A novelty search and power-law-based genetic algorithm for exploring harmonic spaces in J.S. Bach chorales. In: Romero, J., McDermott, J., Correia, J. (eds.) EvoMUSART 2014. LNCS, vol. 8601, pp. 95–106. Springer, Heidelberg (2014). https://doi.org/10.1007/978-3-662-44335-4_9
18. Navarro-Cáceres, M., Olarte-Martínez, M., Amílcar Cardoso, F., Martins, P.: User-guided system to generate Spanish popular music. In: Novais, P., et al. (eds.) ISAmI2018 2018. AISC, vol. 806, pp. 24–32. Springer, Cham (2019). https://doi.org/10.1007/978-3-030-01746-0_3
19. Pachet, F., Roy, P., Barbieri, G.: Finite-length Markov processes with constraints. In: Proceedings of the International Joint Conference on Artificial Intelligence (IJCAI), pp. 635–642 (2011)
20. Rueda, C., Assayag, G., Dubnov, S.: A concurrent constraints factor oracle model for music improvisation. In: Proceedings of the XXXII Conferência Latinoamericana de Informática (CLEI), Santiago, Chile, p. 1 (2006)
21. Sears, D.R.W., Arzt, A., Frostel, H., Sonnleitner, R., Widmer, G.: Modeling harmony with skip-grams. In: Proceedings of the 18th International Society for Music Information Retrieval Conference (ISMIR), pp. 332–338 (2017)
22. Tatar, K., Pasquier, P.: Musical agents: a typology and state of the art towards Musical Metacreation. J. New Music Res. **48**(1), 56–105 (2019)
23. Wang, C., Dubnov, S.: Guided music synthesis with variable Markov Oracle. In: Proceedings of the Artificial Intelligence and Interactive Digital Entertainment Conference, vol. WS-14-18, pp. 55–62 (2014)
24. Wang, C.I., Dubnov, S.: Variable Markov oracle: a novel sequential data points clustering algorithm with application to 3D gesture query-matching. In: Proceedings of the IEEE International Symposium on Multimedia, pp. 215–222 (2014)

Identification of Pure Painting Pigment Using Machine Learning Algorithms

Ailin Chen[1,2](✉), Rui Jesus[2,3], and Márcia Vilarigues[1]

[1] Departamento de Conservação e Restauro, Faculdade de Ciências e Tecnologia, Universidade NOVA de Lisboa, 2825-149 Caparica, Lisboa, Portugal
ailin.chen@campus.fct.unl.pt

[2] NOVA LINCS, Faculdade de Ciências e Tecnologia, Universidade NOVA de Lisboa, 2825-149 Caparica, Lisboa, Portugal

[3] M2A/ADEETC, Instituto Superior de Engenharia de Lisboa (ISEL), IPL, Rua Conselheiro Emídio Navarro, nº1, 1959-007 Lisboa, Portugal

Abstract. This paper reports the implementation of machine learning techniques in the identification of pure painting pigments applying spectral data obtained from both the paint tubes used and the paintings produced by Portuguese artist Amadeo de Souza Cardoso. It illustrates the rationales and advantages behind the application of more accurate artificial mixing by subtractive mixing on the reference pigments as well as the use of Root Mean Square Error (RMSE) for distinguishing especially the mixtures that contain white and black, so that a more holistic machine learning approach can be applied; notably, the experiment of neural network for discerning black and white pigments, which later could be applied for both pure and mixed pigment identification. Other machine learning techniques like Decision Tree and Support Vector Machine are also exploited and compared in terms of the identification of pure pigments. In addition, this paper proposes the solution to the common problem of highly-imbalanced and limited data in the analysis of historical artwork field.

Keywords: Machine learning · Neural network · Artificial intelligence · Pigment unmixing · Pigment identification · Painting reconstruction · Hyperspectral imaging · Visualization · Restoration

1 Introduction

The identification of painting pigments has been studied to a certain extent by various researchers; the majority of the methods focuses on the in-situ or sample identification and analysis of paint pigments implementing different hardware technologies which oftentimes can be invasive to the artworks. The combination of applying the different chemical analysis results from the pigment on the paintings and the spectral images of the paintings, especially, the employment of non-invasive machine learning method, and to be more precise, the application of Neural Network (NN) on the identification and visualization of the pigments of the paintings has been relatively less. This work utilizes

J. Romero et al. (Eds.): EvoMUSART 2021, LNCS 12693, pp. 52–64, 2021.
https://doi.org/10.1007/978-3-030-72914-1_4

the previously acquired data by the Conservation and Restoration Department (DCR) of Universidade NOVA de Lisboa (NOVA) [1, 2] and, intends to optimize the identification of the pigments, so that the results can be better integrated with the previously-reported non-invasive brushstroke analysis not only for the purposes of authentication or restoration of works by late Portuguese artist Amadeo de Souza Cardoso, but also for the aim of its application and generalization on paintings and drawings by other artists and other genres. This paper centers around the major improvements on the work by Montagner [1, 2] and further developments in terms of the recognition and visualization of the pure pigments of Amadeo paintings, particularly, the first attempt on the use of NN in the identification of white and black pigments, which is applied for the identification of mixed pigments at a later stage.

2 Related Work

As noted previously, the investigation of paint pigment comprises of both destructive and non-destructive techniques. Due to the unique cultural and historical values of heritages, even the most careful sampling of the materials from the artworks can only be performed when further information needed is believed to be essential and repeated sampling is not favored [3]; therefore, it's preferable that such examination be done without additional intervention on the artwork itself.

The non-invasive examination and identification of paint pigment have been studied extensively in recent years, in particular, with the aid of hardware technologies like hyperspectral imaging. For instance, Polak et al. [4] reported a non-destructive novel mid-infrared tunable hyperspectral imager extending the electromagnetic spectrum to as wide as between 2500 nm and 3750 nm, which normally the conventional hyperspectral imaging cannot attain, in order to facilitate a more extensive analysis of painting pigments. Cosentino [5] also designed a more chemistry- and physics-oriented flowchart method employing hyperspectral imaging for the recognition of historical pigments, which is deemed to be more suitable for artworks with one single non-mixed layer; in reality it's often not the case with cultural heritages. Conroy et al. [6] appears to be one of the earliest attempts that combine machine learning with imaging technology Raman spectroscopy to identify chemical materials; such application used in waste disposal, chemical or pharmaceutical manufacturing however differs from the implementation of machine learning algorithms on paint pigment identification in cultural heritage settings.

On the other hand, Grabowski et al. [7] intended to integrate hyperspectral information and photography pixel data for the identification of paint pigments with a high-degree unsupervised learning; however, such model has its limitation of the minimum size needed as input for small homogenous regions, hence additional requirement during the acquisition of hyperspectral data. In the research by Deborah et al. [8], the techniques of Spectral Angle Mapper (SAM), Spectral Correlation Mapper (SCM), and a fully constrained unmixing model together with pre-determined parameters are implemented and compared in a case study, with the conclusion that all methods have both advantages and disadvantages. In the recent work done by Fan et al. [9] and Wang et al. [10], machine learning algorithms like SAM, Decision Tree, Support Vector Machine (SVM), and Convolutional Neural Network (CNN) have been implemented and compared accordingly on the hyperspectral data for the purpose of either pigment or painting classification. Their results seem to be effective but none of them have incorporated the systematic artificial mixtures of the reference pigment data with the hyperspectral imaging data, especially with the consideration of Deep Neural Network (DNN), the effect of lighter and darker colors on the identification of pigments and the typical problem with limited and imbalanced data in such field. Our research addresses the afore-mentioned issues.

3 Data

The material analysis data applied in our work was obtained from Montagner [1, 2]; they include both the tube data and the painting data from the late Portuguese artist Amadeo de Souza Cardoso.

In terms of the data acquired from the pigments of the 17 tubes that Amadeo used for painting, they provide the following information for each pigment: color name, RGB and CIELab color components, list of the most significant chemical elements collected via Energy Dispersive X-ray Fluorescence (μ-EDXRF) and reflectance spectra attained via Fiber Optic Reflectance Spectroscopy (FORS). The reference pigment spectral database consists of pigments from the 17 painting tubes including black, white and 15 colors and is shown in Fig. 1.

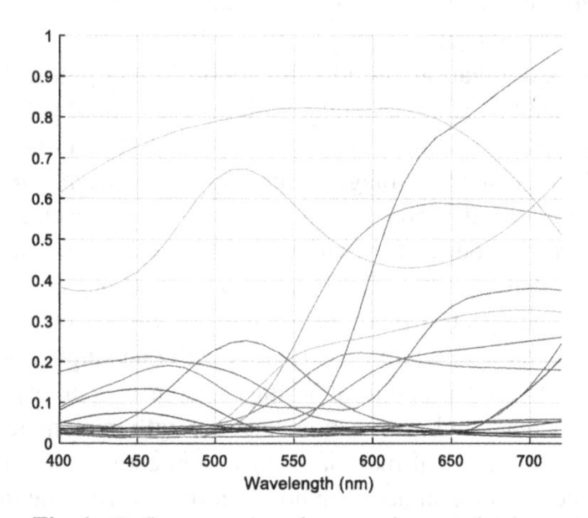

Fig. 1. Reflectances in reference pigment database.

As for the painting data used, there are two sets: one set is the hyperspectral data of 11 Amadeo originals. For each painting to be evaluated, a hyperspectral set of data covering the entire painting is obtained using hyperspectral camera. The hyperspectral data used comprises wavelengths from 400 nm to 720 nm in steps of 10 nm. An example of the hyperspectral image at wavelength 400 nm is shown below (see Fig. 2).

Next, the painting under evaluation is sampled at multiple points with non-destructive μ-EDXRF techniques to obtain information related to the chemical composition at the points sampled. In addition, the RGB and CIELab color components of the point evaluated are obtained. This set of data is referred to as the XRF samples database. For the current research, no further data has been acquired.

Fig. 2. Original image (left) and example of hyperspectral image at 400 nm (right)

3.1 The Challenge

The main challenge faced in the implementation of a machine learning algorithm to this particular field is the quality and quantity of data available. While the former is fixed and specific to the technique used to acquire data the latter remains the major issue as only 11 hyperspectral images are available along with the reference pigment database and existing XRF samples.

Training a machine learning algorithm to identify any of the reference pigments in the database would require a large number of different samples of each pigment ideally under different illumination, bases and of different ages. Yet, only a single sample (single reflectance) of each pigment in the database is available.

A second challenge for the machine learning algorithm is the number of classes required. For a data set of two classes a binary classifier such as Regularized Least Squares Classifier (RLSC) or Support Vector Machine (SVM) could be used. However, to identify 3 or more classes can increase the design complexity of binary classifiers. Using a one-to-one approach would require $m(m - 1)/2$ classifiers where m is the number of classes. Considering the number of pigments in the reference database and all the possible artificial mixtures therefore requires careful evaluation and selection of a suitable machine learning technique.

3.2 Data Augmentation

A way to overcome the data limitation challenge is the use of data augmentation. In this research, to increase the number of samples in the following sections, data has been augmented, where stated, using white Gaussian noise with a signal to noise ratio of 10 dB. This allows generating samples within the standard deviation of the reference signal to avoid overfitting when training. An example of data augmentation is shown in Fig. 3 where 7064 samples have been generated for the reflectance of Cerulean Blue in the reference pigment database. For comparison, the reflectance of pixels identified with the improved version on the original Montagner method [1, 2] as Cerulean Blue in multiple paintings are also shown.

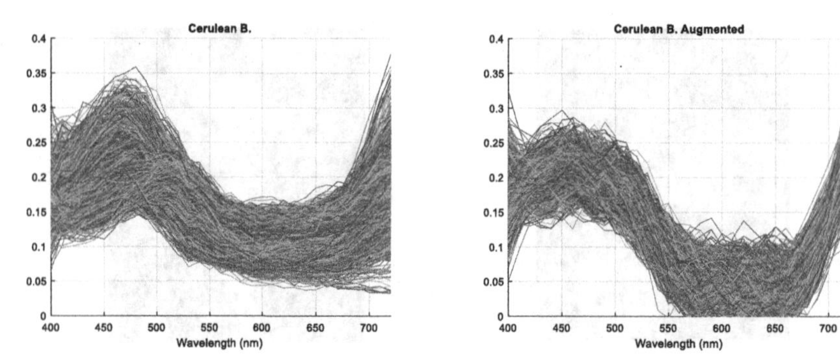

Fig. 3. Comparison of true reflectances for Cerulean Blue (left) and augmented data of Cerulean Blue (right). (Color figure online)

4 Algorithm

The algorithm proposed for pure pigment identification is based on the method described in [1, 2]: in a hyperspectral image, first, black and white pixels are detected by thresholding the Euclidean distance of the hyperspectral image reflectances to black and white reference reflectances. Next, the rest of the hyperspectral image points are compared against other reference pure pigments using SAM. A point is marked as a potential pure pigment if the spectral distance to the closest pigment match is significantly different from the second closest match. Finally, potential pigments are confirmed by verifying the chemical content of the matching pigment and the potential pigment.

To improve this algorithm, it is proposed using Subtractive Mixing of Reflectance and the Root Mean Square Error (RMSE) as metric. It is also analyzed the use of machine learning algorithms, namely, classification methods for pure pigment identification.

4.1 Additive vs Subtractive Mixing of Reflectance

Montagner [1, 2] proposed additive mixing of reflectance in order to artificially produce a series of mixtures based on the available hyperspectral data; it suggests that artificial mixture of colors can be done as the sum of the color reflectances:

$$R_{mix} = \sum_{i=1}^{n} p_i R_i \quad (1)$$

where R_{mix} is the artificial mix reflectance and p_i is the percentage of reflectance R_i with the sum of all percentages equal to 100%.

The above method has the advantage of preserving the shape of the reflectances when added up. This is particularly useful when using matching techniques like Spectral Angle Mapper (SAM) which are particularly insensitive to illumination; this is, SAM performs well when analyzing the overall shape of reflectances where the offset is not important. This method is adequate when applied to color mixing using lights; this is, the superposition of multiple color lights would result in a bright white light. For paints however, this does not apply as the effect of mixing colored paints does not yield white; for a more accurate, non-linear mix of colors, subtractive mixing should be used

[11, 12]. This method allows mixing reflectances in different quantities considering the subtractive effect of amplitudes while maintaining the shape of the resulting reflectance as follow:

$$R_{mix} = \prod_{i=1}^{n} R_i^{p_i} \quad (2)$$

Additive mixing increases the reflectances linearly without correctly handling the amplitude mix. This effectively distorts the color as it increases the overall amplitude of the mix at all wavelengths. The following images, in Fig. 4, illustrate the mixtures obtained with Cerulean Blue/White. As it can be seen, for subtractive mixing the underlying blue tone and reflectance shape is preserved even for mixes containing a large quantity of white, for example 10% blue and 90% white. While for additive mixing the incorrect handling of the amplitude forces the blue tone to gray/white quickly.

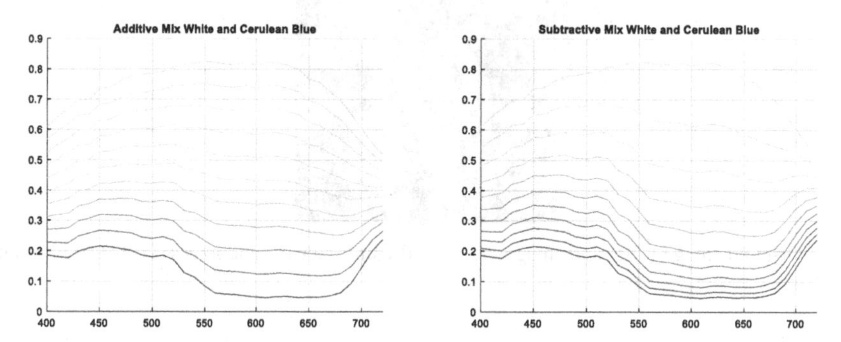

Fig. 4. Effect of mixing colors using additive (left) and subtractive (right) color mixing.

The above confirm that subtractive mixing should be used for all hypothetical mixing of colors using reflectance.

4.2 SAM vs RMSE

In the elemental analysis method proposed in [1, 2], black and white pixels were never considered part of artificial mixtures given that SAM was used as the metric function to evaluate the similitude of reflectances; SAM cannot be used to accurately distinguish between a color that has been mixed with black or white. As a better approach, RMSE is proposed as this metric evaluates difference in amplitudes; this also mean RMSE will more accurately distinguish mixtures that contain white and black.

To exemplify this, the hypothetical mixture of up to 3 pigments out of 17 including black and white in steps of 10% has been created. Then, the reflectance of each pixel in the hyperspectral image is matched using SAM and RMSE for comparison. The image is then recolored with the matching reflectances converted to sRGB. The conversion requires some rough gamma correction which when applied, highlights the issues with SAM; it matches reflectance shapes regardless of amplitude so it cannot effectively discern between lighter and darker colors (see Fig. 5).

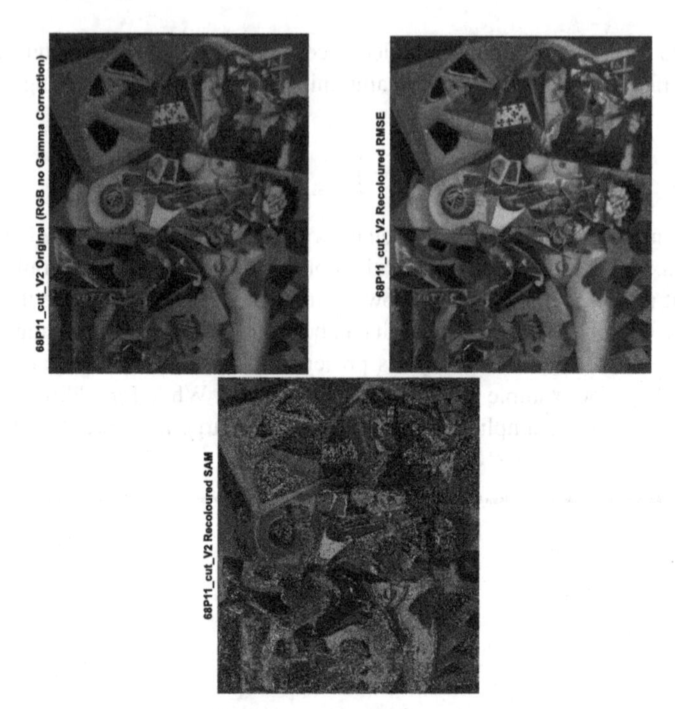

Fig. 5. Original image (left) and images recolored using metric functions RMSE (middle) and SAM (right).

While RMSE shows a promising result, its implementation requires careful use as consistent illumination is required. Yet, RMSE for color matching has the potential to be used to highlight all areas in a painting that contain a certain pigment including black and white colors as shown in the next exemplary images (Fig. 6) where certain pigments are highlighted. The use of RMSE and images like in Fig. 6 could be used in the future to select areas where XRF could be performed to verify the presence of a certain pigment from the reference database for restoration purpose for example.

4.3 Black and White Identification

To identify the simplest tones, black and white, with a classifier three classes are required: black, white or color. To achieve this, a simple NN consisting of an input layer with 33 elements corresponding to a pixel reflectance, a hidden layer of 1 neuron and a single output was used. The training set consisted of three groups of data including augmented samples for white and black pigments and a collection of artificial color mixtures as shown in Table 1.

As the prediction is a numerical value, a threshold is required to classify the prediction into one of the three groups. Thresholds of 0.1 and 0.9 were selected. This is, any predicted value below 0.1 will be classified as black, any value above 0.9 will be classified as white and values between 0.1 and 0.9 inclusive will be classified as color: black (prediction < 0.1), white (prediction > 0.9) or color otherwise.

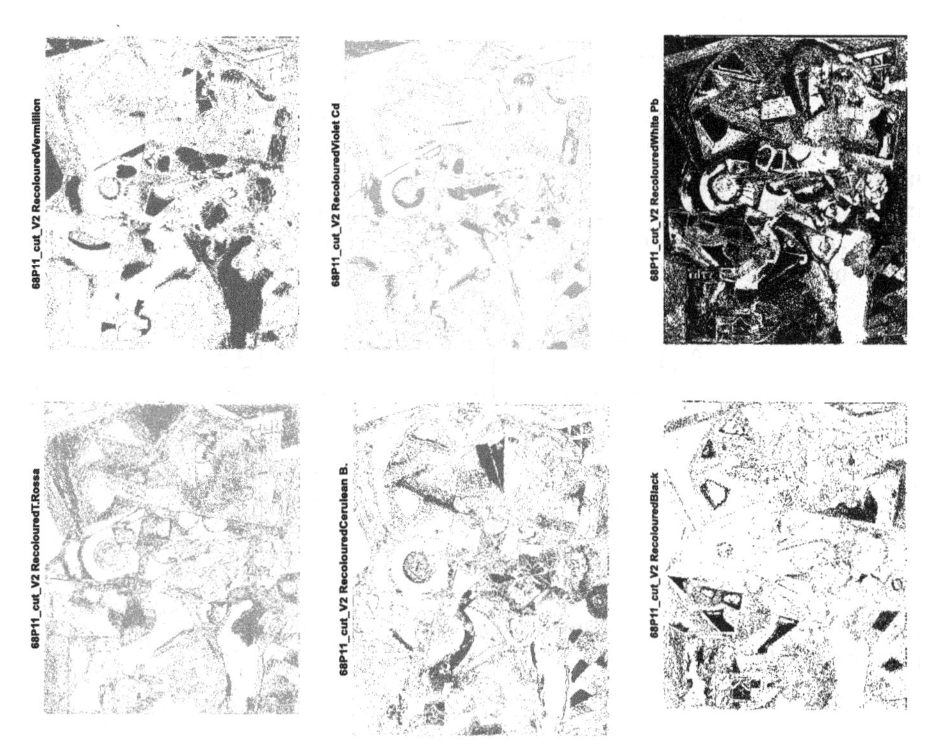

Fig. 6. Areas highlighting the presence of a pigment found through spectral matching with RMSE.

Table 1. Dataset for training of NN with augmented data.

Reflectance	Samples	Augmentation SNR	Label
Black	30000	10 dB	0
White	30000	10 dB	1
3/15 pigment artificial mixture in steps of 10% using subtractive mixing	30030	-	0.5

4.4 Pure Pigment Identification

Using Decision Trees

A second method evaluated for pure pigment identification is decision trees. Three different trees were evaluated: fine, medium and coarse. As opposed to the neural network in the previous section, the trees were trained and tested with reflectance of pixels of a single image classified with the improved original method proposed in [1, 2]. The training set therefore consisted of pure pigments whose label corresponds to the index in the reference pigment database and all other mixed/negative pixels labelled as class 0. The evaluation was performed using k-fold and the results summarized in terms of

Tree Topology and Accuracy are as what follows: Coarse - 93.6%; Medium - 95.2%; Fine - 96.5%.

Although the results seem to show a high accuracy, the training and testing data is highly imbalanced and therefore the accuracy is driven by correct classification of mixed pigments rather than pure pigments as shown in the confusion matrix of the fine tree shown in Fig. 7; the gray diagonal represents the accurate classifications while red entries are misclassifications. The high accuracy is driven by the correct classification of non-pure pixels (class 0).

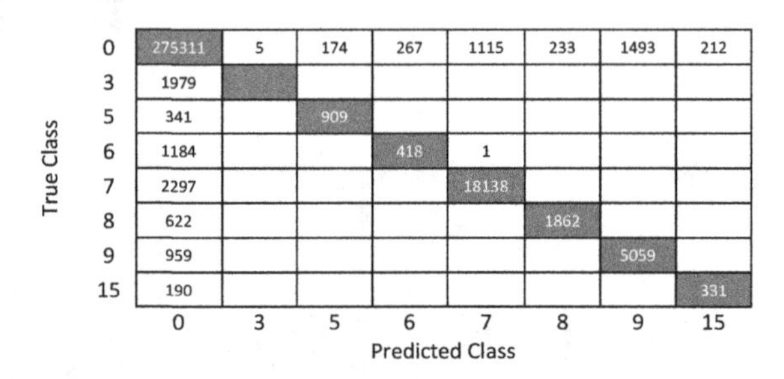

True Class \ Predicted Class	0	3	5	6	7	8	9	15
0	275311	5	174	267	1115	233	1493	212
3	1979							
5	341		909					
6	1184			418	1			
7	2297				18138			
8	622					1862		
9	959						5059	
15	190							331

Fig. 7. Confusion matrix for a fine tree classifier with 15 outputs (pure pigments) and non-pure pigments (class 0).

Using Support Vector Machine Classifiers

A fine Gaussian support vector machine classifier was designed to identify any of the 17 pure pigments (including black and white) in the reference pigment database or a non-pure color. For this particular test the use of true data in the training process was tested. To achieve this, the pixels of all 11 images available were classified using an improved method on the method proposed in [1, 2]. The results of the classification showed that the number of pure pigment pixels is highly imbalanced with some pigments not present in any of the images analyzed regardless of the method used. Take for example, the total number of pixels per pigment in all 11 painting images are: Black - 166014; White Pb - 163165; Emerald - 23909; Ultramarine - 0; Viridian - 0; Green Cd - 0; Prussian B. - 0; Cobalt B. - 5614; Cerulean B. - 47170; Violet Cd - 5732; Yellow Cr - 83046; Y. Ochre - 4576; Orange Cd - 11624; T. Rossa - 0; R. Sienna - 2875; Carmine - 0; Vermillion - 43.

To overcome this problem the following approach was used: set the number of samples to 3000 (average number of reflectances per pure pigment found); if the number of pure reflectances is less than 3000, balance the number using data augmentation; if the number of pure reflectances is greater than 3000, randomly select 3000; randomly select 3000 samples from the non-pure reflectances.

The total number of classes for this test was therefore 18: 17 pure pigments including black and white plus class 0 for non-pure pigments, and the number of samples 54000. For the calculation of the confusion matrix the number of folds was set to 5. The resulting accuracy and confusion matrix, in Fig. 8, showed that the SVM classifier accurately distinguished pure from non-pure pigments.

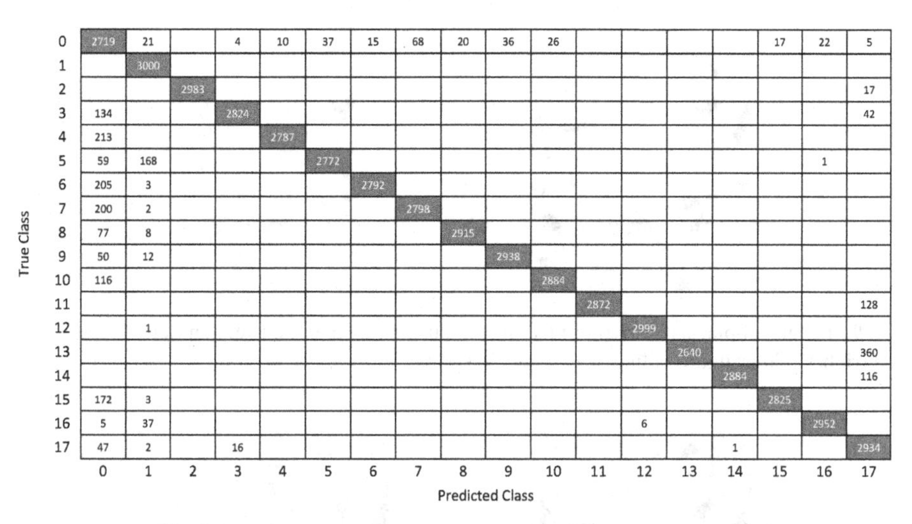

Fig. 8. Confusion matrix for SVM classifier. Accuracy 95.40%.

5 Results and Discussion

In terms of black and white pigment identification, the result of the classification is shown in Table 2. Where the values obtained based on the method proposed by Montagner and applying simple NN are shown for comparison. Figure 9 highlights the classified pixels where a good similarity is found between the improved version and the NN approach. This result shows that NN could be a potential candidate for pixel classification. However, this approach in its current form is limited to three classes and requires further investigation.

Table 2. Comparison of results from NN for black/white/colour identification.

Method	White pixels	Black pixels	Color pixels
Improved Version on the Method proposed in [1, 2]	10588 (7.59%)	9464 (6.78%)	119466 (85.63%)
NN	5136 (3.68%)	2345 (1.68%)	132037 (94.64%)

For multiple pigment identification the SVM classifiers was evaluated using the data from painting 68P11_Coty. This is, the reflectance of each pixel was fed into the classifier to predict whether the reflectance belonged to one of the 17 pure pigments or to a non-pure pigment. To reduce computation time the hyperspectral images were downsized to ensure the maximum number of total reflectances per painting did not exceeded 130000. To enhance visualization of the results, the non-pure pixels were colored with gray to allow white and black to be identified (see Fig. 10). For a true comparison, the images were also re-analyzed using the original method but downsized to match the size of images tested with the classifier.

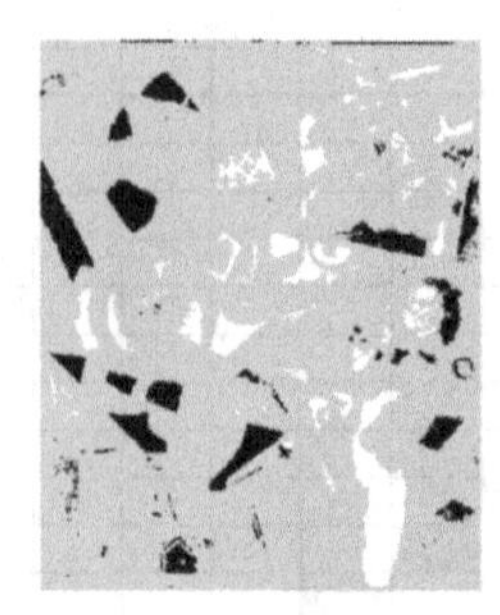
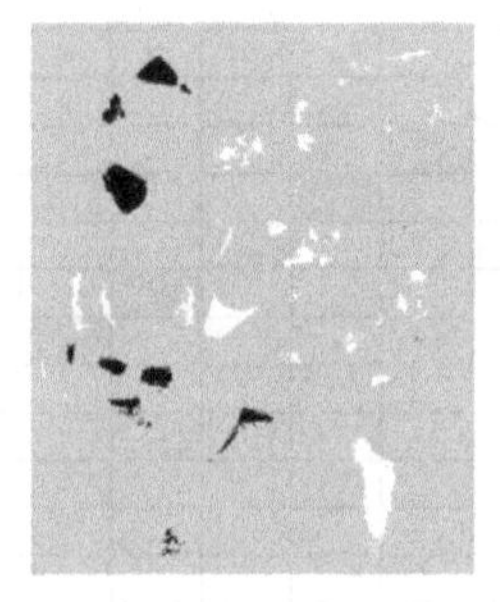

Fig. 9. Results of black/white/color identification using the improved version (left) and a simple NN (right). (Color figure online)

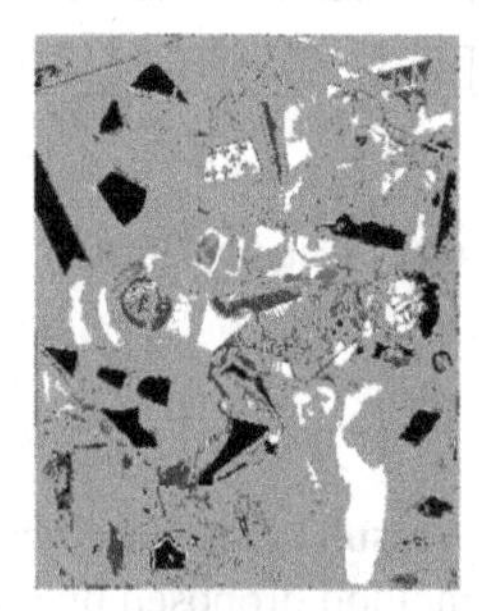
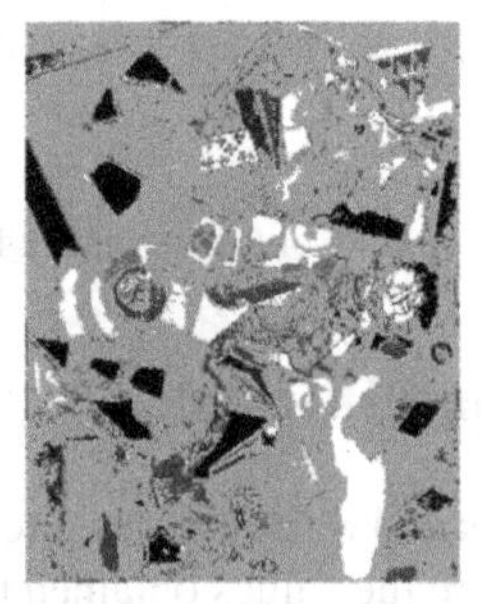

Fig. 10. Comparison of pure pigments identified in 68P11_Coty with original improved method (left) vs SVM classifier (right).

The results of the processing show a good correlation of pigments identified by the SVM suggesting that this SVM model generated is adequate for identification of pure pixels including black and white.

6 Conclusion

Our study explored the use of artificial mixtures based on the reference pigments to be used on the classification of black, white and other pure color pigments. First, applying brain style NN for black and white pigment identification, then conventional machine learning techniques Decision Tree and SVM for pure pigment identification.

Furthermore, our paper highlighted the importance of using the correct method for the generation of hypothetical color mixtures as this has a significant effect when black and white are used to generate lighter and darker colors. This enabled us to successfully use RMSE as the metric function over SAM which was unable to identify colors containing black and white. The difference was clearly visible in the reconstructed images where the color was estimated from the matched reflectances using both methods.

The improved pigment mixing method and metric function are fundamentally important for the implementation of holistic machine learning technique where black and white are considered as pigments like other color pigments.

7 Future Work

The experimental results of NN on black and white pigment showed that NN has the potential to be applied for the identification of other pure pigments as well as mixed pigments. Similarly, SVM used for pure pigment identification can be employed on the classification of black and white pigment as well based on the similar results obtained. In our future work, NN will be applied in the classification of both pure and mixed pigments along with quantitative data indicating percentages of base pigments in mixtures and the visualization of the individual pigments. Correspondingly, the results obtained applying SVM will be used to compare with the counterpart mentioned above to find out which one outperforms the other.

In all, the obtained pigment information resulting from the spectral data analysis will be compared with the chemical components acquired previously and then combined with the formerly-reported brushstroke analysis to determine the authenticity of the artworks.

Additionally, it is intended that the methods proposed be evaluated on other media, e.g., drawing data by the same artist, or artwork by different artists or genres allowing a more in-depth comparison and development of a method of non-intrusive analysis of artworks.

Acknowledgement. This work is supported by NOVA LINCS (UIDB/04516/2020) and VICARTE (UIDB/00729/2020), with the financial support of FCT - Fundação para a Ciência e a Tecnologia, through national grant PD/BD/135223/2017.

References

1. Montagner, C.: The brushstroke and materials of Amadeo de Souza-Cardoso combined in an authentication tool. Ph.D. dissertation, Departmento de Conservação e Restauro, Facul-dade de Ciências e Tecnologia, Universidade NOVA de Lisboa (2015)
2. Montagner, C., Jesus, R., Correia, N., Vilarigues, M., Macedo, R., Melo, M. J.: Features combination for art authentication studies: brushstroke and materials analysis of Amadeo de Souza-Cardoso. In: Multimedia Tools Application, vol. 75, pp. 4039–4063. Springer, Heidelberg (2016)
3. Epitropou, G.: Hyperspectral imaging and spectral classification algorithms for the non-destructive analysis of El Greco's paintings. Technical University of Crete, Kounoupidiana, Chania
4. Polak, A., Kelman, T., Murray, P., Marshall, S., Stothard, D.J.M., Eastaugh, N., Eastaugh, F.: Use of infrared hyperspectral imaging as an aid for paint identification. J. Spectral Imaging **5**(2) (2016)
5. Cosentino, A.: Identification of pigments by multispectral imaging; a flowchart method. Heritage Sci. **2**(1), 1–12 (2014). https://doi.org/10.1186/2050-7445-2-8
6. Conroy, J., Ryder, A.G., Leger, M.N., Hennessey, K., Madden, M.G.: Qualitative and quantitative analysis of chlorinated solvents using Raman spectroscopy and machine learning. In: Proceedings of the SPIE 5826, Opto-Ireland 2005: Optical Sensing and Spectroscopy (2005)
7. Grabowski, B., Masarczyk, W., Głomb, P., Mendys, A.: Automatic pigment identification from hyperspectral data. J. Cult. Herit. **31**, 1–2 (2018)

8. Deborah, H., George, S., Hardeberg, J.Y.: Pigment Mapping of the Scream (1893) Based on Hyperspectral Imaging. In: Elmoataz, A., Lezoray, O., Nouboud, F., Mammass, D. (eds.) ICISP 2014. LNCS, vol. 8509, pp. 247–256. Springer, Cham (2014). https://doi.org/10.1007/978-3-319-07998-1_28
9. Fan, C., Zhang, P., Wang, S., Hu, B.: A study on classification of mineral pigments based on spectral angle mapper and decision tree. In: Proc. SPIE 10806, Tenth International Conference on Digital Image Processing (ICDIP 2018) (2018).
10. Wang, Z., Lu, D., Zhang, D., Sun, M., Zhou, Y.: Fake modern Chinese painting identification based on spectral–spatial feature fusion on hyperspectral image. Multidimension. Syst. Signal Process. **27**(4), 1031–1044 (2016). https://doi.org/10.1007/s11045-016-0429-9
11. Interactive Sensory Laboratory Exercises (ISLE). https://isle.hanover.edu/. Accessed 12 October 2020
12. Schwartz, B.L., Krantz, J.H.: Sensation and Perception. SAGE Publications, Inc (2015)

Evolving Neural Style Transfer Blends

Simon Colton[1,2](✉)

[1] SensiLab, Monash University, Melbourne, Australia
[2] Game AI Group, EECS, Queen Mary University of London, London, UK
s.colton@qmul.ac.uk

Abstract. Neural style transfer is an image filtering technique used in both digital art practice and commercial software. We investigate blending the styles afforded by neural models via interpolation and overlaying different stylisations. In order to produce preset stylisation filters for the development of a casual creator app, we experiment with various MAP-Elites quality/diversity approaches to evolving style transfer blends with particular properties, while maintaining diversity in the population.

Keywords: Neural style transfer · MAP-Elites · Casual creators

1 Introduction

Neural style transfer produces *pastiche* versions of a *content image* by projecting the textures, colours and patterns of a *style image* onto the content image in a non-uniform and often impressive way. The trained neural model performs much processing of the content image's data, so the technique exhibits a level of intelligence by identifying and manipulating high and low level aspects of the content image (such as edges, curves, regions, colour gradients, textures, etc.) This can produce a visual transfer from the style image in a way that would be difficult or impossible to do with other techniques. As examples, in Fig. 1 below, the patterns and textures of the style images are adapted to preserve the edges, regions and points of contrast in the content image.

The production of pastiche images exhibiting a particular style has a long history in non-photorealistic rendering [19]. An influential approach for this task was described in [10], where a single neural model is trained for a given pair of (content, style) images. When run, the model produces an image with the low-level textural features of the style image and the high-level features of the content image. Models are trained using loss functions with two terms relating to representations of style and content, so users can weight each term and observe weaker or stronger applications of the style in the pastiches. A key advance involved training models over a given database of content images for a single style image [13]. Learning optimises perceptual loss over the database, so the trained model can be used in a feed-forward way to stylise any given image efficiently enough for real-time applications. A further advance enabled the training of a single model to represent and apply numerous different styles [9]. Here, feed-forward application of the model can be parameterised by weights, so the user can

J. Romero et al. (Eds.): EvoMUSART 2021, LNCS 12693, pp. 65–81, 2021.
https://doi.org/10.1007/978-3-030-72914-1_5

Fig. 1. Content, three style images and three corresponding generated pastiche images, produced with applications of neural style transfer in the Style Done Quick app.

interpolate between the embedded styles, producing style blends. The authors of [9] demonstrate the application of models trained over 32 different style images. Further discussion of how and why neural style transfer works is given in [15].

In 2016, in a breakthrough moment for the 'CreativeAI' movement of artistic and commercial deep learning practitioners (see [8]), the Prisma style transfer app (prisma-ai.com) won both Google and Apple awards. Prisma enables users to apply a single style to a chosen image, then change some standard settings in the pastiche image, such as saturation. As discussed in Sect. 2, we are building an iOS app similar to Prisma, called *Style Done Quick.* Improving on most consumer style transfer apps, all image manipulation will be performed on-device, rather than in the cloud, and users will have more control over the style manipulations applied. This will be enabled through (a) blending the styles that models embed (b) changing the grain of the texture applied by a model, and (c) using standard image compositing to produce more sophisticated imagery.

The simple user interfaces in apps like Prisma support *casual creation* [6,7], where user enjoyment is emphasised over fine-grained control or professional production. Unfortunately, with more powerful control over the style transfer application comes more complexity in the app and heightened potential for confusion. Hence, to keep the usage of Style Done Quick casual, and to help users navigate the space of possible combinations, we will provide numerous **preset stylisation filters** as easy entry points. As described in Sect. 2, users will be able to opt to simply apply a preset and possibly tweak the settings, and/or to construct filters from scratch. We aim to generate thousands of candidate preset filters, and cherry pick ones to include with the app. We have two main considerations: (i) the presets should cover much of the space of possibilities, to introduce users to a range of possibilities, and (ii) while it's useful for generated presets to be grouped into themes, we also want to be opportunistic, and for the generation to highlight stylisations we didn't know we were looking for.

Given these considerations, we employed a MAP-Elites approach [16] to generate archives of presets. MAP-Elites is a quality-diversity algorithm [17], which uses niching to maintain diversity, while simultaneously evolving solutions with respect to a fitness function. In Sect. 3, we describe aspects of MAP-Elites searches, including a machine vision approach to defining niche boundaries. In Sect. 4, we describe experimental results from MAP-Elites searches, and we conclude with some general observations and some avenues for future work.

2 Blending Neural Styles

After some experimentation with alternatives, the approach we adopted for training neural style transfer models was to use the TuriCreate python/C++ environment based on TensorFlow [1], described at github.com/apple/turicreate. This is supplied by Apple specifically for developers to train models for deployment in iOS applications. The processing is optimised to produce small, fast models which can run in real-time on 256 × 256 images. Our use-case is different to this, but we have found that, with careful choice of style image and training parameters, TuriCreate can produce effective models which run on the 1024 × 1024 images in Style Done Quick in 0.3 s, which satisfies our needs. Models are trained to apply to images of a fixed size, T. If the model is applied to an image I of size greater than T, it is sequentially applied to overlapping tiles of I of size T, with the tiles being feathered together at the end.

We have trained around 1,000 models at sizes ranging from 256 × 256 to 1024 × 1024 pixels. The size is chosen with reference to the style image(s) that the model will transfer. For instance, the three models employed in Fig. 1 were trained on 384 × 384 pixel style images, because we were able to find a patch in the style image of this size which captured the pattern and textures required. Sometimes, however, this requires a larger patch, and care must be taken to train the model at the same size, otherwise scaling down of the style image may distort or remove fine-grained textures that are part of the required style. Training on small style images usually takes less than an hour for 50,000 epochs on a single GPU, but training 1024 × 1024 pixel models can take up to 8 h. We have found that 50,000 epochs is usually sufficient for a model to converge to a suitably low loss value, but this doesn't necessarily mean that the model will work well as a style transfer filter: we have found that only around 1 in 5 style images eventually yield usable models. Note that the time taken for a run-time application of a model to a content image is only dependent on the size of the content image being processed, not the model size.

As Style Done Quick is a mobile app, it is advantageous to have a small download footprint, so we quantize models from 32 to 8-bit versions [12], bringing each model size down to less than 2 Mb. We have found that this doesn't noticeably alter the look of the pastiches generated. The style transfer training regime in TuriCreate enables multiple styles to be captured in a single model, as it is an implementation of the approach in [9]. Hence we can combine multiple styles into single models to further reduce the footprint. However, we have found that when there are more than 10 different styles being trained over, a local minimum is often found where the model transfers the colours of each style, but only a generic hexagon-like texture onto pastiches, regardless of the original style. Note that the authors of [9] tended to train models over multiple similar styles. We have found that training models to each capture 6 relatively similar styles works well for the efficacy of the process and fits with the user interface (see below). Our current version has 36 models, hence exposes 36 × 6 = 216 styles for the user to create with, and adds 72 Mb to the app's footprint.

2.1 Blending Through Model Interpolation

Embedding multiple styles in each model opens up options for their usage. Firstly, a variation, S', of a style image, S, can be used as an additional style image during training, so the user can interpolate between the two styles when applying the model. For homogeneity in Style Done Quick, for each style, S' is generated to be the same size as S by mirroring it three times then scaling it accordingly, as depicted in Fig. 2. The interpolation between the styles afforded by S and S' can then be made available with a single user-interface slider marked as **grain**, the action of which should feel fairly natural, as also depicted in Fig. 2. Secondly, once a user has chosen a model which embeds six styles, sliders can be used to explore the space of interpolated blends between the styles. In practice, we provide only five **style weight** sliders for user control, because this enables one of the styles to be the default, i.e., if all the sliders are set to zero, there is still a style applied. The alternative is for there to be no style applied when all sliders are set to zero, which is confusing, or for the model to be applied with a zero weight vector, which produces an unappealing grey image.

To enable the above functionality, each model is trained over 12 style images, i.e., employing six pairs of images $(S_1, S_1'), \ldots, (S_6, S_6')$. To control the application of the styles at run-time, the model therefore needs a vector of twelve floats to weight the different styles. Supposing that the grain slider and each of the style weight sliders range over 0 to 1 continuously, and the user has chosen a grain value g and style weights $s_2, \ldots, s_6$ (noting that they are not able to choose s_1), then the vector of weights passed to the model is:

$$[g * r, g * s_2, \ldots, g * s_6, (1 - g) * r, (1 - g) * s_2, \ldots, (1 - g) * s_6]$$

where $r = max\left(0, 1 - \sum_{i=2}^{6} s_i\right)$.

In practice, users can choose a pack of styles with a particular theme such as 'Scribble', which loads the neural model for six styles related to that theme. Then, when all sliders including the grain slider are set to zero, default style S_1 is applied. As the grain slider is increased, the pattern/texture that is transferred gets larger in grain, and as the style slider for S_n is increased, the style expressed by S_1 and S_n is blended until the slider reaches 1. When multiple sliders are non-zero, the effect of S_1 on the blend decreases until it reaches zero when the sum of the other slider values becomes 1 or more. We have found that this usage of sliders feels fairly natural to users. Where possible, the default style has been

Fig. 2. Left pair: example style image S and variant S' used in training. Right triple: interpolation between the styles on application to a content image.

Fig. 3. Default scribble style with pen style emerging as the slider is adjusted.

chosen to be somehow quite relevant to the pack theme, so it feels to users like a good starting point. For instance, in the Scribble pack, the default style is particularly messy, so that increasing the value of any weight slider brings more order to the mess, albeit in a scribbly way, as portrayed in Fig. 3.

2.2 Blending Through Graphical Layers

There are three **overlay options** available to Style Done Quick users, labelled *Style*, *Texture* and *Fusion*. When a pastiche image is produced, it is overlaid onto the original content image to produce a **stylised image** using a compositing technique corresponding to the overlay option: *Style* forces on-top compositing; *Texture* forces luminescence compositing; and *Fusion* forces multiplication compositing. Example overlays are given in Fig. 4 and we see that the results are quite different: on-top compositing transfers the whole style; luminescence compositing transfers textures, but largely retains the colours of the content image; and multiplication compositing produces more photorealistic results, e.g., in Fig. 4, it seems as if the style was applied directly to the subject's face.

A screenshot of the user interface to Style Done Quick is given in Fig. 5, showing the sliders for the style weights, which have preview samples on, as a guide to what to expect. There are also five tabs, each exposing the controls of a different neural model that produces a stylised image. These images are alpha composited, to produce more sophisticated imagery, e.g., a base stylised image capturing some texture, overlaid with a stylised image capturing a colour palette, overlaid with a pattern, etc. There is an overlay slider which prescribes the alpha value of each layer's stylised image when alpha composited. There is also a hidden layer containing the content image underneath the first layer, onto which the stylised image in the first layer is overlaid. Finally, there are three

Fig. 4. Results from the *Style*, *Texture* and *Fusion* overlay options.

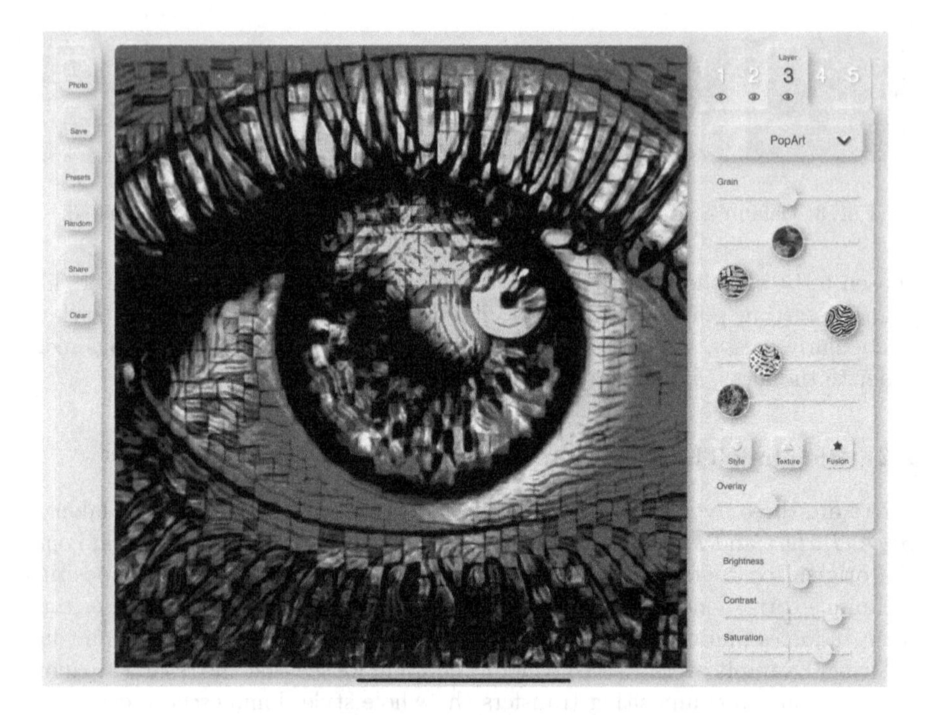

Fig. 5. Screenshot of the Style Done Quick user interface.

sliders for adjusting the brightness, contrast and saturation of the overall image produced, which is presented in the centre of the user interface.

While the stylised images in Fig. 1 were produced by the projection of a single style, those in Figs. 5 and 6 highlight the kinds of more sophisticated images users can produce using blended styles with the multi-layer approach. In both Figs. 5 and 6, three stylised images have been overlaid to produce the final image. Each stylised image blended multiple styles within a neural model, and the final layer of each used the Fusion overlay option, i.e., multiplication composition, to add elements of fine detail over the preceding layers.

Fig. 6. Original content image, example three layer stylised image and detail.

3 MAP-Elites Searches for Stylisation Filters

Adobe Inc. recently announced upgrades to Photoshop that employ deep learning techniques including style transfer [2]. We aim for Style Done Quick to lie between Photoshop and Prisma in terms of power of, and control over, style transfer, but for users to have definitively casual and fun experiences. People are accustomed to style transfer systems which enable them to rapidly trial numerous styles on a chosen image, and the fun lies in finding ones which produce surprising/sophisticated results. Hence we intend for Style Done Quick to enable one-tap stylisation of images, and for this we need a library of preset filters which can be quickly and easily experimented with. In turn, to properly populate such a library, we need to automate the search for potentially valuable stylisation filters. To do this, we use MAP-Elites searches, as described in terms of a genetic representation, niching operations and evolutionary mechanisms below.

3.1 Genetic Representation

With the affordances of the Style Done Quick app described above, we define the genotype of a stylisation filter as an ordered list of between 1 and 5 chromosomes, along with floating point values b, c and s, each between 0 and 1 specifying (b)rightness, (c)ontrast and (s)aturation edits to the overall layered image: $\langle C_1, C_2, C_3, C_4, C_5, b, c, s\rangle$. Each chromosome represents a layer applying a neural model, M, to the content image, parameterised by (g)rain, six (s)tyle weights, an overlay alpha value, α (all of which are between 0 and 1), and finally an overlay type, OT, which is either *style*, *texture* or *fusion*: $\langle M, g, s_1, s_2, s_3, s_4, s_5, s_6, \alpha, OT\rangle$. Note that s_1 is calculated as above, namely as the remainder when the sum of s_2 to s_6 is taken from 1, or 0 if this remainder is negative. Mapping M and OT to numbers, the genotype becomes a non-empty ordered list of up to five vectors of floats with well defined ranges.

Filter phenotypes are relative to a given content image, I, and are generated from a genotype, G, by alpha compositing, in order, the stylised images resulting from each of G's chromosomes, using the α values in each chromosome. The stylised image for a chromosome, C, is generated using its neural model, M, parameterised with the 12-point weight vector derived from s_1 to s_6 as above. This produces a pastiche which is combined through either on-top, luminescence or multiplication compositing with I, dependent on the value of OT in C. For the experiments described here, we work only with 384×384 pixel images which a neural model takes around 0.1 s to process. The generation of a phenotype from a genotype takes around 0.3 s on average (on a 2018 iPad Pro).

3.2 Mapping Phenotypes to Unique Cells

The MAP-Elites algorithm described in [16] requires a pre-defined way to place a (genotype/phenotype) pair into a unique *cell* (or *niche*), and a pre-defined way to determine whether one phenotype is fitter than another, usually done by calculating a concrete fitness value for both and comparing. An *archive* of cells is maintained throughout the process with each cell containing either zero or one (genotype/phenotype) pairs. Genotypes are initially generated randomly and their associated phenotypes calculated accordingly. The cell, C, for a randomly generated pair (g, p) is determined, and if there is no occupant of C, or the phenotype of the current occupant is less fit than p, then (g, p) is placed into C.

After a certain number of pairs have been produced, or a time limit exceeded, etc., random generation stops. Then, for a certain number of steps, the occupants (g, p) of a randomly chosen non-empty cell are selected, a variant, g', of g is produced (usually through mutation and/or crossover with other genotypes in the archive), and p' is calculated accordingly. Then the cell for (g', p') is determined, and as before, the pair is placed in the cell if it is empty or has a less fit pair in it. Given that any pair in a cell is always the best seen so far (with respect to fitness) for that cell, archive members in the cells are called *elites*. Under certain circumstances, MAP-Elites can outperform standard evolutionary search techniques and other quality/diversity algorithms for optimisation problems, with details given in [16]. However, for us, its main value is in maintaining a diverse set of stylisation filter genotypes that we can peruse after the search concludes, to cherry pick suitable ones for inclusion in Style Done Quick. We found the additional advantage of using fitness evaluations for more focused searches which converge on visually similar phenotypic outputs, but, due to the nature of MAP-Elites, still guaranteed a level of diversity in those outputs.

To harness MAP-Elites, we first derived a method for categorising a stylised image (phenotype) into a unique cell. To do this, we first randomly generated 1,000 stylised images for each of 10 representative content images (covering faces, landscapes, cityscapes and still life objects). The output from a headless application of the ResNet50 neural model for each of the 10,000 stylised images was then recorded. ResNet50 is a pre-trained model which is normally used for classifying images [14]. However, the headless version, which outputs vectors of 2,048 floats for each input image, has also been used successfully in [4] and [5] to provide a visual analysis of artistic images, e.g., for clustering purposes.

We performed a principal component analysis over the outputs of ResNet on the 10,000 stylised images, and examined the first 10 principal components. From each one, we selected the ResNet output feature most highly weighted, in order to find approximately orthogonal ways to separate images according to their visual properties, in order to assign them to cells. After some experimentation, we decided to use the first four ResNet features selected in this way, and we performed a number of analyses to confirm that images achieving different values on any of the features do indeed look different. For each of the features, f, its mean and standard deviation over the 10,000 stylised images were calculated and recorded. Then, given a stylised image, its output, v, for ResNet feature f with mean μ and standard deviation σ is discretised via a map to:

$$\begin{cases} 0 \text{ if } v - \mu < -2\sigma \\ n - 1 \text{ if } v - \mu > 2\sigma \\ \lfloor \frac{n(v-\mu+2\sigma)}{4\sigma} \rfloor \text{ otherwise} \end{cases}$$

where n is a number of divisions, which we set to 5 for our purposes here.

Informally, this calculation looks at how far an image's feature value is from the mean for that feature and assigns it an integer between 0 and $n-1$ based on this distance relative to the standard deviation. Doing this for all the selected ResNet features generates a **signature** for image I as a list of four integers, each between 0 and 4 inclusive. If (genotype/phenotype pairs are assigned to the cell in the MAP-Elites archive corresponding to the ResNet signature of the phenotype, there will potentially be $5^4 = 625$ cells in the archive. To test this approach, we assigned 1,000 randomly generated stylised images to cells and looked at the entries in different cells. We found that while it is possible for two images in adjacent cells to look overly similar, we found that this gradation of differences sufficiently splits stylised images according to their appearance. This means that when scrutinising the elite stylised images in an archive after a search has completed, we would expect a suitable level of visual difference among them.

3.3 Mutation, Crossover, Fitness and Selection Mechanisms

In the experiments described below, random generation of genotypes involved randomly choosing a number between 1 and 5 (inclusive) of chromosomes, then randomly generating values within the appropriate range for each of the floats in each chromosome. Moreover, given a genotype, G, and a mutation rate, r, and referring to the chromosome representation described above, we implemented the following genotype mutation operators in MAP-Elites searches:

- **Layer mutation**: take r randomly chosen chromosomes from G and swap them for randomly generated ones.
- **Weight mutation**: take r randomly chosen chromosomes from G and for each, alter g, $s2$, $s3$, $s4$, $s5$ and $s6$ to random values between 0 and 1, calculating s_1 based on $s2$ to $s6$ as usual.
- **Overlay mutation**: take r chromosomes and swap α and OT for randomly generated values.
- **BCS mutation**: alter the brightness, contrast and saturation of the overall layered image to random values.

We also implemented a crossover mechanism whereby the first p chromosomes of parent genotype G_1 are taken and the final q chromosomes of parent genotype G_2 are taken as the starting and ending chromosomes of child C respectively. Note that p and q are chosen randomly, but constrained such that $p > 1$, $q > 1$ and $p + q \leq 5$. We reasoned that, due to the overlaying procedure, the final layers of a stylisation filter project most of the visual effect onto a styled image, with the first layers fine-tuning this effect. Hence we implemented the crossover mechanism to preserve these contributions in the offspring of two parents.

To systematically find preset stylisation filters for Style Done Quick, we vary the fitness function which guides MAP-Elites searches. In particular, we implemented a fitness function that uses the ResNet output from a given **target** image, and calculates the reciprocal of the Euclidean distance from this to the

ResNet output for a stylised image (phenotype). In this way, the search should converge on a set of solutions that resembles the target image, while – thanks to the niching inherent in MAP-Elites – maintaining diversity somewhat. To best use this fitness function, the target image and the stylised image phenotypes should ideally contain the same subject material, so that the ResNet distance is based on the stylisation of the content, rather than the content itself. Hence we only used target styled images where we also had a matching content image.

We describe experiments with different evolutionary setups and target images in the next section. In order to potentially increase the yield of high-scoring elite phenotypes, we adapted the MAP-Elites approach to use a mechanism for selecting which elites to crossover and mutate in any step, as opposed to randomly choosing them. This brings the approach more in line with standard evolutionary search, but the niching should maintain diversity. To do this, we tried two different selection mechanisms. Firstly, note that the target image can be assigned a unique signature in the same way as stylised images, described in Subsect. 3.2. As this signature is a list of four integers, the Euclidean distance between the signature, $sig(T)$, of target image T and the signature, $sig(E)$, of an elite phenotype, E, can be calculated and denoted $dist(sig(T), sig(E))$.

Our first selection mechanism, entitled **target distance selection** (TDS) chooses a cell randomly and then mutates it (possibly via crossover with another elite) with probability $p = |dist(sig(T), sig(E))|^{-1}$. Hence elites which are closer to the cell that the target image would naturally be placed in are selected for mutation with higher probability than those further away. Our second selection mechanism, entitled **score-based selection** (SS) works similarly by choosing an elite randomly, and mutating it with a probability, but this time based on the fitness that the elite already has. If, at the time the selection is to be carried out, the archive has a elites in it, these are ranked from fittest (with rank a) to least fit (with rank 0). Then a randomly chosen elite with rank k is mutated with probability $p = k/a$, so that low ranking elites are mutated with smaller probability than higher ranking ones. Note that partners for a crossover operation can be similarly chosen with either of these selection mechanisms.

3.4 Many-Archive MAP-Elites Searches

As discussed in the next section, we have found that seeding a MAP-Elites archive with the elites derived from 1,000 randomly generated phenotypes, followed by 2,000 mutation/crossover steps usually produces valuable search results in around 15 min on an iPad Pro. Our plan is to undertake dozens of such searches with different target images to systematically attempt to find preset stylisation filters for Style Done Quick. In these searches, the majority of time is spent using neural style transfer to generate a phenotype from a genotype (on average, 0.25 s) and on analysing the resulting image with headless ResNet (on average, 0.05 s). The time taken to assign a phenotype to a cell and to compare fitness to a current occupant is negligible. Moreover, the styled images for the phenotype don't need to be stored, only a set of ResNet feature outputs, which occupy tiny amounts of memory, as do the genotype representations.

In light of these long processing times and low storage requirements, we decided to test whether it is beneficial to run multiple searches at once. That is, we implemented a version of MAP-Elites whereby $n \geq 1$ archives are compiled at the same time, each with their own fitness function. In the initial random generation stage, each phenotype is assigned a cell with its signature, then is added to that cell in each archive if the cell is empty or the phenotype is fitter than the current occupant. Then an elite is selected from an archive, A, and mutated. The mutation is then potentially added not just to the appropriate cell in A, but to the same cell in each of the other archives. In the next step, the next archive is chosen in a round-robin fashion, and the process is repeated. Our thinking is that, due to the elites in different archives maximising different fitness functions, the chances of a mutated elite from one archive being elite in another archive are small, but they are not non-existent. In such a many-archive MAP-Elites search, each archive is exposed to n times as many phenotypes as in a single-archive search, which may drive up the overall fitness and diversity.

We also experimented with substituting the random generation initial phase with generation of all styled images arising from an exhaustive set of genotypes with a single *one-hot* chromosome, i.e., with a single one of the six style weights set to 1, with all others set to 0. To increase the spread of genotypes, we also varied the grain parameter over the set $\{0, 0.5, 1\}$, the overlay parameter over the set $\{0.5, 1\}$ and the OT (overlay type) parameter over its three possible values. With 36 neural models available, each embedding 6 styles, this exhaustive search produced $36 \times 6 \times 3 \times 3 \times 2 =$ 3,888 phenotypes, but took roughly the same time as producing 1,000 phenotypes randomly. This is because the exhaustive genotypes only have one layer, hence require only one application of the neural model, rather than, on average, 2.5 applications required to produce phenotypes from the randomly generated genotypes.

4 Experimental Results

In general, we have found that MAP-Elites searches work well for our purposes. With the separation of images into one of potentially 625 cells via the signature calculation, the searches always yield more than 200 elites to inspect. The fittest individuals do tend to look like the target images, and the visual diversity increases as the fitness decreases, which in turn heightens the potential for opportunistically finding stylisation filters that we didn't know we were looking for. In many cases, we have identified and kept filters in this opportunistic way.

4.1 Finding a Suitable Evolutionary Setup

Given that we will be using dozens of targeted searches to compile a set of preset filters for Style Done Quick, it was sensible to find a good setup for the MAP-Elites search through experimentation. We did this in two stages, firstly tweaking the evolutionary setup for a single-archive approach, and then taking the best setup forward to use in a comparison of single versus many-archive MAP-Elites

Fig. 7. Centre and right: content image and target image for first experiments with MAP-Elites searches. Left: additional content image for second experiments comparing single-archive with many-archive MAP-Elites searches.

searches. After each search concluded, we measured two values of the final archive it produced: quality and diversity. Given our use case, we calculated quality as the average fitness of the highest-scoring 50 elites in the archive. Diversity was measured as the number of occupied cells in the archive.

The fitness results from 22 evolutionary setups is given in the bar charts of Fig. 8. The content and target images were in the centre and on the right respectively of Fig. 7. Fitness of a phenotype was measured as the reciprocal of the Euclidean ResNet distance of the stylised image from the target image. In most of the search sessions, the same set of 1,000 randomly generated genotypes were used to initialise the archive, and then 2,000 mutation/crossover steps were undertaken. However, in the setups marked *Ex+* on the charts, the exhaustive genotype initialisation method was employed. The setups varied over the selection and mutation mechanisms described in the previous section. Rather than undertake a systematic exhaustion of all possible setups, we opportunistically took well-performing setups and varied them to see if this improved matters.

The charts in Fig. 8 record the percentage increase/decrease in the average fitness of the highest scoring 50 phenotypes over the baseline of simply generating another 2,000 phenotypes randomly. The charts also record the number of occupied cells in the final archive (noting that the maximum this could be is 625). In the first batch of experiments, we tested mutation-only search, and in the second batch, we introduced crossover and crossover+mutation searches. We see that the overlay and BCS (brightness, contrast and saturation) searches populated more cells than the random one did, but this lowered the average fitness of the top 50 elites. We found that the best mutation-only search setup for fitness (and second-best for diversity) used weight mutation and the target-distance selection mechanism, as per Subsect. 3.3. This also used an additional mild BCS mutation of the genotype, whereby the three default values of 0.5 were varied to within the range 0.45 to 0.55.

In general, we found that crossover searches outperformed mutation-only ones, and that the target-distance and score-based selection mechanisms improved the overall fitness of the top elites in the final archives. In the second batch of experiments, namely those involving crossover, we found that the best in terms of fitness employed score-based selection for both parents in the crossover operation, then applied weight mutation with mutation rate 2 and mild BCS mutation to the offspring. We further found that using the exhaustive rather than random initialisation method substantially improves the diversity of the final archive, although it may slightly decrease the average fitness. The two

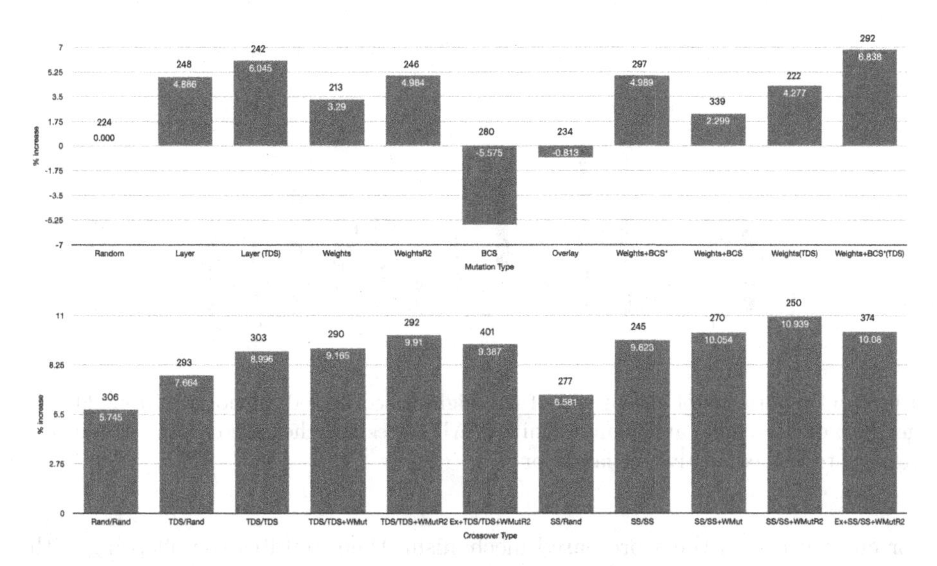

Fig. 8. Top chart: quality and diversity results from mutation-only single-archive MAP-Elite searches. Bottom chart: results from searches with crossover and mutation. The figures inside the bars are the percentage increase in average fitness of the top 50 elites over that gained from generating 2,000 additional genotypes randomly. The figures above the bars are the number of occupied cells in the final archives. *Rand*, *TDS* and *SS* denote random, target-distance and score-based selection mechanisms, with X/Y on crossover setups indicating the mechanism used for each parent. *BCS** denotes a milder version of BCS (brightness, contrast, saturation) mutation. Ex+ indicates an exhaustive initialisation rather than random. +WMut indicates that weight mutation of an offspring was undertaken after crossover, with mutation rate 1, followed by mild BCS mutation; +WMutR2 is the same, but with mutation rate 2.

most diverse archives, containing 374 and 401 occupied cells, both used this approach, with the next best crossover search producing only 306 elites. Note that our aim here was to produce a workable search setup for finding candidate preset stylisation filters. In future work, we will further test the evolutionary setups to validate results over multiple runs, multiple content and multiple target images.

4.2 Comparing Single and Many-Archive MAP-Elites Searches

In a final experiment, we tested whether the searches do indeed produce a diverse set of image filters that, when applied to an appropriate content image, produce pastiche images with visual similarity to a given target image. We also used this to experiment to test the hypothesis that, for the same amount of processing, using a many-archive approach improves the fitness and diversity of the archives produced. The base evolutionary setup for the MAP-Elites searches in this experiment was the final one recorded in Fig. 8, which, as second-best for both quality and diversity was a good all-rounder to use. This initialises the archive with the exhaustive rather than random approach, selects both parents

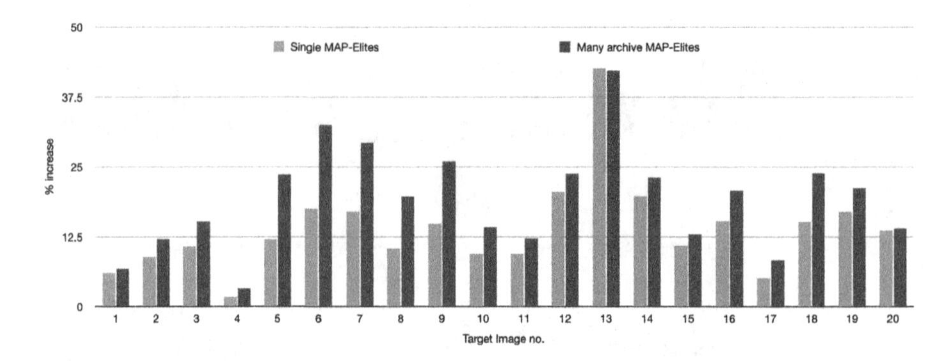

Fig. 9. Comparisons of the fitness of the highest scoring 50 phenotypes over 20 target images, for single and many-archive MAP-Elites searches. Percentage increase is relative to the exhaustive initialisation.

for crossover with the score-based mechanism, then mutates the offspring with rate 2, using the weight mutation scheme followed by mild BCS mutation.

For two separate content images, given in Fig. 7, we found online 10 different stylised version of each, produced by: fan-artists by hand as real paintings/drawings; digital editing tools and other style transfer projects. The stylised versions are given in the first columns of Fig. 10. We ran 20 separate single-archive MAP-Elites searches, each initialised with the exhaustive method, each performing 2,000 crossovers, and each using a single stylised image as a target. We also ran two many-archive searches, one per content image, each with ten stylised images as targets. These both used 20,000 crossover steps, so 2,000 elites from each of the 10 archives were mutated, to provide a fair comparison.

The quality results are given in Fig. 9. Note that the archives produced are compared not to one generated with random additions, as in Subsect. 4.1, but rather to the archive achieved with the exhaustive initialisation. We found that the random archives (of 3,000 genotypes with average 2.5 layers) and the exhaustive archives (of 3,888 genotypes with 1 layer) were similar for comparison against, but the latter were much quicker to generate. In Fig. 9, we see that percentage increases over the exhaustive search vary a lot over the different target images, from almost nothing to nearly 45%. The same is true for any advantage gained from the many-archive approach, where in some cases, the many-archive approach barely improves matters, but in others, it roughly doubles the percentage increase. In all cases, though, the crossover searches improved upon the exhaustive search, and, given that they only take around 15 min to complete, they are clearly worth doing. In all but one case, the many archive approach did produce fitter archives, but in general the improvement is modest. However, this doesn't involve more computation time and we found it to be more convenient to run two rather than twenty searches. Moreover, while the single-archive approach produced archives with an average of 339 occupied cells, the many-archive approach produced 483 occupied cells in the first search and 433 in the second, hence bringing clear improvements in diversity.

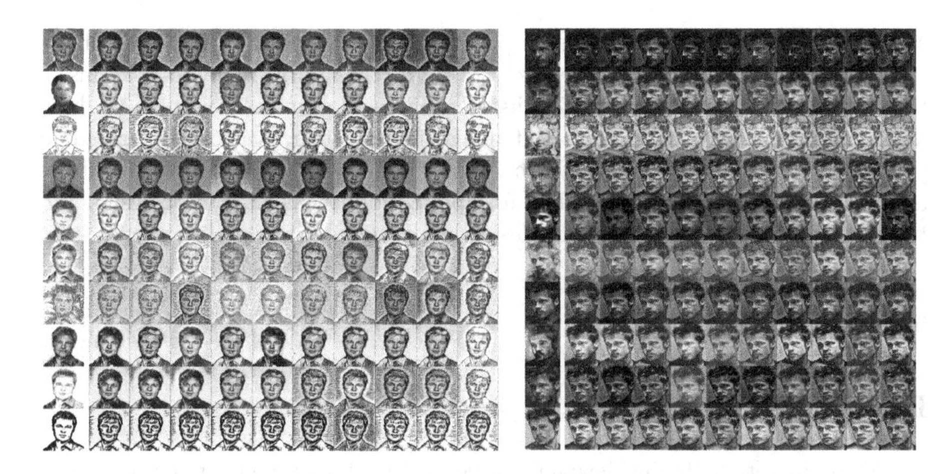

Fig. 10. Left column in both sets: target images in a many-archive MAP-Elites search. Other columns: the 10 highest scoring elites, with decreasing scores left to right.

The stylised images resulting from the top ten fittest phenotypes in all 10 archives of both many-archive searches are given in Fig. 10, along with the respective target images. We see that the stylised images have indeed converged to visual similarity with the targets, with some having more success than others. These images only tell half the story, however, as the diversity of the archives meant they also contained numerous interesting stylisations which had some but not huge resemblance to the target, and many low fitness stylisations which could be mined for interesting and inspiring outliers.

5 Conclusions and Future Work

We have described how neural style transfer models can be used to produce blended styles both through interpolation in a single model and using standard graphics overlaying techniques, under the control of a user in a casual creator setting. In the development of the Style Done Quick app, we need to produce a diverse set of thousands of potential stylisation filters in order to provide users with easy to use presets that introduce them to a broad swathe of the space of possibilities. We've shown how that can be achieved systematically with directed MAP-Elites searches for stylisations measured against a target image, and have experimented to find an evolutionary setup which improves the efficacy of the approach. While our experiments were on 384 × 384 pixel images, we've undertaken further experimentation (not reported here) which confirmed that the above approach will scale up to finding filters for 1024 × 1024 pixel images, as required for Style Done Quick, although the searches will take longer.

While the focus here has been on a particular application, there are some general findings that have arisen. For instance, calculating visual signatures using machine vision models like ResNet might find further application in other generative projects where separating outputs visually is required. The many-archive

MAP-Elites approach also has some promise, as it makes good use of available computation time, and there are likely to be further improvements in terms of reducing redundancy and cross-fertilisation between archives. We plan to draw from similar projects which evolved image filters, such as those described in [3,11] and [18]. We also plan to further experiment with and improve the above and other quality/diversity searches, for application to casual creator design and to provide best practice advice for designers of such apps.

Acknowledgements. We would like to thank the anonymous reviewers for their insightful suggestions.

References

1. Abadi, M., et al.: TensorFlow: large-scale machine learning on heterogeneous systems (2015). tensorflow.org
2. Clark, P. (2020). https://blog.adobe.com/en/2020/10/20/photoshop-the-worlds-most-advanced-ai-application-for-creatives.html
3. Colton, S., Torres, P.: Evolving approximate image filters. In: Giacobini, M., et al. (eds.) EvoWorkshops 2009. LNCS, vol. 5484, pp. 467–477. Springer, Heidelberg (2009). https://doi.org/10.1007/978-3-642-01129-0_53
4. Colton, S., McCormack, J., Berns, S., Petrovskaya, E., Cook, M.: Adapting and enhancing evolutionary art for casual creation. In: Romero, J., Ekárt, A., Martins, T., Correia, J. (eds.) EvoMUSART 2020. LNCS, vol. 12103, pp. 17–34. Springer, Cham (2020). https://doi.org/10.1007/978-3-030-43859-3_2
5. Colton, S., McCormack, J., Cook, M., Berns, S.: Creativity theatre for demonstrable computational creativity. In: Proceedings of 11th ICCC (2020)
6. Compton, K.: Casual creators: AI supported creativity for casual users. Ph.D. thesis, University of California, Santa Cruz (2019)
7. Compton, K., Mateas, M.: Casual creators. In: Proceedings of the 6th International Conference on Computational Creativity (2015)
8. Cook, M., Colton, S.: Neighbouring communities: interaction, lessons and opportunities. In: Proceedings of 9th International Conference on Computational Creativity (2018)
9. Dumoulin, V., Shlens, J., Kudlur, M.: A learned representation for artistic style. In: Proceedings of the International Conference on Learning Representations (2017)
10. Gatys, L., Ecker, A., Bethge, M.: A neural algorithm of artistic style. arXiv:1508.06576 (2015)
11. Harding, S.: Evolution of image filters on graphics processor units using Cartesian genetic programming. In: IEEE Congress on Evolutionary Computation (2008)
12. Jacob, N., et al.: Quantization and training of neural networks for efficient integer-arithmetic-only inference. In: Proceedings of CVPR (2017)
13. Johnson, J., Alahi, A., Fei-Fei, L.: Perceptual losses for real-time style transfer and super-resolution. In: Leibe, B., Matas, J., Sebe, N., Welling, M. (eds.) ECCV 2016. LNCS, vol. 9906, pp. 694–711. Springer, Cham (2016). https://doi.org/10.1007/978-3-319-46475-6_43
14. Krizhevsky, A., Sutskever, I., Hinton, G.: ImageNet classification with deep convolutional neural networks. In: Advances in Neural Information Processing Systems (2012)

15. Li, Y., Wang, N., Liu, J., Hou, X.: Demystifying neural style transfer. In: Proceedings of the Twenty-Sixth International Joint Conference on Artificial Intelligence (2017)
16. Mouret, J.-B., Clunes, J.: Illuminating search spaces by mapping elites. arXiv: 1504.04909 (2015)
17. Pugh, K., Soros, L., Stanley, K.O.: Quality diversity: a new frontier for evolutionary computation. Front. Robot. AI **3**, 40 (2016)
18. Smith, S.L., Leggett, S., Tyrrell, A.M.: An implicit context representation for evolving image processing filters. In: Rothlauf, F., et al. (eds.) EvoWorkshops 2005. LNCS, vol. 3449, pp. 407–416. Springer, Heidelberg (2005). https://doi.org/10.1007/978-3-540-32003-6_41
19. Strothotte, T., Schlechtweg, S.: Non-Photorealistic Computer Graphics. Morgan Kaufmann, Burlington (2002)

Evolving Image Enhancement Pipelines

João Correia[1(✉)], Leonardo Vieira[1], Nereida Rodriguez-Fernandez[2], Juan Romero[2], and Penousal Machado[1]

[1] CISUC, Department of Informatics Engineering, University of Coimbra, 3030 Coimbra, Portugal
{jncor,machado}@dei.uc.pt, lmavieira@student.dei.uc.pt

[2] Faculty of Computer Science, University of A Coruña, Coruña, Spain
{nereida.rodriguezf,jj}@udc.es

Abstract. Image enhancement is an image processing procedure in which the original information of the image is improved. It alters an image in several different ways, for instance, by highlighting a specific feature in order to ease post-processing analyses by a human or machine. In this work, we show our approach to image enhancement for digital real-estate-marketing. The aesthetic quality of the images for real-estate marketing is critical since it is the only input clients have once browsing for options. Thus, improving and ensuring the aesthetic quality of the images is crucial for marketing success. The problem is that each set of images, even for the same real-estate item, is often taken under diverse conditions making it hard to find one solution that fits all. State of the art image enhancement pipelines applies a set of filters that solve specific issues, so it is still hard to generalise that solves all types of issues encountered. With this in mind, we propose a Genetic Programming approach for the evolution of image enhancement pipelines, based on image filters from the literature. We report a set of experiments in image enhancement of real state images and analysed the results. The overall results suggest that it is possible to attain suitable pipelines that visually enhance the image and according to a set of image quality assessment metrics. The evolved pipelines show improvements across the validation metrics, showing that it is possible to create image enhancement pipelines automatically. Moreover, during the experiments, some of the created pipelines create non-photorealistic rendering effects in a moment of computational serendipity. Thus, we further analysed the different evolved non-photorealistic solutions, showing the potential of applying the evolved pipelines in other types of images.

Keywords: Image enhancement · Image processing · Computer vision · Evolutionary computation · Genetic programming

1 Introduction

Digital images are, now more than ever, an essential element in our daily lives, considering that almost all online activities depend, in one way or another, on

J. Romero et al. (Eds.): EvoMUSART 2021, LNCS 12693, pp. 82–97, 2021.
https://doi.org/10.1007/978-3-030-72914-1_6

this type of resource. Every image we see in our daily life has a set of attributes that define the way it is perceived. Often, these attributes are not well balanced or optimised for the context of the image and affect image quality.

Image Enhancement (IE) is an image processing approach that aims to improve the perception of a feature or the overall quality of the image, by a person or a computer. Although by definition IE can be done manually, in scenarios involving a large number of images and under different conditions and constraints, the task becomes complex and under specific scenarios unfeasible. In this work, we focus on creating an automatic IE approach. More specifically, we are concerned with creating an approach to automatic enhance images for the context of real estate marketing. Automatic IE brings significant challenges, especially when it comes to manipulating multiple aspects of the image simultaneously since the individual features are not independent of each other. There are multiple types of IE techniques with different purposes and characteristics. Some are more detailed static filters applied to the spatial domain, and others seek to adapt to the image context, avoiding heterogeneous results across multiple images. Based on this research, we explore IE methods focused on improving image aesthetics.

We propose a Genetic Programing (GP) approach that generates pipelines for image enhancement based on image processing filters, with decision components, which aim to alter the pipeline's output, depending on the input image state and features. We resorted to automatic fitness assignment schemes based on the response of an aesthetic evaluator, the Neural Image Assessment (NIMA) classifier [1]. We tested it using a provided dataset of various real-estate pictures of different quality. Furthermore, the outputted enhanced images are evaluated using image quality assessment tools to assess and validate the outputs' quality. We were able to evolve image enhancement pipelines that successfully enhance input images according to the aesthetic evaluate that assigned fitness and other image quality metrics that were only used for validation. However, some of the solutions that the GP approach was optimising were creating non-photorealistic renderings of the input images. The renderings resulted in aesthetically appealing images of arguably artistic merit. In the context of Computational Creativity, this can be viewed as a moment of serendipitous discovery [2], where a system was prepared with an objective in mind, and partially due to chance, an exciting and unexpected value output occurs. In this work, we also explore how the pipelines that create non-photorealistic renderings affect other types of images, further showing those solutions' value.

In this way, the contribution of our work can be summarised in the following main points: (i) the design of an approach that creates a sequence of image filters for image enhancement; (ii) analysis of results obtained with automatic fitness assignment schemes that quantify image aesthetics; (iii) comparison between evolved filters pipeline and a baseline subset of the state of the art image enhancement filters; (iv) analysis of the non-photorealistic effect detected during the experiments; and (v) exploration of the application of the non-photorealistic rendering pipelines on other types of images. The remainder of this paper is organised as follows. Section 2 presents related work. Section 3 present commonly used image filters in the area. Section 4 presents the approach. Section 5 lays the experimental setup and in Sect. 6 we present and discuss the experimental results. Finally, Sect. 7 draws final conclusions and points future work.

2 Related Work

There are multiple types of IE techniques with different purposes and characteristics. The research around this field revolves around machine learning models, computer vision pipelines by applying filters, or both. This section reviews the works related to image filters that are mostly related to our approach.

In terms of image processing and filter approaches, W. Wencheng et al. proposed an IE pipeline that aims to improve the overall brightness and contrast of low-illumination images [3]. C. Y. Wong et al. proposed another pipeline that tries to bridge the problem where approaches that are only based on intensity enhancement may produce artefacts in "over-enhance" regions, and lack enrichment on colour-based features [4]. Moreover, H. Talebi et al. proposed in [5] a novel way of improving an image detail and contrast by expanding on Laplacian operators of edge-aware filter kernels. Closing the classical techniques, we want to mention S. Zhuo et al. in [6], where a noise reduction pipeline is proposed.

Some works integrate evolutionary computation in IE. L. Rundo et al. proposes an evolutionary method based on genetic algorithms to improve medical imaging systems [7]. C. Munteanu also proposed an IE method that relies on evolutionary techniques to improve grey-scale images by evolving the shape of the contrast curve [8]. The work of Shan et al. [9] used an Immune Clone Algorithm (ICA) which make the enhancement method suppress noise and increase the visibility of the underlying signal at the same time on grayscale images.

Most of the approaches mentioned above are mostly non-modular pipelines with fixed parameterisation and applied to solve specific issues with the input image to the best of our knowledge. Based on the review, we moved to implement a set of image filters that will be included in our approach to providing flexibility to the pipelines that we aim to evolve. Thus, the selected filters are described in the next section.

3 Image Filters

A set of 7 previously reviewed image filters were implemented and used during this work. The implemented methods focus on five main aspects of IE approaches contrast adjustment, brightness adjustment, colour balance, noise removal and edge enhancement (also referred as *sharpening*). In this section, we present and explain each one individually.

The contrast in image processing is the range of intensity values available to an image. The contrast stretching is a point operation method that, as the name implies, tries to improve the image contrast by linearly increasing the difference between the maximum intensity value and the minimum intensity value in an image, therefore increasing the contrast level. The work of Bazeille et al.[10] discuss and report of such filter for IE.

Histogram Equalisation (HE) is another method that tries to improve an image's quality by manipulating the contrast. It does this by spreading the most common intensity values by the less common ones, increasing the global contrast of an image. This method is highly used, and there are multiple iterations and discussion about its results [11].

Contrast Limited Adaptive Histogram Equalization (CLAHE) is yet another contrast enhancement method and adaptable for different use cases [12]. It is an iteration of the Adaptive Histogram Equalization (AHE) technique that is an improved version of the regular histogram equalisation. CLAHE improves upon the AHE by clipping the maximum intensity values of each region and redistributing the clipped values uniformly throughout the histogram before applying the equalisation.

Gamma Correction (CB) tries to accommodate the fact that the Visual System (HVS) perceives brightness in a non-linear way. This is done by scaling each pixel brightness from *[0–255]* to *[0–1]* and applying an expression to map the original values.

Non-local Means Denoising (NLMD) [13], as the name implies, tries to reduce the existing noise in an image. It replaces the value of each pixel in each channel to the average of similar pixels.

Unsharp Masking (UM) is an IE technique that sharpens the edges of an image [14]. It does that by subtracting a blurred version of the original image from the original image to create a mask. This mask is then applied to the original image, enhancing edges and details.

Simplest Color Balance (SCB) was proposed by *N. Limare* et al. in [15]. The algorithm tries to remove incorrect colour cast by scaling each channel histogram to the complete *0–255* range via affine transform.

4 Evolving Image Enhancement Pipelines

The analysis of the related work and preliminary work with the implemented filters shown that individually the filters can perform well if under the right conditions but may struggle with versatility, i.e. input images under different conditions may require adjustments on parameters. Moreover, it is clear that applying different filters sequentially can produce unique results and that slight adjustments in the order of the filters may cause significant changes to the output. Furthermore, some images may require the application of one or two filters depending on different conditions. This led us to conclude, that for automatic IE we need to find a generic pipeline for application of image filters suitable to different input images. Contemplating these insights, we developed a way to automatically generate pipelines that compute image filters to be applied to the input images. We opted for GP using a tree representation since the problem inherently can be viewed as a program, a succession of steps and decisions of what filters should be applied on the input image and by what order. GP provides us with a representation suitable for exploring solutions in a structured and flexible way, with variation operators well-defined and adaptable to our problem.

Making use of GP requires the definition of the primitives and terminals that will be available to the population during the evolution. In our scenario, we want to evolve a sequence of filter functions that generally received at least an image and a numeric value as input, i.e. the filter's parameters. We defined a primitive set containing all the seven classical functions previously implemented, a terminal set containing the input image, and *ephemeral constants* ranging from -1 to 1. Each function then mapped the defined range in order to adapt it to the desired magnitude. Each function parameter range was empirically defined so that the function provided acceptable results. Additionally, we introduced an "*if-then-else*" function that, depending on the boolean value of a condition, returns the output of the "*then* tree" or the "*else* tree", allowing the same program solution to behave differently according to the input characteristics.

Since one of the primitives is the whole image, we required values that could be used for comparison to make conditionals. To make this possible, a set of "conditional functions" were introduced. Thus, we added to the primitive set, five functions that extracted relevant features from the image. The features are image-related features that capture characteristics of the perceived quality of the image: noise, contrast, saturation, brightness and sharpness. To extract the noise, we used the work proposed in [16] to quickly estimate the images Gaussian noise. For contrast, we calculated the *RMS* contrast [17], meaning standard deviation of pixel intensities. For saturation, we averaged the pixels' intensity in the S channel of the *HSV* colour system. For brightness, we used the *HSP* colour system [18], as it grants a brightness value closer to the real human perception when compared to the luminance (L) channel of the *HSL* or the value (V) channel from the *HSV*. We then averaged the perceived brightness (P) channel to obtain a final value. Finally, for sharpness, we applied a Laplacian filter, calculated the variance of the output and used that as a sharpness score, as proposed in [19].

All these functions were modified to expect an image as input and an ephemeral constant, which serves as a threshold for that condition. Figure 1 shows a graphical example of a possible individual. All the implementation was done using *DEAP* [20] for *Python* as the base evolutionary engine.

5 Experimental Setup

In this section, we present the experimental setup of our work. The GP approach creates image enhancement pipelines to create enhanced outputs of an input image. Without a lack of generality, we deploy the approach in online real-estate marketing scenarios, where the images should be aesthetically appealing to the audience. To validate the outputs we used a set of Image Quality Assessment (IQA) tools that are presented in Section . Afterwards, we present the test sample used in this work's experiments in Sect. 5.2. Furthermore, in the last section, we present the setup for our evolutionary approach in Sect. 5.3.

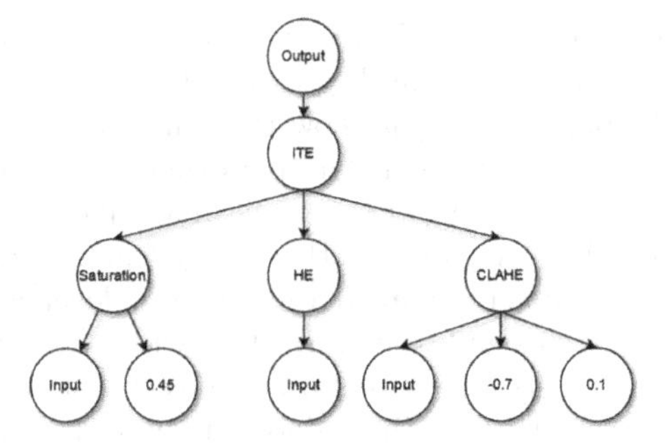

Fig. 1. Graphical example of a possible individual. The numbers represent the ephemeral constants, the *ITE* node represents the *if-then-else* primitive and the Saturation node represents the conditional function.

5.1 Image Quality Assessment

Although our work has its focus on IE, it is essential to understand how an image's quality can be measured. L. He et al. proposes the definition of image quality in three levels: fidelity, perception and aesthetics [21]. Fidelity is how well the image preserved the original information; Perception is how well the image is perceived according to every part of the HVS. Lastly, the image's aesthetic is the most subjective level because it varies from person to person. It is also the most difficult to measure objectively because *"aesthetics is too nonrepresentational to be characterised using mathematical models"* [21]. Since we are dealing with image quality measurements, which most often derives from a subjective appreciation, we must have a deterministic and automated way of qualifying an image quality. To tackle this problem, we made use of 3 distinct *no-reference* IQA tools, where *no-reference* means that the evaluation does not depend on a target or reference image to evaluate the quality of an input image.

PhotoILike (*PHIL* for short) is an IQA service provided by an external company *omitted for blind review.* This service is a closed source, third-party, black-box software that receives an image and returns a value from 1 to 10, where 1 means the worst quality and 10the best quality. Note that the calculated score is not solely based on the images aesthetic but also on multiple features considered relevant for real-estate marketing. For instance, the baseline score of a pool picture is much higher than the bathroom's baseline.

Blind Image Spatial Quality Evaluator [22] (*BRISQUE* for short), is a *no-reference* IQA tool, proposed by *A. Mittal* et al., used in image enhancement contexts [23,24]. As opposed to the previous methods, *BRISQUE* is based on a set of classical feature extraction procedures that computes a collection of 36 features per image. This tool originally outputs a value between 0 and 100, where 0 represents the best quality and 100 the worst. However, in order for the outputs to be in concordance with the previously presented methods, the output was mapped to a 1 to 10 range where 1 means is the worst score and 10 the best.

Neural Image Assessment [1] (NIMA for short), is a *no-reference* IQA tool based on a deep Convolutional Neural Network (CNN), proposed by H. Talebi and P. Milanfar. The paper highlights how the same architecture, trained with different datasets, leads to state-of-the-art performance predicting both technical and aesthetic scores. As the paper states, technical judgment considers noise, blur, and compression artefacts, among other image features. On the other hand, the aesthetic evaluation aims to quantify the semantic level characteristics associated with images' emotions and beauty. Both provided models predict the final score as an average of a distribution of scores between 1 and 10, where 1 means the worst score and 10 the best. Both models were used during the experiments, and require each input image to have a resolution of 224 by 224 pixels. The NIMA aesthetic response was used for evaluation of the individuals during the evolutionary process.

5.2 Test Dataset

To examine our results in an unbiased way, we separated in an early stage, 10% of the 12,090 images from the rest of the dataset. Thus, 1,209 images represent the test set. The rest served as a training set to our approaches, i.e. the dataset used by the evolutionary approach to be selected along the evaluation's evolutionary process. As a way of having a baseline for our measurements, we examined the original images of this test set, using all the IQA metrics presented in Sect. 5.1.

5.3 Evolutionary Setup

All the GP configuration used is presented in sum in Table 1, taking into account the standard GP operators and probabilities [25].

Table 1. Summary of the GP configuration used during the experiments.

Cross-over	One-point
Mutation	Sub Tree mutation - adds a tree with depth between 0 and 2
Selection	Tournament Size 3
Tree generation	Ramped half-and-half
Population size	80
Number of generations	150
Crossover probability	75%
Mutation probability	5%
Elite size	1
Max depth	10

As mentioned in Sect. 4, it is not the objective of these experiments to produce a solution that improves upon a specific image. Instead, we endeavour in searching for a solution as generalist as possible. To achieve this, it is necessary to do a fundamental change to the typical GP evaluation. Each solution will be evaluated on a set of 10 randomly selected images from the training subset. The average fitness of all images will be considered the fitness of the individual. In addition to that, to further prevent overfitting to a specific group of images, a new set of 10 images is selected in each generation. For this reason, it is expected a significant variation in fitness from one generation to another. In all performed experiments, the individual with the overall highest fitness was selected as the test subject for validation.

Defining a fitness function is one of the most crucial steps of building an evolutionary approach. In our case, we wanted a fitness function that evaluated each individual based on its output visual quality. We have selected the *NIMA* classifier tool as it produces results closer to *state-of-the-art* on visual quality evaluation. In the set of experiments presented in this work, we used values of the response from the *NIMA* classifier to assess the visual quality of the images produced by the individuals being evolved.

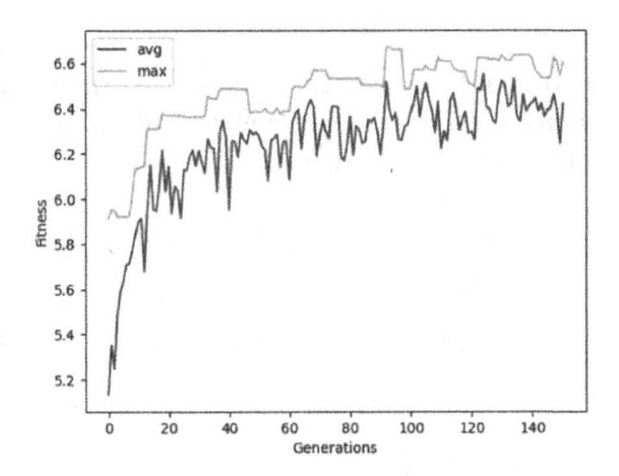

Fig. 2. Fitness evolution during 150 generations using the aesthetic model, with a depth of 10. The results are averages of 15 runs.

6 Experimental Results

In this subsection, we explore the experiments' results in a sub-dataset of real estate image, intending to enhance the image quality. Starting with the evolutionary approach, we can observe in Figure 2 that we can maximise the fitness function and the iterations. Note that since the set of images for evaluation is changing at each generation, the maximum value can oscillate even with elitism. Based on these results, we can see that the process is creating pipeline solutions that enhance the aesthetic score of images according to the NIMA aesthetic classifier output.

To further evaluate the pipeline solutions, we created a validation process where the best pipeline is applied to the test set, i.e. images not used in the evolutionary process. In this way, we can analyse the generalisation of the evolved pipelines. To establish a baseline approach, we conducted the same validation with a manually created pipeline, arranged based on expertise and state of the art references, with the default parameterisation. The pipeline by order of application is as follows: *Contrast Balance (CB), CLAHE, Unsharp Masking (UM, Non-local means denoising (NLMD) and Contrast Stretching (CS).* The best pipeline encountered uses the following pipeline (using to image filters acronyms from Sect. 3): $ITE(Sharp(x, -0.29), SCB(GC(x, 0.41), 0.69), CB(NLMD(GC(GC(HE(CS(CLAHE(x, 0.36, 0.39))), 0.03), -0.97), 0.75), -0.97))$.

The original scores according to the IQA tools are presented in Table 2. In Table 3, and Fig. 3 we can see the results after applying the manual and the best evolutionary pipeline. In benefit of readability, *NIMA* models are abbreviated to respective initials and *PhotoILike* is abbreviated to *PHIL*. All the results are presented in Table 3 and show the metrics improvement over the original images scores. A negative *improvement* score means the degradation of quality according to the respective metric.

Table 2. Average (μ) and standard deviation (σ) of the original images from the test dataset using 4 *no-reference* metrics. All the metrics values range from 1 to 10 where 1 means the lowest quality and 10 highest quality.

	NIMA A.	*NIMA T.*	*BRISQUE*	*PHIL*
Original - μ	4,91	5,36	7,77	5,56
Original - σ	0,45	0,43	1,15	1,41

Table 3. Average (μ) and standard deviation (σ) of the improvement a list of the manual pipeline and the evolutionary pipeline, on the test dataset. All the metrics values range from 1 to 10 where 1 means the lowest quality and 10 the highest quality, and the improvement is calculated by subtracting the original score from the resulting one.

	NIMA A.	*NIMA T.*	*BRISQUE*	*PHIL*
Manual - μ	0,43	−0,06	−0,08	0,25
Manual - σ	0,31	0,26	1,86	0,64
Evolutionary - μ	1,30	−0,55	0,40	0,47
Evolutionary - σ	0,45	0,43	0,10	0,96

The results in Table 3 and Fig. 3 indicate that the manual pipeline resulted in small alterations across all IQA scores, with the NIMA aesthetic response showing some improvements. In average the manually defined pipeline improves

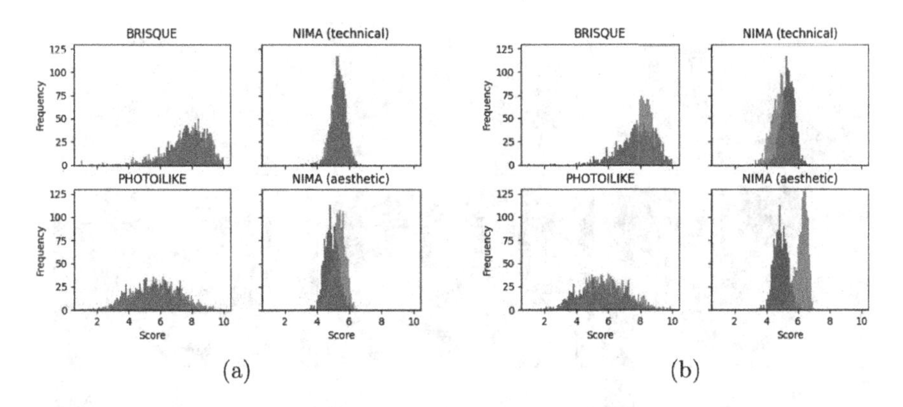

(a) (b)

Fig. 3. (a) Graphical comparison between the original test dataset (blue) and the results from the list of classical functions manually selected (orange), computed by all four IQA tools used during the experiments. (b) Graphical comparison between the original test dataset (blue) and the results from using the best evolved solution using *NIMA* aesthetic as fitness, computed by all four *no-reference* IQA tools (Color figure online)

the test dataset in the NIMA and *PHIL*, with the other two metrics showing that it worsens the images. Evaluating the best-evolved pipeline, we can observe that all the metrics show greater increases when compared with the manual pipeline except for the NIMA technical, where for some images tend to be worse than the original and the manual pipeline on that metric, averaging a −0.55 difference for the original and −0.49 to the manual pipeline. Overall, we have significant improvements to the images' aesthetic component, considering the initial distribution of the scores, when using the aesthetic model as fitness. Based on the results, we can also say that the best-evolved pipeline has better results than the manual pipeline, the baseline approach.

Figure 4 show output examples for the experiments using the *NIMA* aesthetic model. All the images presented show an improvement in metrics and subjectively visual improvements compared with the original ones and the baseline. Using the *NIMA* aesthetic model can produce fascinating results, that almost resemble over stylised photos, almost like paintings in some details, instead of real pictures. The model's evaluation tends to associate higher scores with over-edited and saturated images. Another conclusion we can extrapolate from the results is the fact that improvements of the same magnitude as those of this test will also mean extreme adulteration of the original image even if they are considered good for the aesthetic model and *PhotoILike*, showing that this tool is also conductive to high aesthetic scores.

Besides the relevant results, we obtained pipelines where a non-photorealistic rendering effect occurs. The results shown in Fig. 5 were unexpected but a pleasant surprise, indicating a moment of computational serendipity [2]. The overall

Fig. 4. Examples of images from the test set (left), processed from the manual pipeline (middle) and from the evolved pipeline (right).

idea and goal were to create a filter application pipeline that maximised aesthetic response by using NIMA aesthetic response to guide evolution. Indeed we can maximise the response, and the results indicate that the images are being enhanced across the IQA tools that take aesthetics into account. More importantly, we are improving the PHIL metric, which estimates aesthetics and value from a real-estate marketing perspective. So the system tends to evolve further these solutions which maximise the response of *PHIL* which is the metric more closely related to the problem that we are trying to solve. Moreover, a fascinating fact is that NIMA aesthetic was not trained with paintings; however, it tends

Fig. 5. Examples of images with non-photorealistic rendering effects.

to score higher for the more non-realistic images, which tend to be more like paintings.

Based on these results, we tested evolved pipelines in another type of imagery to evaluate the effect and assess if the results would render aesthetically pleasing images. Figure 6 show some of those outputs. As we can see, the pipelines are generic enough to alter the different type of images, abstract, minimalist and raw photographs. From a subjective standpoint, some of the outputs are more interesting and aesthetically pleasing than the inputs. The results further indicate our approach's potential to create non-photorealistic rendering pipelines for the creation of different artefacts.

For real estate marketing, based on these insights and results, we are convinced that we should incorporate the NIMA technical into the fitness function and some control mechanism to prevent the image from suffering many alterations compared with the original. This line of experimentation and research is already being pursued.

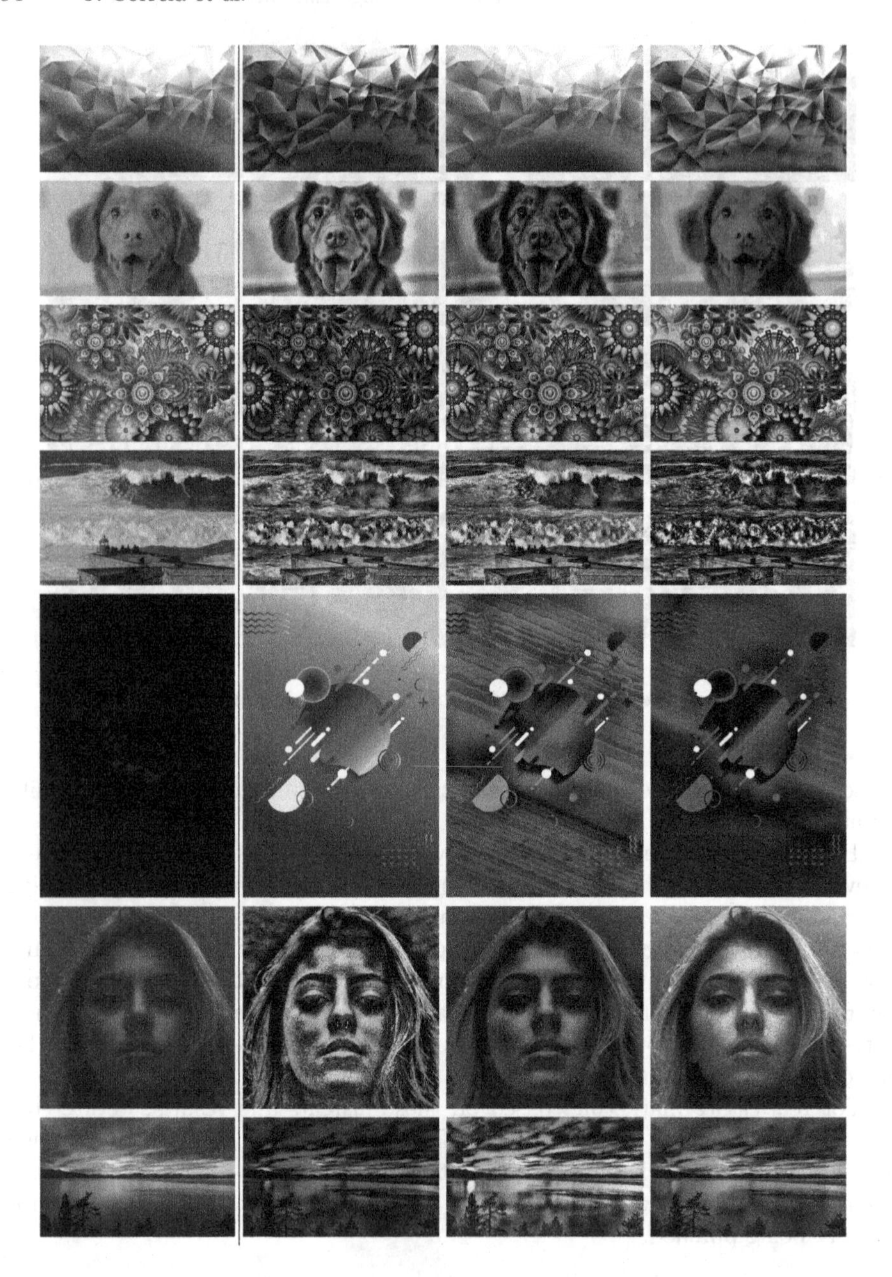

Fig. 6. Samples of images created with a selection of different evolved non-photorealistic pipelines. The original image is the first one on the left.

7 Conclusions

In this work, we presented an approach for automatic image enhancement using GP. The approach is instantiated in real estate marketing, to improve images from different types of real estate under diverse conditions, which requires a more modular approach. We propose an approach that relies on a set of 7 filters from the literature linked to image enhancement and image quality assessment. The context of the problem indicates that the image enhancement should aim towards aesthetically pleasing images. We explore that characteristic by resorting to the aesthetic classifier NIMA to evaluate the individuals. After evolving pipeline solutions, we tested the best solution in a subset of test images compared with the initial results and manual pipeline as the baseline. We show that the system can create filter application pipelines that improve the image quality in all the metrics chosen for image quality assessment. The technical IQA tool metric was the only metric that suffered a negative effect. We argue that the images tend to be altered much to the point that some parts can become more stylised than the original. We also argue that some mechanism to prevent the image from suffering many alterations could mitigate this effect.

During the experimentation, some of the results resulted on a non-photorealistic rendering effect that came to be unexpected and a computational serendipity phenomenon. The system maximised the objective function and the validation metrics response while creating effects that manipulate the images to the point that they became almost paintings. We moved to explore the application in another type of imagery and analysed the effect, showing this approach's potential to create non-photorealistic rendering and transforming effects.

As future work, we plan to expand and alter the set of functions available to the GP algorithm, fine-tuning evolution parameters, more experimentation with the fitness assignment alternatives and changing the evolution dataset, and ways of enlarging the scope of this work. We also plan to incorporate machine learning approaches to the pipeline to be used as filters for the pipeline's input images. Furthermore, in the context of the problem at hand, the results suggest that we should not exaggerate and alter the images' feature that much. We plan on doing a set of experiments using similarity metrics to control the pipelines and prevent them from altering the input images to much.

Acknowledgments. This work is funded by national funds through the FCT - Foundation for Science and Technology, I.P., in the scope of the project CISUC - UID/CEC/00326/2020 and by European Social Fund, through the Regional Operational Program Centro 2020. This work is also funded by the INDITEX-UDC Program for predoctoral research stays through the Collaboration Agreement between the UDC and INDITEX for the internationalization of doctoral studies.

References

1. Esfandarani, H.T., Milanfar, P.: NIMA: neural image assessment. CoRR abs/1709.05424 (2017). http://arxiv.org/abs/1709.05424
2. Pease, A., Colton, S., Ramezani, R., Charnley, J., Reed, K.: A discussion on serendipity in creative systems. In: Maher, M., Veale, T., Saunders, R., Bown, O. (eds.) Proceedings of the 4th International Conference on Computational Creativity, ICCC 2013, 12 June 2013 Through 14 June 2013, pp. 64–71. University of Sydney, Faculty of Architecture, Design and Planning (2013). http://www.computationalcreativity.net/iccc2013/
3. Wang, W., Chen, Z., Yuan, X., Wu, X.: Adaptive image enhancement method for correcting low-illumination images. Inf. Sci. **496**, 25–41 (2019)
4. Wong, C.Y., et al.: Histogram equalization and optimal profile compression based approach for colour image enhancement. J. Vis. Commun. Image Represent. **38**, 802–813 (2016) http://dx.doi.org/10.1016/j.jvcir.2016.04.019
5. Talebi, H., Milanfar, P.: Fast multi-layer laplacian enhancement. IEEE Trans. Comput. Imaging (2016)
6. Zhuo, S., Zhang, X., Miao, X., Sim, T.: Enhancing low light images using near infrared flash images. In: Proceedings - International Conference on Image Processing, ICIP, pp. 2537–2540 (2010)
7. Rundo, L., et al.: MedGA: a novel evolutionary method for image enhancement in medical imaging systems. Expert Syst. Appl. **119**, 387–399 (2018)
8. Munteanu, C., Rosa, A.: Evolutionary image enhancement with user behaviour modeling. ACM SIGAPP Appl. Comput. Rev. **9**, 8–14 (2000)
9. Shan, T., Wang, S., Zhang, X., Jiao, L.: Automatic image enhancement driven by evolution based on ridgelet frame in the presence of noise. In: Rothlauf, F., et al. (eds.) EvoWorkshops 2005. LNCS, vol. 3449, pp. 304–313. Springer, Heidelberg (2005). https://doi.org/10.1007/978-3-540-32003-6_31
10. Bazeille, S., Quidu, I., Jaulin, L., Malkasse, J.P.: Automatic underwater image pre-processing. In: Proceedings of CMM 2006, October 2006
11. Xie, Y., Ning, L., Wang, M., Li, C.: Image enhancement based on histogram equalization. J. Phys.: Conf. Ser. **1314**, 012161 (2019)
12. Chang, Y., Jung, C., Ke, P., Song, H., Hwang, J.: Automatic contrast-limited adaptive histogram equalization with dual gamma correction. IEEE Access **6**, 11782–11792 (2018)
13. Buades, A., Coll, B., Morel, J.M.: Non-local means denoising. Image Process. Line **1**, 208–212 (2011)
14. Deng, Y., Loy, C.C., Tang, X.: Aesthetic-driven image enhancement by adversarial learning. In: MM 2018 - Proceedings of the 2018 ACM Multimedia Conference, pp. 870–878 (2018)
15. Limare, N., Lisani, J.L., Morel, J.M., Petro, A.B., Sbert, C.: Simplest color balance. Image Process. Line **1**, 297–315 (2011)
16. Immerkær, J.: Fast noise variance estimation. Comput. Vis. Image Underst. **64**(2), 300–302 (1996). https://doi.org/10.1006/cviu.1996.0060
17. Peli, E.: Contrast in complex images. J. Opt. Soc. Am. A **7**(10), 2032–2040 (1990)
18. Rex Finley, D.: HSP color model - alternative to HSV (HSB) and HSL (2006). http://alienryderflex.com/hsp.html
19. Pech-Pacheco, J.L., Cristobal, G., Chamorro-Martinez, J., Fernandez-Valdivia, J.: Diatom autofocusing in brightfield microscopy: a comparative study. In: Proceedings 15th International Conference on Pattern Recognition, ICPR-2000, vol. 3, pp. 314–317 (2000)

20. Fortin, F.A., De Rainville, F.M., Gardner, M.A., Parizeau, M., Gagné, C.: DEAP: evolutionary algorithms made easy. J. Mach. Learn. Res. **13**, 2171–2175 (2012)
21. He, L., Gao, F., Hou, W., Hao, L.: Objective image quality assessment: a survey. Int. J. Comput. Math. **91**(11), 2374–2388 (2014). https://doi.org/10.1080/00207160.2013.816415
22. Mittal, A., Moorthy, A.K., Bovik, A.C.: No-reference image quality assessment in the spatial domain. IEEE Trans. Image Process. **21**(12), 4695–4708 (2012)
23. Lim, J., Heo, M., Lee, C., Kim, C.S.: Contrast enhancement of noisy low-light images based on structure-texture-noise decomposition. J. Vis. Commun. Image Represent. **45**, 107–121 (2017). http://www.sciencedirect.com/science/article/pii/S1047320317300603
24. Wang, G., Li, L., Li, Q., Gu, K., Lu, Z., Qian, J.: Perceptual evaluation of single-image super-resolution reconstruction. In: 2017 IEEE International Conference on Image Processing (ICIP), pp. 3145–3149 (2017)
25. Banzhaf, W., Francone, F.D., Keller, R.E., Nordin, P.: Genetic Programming: An Introduction: On the Automatic Evolution of Computer Programs and Its Applications. Morgan Kaufmann Publishers Inc., San Francisco (1998)

Genre Recognition from Symbolic Music with CNNs

Edmund Dervakos, Natalia Kotsani(✉), and Giorgos Stamou

School of Electrical and Computer Engineering,
National Technical University of Athens, Athens, Greece
eddiedervakos@islab.ntua.gr, nkotsani@corelab.ntua.gr, gstam@cs.ntua.gr

Abstract. In this work we study the use of convolutional neural networks for genre recognition in symbolically represented music. Specifically, we explore the effects of changing network depth, width and kernel sizes while keeping the number of trainable parameters and each block's receptive field constant. We propose an architecture for handling MIDI data which makes use of multiple resolutions of the input, called MuSeReNet - Multiple Sequence Resolution Network. Through our experiments we significantly outperform the state-of-the-art for MIDI genre recognition on the topMAGD and MASD datasets.

Keywords: Artificial intelligence · Machine learning · Music information retrieval · Genre classification

1 Introduction

Music is a domain which in recent years has facilitated important research on artificial intelligence, including information retrieval and generative models.

The two most common forms of musical data are symbolic representations - such as MIDI messages, pianorolls and scores - and audio recordings. Each of the two representations is most suited for different tasks of interest for the artificial intelligence community. For instance, algorithmic composition of music is almost exclusively tackled with symbolic representations of music [1–3], while music classification tasks, such as genre recognition [4–7], or mood classification [8–10] are typically approached in the audio domain.

There are certain aspects of musical data which make it difficult to develop new methodologies for processing and extracting information. One of the main challenges in handling musical data is the existence of information across multiple time-scales. In audio recordings for example there exists information which ranges from very high frequencies (>10 kHz) mainly effecting the timbre of the sound to very low frequency information which may pertain to structural elements of a piece of music. This is also true, to a lesser extent, for symbolic representations of music, and is one of the main reasons for which many deep learning based composition models cannot generate pieces of music longer than a few bars while maintaining musicality, prosody and structure. On the other

J. Romero et al. (Eds.): EvoMUSART 2021, LNCS 12693, pp. 98–114, 2021.
https://doi.org/10.1007/978-3-030-72914-1_7

hand, symbolic music does not fully capture information which might be important for solving a task such as genre recognition, for instance the different sound of various instruments or nuanced aspects of human performance. These difficulties, among others, have led to the development of new approaches for processing music in both the audio and the symbolic domain, many of which take the form of specialized neural network architectures such as WaveNets [11]. In this work one of our main goals was to develop a methodology for classification of symbolic music into genres, based on neural networks.

Specifically convolutional neural networks (CNNs) have seen widespread use in multiple domains, spearheaded by their success in computer vision. However, finding an optimal architecture and set of hyper-parameters, such as number of layers, kernel size, number of kernels etc. for a given task remains a difficult problem which is still being tackled mostly experimentally. In this work, we focus our experiments on comparing CNNs which have the same receptive field and number of trainable parameters, but differ in depth, width and kernel size to explore their effectiveness for genre recognition in symbolically represented music.

2 Related Work

Most works related to genre classification involve music in audio (raw) format. The following are some of the most up-to-date techniques on the classification of music through neural networks. Oramas et al. [4] propose a convolutional neural network architecture (a deep residual network) for the genre classification task from the audio spectrogram of the songs. Their neural model learns to embed different modalities in a new multimodal space. Yu et al. [5] incorporating an attention mechanism into representation learning, proposing a parallelized attention framework in a Recurrent Neural Network architecture. Medhat et al. [6] introduce the ConditionaL Neural Network (CLNN) and the Masked ConditionaL Neural Network (MCLNN), designed to exploit the nature of sound in a time–frequency representation. The CLNN preserves the time and frequency specificity of the features as the MCLNN combines frequencies in a controlled fashion, subdividing the frequency bins of a spectrogram into bands. Yang et al. [7] propose a hybrid architecture, the parallel recurrent convolutional neural network (PRCNN), that combines feature extraction and time-series data classification in one stage showing that this architecture outperforms the previous approaches applied to the same datasets.

The abundant availability of MIDI files online, from multiple sources, has given rise to the challenge of automatically organizing such large collections of MIDI files. One criterion for organization is music genre, among others such as music style, similarity and emotion. In [12] McKay and Fujinaga argue in favor of genre classification, despite inherent difficulties such as ground truth reliability.

There are many different approaches in the literature for genre recognition in the symbolic music domain. Dannenberg et al. [13] used a machine learning approach, including a simple neural network, on a custom dataset for successful genre recognition in the symbolic domain. In [14], Karydis et al. combine pattern recognition with statistical approaches to successfully achieve genre recognition for five subgenres of classical music. Kotsifakos et al. in [15] compute a sequence similarity between all pairs of channels of two MIDI files and then use a k-NN classifier for genre recognition. They experiment on a dataset of 100 songs and four genres. Zheng et al. [16] extract features related to melody and bass through a musicological perspective, incorporating text classification techniques and using Multinomial Naive Bayes as the principal probabilistic classifier. Their self-collected dataset consist of 273 records.

These approaches were experimentally validated on relatively small datasets compared to for example the openly available Lakh MIDI dataset [17]. For large scale datasets, Ferraro and Lemström in [18] utilize pattern recognition algorithms SIA [19] and P-2 [20] in addition to a logistic regression classifier to solve the task. The benefit of this approach is interpretability since the authors have created a large corpus of genre-specific patterns, which could also be utilized for other music related tasks. Duggirala and Moh [21] apply Hierarchical Attention Networks in music genre classification, after converting the audio files into a word embedding representation. Liang et al. [22] propose four word embedding models consisting of three vocabularies (chroma, velocity and note state) and apply these models in three MIR tasks: melody completion, accompaniment suggestion, and genre classification, concluding the robustness and effectiveness of their embeddings.

Recently in multiple domains, there is a tendency to forgo feature extraction stages of an information retrieval pipeline, instead using a more complex neural network architecture, in which the first layers act as feature extractors. In computer vision for example, this end-to-end approach has surpassed most previous feature-based works in performance, which motivates us to implement such a system for MIDI genre recognition.

3 Preliminaries

Genre recognition from symbolic music can be described as a multi-label sequence classification task, in which a set of labels $Y = \{y_1 \dots y_m\}$ is assigned to a sequence of vectors $X = (x_1, x_2 \dots x_t)$.

Given a large enough dataset of n such sequences, along with their ground truth labels $D = \{(X^1, Y^1), \dots, (X^n, Y^n)\}$, in the context of machine learning, the goal is to train an algorithm $F(X; \theta)$ to model the conditional distribution of labels with respect to input sequences.

$$F(X; \theta)_m = P(y_m | X) \tag{1}$$

where m is the index of a label.

This is approached as an optimization problem of finding parameters $\hat{\theta}$ which minimize the cross-entropy between the modelled distribution and the distribution in the train-set for each label separately. Each classifier's performance is then measured on a test dataset which does not overlap with D.

3.1 1D Convnets

The state-of-the-art approach for MIDI genre classification presented by Ferraro and Lemström in [18] involves using an algorithm for recognizing patterns of notes in an input sequence and then performing classification based on recognized patterns. These patterns are local, with the best results achieved from extracting four or five note patterns. This, along with the success of 1D CNNs for other tasks in the symbolic music domain [1] and their suitability for sequence pattern recognition, as has been shown in multiple domains, such as pattern recognition in DNA sequences by Lanchantin et al. in [23], motivates us to explore 1D CNNs for the task of recognizing genres in symbolic music.

A one-dimensional convolution with kernels of size k is an operation, which acts on a sequence of T vectors X of size N to produce a sequence of vectors Y, where:

$$Y_{i,j} = \sigma(\sum_{n=1}^{N}\sum_{m=1}^{k} W_{j,n,m} X_{i+m-1,n}) \tag{2}$$

The matrix W consists of the convolution's trainable parameters, while the function σ enforces non-linearity. Note that there is also a bias term which has been omitted. Each element of the output sequence Y only depends on k consecutive elements of the input sequence X, leading to the effectiveness of the convolutional operation for capturing local structures and patterns in the data. A deep neural network may then be constructed by stacking such operations in a depth-wise fashion. This results in the first convolutional operations capturing low-level features in the data, while deeper operations capture more complex high-level features.

3.2 Receptive Field and Trainable Parameters

An important attribute of such a network is its receptive field, which is defined as the number of elements in the input sequence which effect a single element of the output sequence. A single convolution C_1 with kernels of size k_1 has a receptive field of k_1. A second convolution with kernels of size k_2 which is fed C_1's output will depend on k_2 consecutive elements of said output, leading to its dependence on $k_2 + k_1 - 1$ elements of the original input sequence. Another important attribute of a CNN to keep track of is the number of trainable parameters, equivalent to the size of all weight matrices W. The number of trainable parameters is the primary factor for the memory requirements of a network which is often a bottleneck for network design.

Efficiently increasing a network's receptive field is crucial for effectively capturing features across multiple time-scales. By stacking convolutional layers, receptive field increases linearly with network depth and with kernel size k. However, increasing depth could give rise to difficulties during training such the exploding and vanishing gradients problem (EVGP) [24,25] among others in addition to increasing the number of trainable parameters, while increasing k by dk leads to a $f_{in} \times f_{out} \times dk$ increase in trainable parameters, where f_{in} and f_{out}

are the numbers of input features and output features (number of kernels in the layer). There exist many methods for increasing a network's receptive field more efficiently, for instance using dilated convolutions such as in [11], using strided convolutions, or the most common approach: pooling layers.

3.3 Pooling

A pooling layer of stride S and kernel size K acts on a sequence of vectors X of dimensions $T \times N$ and outputs a sequence Y of dimensions $(\frac{T-K}{S}+1) \times N$. The stride parameter S determines how many input samples are skipped in between applications of the pooling kernel. A common pooling operation is max pooling with a stride equal to kernel size $K = S$. In this case for the output sequence Y:

$$Y_{i,j} = \max_{m<K} X_{(i*K+m),j} \tag{3}$$

Another common pooling operation is average pooling with $K = S$, for which case:

$$Y_{i,j} = \mathbb{E}_{m<K}(X_{(i*K+m),j}) \tag{4}$$

A pooling operation has a receptive field K and does not have any trainable parameters. Each of two consecutive samples in the pooling operation's output depend on K input samples, however these samples are spaced apart by on average S samples in the input sequence. This means that feeding the pooling operation's output to another layer effectively increases receptive field by a factor of S without an increase in trainable parameters. Finally, pooling operations introduce invariance to local translations of an input sequence which could be useful for the learning process but they entail information loss which could be detrimental for learning.

4 Architecture

Trees are an effective representation of music, since they are able to capture information, patterns and structures at multiple different time-scales. There are however many different ways of constructing such trees from a piece of music, in addition to different approaches for feeding tree representations to neural networks.

In this work we set a baseline for tree representation utilization for information retrieval from symbolic music, by using full binary trees as input structures and by then treating each level of the tree as a separate input. Each node of the binary tree has as a value the average of its children, thus each level of the tree is equivalent to the original sequence - which is represented by the leaves of the tree - at a lower resolution. This means that MuSeReNets presented in this paper are similar to multiple resolution CNNs which have been successful for some computer vision tasks such as skin lesion recognition by Kawahara and Hamarneh in [26].

Intuitively, the first levels of a CNN detect low-level local features in the data, and more complex higher level features are captured in deeper layers. However there could exist high level features which may be simply extracted from a lower-resolution representation of the input without requiring increasing the depth of the network. For instance in the case of symbolic music, a simple feature which could be extracted from higher levels of a tree (closer to the root) would be the key signature of a large segment of the input - which relates to the set of notes which appear in the segment. Such features may then be combined with lower-level features extracted from levels closer to the leaves of the tree and fed to deeper layers of the network for further feature extraction and eventually solving a task, which in our case is genre classification.

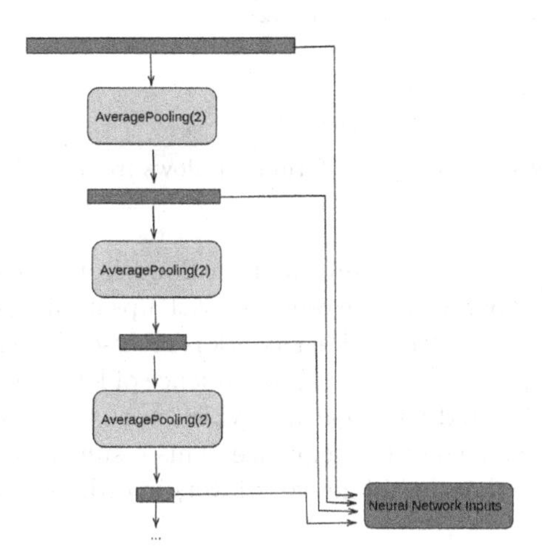

Fig. 1. Constructing a binary tree where levels are equivalent to the input sequence at lower resolutions

The first module of a MuSeReNet is a set of average pooling operations which act on the original sequence, each producing a version of the original sequence at a different resolution (Fig. 1). Each of these is treated as a separate input for the neural network.

There are many different ways to make use of these inputs. For MuSeReNets we distinguish between two cases: When information flows from the leaves to the root (Fig. 2) and when information flows from the root to the leaves (Fig. 3).

In the first case, in which information flows from the leaves to the root (Fig. 2), each input is fed through a block which consists of convolutional layers followed by a max pooling operation with the same stride and kernel size as the average pooling operation which generated the specific input from its higher resolution counterpart. This way, and by using 'same' padding for convolutional operations, the output of a specific block has the same sequence length as the original input at the previous resolution level and may be concatenated along their second axis, producing a sequence of the same length. The result of concatenation is the original sequence at a lower resolution augmented with features extracted from the convolutional block which processed the input at a higher resolution. This process is repeated until we reach the root of the tree, where the sequence length is 1, and the vector consisting of the root and features extracted from the previous convolutional block are fed to a fully connected layer with the goal of solving a specific task.

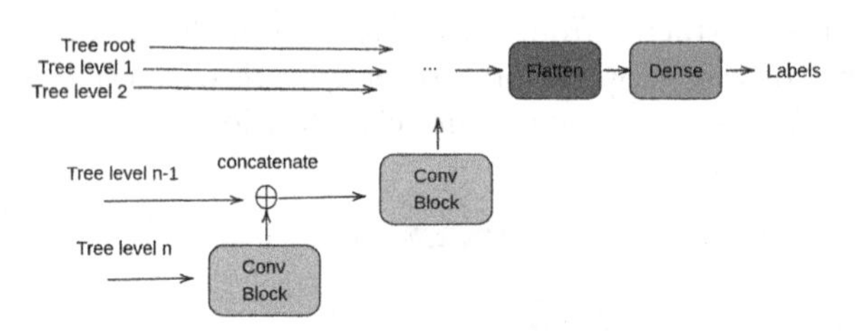

Fig. 2. A MuSeReNet where the information flows from the leaves to the root.

In the second case, in which information flows from the root to the leaves (Fig. 3), max pooling operations are replaced with upsampling operations and the order with which inputs are fed to the network is reversed (the root first instead of the leaves first). In this case the result is a sequence of length equal to the original sequence, but is augmented with features which were extracted by convolutions on lower resolution versions of the sequence. This resulting augmented sequence may then be fed to further neural network layers in order to solve a specific task.

5 Experiments

In order to explore the effectiveness of our architecture for information retrieval from symbolic music and to check the compatibility of 1D CNNs for the task, along with the effect of allocating resources to network depth or to kernel size we conducted the following experiments[1].

5.1 Data

For our experiments we use the Lakh Pianoroll Dataset as presented by Dong et al. in [1], specifically the LMD-matched subset. This dataset consists of pianoroll representations of MIDI files in the Lakh MIDI Dataset presented by Raffel in [17]. The pianoroll is an array representation of music in which columns represent time at a sample rate of n samples per quarter note and rows represent pitch in the form of MIDI note numbers. The LMD-matched subset contains pianorolls which have been linked with the Million Song Dataset (MSD) [27]. We use labels

[1] The code is available in the following github repository: https://github.com/kinezodin/cnn-midi-genre.

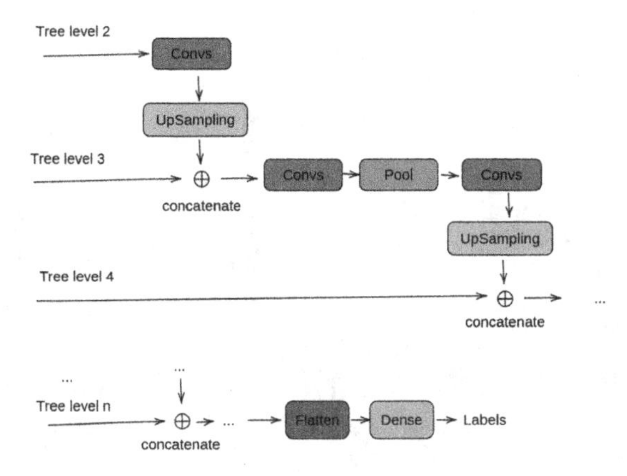

Fig. 3. A MuSeReNet where information flows from the root to the leaves.

acquired by MSD to construct the MASD and top-MAGD datasets presented by Schindler et al. in [28], so we can compare our results with existing work. To our knowledge Ferraro and Lemström in [18] have achieved the best results with regards to genre classification of symbolic music for the MASD and top-MAGD datasets. Finally we randomly split each dataset into a train and test set (.75/.25), we use the train set for training our models and the test set for evaluating them.

Both datasets are imbalanced with regard to the number of files corresponding to each label (Tables 1 and 2). Methods such as over-sampling rare classes and under-sampling common classes could be used to potentially improve the generalization ability of trained models, but it is left for future work since we are interested in observing the behaviour of different network configurations for this task and to what extent this extreme imbalance is detrimental to performance for each case.

5.2 Models

We construct neural networks by using blocks of 1-D convolutions followed by max-pooling operations of kernel size and stride 2. Specifically, we use a shallow block, which consists of only one convolutional layer prior to the pooling operation, and a deep block which consists of three convolutional layers before each pooling operation (Fig. 4). A network built with shallow blocks will be referred to with the prefix 'shallow', and those built with deep blocks 'deep'.

In addition, each network has a 'Sequence' version in which blocks are stacked depth wise and the first block receives as its input the original sequence, and a 'MuSeRe' (Multiple Sequence Resolution) version in which a level of a tree constructed from the input sequence is concatenated to the output of each block. The first case represents a traditional CNN architecture, while the second represents MuSeReNets in which information flows from the leaves to the root (Fig. 3).

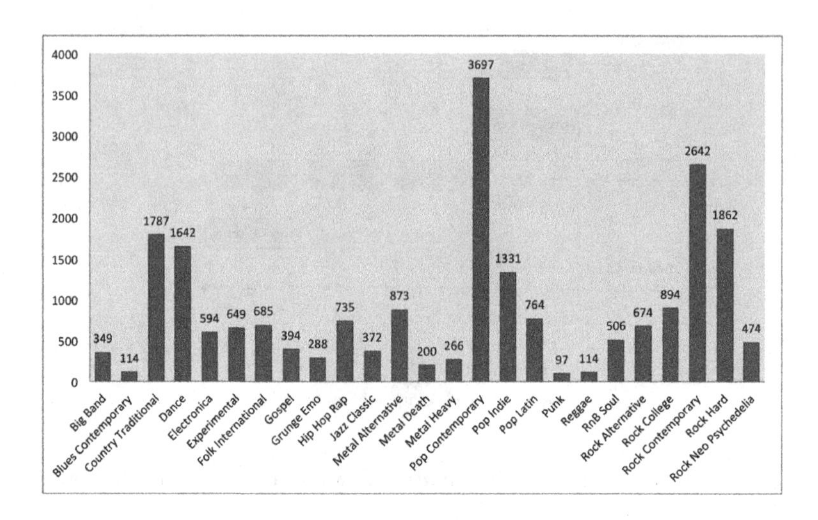

Fig. 4. Number of files in the LPD dataset per label of the MASD dataset

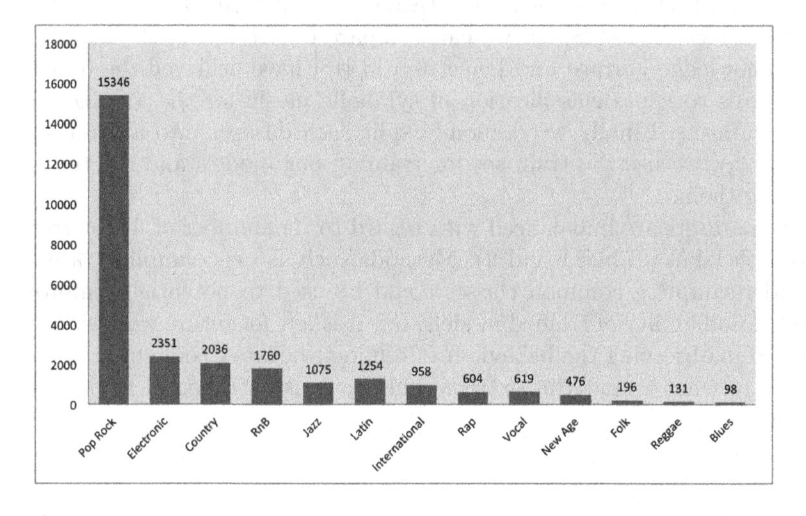

Fig. 5. Number of files in the LPD dataset per label of the topMAGD dataset

Shallow vs Deep. All blocks are set to have a similar receptive field of 24 samples. In the shallow block case this implies a kernel size of 24. For the deep block case, assuming all convolutions have the same kernel size k, if $k = 9$ the receptive field at the output of the third layer is 25 input samples. We arbitrarily chose the smallest kernel size $k = 9$ which has the same number of trainable parameters as a 3×3 kernel which is popular for computer vision two-dimensional CNNs.

All blocks are also set to have a similar number of trainable parameters. Given fixed input dimensions of $(|x|, f_{in})$, a single convolutional layer with f_{out} kernels of size k will have n_p trainable parameters:

$$n_p = (f_{in} * k + 1) * f_{out}$$

We set $f_{out} = f_{in} = 128$ for all blocks, thus the number of trainable parameters for a shallow block is:

$$n_{shallow} = 393,344$$

For the deep block, we set the number of kernels of the third layer $f_{out_3} = 128$, so that the output of each block is the same shape as with the shallow block. In order to satisfy the condition of having an equal number of trainable parameters to the shallow case

$$n_{deep} = n_{shallow}$$

$$1153 f_{o_1} + 9 f_{o_1} f_{o_2} + f_{o_2} + 1153 f_{o_2} + 128 = n_{shallow},$$

where f_{o_1} and f_{o_2} are the number of kernels for the first and second layers of the deep block. By arbitrarily setting $f_{o_1} = f_{o_2}$ we get 117 kernels per layer. This way we end up with the two blocks shown in Fig. 4.

For the different models, we use powers of 2 as input sequence lengths l, ranging from $l = 64$ to $l = 2048$. In the context of our dataset, these lengths represent musical time from approximately 5 quarter notes to 170 quarter notes, or 42 bars for a $\frac{4}{4}$ time signature (around one to two minutes for typical values of a song's tempo). Then each network will consist of $\log_2 l$ blocks stacked depth-wise, followed by a fully connected layer at the output, with as many sigmoid activated units as there are different labels in each dataset. For all convolutional layers we used ReLu activations (Fig. 6).

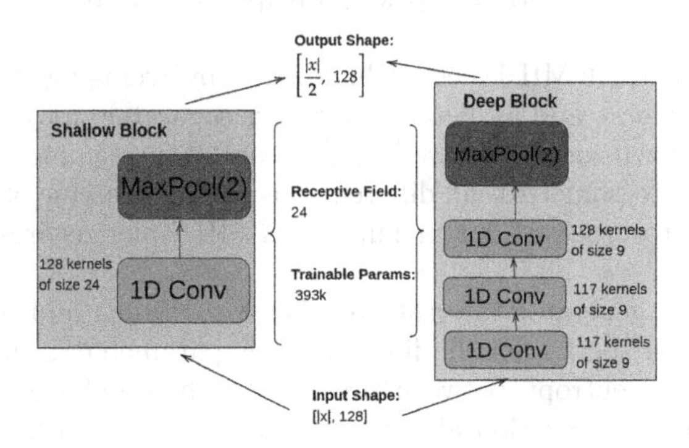

Fig. 6. The convolutional blocks used to construct the CNNs for use in our experiments

Sequence vs MuSeRe. The two versions of each network (Fig. 5) differ with regard to the inputs of each block which are sequences of 128-dimensional vectors in the 'Sequence' case and 256-dimensional vectors in the 'MuSeRe' case, which are a result concatenation of a previous block's output with the original sequence at a lower resolution. This leads to an increase in trainable parameters for the first layer of each block (Fig. 7).

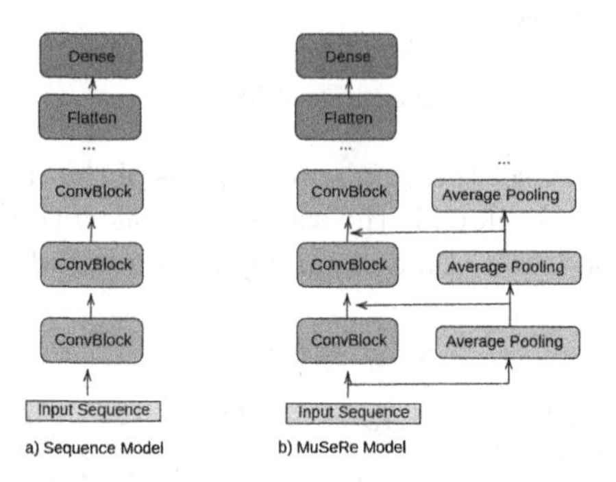

Fig. 7. a) Sequence architecture and b) MuSeRe architecture used in experiments

5.3 Data Preparation and Training

Every piano-roll is a fixed length sequence of vectors $\mathbf{x} = [\mathbf{x_1}, \mathbf{x_2}, ...\mathbf{x_t}]$, where each vector x_i has 128 dimensions representing MIDI note numbers. During training, before feeding a sequence to a network, we perform a random transposition by shifting elements of every vector of a sequence by a random integer in $[-6, 6]$. This corresponds to transpositions up to a tritone below or above and is done as a data augmentation step which helps to avoid bias with respect to a particular tonal center.

The way multi-track MIDI files are handled, is by averaging the piano-rolls of all instrumental tracks and all percussive tracks separately into two cumulative pianorolls. The downside is that we lose information pertaining to the different instruments and to some extent different voices become convoluted, but this way we are able to process a larger number of MIDI files regardless of number of tracks.

We use a data generator to create batches for training and apply any data transformations on the fly. For finding trainable parameters of networks which minimize the cross entropy between predicted genres and real genres we use Adam [29] as the optimization algorithm during training with a learning rate of $\alpha = 10^{-5}$ and for the other hyper-parameters of the optimizer $\beta_1 = 0.9$, $\beta_2 = 0.999$ and a batch size of 32. Before training a model, we further split the original train set into a validation set and a train set (0.2/0.8) randomly. We then trained our models for up to 300 epochs while using an early stopping criterion for each model's F1-score on the validation set.

5.4 Evaluation and Post Processing

We evaluate our models on the held-out test set for each of the MASD and top-MAGD datasets by computing precision, recall, micro f1 metric. Due to the varying sequence lengths of pianorolls in the test set, we use the post-processing procedure described below to aggregate a model's prediction across whole sequences, which are of greater length than the expected neural network inputs.

Given a pianoroll x_p of sequence length N and a model with input length $N_m < N$, we retrieve sequences of length N_m from x_p by using a sliding window of N_m samples and a hop size of $\frac{N_m}{2}$ samples on the sequence. Each window is then fed to the model for inference, which returns a vector where each element represents a probability that a specific label is assigned to x_p. Then we assign as predicted labels those with a probability value greater than 0.5. If no labels have a probability greater than 0.5 then we assign as a single label the element of the vector which has the maximum probability, since there are no unlabeled samples in the dataset. These predictions are then used to calculate false and true positives and negatives, recall, precision and F1 score.

5.5 Results

The results of our experiments are shown in Table 1 which list micro F1 scores of each trained model on the test set. In general, all CNNs which were trained on sequences longer than 256 samples surpassed Ferraro and Lemström's pattern recognition approach as well as Liang et al. model with regards to F1 metric [18,22]. Increasing input length by a factor of two along with increasing the number of blocks by one in most cases improved performance, with a notable exception for the longest sequence lengths that we experimented on (1024 vs 2048). In general, tree models outperform sequence models for shorter input lengths. The poor performance of tree models for larger inputs could be a result of overfitting, but requires further experimentation.

In addition we present precision, recall and F1 scores for each label in the topMAGD dataset for the best performing model (Table 2). On the one hand the effect of the imbalanced dataset is apparent in the network's performance for the most common label (Pop Rock) when compared to those with fewer files in the dataset such as Blues, Reggae and Folk. It is interesting that genres such as Jazz which have little representation in the dataset are better classified than genres such as Electronic which has almost double the support. This could be due to distinguishing musical characteristics of each genre, which are apparent in symbolic representations of music - for instance jazz music tends to have complex harmony and utilize more notes, while electronic music tends to contain loops of very few notes.

Table 1. Micro F1 scores on the test sets of the MASD and topMAGD dataset for each of our architectures P2-4 and P2-5 refer to the best performing configuration of those presented in [18] and PiRhDy_GM refer to the best performing configuration of those presented in [22]

Length	Block	Input	MASD	topMAGD
64	Deep	Sequence	0.258	0.620
		MuSeRe	0.265	0.622
	Shallow	Sequence	0.295	**0.623**
		MuSeRe	**0.308**	0.622
128	Deep	Sequence	0.315	0.624
		MuSeRe	0.317	0.631
	Shallow	Sequence	0.361	0.632
		MuSeRe	**0.407**	**0.639**
256	Deep	Sequence	0.411	0.654
		MuSeRe	0.404	0.639
	Shallow	Sequence	0.335	0.663
		MuSeRe	**0.491**	**0.668**
512	Deep	Sequence	0.456	0.661
		MuSeRe	0.374	0.653
	Shallow	Sequence	**0.545**	**0.711**
		MuSeRe	0.525	0.703
1024	Deep	Sequence	0.507	0.673
		MuSeRe	0.337	0.641
	Shallow	Sequence	**0.581**	**0.777**
		MuSeRe	0.526	0.737
2048	Deep	Sequence	0.456	0.696
		MuSeRe	0.264	0.627
	Shallow	Sequence	**0.593**	**0.759**
		MuSeRe	0.444	0.733
P2-4			0.468	0.662
P2-5			0.431	0.649
PiRhDy_GM			0.471	0.668

Table 2. Per label precision recall and F1 score on the test set for Shallow Sequence model with input length 1024 (best performing model) on the topMAGD dataset

Label	F1	Precision	Recall	Support
Pop Rock	0.86	0.81	0.96	3705
Electronic	0.58	0.74	0.47	557
Country	0.67	0.83	0.56	502
RnB	0.61	0.92	0.45	432
Jazz	0.76	0.91	0.65	281
Latin	0.45	0.78	0.32	338
International	0.53	0.77	0.41	236
Rap	0.34	0.78	0.22	133
Vocal	0.65	0.90	0.51	150
New Age	0.66	0.94	0.51	116
Folk	0.48	1.00	0.32	44
Reggae	0.48	1.00	0.31	38
Blues	0.55	0.73	0.44	18
Micro avg	0.78	0.81	0.74	6550

6 Conclusions and Future Work

We have demonstrated the effectiveness of CNNs for information retrieval from symbolically represented music. From our results it is apparent that for our definitions of 'shallow' and 'deep' CNNs, the 'shallow' ones are most suited for the task at hand. We plan to further explore this idea, by using network architecture search to find a good baseline network and similarly to [30] find an efficient way to scale up neural networks for MIDI classification tasks.

In addition, even though 'MuSeRe' networks performed poorly when compared to 'Sequence' networks in most cases, it is interesting that for the smallest models and shortest input lengths 'MuSeRe' models tended to outperform 'Sequence' ones. In our experiments we used simple tree representations, which do not hold a lot of musical significance. In the future we will incorporate more musically meaningful tree representations of music which have been proposed in literature over the years.

Furthermore, increasing input sequence length effectively reduces the number of (non-overlapping) sequences in the dataset while network size increases, which makes models more prone to overfitting. This is hinted when comparing models trained on sequences of 1024 samples and those trained on sequences of 2048 samples. For future work we will explore ways in which our networks will dynamically accept multiple different sequence lengths and will be fed entire MIDI files instead of fixed length sequences.

In addition, even though we used one of the largest available MIDI genre annotated datasets for training and evaluating our models, the dataset is by no means representative of all available music and suffers from poor class balance. For future work we plan to augment our dataset by including files from other sources, such as the reddit[2] MIDI dataset and automatically acquire additional labels and annotations from online sources such as The Echo Nest[3] and Spotify APIs.

Finally, music genres themselves are part of a complex domain with hierarchical structures and different relationships between elements, such as sub-genres, fusions of genres etc. Utilizing these relationships could lead to more robust genre recognition, similarly to how utilizing relationships between chords by Carsault et al. in [31] improved performance for chord recognition.

Acknowledgment. This research is carried out/funded in the context of the project "Automatic Music Composition with Hybrid Models of Knowledge Representation, Automatic Reasoning and Deep Machine Learning" (5049188) under the call for proposals "Researchers' support with an emphasis on young researchers- 2nd Cycle". The project is co-financed by Greece and the European Union (European Social Fund-ESF) by the Operational Programme Human Resources Development, Education and Lifelong Learning 2014–2020".

References

1. Dong, H.-W., Hsiao, W.-Y., Yang, L.-C., Yang, Y.-H.: MuseGAN: multi-track sequential generative adversarial networks for symbolic music generation and accompaniment. In: Thirty-Second AAAI Conference on Artificial Intelligence (2018)
2. Mao, H.H., Shin, T., Cottrell, G.: DeepJ: style-specific music generation. In: 2018 IEEE 12th International Conference on Semantic Computing (ICSC), pp. 377–382. IEEE (2018)
3. Brunner, G., Konrad, A., Wang, Y., Wattenhofer, R.: Midi-VAE: modeling dynamics and instrumentation of music with applications to style transfer. In: 19th International Society for Music Information Retrieval Conference (ISMIR 2018) (2018)
4. Oramas, S., Barbieri, F., Nieto, O., Serra, X.: Multimodal deep learning for music genre classification. Trans. Int. Soc. Music Inf. Retrieval **1**(1), 4–21 (2018)
5. Yu, Y., Luo, S., Liu, S., Qiao, H., Liu, Y., Feng, L.: Deep attention based music genre classification. Neurocomputing **372**, 84–91 (2020)
6. Medhat, F., Chesmore, D., Robinson, J.: Masked conditional neural networks for sound classification. Appl. Soft Comput. **90**, 106073 (2020)
7. Yang, R., Feng, L., Wang, H., Yao, J., Luo, S.: Parallel recurrent convolutional neural networks-based music genre classification method for mobile devices. IEEE Access **8**, 19 629–19 637 (2020)
8. Ren, J.-M., Wu, M.-J., Jang, J.-S.R.: Automatic music mood classification based on timbre and modulation features. IEEE Trans. Affect. Comput. **6**(3), 236–246 (2015)

[2] https://www.reddit.com/r/WeAreTheMusicMakers/comments/3ajwe4/the_largest_midi_collection_on_the_internet/.

[3] http://static.echonest.com/enspex/.

9. Xue, H., Xue, L., Su, F.: Multimodal music mood classification by fusion of audio and lyrics. In: He, X., Luo, S., Tao, D., Xu, C., Yang, J., Hasan, M.A. (eds.) MMM 2015. LNCS, vol. 8936, pp. 26–37. Springer, Cham (2015). https://doi.org/10.1007/978-3-319-14442-9_3
10. Padial, J., Goel, A.: Music mood classification (2018)
11. Oord, A.V.D., et al.: WaveNet: a generative model for raw audio. arXiv preprint arXiv:1609.03499 (2016)
12. McKay, C., Fujinaga, I.: Musical genre classification: is it worth pursuing and how can it be improved? In: ISMIR, pp. 101–106 (2006)
13. Dannenberg, R.B., Thom, B., Watson, D.: A machine learning approach to musical style recognition (1997)
14. Karydis, I., Nanopoulos, A., Manolopoulos, Y.: Symbolic musical genre classification based on repeating patterns. In: Proceedings of the 1st ACM Workshop on Audio and Music Computing Multimedia, pp. 53–58 (2006)
15. Kotsifakos, A., Kotsifakos, E.E., Papapetrou, P., Athitsos, V.: Genre classification of symbolic music with SMBGT. In: Proceedings of the 6th international conference on PErvasive technologies related to assistive environments, pp. 1–7 (2013)
16. Zheng, E., Moh, M., Moh, T.-S.: Music genre classification: a n-gram based musicological approach. In: 2017 IEEE 7th International Advance Computing Conference (IACC), pp. 671–677. IEEE (2017)
17. Raffel, C.: Learning-based methods for comparing sequences, with applications to audio-to-midi alignment and matching. Ph.D. dissertation, Columbia University (2016)
18. Ferraro, A., Lemstöm, K.: On large-scale genre classification in symbolically encoded music by automatic identification of repeating patterns. In: 5th International Conference on Digital Libraries for Musicology, Paris (2018)
19. Meredith, D., Lemström, K., Wiggins, G.A.: Algorithms for discovering repeated patterns in multidimensional representations of polyphonic music. J. New Music Res. **31**(4), 321–345 (2002)
20. Ukkonen, E., Lemström, K., Mäkinen, V.: Sweepline the music. In: Klein, R., Six, H.-W., Wegner, L. (eds.) Computer Science in Perspective. LNCS, vol. 2598, pp. 330–342. Springer, Heidelberg (2003). https://doi.org/10.1007/3-540-36477-3_25
21. Duggirala, S., Moh, T.-S.: A novel approach to music genre classification using natural language processing and spark. In: 2020 14th International Conference on Ubiquitous Information Management and Communication (IMCOM), pp. 1–8. IEEE (2020)
22. Liang, H., Lei, W., Chan, P.Y., Yang, Z., Sun, M., Chua, T.-S.: PiRhDy: learning pitch-, rhythm-, and dynamics-aware embeddings for symbolic music. In: Proceedings of the 28th ACM International Conference on Multimedia, pp. 574–582 (2020)
23. Lanchantin, J. Singh, R., Lin, Z., Qi, Y.: Deep motif: visualizing genomic sequence classifications. arXiv preprint arXiv:1605.01133 (2016)
24. Hanin, B.: Which neural net architectures give rise to exploding and vanishing gradients? In: Advances in Neural Information Processing Systems, pp. 582–591 (2018)
25. Xie, D., Xiong, J., Pu, S.: All you need is beyond a good init: exploring better solution for training extremely deep convolutional neural networks with orthonormality and modulation. In: Proceedings of the IEEE Conference on Computer Vision and Pattern Recognition, pp. 6176–6185 (2017)

26. Kawahara, J., Hamarneh, G.: Multi-resolution-Tract CNN with hybrid pretrained and skin-lesion trained layers. In: Wang, L., Adeli, E., Wang, Q., Shi, Y., Suk, H.-I. (eds.) MLMI 2016. LNCS, vol. 10019, pp. 164–171. Springer, Cham (2016). https://doi.org/10.1007/978-3-319-47157-0_20
27. Bertin-Mahieux, T., Ellis, D.P., Whitman, B., Lamere, P.: The million song dataset (2011)
28. Schindler, A., Mayer, R., Rauber, A.: Facilitating comprehensive benchmarking experiments on the million song dataset. In: ISMIR, pp. 469–474 (2012)
29. Kingma, D.P., Ba, J.: Adam: a method for stochastic optimization. In: Bengio, Y., LeCun, Y., (eds.) 3rd International Conference on Learning Representations, ICLR 2015, San Diego, CA, USA, 7–9 May 2015, Conference Track Proceedings (2015). http://arxiv.org/abs/1412.6980
30. Tan, M., Le, Q.V.: EfficientNet: rethinking model scaling for convolutional neural networks. arXiv preprint arXiv:1905.11946 (2019)
31. Carsault, T., Nika, J., Esling, P.: Using musical relationships between chord labels in automatic chord extraction tasks. arXiv preprint arXiv:1911.04973 (2019)

Axial Generation: A Concretism-Inspired Method for Synthesizing Highly Varied Artworks

Edward Easton(✉), Anikó Ekárt, and Ulysses Bernardet

Computer Science, Aston University, Birmingham, UK
{eastonew,a.ekart,u.bernardet}@aston.ac.uk

Abstract. Automated computer generation of aesthetically pleasing artwork has been the subject of research for several decades. The unsolved problem of interest is how to automatically please any audience without too much involvement of the said audience in the process of creation. Two-dimensional pictures have received a lot of attention however, 3D artwork has remained relatively unexplored. This paper introduces the Axial Generation Process (AGP), a versatile generation algorithm that can be employed to create both 2D and 3D items within the Concretism art style. A range of items generated through the AGP were evaluated against a set of formal aesthetic measures. This evaluation shows that the process is capable of generating visually varied items which generally exhibit a diverse range of values across the measures used, in both two and three dimensions.

Keywords: Evolutionary computation · 2D and 3D art generation · Concretism

1 Introduction

The abstraction of real-world phenomena is a key element to art. The level of abstraction depends on many different factors, such as the subject of the artwork, the medium the artwork is generated in and the style it is created in. 3D media allows more accurate representations of real-world objects to be created than 2D media, which often abstracts by flattening 3D items onto a 2D canvas. The style of a piece of art transcends across both 2D and 3D media and heavily influences the subject. A style can be thought of as the combination of aspects which can be used to identify and categorise pieces of art. These categories are used as a method of identifying and comparing both artists and their work. For paintings and sculptures, these styles are often well defined, Van Gogh is known as a Post-Impressionist, Andy Warhol for Pop Art and Banksy for Street Art. This categorisation process offers many benefits, such as making comparisons between artwork and artists more straightforward, for example comparing Georgette Chen to Vincent Van Gogh is easier than to Banksy. Styles can also help people understand and describe their aesthetic preferences.

J. Romero et al. (Eds.): EvoMUSART 2021, LNCS 12693, pp. 115–130, 2021.
https://doi.org/10.1007/978-3-030-72914-1_8

In art generation systems, the style in which the items are created is often not well defined due to the system itself being minimally constrained, searching through all possible permutations of colours on a canvas, meaning any style of artwork may be produced. A person's aesthetic preferences only represent a minute portion of the items which can potentially be created, the size of the search space compared to the region containing an individual's preferences makes finding aesthetically pleasing artwork difficult. Two methods exist for navigating to the region containing someone's preferences: using formalised measures to automatically navigate or asking the person themselves to direct the search. Formal measures are often limited in their effectiveness to accurately describe aesthetically pleasing artwork and on the other hand, a human user, who despite being instinctively more proficient at assessing the aesthetic appeal, hinders the search in other ways such as through user fatigue [38]. One possible explanation for this is due to the extensive size of the search space, the system is not capable of locating sufficiently interesting artwork sufficiently quickly to continually engage the user.

Attempts have been made to address these issues by giving the user more control over the direction of the search [13] or allowing the generation of artwork to be shared across multiple users [36] however, these methods also have their limitations. An effective, alternative method would be to constrain the search space to include only the items someone finds attractive, but due to the lack of exact formalisms for individual aesthetic preferences, this would be a highly complex task. Another way the search space can be constrained is to apply a specific style, this can make the search for more interesting items easier and consequently reduce user fatigue and improve the understanding of and ability to measure different aesthetic aspects. Examples of systems which have tried this approach include generating artwork in the Pop-Art style [7] or within the style of Mondrian [17].

This paper introduces a versatile generation process which produces artwork within the Concretism style, that can be rendered in both three dimensions and two dimensions. As defined by Tate: "Concrete art is abstract art that is entirely free of any basis in observed reality and that has no symbolic meaning" [39]. It was chosen as the style of the generated artwork due to three main reasons: (1) it is difficult to assess and generate representational artwork, this style allows this to be avoided; (2) the style works within simple boundaries which are easily translated into an Evolutionary Art context and finally (3) the style is applicable for both 2D and 3D artwork. The remainder of the paper is organised as follows: Sect. 2 provides a brief overview of existing 2D and 3D generation methods, Sect. 3 details how the proposed Axial Generation Process generates its content, Sect. 4 provides an evaluation on how effective the process is at generating interesting content and finally the conclusions and potential extensions are discussed in Sect. 5.

2 Existing Generation Methods

A wide variety of methods have been used for generating artwork within Evolutionary Art systems [15,18,25,36,41,42], these different processes can be split across multiple categories. For the purpose of this work, the categorisation into methods that generate 2D and 3D items is the most relevant.

2.1 2D Items

Two-dimensional items are widely studied and the generation of images based on mathematical expressions is the most popular approach. These can be represented by lisp-style expressions [18,37] or in the form of expression trees [10,15,21,23,28,42]. The way these expressions are used to generate an image takes a variety of forms, the most common procedure is to use a value generated by the expression to set a property of each pixel in an image such as the luminosity [16,23], using multiple expressions to calculate the RGB value of a pixel [9,15,21] or mapping the value to a colour lookup table [18]. Expressions have also been used to calculate the position and colour of a line over a series of time-steps [10,13] and to create animated content [28,41]. Other underlying data types are routinely used, such as Compositional Pattern Producing Networks (CPPNs) [20,30,36,41]; Context-Free Design Grammars (CFDGs) [42]; encoding the parameters to use in another generation process [11], using SVG [17], or maintaining a list of shapes to be placed on the canvas [8]. Entirely custom representations have also been designed to solve particular problems for example to efficiently hold the data to create a piece of art in the style of Mondrian [7].

2.2 3D Items

3D item generation is not as widely studied however, the range of representations and their use is just as diverse as their 2D counterparts. This includes manipulating the colour and visibility of voxels using CPPNs [19] or Context-Free Grammars [2,27,33], using Graph Grammars to generate a set of points in 3D space [26], using Shape Grammars to create 3D items [5,29,32] and evolving parameters to use within an external generation system [25,31]. Other non-direct generation methods have also been used, such as by taking 2D content and adding the extra dimension [12]. The process presented in this paper works in the opposite direction, converting from 3D to 2D allowing the same content to be generated without losing any information about the original item.

3 Axial Generation Process

The Axial Generation Process (AGP) uses three values to position geometric items around a central axis: the height, the angle and the radius, representing the precise location in 3D space. Figure 1 shows how these values are used to

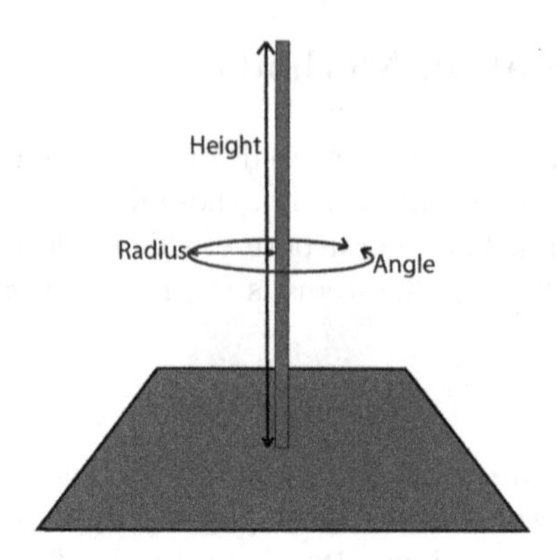

Fig. 1. The use of the height, the radius and the angle to place the geometric objects around the axis

place the items. The process maintains three expression trees which are used to calculate each value, as the process does not amend properties of individual pixels/voxels, only two values are used for the leaf nodes, the index of the item and a randomly generated value. This is different from the expressions within most existing systems, which commonly use the position of the pixel/voxel as well as a random value. Even though the process does not work on pixels directly, this does not affect its ability to generate high resolution images or large sculptures. The calculation of the position values is shown in Algorithm 1. This process is the same whether generating 2D or 3D content and therefore allows the same measures to be applied to both types allowing for easier comparison between all items.

Algorithm 1. Generation method

1: $totalItems \leftarrow n$
2: $heightTree \leftarrow Tree$
3: $angleTree \leftarrow Tree$
4: $radiusTree \leftarrow Tree$
5: $Points[] \leftarrow [totalItems]$
6: **for** $k \leftarrow 1$ to $totalItems$ **do**
7: $\quad y \leftarrow heightTree.getValue(k)$
8: $\quad angle \leftarrow angleTree.getValue(k)$
9: $\quad radius \leftarrow radiusTree.getValue(k)$
10: $\quad x \leftarrow radius \times Sin(angle)$
11: $\quad z \leftarrow radius \times Cos(angle)$
12: $\quad Points[k] \leftarrow (x, y, z)$
13: **end for**

The process is controlled by eight parameters, shown in Table 1, including the number of items to be added, the bounds the items should be placed within and how the items should be rendered. Amending these parameters can add additional complexity or change the visual structure of the generated items.

Table 1. Parameters used by the AGP

Total Items	The number of items which will be placed in the artwork
Maximum Height	The furthest distance away from the base the items can be placed
Maximum Radius	The furthest distance from the axis the items can be placed
Element Size	The radius of the circle to be placed on the canvas (2D only)
Background Colour	The canvas colour to be used when drawing the image (2D only)
Is 3D	Indicates whether 2D or 3D items should be generated
Canvas Width	The width of the bounds to render the items in
Canvas Height	The height of the bounds to render the items in

To allow for greater versatility, the AGP abstracts the rendering process away from the generation method, allowing the same output to be rendered in many different ways, the methods used here involve rendering the output in 2D and 3D. Rendering the 3D content is a simple process and involves just placing a geometric shape at the specified location. Examples of 3D content, rendered using the Unity3D engine, are shown in Fig. 3. For 2D content, the rendering process works in a similar way to creating 2D art, where a 3D object is abstracted down onto a 2D plane. This plane shows the top-down view of the items where their size is related to their distance from the canvas. This means that items closer to the max height will be larger when drawn onto the canvas, this size is calculated using Eq. 1. To ensure all items are visible, the element size parameter is used as the minimum size. This means that the size of each item can vary by up to a factor of the specified max height. An example of an image generated with the height taken into account is shown in Fig. 2. The perspective-driven flattening is optional within the process and depends on the max height specified. Example images which have been created with no perspective are shown in Fig. 4. The perspective will not be used for the evaluation of the 2D items within this paper, as many of the measures will not take this information into account. The parameters used for both the generation and rendering of the sculptures and images are shown in Table 2.

$$\frac{elementSize * maxHeight}{Clamp(maxHeight - itemHeight, 1, maxHeight)} \tag{1}$$

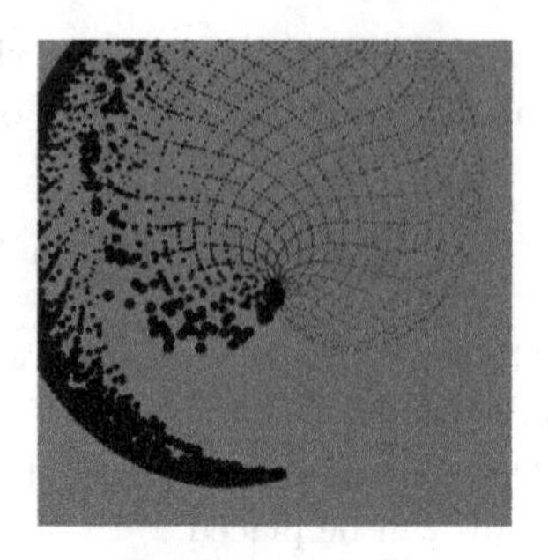

Fig. 2. 2D item with variable object sizes

Fig. 3. A variety of randomly generated sculptures using the AGP

Fig. 4. A variety of images generated using the AGP

4 Evaluation of the Axial Generation Process

In order for the AGP to be useful within an Evolutionary Art context, it needs to produce artefacts of interest. A common method is to classify artefacts based on their "novelty" and "value" [4,6,43]. However, because novelty is a context specific measure, it cannot be directly assessed without knowing how the output

Table 2. Output generation parameters

	Images	Sculptures
Total items	1000	750
Max height	1	10
Max radius	764	10
Element size	15	1
Background colour	#F5DE5D83	N/A
Geometric shape	Circle	Sphere
Canvas height	1080	N/A
Canvas width	1080	N/A

might be used. Alternatively, the process can be evaluated for its versatility, i.e. the range of novel items that can be generated which show a variety of values across different measures. There are many formal measures which aim to describe the aesthetic appeal and can be used to evaluate these items. To ensure that both 2D and 3D items can be compared, only measures which are applicable for both were considered. As the AGP does not currently calculate a value for the colour of the items, the selection of candidate measures was limited to those which are solely based on the form of the output.

Hence, several established measures such as the Global Contrast Factor [24], the colour gradients present within an image [35] or the naturalness of the image [1] were not considered. We are interested in how varied the artefacts generated by AGP are, therefore for the evaluation of AGP we focus on three canonical properties, namely level of symmetry, complexity, and compressibility, as these are all considered to contribute to the aesthetic appeal of the generated artefact.

Firstly, symmetry is known to be an important factor in the aesthetic judgement [40], the AGP places items around a central axis, often yielding highly symmetrical items. There are many choices of methods to calculate symmetry, however, the measure presented by [14] is used as it can calculate the level of symmetry for both 2D and 3D items. This measure works by finding multiple candidate symmetry planes and then refining their positions using the Levenberg Marquardt algorithm that minimises the error between points reflected in the plane. If the error is lower than a specified threshold, the item is considered symmetrical around it. The score calculated for each item is the average distance error across all detected planes for an item.

Secondly, the complexity of an item is also known to impact the aesthetic appeal [3,22]. The method we employ is based on the measure utilised by Birkhoff [3], involving counting the vertices in an item. Instead of fully rendering each item and counting the vertices, the value was estimated by merging items with similar positions, using a distance threshold of 0.2. The remaining geometric items in the artwork are then counted and divided by the original total, the intention is if items have many shapes in the same position, they are less complex. This process does have a small limitation by not accounting for the variation in size of the 2D items however, it is still suitable for this evaluation.

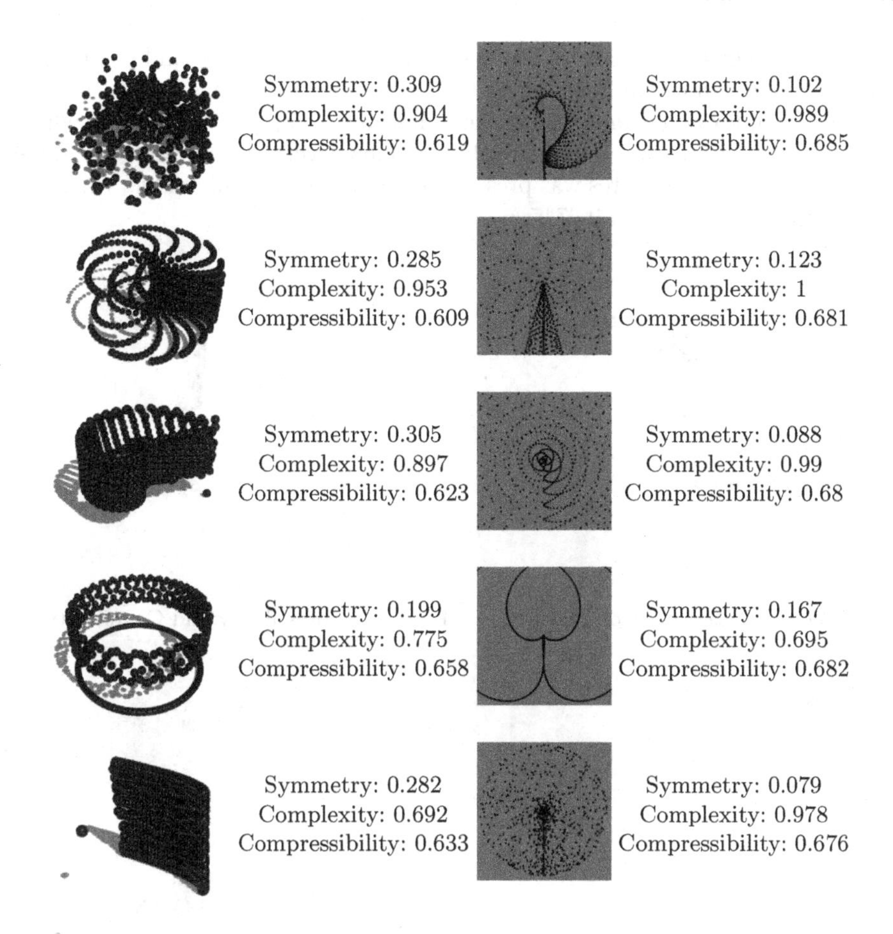

Fig. 5. Example output and associated values

Finally, we use the Global Normalised Kolmogorov Complexity [34], shown in Eq. 2, calculating the compressibility by encoding all points into text and then compressing this string using ZIP compression, similar to the process used in [15].

To perform the assessment, 500 items in both 2D and 3D were generated using the parameters shown in Table 2. A selection of generated items and their calculated values are shown in Fig. 5.

$$\frac{originalSize - compressedSize}{originalSize} \tag{2}$$

The results for the sculptures, shown in Figs. 6a and 6b, indicate that they were relatively well spread out with the levels of symmetry and complexity. However, even though the range of values covered the entire spectrum, as the level of symmetry increased the complexity tended to fall. The results for the Normalised Kolmogorov measure were relatively consistent, however another noticeable gap was present with no item rating lower than 0.6067.

A wide range of values were present for both the symmetry and complexity of the 2D items, shown in Figs. 6c and 6d. Unlike the 3D items, examples existed of highly complex and symmetrical items. However, a gap was still present where items with a mid-level of symmetry only had a low level of complexity. The same pattern as for the sculptures was present for the compressibility value, where no item had a value less than 0.6747.

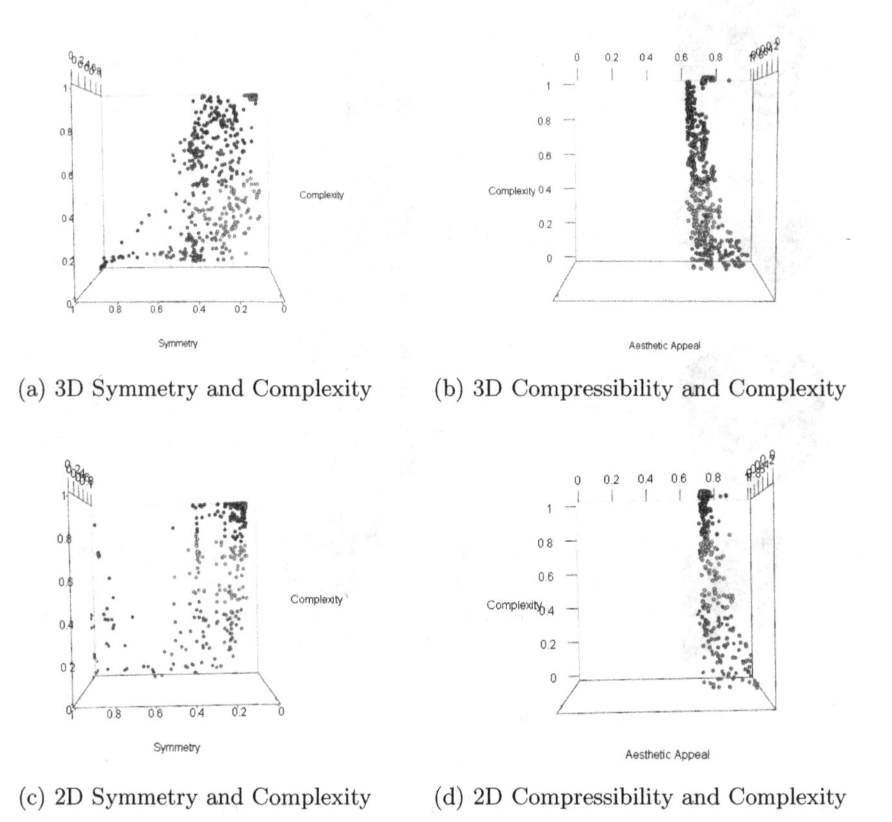

(a) 3D Symmetry and Complexity (b) 3D Compressibility and Complexity

(c) 2D Symmetry and Complexity (d) 2D Compressibility and Complexity

Fig. 6. The distribution of the randomly generated items

These major gaps potentially limit the process when attempting to generate novel items, as these regions of the search space may not contain sufficiently varied artefacts. There may exist more varied artefacts which are not reached when initially generating these items.

4.1 Filling the Gaps

To determine whether items existed within the noted gaps in the search space, two evolutionary algorithms were used. These were maximisation of symmetry and complexity as two objectives and the minimisation of compressibility as a single objective, for both 2D and 3D items. The evolutionary parameters are outlined in Table 3.

Once the process had finished the analysis was run again, the results for the newly generated items are shown in Fig. 7. A visual inspection shows that the evolutionary optimisation was not able to fill the gaps in the measured values. There are two possible explanations for this: as the measures are highly related to the number of geometric shapes included in the items, increasing this number would potentially improve the range of the values generated. The measures themselves may also not accurately represent the details of the generated artwork, highlighting the need for more appropriate formalised measures to be created.

Table 3. Evolutionary Algorithm parameters

Total Generations	30
Mutation Probability	0.7
Initialisation Method	Full
Selection Method	Tournament (k = 3)
Population Size	150

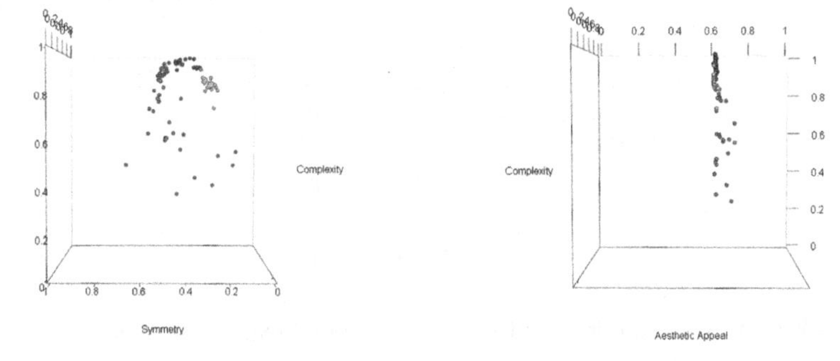

(a) 3D Symmetry and Complexity (b) 3D Compressibility and Complexity

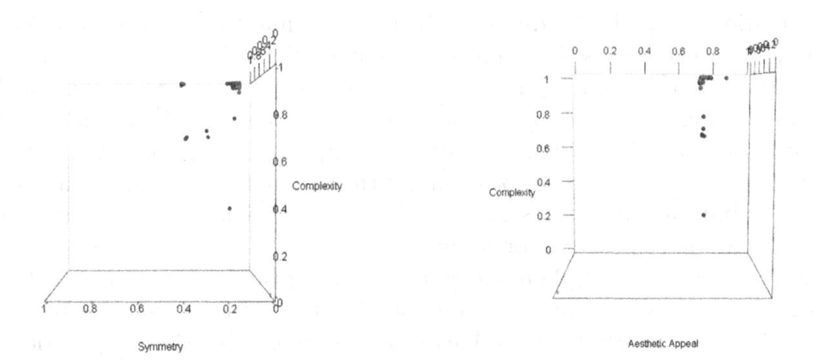

(c) 2D Symmetry and Complexity (d) 2D Compressibility and Complexity

Fig. 7. Distribution of the items after the Evolutionary run

5 Conclusion

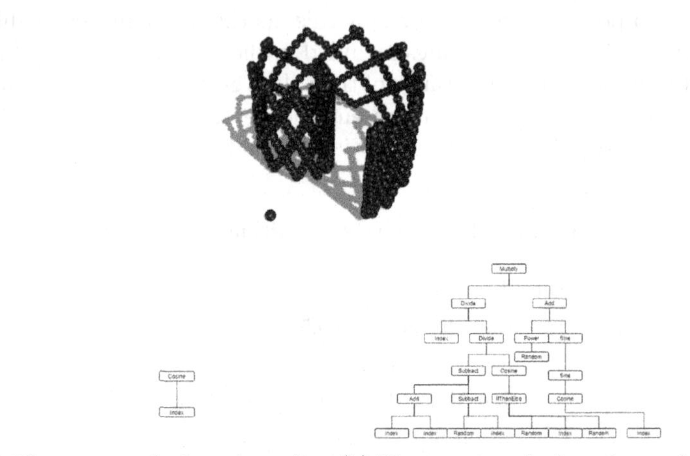

(a) The tree to calculate the radius (b) The tree to calculate the angle

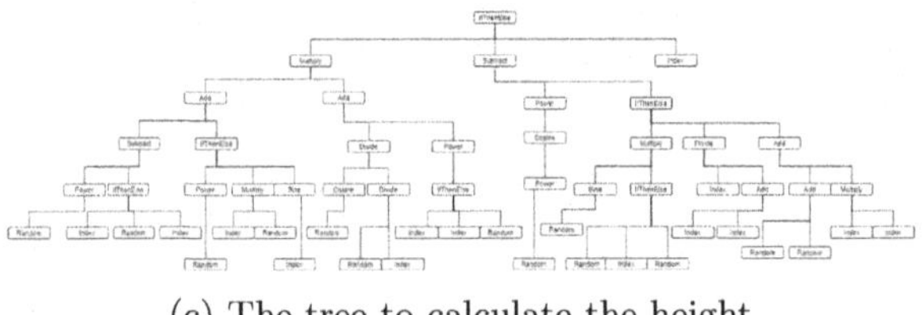

(c) The tree to calculate the height

Fig. 8. An example sculpture and associated expression trees

The generation method introduced in this paper aims to improve on aspects of generation processes used within existing Evolutionary Art systems. These improvements stem from applying the Concretism style to constrain the search space whilst being versatile and capable of generating interesting, visually varied and novel content. The process can easily be applied to existing and future Evolutionary Art systems and constraining the available search space has the potential to improve on various areas such as user fatigue, by improving how quickly interesting content can be found.

The AGP exposes multiple parameters improving the variety of items which can be generated and gives another dimension for the output to be explored. As the algorithm does not calculate a value for each pixel or voxel, the computational intensity of generating large or high-resolution items is reduced. This means that it is suitable for use in performance-intensive environments such as Augmented and Virtual Reality as well as online systems.

It was found that the process can produce varied content in 2D and 3D, with items exhibiting a range of values across symmetry and complexity measures. As the example in Fig. 8 shows, this complexity can be unrelated to the expressions used to generate the items. The set of functions used within these examples is limited and the trees can be very simple, this is a good indication that the restriction of the search space is beneficial for finding interesting artwork.

The adherence to the Concretism style potentially limits the versatility, as the style may not be considered to be aesthetically pleasing by everyone. Whilst the process may not be suitable for all scenarios, the benefits it does possess indicate that it may be useful to help improve the search for highly aesthetic items, in both automatic and individual-led systems.

The analysis of the items generated by the AGP exposed potential gaps where the location in the output parameter space was unattainable, this can be attributed to the measures used, indicating that more accurate measures are required. As the process generates a wide range of varied items it is a good candidate to help identify and formalise the contributing aspects of 3D aesthetic judgement, this can include measures describing stable aspects such as the complexity as well as less stable aspects that vary between individuals. These aspects will be assessed by performing a user study, which will also give a more complete indication of how versatile and effective the generation process is.

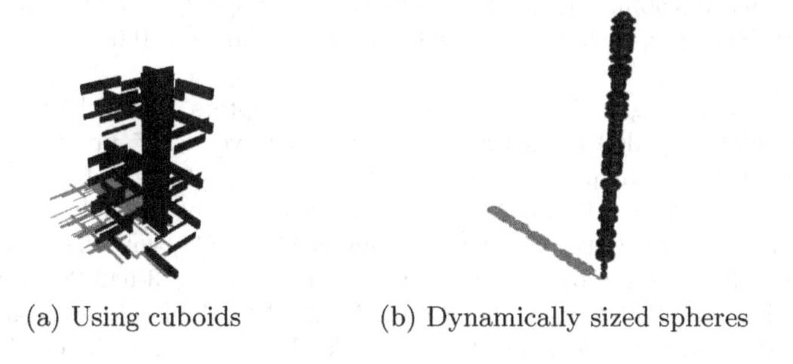

(a) Using cuboids (b) Dynamically sized spheres

Fig. 9. Potential extensions for the AGP

Developing more accurate measures of the aesthetic value of both 2D and 3D content will eventually require the output to be empirically tested either with participants or classifying using ML techniques. It may also be required to extend the AGP, for example increasing the function set used within the expression trees, amending the base geometric shape used to generate the items, such as the item shown in Fig. 9a, generating a colour for each item and amending the 3D process to allow the geometric items to be dynamically sized as shown in Fig. 9b. These extensions will expand the range of items generated by the process, increasing the likelihood of creating items within the unreachable locations in the parameter space.

References

1. Acebo, E., Sbert, M.: Benford's law for natural and synthetic images. In: First Eurographics conference on Computational Aesthetics in Graphics, Visualization and Imaging (2005)
2. Bergen, S., Ross, B.J.: Aesthetic 3D model evolution. Genet. Program Evolvable Mach. **14**(3), 339–367 (2013)
3. Birkhoff, G.D.: Aesthetic Measure. Cambridge (1933)
4. Boden, M.A., et al.: The Creative Mind: Myths and Mechanisms. Psychology Press (2004)
5. Byrne, J., Hemberg, E., O'Neill, M., Brabazon, A.: A methodology for user directed search in evolutionary design. Genet. Program. Evolvable Mach. **14**(3), 287–314 (2013)
6. Canaan, R., Menzel, S., Togelius, J., Nealen, A.: Towards game-based metrics for computational co-creativity. In: 2018 IEEE Conference on Computational Intelligence and Games (CIG), pp. 1–8. IEEE (2018)
7. Cohen, M.W., Cherchiglia, L., Costa, R.: Evolving mondrian-style artworks. In: Correia, J., Ciesielski, V., Liapis, A. (eds.) EvoMUSART 2017. LNCS, vol. 10198, pp. 338–353. Springer, Cham (2017). https://doi.org/10.1007/978-3-319-55750-2_23
8. Colton, S.: Automatic invention of fitness functions with application to scene generation. In: Giacobini, M., et al. (eds.) EvoWorkshops 2008. LNCS, vol. 4974, pp. 381–391. Springer, Heidelberg (2008). https://doi.org/10.1007/978-3-540-78761-7_41
9. Colton, S.: Evolving a library of artistic scene descriptors. In: Machado, P., Romero, J., Carballal, A. (eds.) EvoMUSART 2012. LNCS, vol. 7247, pp. 35–47. Springer, Heidelberg (2012). https://doi.org/10.1007/978-3-642-29142-5_4
10. Colton, S., Cook, M., Raad, A.: Ludic considerations of tablet-based evo-art. In: Di Chio, C., et al. (eds.) EvoApplications 2011. LNCS, vol. 6625, pp. 223–233. Springer, Heidelberg (2011). https://doi.org/10.1007/978-3-642-20520-0_23
11. Davies, E., Tew, P., Glowacki, D., Smith, J., Mitchell, T.: Evolving atomic aesthetics and dynamics. In: Johnson, C., Ciesielski, V., Correia, J., Machado, P. (eds.) EvoMUSART 2016. LNCS, vol. 9596, pp. 17–30. Springer, Cham (2016). https://doi.org/10.1007/978-3-319-31008-4_2
12. Easton, E.: Investigating user fatigue in evolutionary art. Master's thesis, Aston University (2018)
13. Easton, E., Bernardet, U., Ekart, A.: Tired of choosing? Just add structure and virtual reality. In: Ekárt, A., Liapis, A., Castro Pena, M.L. (eds.) EvoMUSART 2019. LNCS, vol. 11453, pp. 142–155. Springer, Cham (2019). https://doi.org/10.1007/978-3-030-16667-0_10
14. Ecins, A., Fermuller, C., Aloimonos, Y.: Detecting reflectional symmetries in 3D data through symmetrical fitting. In: Proceedings of the IEEE International Conference on Computer Vision Workshops, pp. 1779–1783 (2017)
15. Ekárt, A., Sharma, D., Chalakov, S.: Modelling human preference in evolutionary art. In: Di Chio, C., et al. (eds.) EvoApplications 2011. LNCS, vol. 6625, pp. 303–312. Springer, Heidelberg (2011). https://doi.org/10.1007/978-3-642-20520-0_31
16. Gircys, M., Ross, B.J.: Image evolution using 2D power spectra. Complexity **2019** (2019)

17. den Heijer, E., Eiben, A.E.: Evolving pop art using scalable vector graphics. In: Machado, P., Romero, J., Carballal, A. (eds.) EvoMUSART 2012. LNCS, vol. 7247, pp. 48–59. Springer, Heidelberg (2012). https://doi.org/10.1007/978-3-642-29142-5_5
18. den Heijer, E., Eiben, A.E.: Comparing aesthetic measures for evolutionary art. In: Di Chio, C., et al. (eds.) EvoApplications 2010. LNCS, vol. 6025, pp. 311–320. Springer, Heidelberg (2010). https://doi.org/10.1007/978-3-642-12242-2_32
19. Hollingsworth, B., Schrum, J.: Infinite art gallery: a game world of interactively evolved artwork. In: 2019 IEEE Congress on Evolutionary Computation (CEC), pp. 474–481. IEEE (2019)
20. Lehman, J., Stanley, K.O.: Exploiting open-endedness to solve problems through the search for novelty. In: ALIFE, pp. 329–336 (2008)
21. Li, Y., Hu, C., Chen, M., Hu, J.: Investigating aesthetic features to model human preference in evolutionary art. In: Machado, P., Romero, J., Carballal, A. (eds.) EvoMUSART 2012. LNCS, vol. 7247, pp. 153–164. Springer, Heidelberg (2012). https://doi.org/10.1007/978-3-642-29142-5_14
22. Machado, P., Cardoso, A.: Computing aesthetics. In: de Oliveira, F.M. (ed.) SBIA 1998. LNCS (LNAI), vol. 1515, pp. 219–228. Springer, Heidelberg (1998). https://doi.org/10.1007/10692710_23
23. Machado, P., Vinhas, A., Correia, J., Ekárt, A.: Evolving ambiguous images. AI Matt. **2**(1), 7–8 (2015)
24. Matkovic, K., Neumann, L., Neumann, A., Psik, T., Purgathofer, W.: Global contrast factor-a new approach to image contrast. Comput. Aesthetics **2005**, 159–168 (2005)
25. McCormack, J., Lomas, A.: Understanding aesthetic evaluation using deep learning. In: Romero, J., Ekárt, A., Martins, T., Correia, J. (eds.) EvoMUSART 2020. LNCS, vol. 12103, pp. 118–133. Springer, Cham (2020). https://doi.org/10.1007/978-3-030-43859-3_9
26. McDermott, J.: Graph grammars as a representation for interactive evolutionary 3D design. In: Machado, P., Romero, J., Carballal, A. (eds.) EvoMUSART 2012. LNCS, vol. 7247, pp. 199–210. Springer, Heidelberg (2012). https://doi.org/10.1007/978-3-642-29142-5_18
27. McDermott, J., et al.: String-rewriting grammars for evolutionary architectural design. Environ. Plan. B: Plan. Design **39**(4), 713–731 (2012)
28. Mills, A.: Animating typescript using aesthetically evolved images. In: Johnson, C., Ciesielski, V., Correia, J., Machado, P. (eds.) EvoMUSART 2016. LNCS, vol. 9596, pp. 126–134. Springer, Cham (2016). https://doi.org/10.1007/978-3-319-31008-4_9
29. Muehlbauer, M., Burry, J., Song, A.: Automated shape design by grammatical evolution. In: Correia, J., Ciesielski, V., Liapis, A. (eds.) EvoMUSART 2017. LNCS, vol. 10198, pp. 217–229. Springer, Cham (2017). https://doi.org/10.1007/978-3-319-55750-2_15
30. Nguyen, A.M., Yosinski, J., Clune, J.: Innovation engines: automated creativity and improved stochastic optimization via deep learning. In: Proceedings of the 2015 Annual Conference on Genetic and Evolutionary Computation, pp. 959–966 (2015)
31. Nicolau, M., Costelloe, D.: Using grammatical evolution to parameterise interactive 3D image generation. In: Di Chio, C., et al. (eds.) EvoApplications 2011. LNCS, vol. 6625, pp. 374–383. Springer, Heidelberg (2011). https://doi.org/10.1007/978-3-642-20520-0_38
32. O'Neill, M., et al.: Evolutionary design using grammatical evolution and shape grammars: designing a shelter. Int. J. Design Eng. **3**(1), 4–24 (2010)

33. O'Reilly, U.M., Hemberg, M.: Integrating generative growth and evolutionary computation for form exploration. Genet. Program. Evolvable Mach. **8**(2), 163–186 (2007)
34. Rigau, J., Feixas, M., Sbert, M.: Conceptualizing Birkhoff's aesthetic measure using Shannon entropy and Kolmogorov complexity. In: Computational Aesthetics, pp. 105–112 (2007)
35. Ross, B., Ralph, W., Zong, H.: Evolutionary image synthesis using a model of aesthetics. In: IEEE International Conference on Evolutionary Computation, pp. 1087–1094. IEEE (2006)
36. Secretan, J., Beato, N., D Ambrosio, D.B., Rodriguez, A., Campbell, A., Stanley, K.O.: Picbreeder: evolving pictures collaboratively online. In: Proceedings of the SIGCHI Conference on Human Factors in Computing Systems, pp. 1759–1768 (2008)
37. Sims, K.: Artificial evolution for computer graphics. In: Proceedings of the 18th Annual Conference on Computer Graphics and Interactive Techniques, pp. 319–328 (1991)
38. Takagi, H.: Interactive evolutionary computation: fusion of the capabilities of EC optimization and human evaluation. Proc. IEEE **89**(9), 1275–1296 (2001)
39. Tate: Concrete art (2017). https://www.tate.org.uk/art/art-terms/c/concrete-art/. Accessed 20 Nov 2020
40. Tinio, P.P., Leder, H.: Just how stable are stable aesthetic features? Symmetry, complexity, and the jaws of massive familiarization. Acta Physiol. (Oxf) **130**(3), 241–250 (2009)
41. Tweraser, I., Gillespie, L.E., Schrum, J.: Querying across time to interactively evolve animations. In: Proceedings of the Genetic and Evolutionary Computation Conference, pp. 213–220 (2018)
42. Vinhas, A., Assunção, F., Correia, J., Ekárt, A., Machado, P.: Fitness and novelty in evolutionary art. In: Johnson, C., Ciesielski, V., Correia, J., Machado, P. (eds.) EvoMUSART 2016. LNCS, vol. 9596, pp. 225–240. Springer, Cham (2016). https://doi.org/10.1007/978-3-319-31008-4_16
43. Wiggins, G.A.: A preliminary framework for description, analysis and comparison of creative systems. Knowl.-Based Syst. **19**(7), 449–458 (2006)

Interactive, Efficient and Creative Image Generation Using Compositional Pattern-Producing Networks

Erlend Gjesteland Ekern[1] and Björn Gambäck[1,2](✉)

[1] Department of Computer Science,
Norwegian University of Science and Technology, 7491 Trondheim, Norway
erlendekern@gmail.com, gamback@ntnu.no
[2] Digital Systems, RISE, Research Institute of Sweden AB, 164 29 Kista, Sweden

Abstract. In contrast to most recent models that generate an entire image at once, the paper introduces a new architecture for generating images one pixel at a time using a Compositional Pattern-Producing Network (CPPN) as the generator part in a Generative Adversarial Network (GAN), allowing for effective generation of visually interesting images with artistic value, at arbitrary resolutions independent of the dimensions of the training data. The architecture, as well as accompanying (hyper-) parameters, for training CPPNs using recent GAN stabilisation techniques is shown to generalise well across many standard datasets. Rather than relying on just a latent noise vector (entangling various features with each other), mutual information maximisation is utilised to get disentangled representations, removing the requirement to use labelled data and giving the user control over the generated images. A web application for interacting with pre-trained models was also created, unique in the offered level of interactivity with an image-generating GAN.

Keywords: Compositional pattern-producing networks · Image generation · Generative adversarial networks

1 Introduction

Interest in generative models such as Generative Adversarial Networks, GANs [8] has grown enormously in recent years. In a GAN, two networks are pitted against each other: a *discriminator* is tasked with distinguishing between instances from the dataset and synthetic ones from a *generator*, which in turn tries to generate samples that the discriminator classifies as coming from the training data. New and improved methods, architectures and applications are frequently being introduced, and the quality of the generated results is constantly being improved, e.g., with StyleGAN [21,22] being able to generate large images of, among other things, cars, human faces and cats exhibiting a level of realism making it almost impossible to separate the real from the computer-generated.

J. Romero et al. (Eds.): EvoMUSART 2021, LNCS 12693, pp. 131–146, 2021.
https://doi.org/10.1007/978-3-030-72914-1_9

Most mainstream research on high-resolution image generation revolve around approaches aimed at generating an entire image at once. An alternative approach is offered by Stanley's Compositional Pattern-Producing Networks, CPPN [34] that generate the value of one pixel at a time. While some previous work has experimented with using CPPNs in GANs [1,12,14,15,28], this approach to image generation has not been studied sufficiently. CPPNs do, however, exhibit properties that might have merits in other areas such as computational creativity and efficiency. To this end, the paper reports a thorough study on how different architectures and parameters materialise in distinct visual features in the images generated by a CPPN, and introduces an Information Maximising and Pattern-Producing Generative Adversarial Networks architecture (IMP-GAN), which performs well across a wide range of datasets and configurations, improving on related work by combining techniques for training stabilisation with methods achieving disentangled latent representations, not with the goal of generating images with the highest degree of realism, but rather to explore the creative properties of CPPNs and their generalisability.

While early GANs usually were limited to training on, and generation of, images with small dimensionality (e.g., 28×28 pixels), recent innovations have enabled generation of images of increasingly higher resolution, by introducing more complex architectures and techniques, and by training networks on larger and more high-resolution datasets for longer time. However, as architectures and datasets become larger, hardware and data requirements also increase. In contrast, CPPNs can generate images of infinite resolution (as each pixel is generated independently) regardless of the dimension of the images they were trained on, thus decreasing both the data and hardware requirements. A CPPN can be trained on a low-resolution image set for a fraction of the time, and afterwards be able to generate not only images approximating the visual characteristics of the training data at the original resolution, but also produce results at very high resolutions that exhibit novel visual properties not apparent at lower resolutions, while still preserving the chief characteristics of the training data.

Section 2 details CPPNs and previous work on using them for image generation. Section 3 describes the system architecture and experimental setup. Results are reported in Sect. 4 and discussed in Sect. 5, detailing how various setup changes affect the produced images. Section 6 evaluates the results in terms of visual quality, efficiency, and computational creativity, and Sect. 7 concludes.

2 Related Work

A CPPN in its simplest form takes a coordinate as input, e.g., (x, y), and outputs a value, e.g., a pixel intensity, for that coordinate. The output values in an arbitrary coordinate system can be calculated by entering each coordinate in turn into the CPPN. CPPNs are similar to Artificial Neural Networks (ANNs) in that they utilise inputs and interconnected units with weights and activation functions to calculate an output; however, while ANNs represent a simplified model of mechanisms in the brain, CPPNs are related to natural development, with complex networks commonly iteratively evolved from simple ones using evolutionary algorithms [34]. Each new generation is a mutated crossover of several parent networks, with new units, edges and activation functions added at random. Such *evolved* CPPNs often get sparsely-connected network structures, much less organised than in ANNs. In contrast, *fixed* CPPN architectures will be explored here, making them more akin to traditional neural networks.

Unlike ANNs, each node in a CPPN can compute a different type of function, so a CPPN is effectively an ANN with heterogeneous activation functions at its nodes. These functions are *composed* to establish increasingly refined coordinate frames, with the final outputs the results of all transformations taking place before them. Different hidden node activation functions in a CPPN give rise to various output patterns: a Gaussian or absolute function leads to symmetric structures, while a sin function gives repeating structures. Various other functions contribute to sharp edges, circles, etc., in the output image. The distance between a given coordinate and the coordinate system centre can be included as an additional input to provide a bias towards symmetry in the output [34]. One of the most interesting properties of CPPNs is that they can generate images of infinite resolution by transforming pixel coordinates to fixed continuous space. Thus, to increase the resolution of a generated image, only the granularity of the input coordinates needs to be increased [34].

Already Sims [33] introduced models in interactive genetic art that were CPPN-like, and with applications where a user was presented with a set of initial images generated by different models and then being allowed to select which images, and thus implicitly models, to evolve. Picbreeder [32] and Neurogram [11] then also utilised CPPNs in such a manner, with NeuroEvolution of Augmenting Topologies, NEAT [35] employed to evolve user-selected models. Ha conducted several further experiments on CPPNs in the context of interactive, generative and genetic art using both NEAT and randomly initialised CPPNs [10,13,16], and introduced a latent noise vector $\mathbf{z}$ as an input to the CPPN as a way to generate different images using a fixed network, and the application of a scaling factor s on all inputs to the CPPN [13], effectively allowing for generation of zoomed-in/-out image versions. Ha also experimented with training a CPPN as the generator in a GAN, i.e., a CPPN-GAN [12,14,15]. Metz and Gulrajani [28] extended this by employing new training techniques and introducing new applications of CPPN-GAN architectures as creative tools. They used multiple discriminators trained on different datasets, and trained the CPPN to approximate multiple distributions at different scales or resolutions.

After the present work, Anokhin et al. [1] independently introduced a CPPN-GAN architecture ("CIPS") achieving quality similar to state-of-the-art convolutional generators at resolutions up to 256×256 pixels. However, Anokhin et al. do not address in depth how different hyperparameters and CPPN architectures affect the output images, nor offer an interactive system or an evaluation of the creative value of such networks; all of which are main contributions here.

Apart from that, there is no directly comparable work, since previous studies have been limited to a small subset of the datasets used by the IMPGAN, and have not evidenced the same degree of realism. The work by Ha [10,13,16] mainly pertained to training a CPPN on MNIST and CIFAR-10 using GAN architectures, and failed to closely approximate the distribution of the training data, although showing interesting visual features when scaled up. Metz and Gulrajani [28] improved Ha's work by using the WGAN-GP objective coupled with batch renormalisation for the CPPN and layer normalisation for the discriminator, training on CIFAR-10 and ImageNet, and focusing on training a single CPPN to approximate different data distributions at different scales and resolutions. It would be hard to reproduce their experiments, since many key implementation details are unclear, but they state that they were "unable to reach the same visual quality of existing convolution transposed models", which the architecture presented here is able to do.

3 Experimental Setup

To experiment with using a CPPN as the generator part of an image-producing GAN, a general architecture was developed using Python and TensorFlow. Most experiments were run on an NVIDIA Tesla P100 GPU with 16 GB memory. A common (default) set of hyperparameters and setups were used for the majority of the experiments, with initial parameter values derived from related work, but subsequently tweaked to improve the stability of the training process.

The architecture supports various GAN **training objectives:** non-saturating; non-saturating with gradient penalty (GP); Wasserstein GAN, WGAN [2]; and WGAN-GP [9]—with WGAN-GP as default objective. The output of the discriminator is either a probability or scalar depending on which GAN objective is used: WGAN and WGAN-GP models output a scalar approximating the Earth mover's distance (EMD) between distribution of the training dataset and the implicit generator distribution; networks using non-saturating objectives output the probability of a given image coming from the generator or from the training data. The objectives for the generator G and the discriminator D for a given input i in a WGAN-GP are defined by their loss functions:

$$\begin{aligned} L_G = & -\mathbb{E}_{g\sim P_g}\left[D(G(g))\right] \\ L_D = & \ \ \mathbb{E}_{g\sim P_g}\left[D(G(g))\right] - \mathbb{E}_{i\sim P_d}\left[D(i)\right] \\ & + \lambda\,\mathbb{E}_{\hat{\imath}\sim P_{\hat{\imath}}}\left[(\|\nabla_{\hat{\imath}} D(\hat{\imath})\|_2 - 1)^2\right] \end{aligned}$$

where P_g is the generator distribution, P_d the actual (true) data distribution, and the last term acts as the gradient penalty, drawn over random samples $\hat{\imath} \sim P_{\hat{\imath}}$, with λ being a constant weighting the penalty (the experiments below use $\lambda = 10$). The discriminator loss function L_D approximates the EMD between P_d and P_g, while the generator loss L_G aims to maximise the discriminator loss.

The **inner network architectures** can be arbitrary. The setup introduced here allows for constructing GANs using either a multilayer perceptron (MLP), Convolutional Neural Network (CNN) or CPPN as the generator. The CPPN can

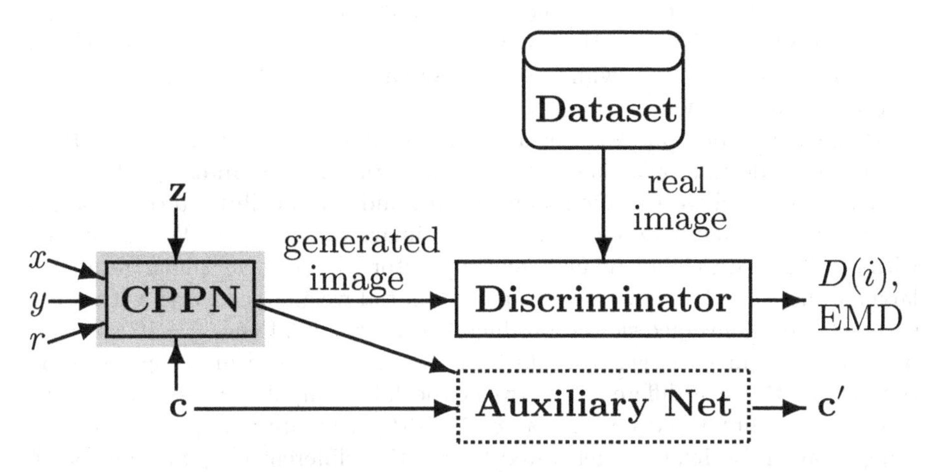

Fig. 1. Information maximising and pattern-producing GAN. (x, y) are pixel coordinates; $\mathbf{z}$ and $\mathbf{c}$ latent noise and code vectors; r an optional distance between (x, y) and the image centre. The discriminator outputs either the probability D_i of an image i belonging to the real data or the Earth mover's distance between the true dataset distribution and the implicit generator distribution. When maximising mutual information between $\mathbf{c}$ and i, an auxiliary net is included to reconstruct $\mathbf{c}$.

be configured as an MLP or a residual neural network [17]. The discriminator can be an MLP or a CNN with customisable layer numbers and sizes. Four common **normalisation schemes** can be applied to the layers: spectral [29], layer [3], batch normalisation [20], and batch renormalisation [19].

The default **discriminator** architecture includes three convolutional layers (kernel size 4, stride 2) with LReLU activation and respective output sizes of 64, 128, and 256, followed by a fully connected layer with size 1 output, using spectral normalisation in all layers. The discriminator's input images are normalised and represented by matrices with values in the range $[-1, 1]$. The **generator** produces images in this range utilising a tanh output layer activation function. In generators using residual blocks, a fully connected layer without activation function is added after the input layer to resize the inputs to dimensions matching the size of the block's final layer.

The size of the **latent noise** input vector $\mathbf{z}$ was set to 128 as done by Miyato et al. [29] and Kurach et al. [25]. An idea borrowed from Information Maximizing GANs, InfoGAN [6] is to encapsulate different features in a separate vector $\mathbf{c}$, called the **latent code**, that is input to the generator in addition to $\mathbf{z}$. $\mathbf{c}$ contains a preselected number of variables that are either categorical or continuous, and can be considered "placeholders" for features that the generator is tasked to learn. The size of $\mathbf{c}$ was inspired by a configuration used in InfoGAN, with one size-10 categorical variable and two continuous variables. This choice is further motivated by the intuition that a larger latent code can be harder or impossible for the auxiliary net to reconstruct, thus leading to a more unstable training process. Both **z** and the continuous variables of **c** are sampled from a uniform distribution $U \sim [-1, 1]$, while the categorical variables are sampled from a categorical distribution.

If the generator is an MLP or a CNN, the input is only **z** and **c**. A CPPN generator in addition takes as input the x and y **pixel coordinates**, both in the interval $[-s, s]$, where s is the scaling factor; and (optionally) a term r for the Euclidean **distance** between (x, y) and the image centre, $(0, 0)$. When training with small initial weights, a small **scaling** factor ($s = 1$) gave similar results as larger values ($s = 10$); however, using large initial weights, networks with lower scaling factors converged slower, producing worse results. Hence $s = 10$ was used per default. A much larger s would likely not be beneficial in an optimisation context, as the size differences between the latent inputs (**z** and **c**) and the coordinate inputs would be very large. A CPPN generates one pixel value at a time, so an entire image is generated by entering different (x, y, r) inputs with fixed **z** and **c** vectors. If **z** and **c** are changed, a different image is created, with the absolute size of the changes controlling how similar the new image will appear.

An optional **auxiliary network** is included if one of the objectives is to maximise the mutual information between the latent code and the generated images. The auxiliary net enables the use of unlabelled data, allowing the model to learn various labels (disentangled representations) in an unsupervised fashion with. The auxiliary net takes an image as input, and outputs a prediction of what the original latent code **c** for that image was. The auxiliary net loss then reflects how closely it reconstructs the latent code (i.e., the original **c** that was input to the generator) of a generated image. It sums the latent code categorical variables' loss (calculated by the cross-entropy of the softmaxed outputs) and the continuous variables' loss (calculated using mean-squared error).

The auxiliary net reuses the same three convolutional layers as the discriminator, but followed by two fully-connected layers using spectral normalisation, the first with LReLU activation and size 128 output, and the second with output size and activation function (softmax or none) dependent on the type of latent code used. When the architecture is configured to use a CPPN and an auxiliary net, the general model will be referred to as an **Information Maximising and Pattern-Producing GAN (IMPGAN)**, with three neural networks in the model as shown in Fig. 1; otherwise there will be two nets only. The **weights** of the auxiliary and discriminator network, as well as MLP and CNN generators, are initialised following the Xavier scheme [7], while all biases are initialised to 0. However, all CPPNs are initialised with weights sampled from a Gaussian distribution, either with a smaller ($\mathcal{N}(0, 0.02)$) or larger standard deviation ($\mathcal{N}(0, 1.0)$).

The Adam [23] optimiser was utilised for all experiments involving **loss function minimisation**. It adjusts network parameters after each training step, in the order discriminator, generator and auxiliary net. Adam's β_1 decay rate hyperparameter defaults to 0.9 in TensorFlow; however, Miyato et al. [29] showed that with WGAN-GP as the main GAN objective, smaller values can give better results, so $\beta_1 = 0.5$ was used as default. The **learning rates** were inspired by the success of Heusel et al. [18] in using the two time-scale update rule, when training a GAN using the WGAN-GP objective, with a discriminator rate of 0.0003 and generator rate 0.0001. The auxiliary net's learning rate was set to the same as the generator to avoid letting the mutual information maximisation dominate the generator optimisation, i.e., the generator should primarily aim to approximate the training data distribution.

The **batch size** was set to 256 to strike a balance between the lower [18,29] and higher values [5] used previously. How much larger batch sizes improve GAN training and performance is an open question [30], but larger batches can induce excessive computational requirements as the implementation generates all pixel values for an image independently and in parallel to speed up training: training a traditional generator with a batch size of N implies generating N images, while a CPPN using the same batch size is activated $N \cdot ImgDim^2$ times, once for each pixel in each output image in the (mini-)batch.

4 Experiments and Results

The main experiments revolved around training Information Maximising and Pattern-Producing GANs (IMPGANs) across a wide range of configurations in order to generate images approximating the original dataset at low resolution, but exhibiting different visual properties or higher degrees of realism at higher resolutions. The experiments were carried out by training on different datasets, with various architectures, activation functions and parameter settings, in order to gauge their effects on the high resolution images' visual properties.[1]

Six different datasets, covering various domains and levels of complexity, were used to train the neural networks. Two sets of 70 000 grayscale 32 × 32 images: MNIST [26], a low-complexity set of handwritten digits, and Fashion MNIST [37], a somewhat higher complexity set containing ten classes of fashion items; as well as four significantly harder sets of RGB images, that in addition to the three colour channels add more complex and diverse images: TensorFlow Flowers [36], with approx. 3 000 images of five different flower species, CelebA [27], containing over 200 000 celebrity face 218 × 178 images, CIFAR-10 [24] with 60 000 32 × 32 images of ten mutually exclusive classes (dogs, airplanes, cars, frogs, etc.), and a pre-processed version of the WikiArt 'Painter by Numbers' dataset[2] with some 100 000 paintings by over 2 000 artists from the last 500 years. To decrease training time and computational requirements, CelebA was resized to either 32 × 32 or 64 × 64, Painter by Numbers and TensorFlow Flowers to 32 × 32, while the other datasets were kept at their original, low resolutions.

The **CelebA** dataset is much more diverse than the greyscale MNIST and Fashion MNIST datasets, with three colour channels and wide colour variety. The networks were by default trained on CelebA images resized to 32 × 32,

[1] The supplement gives further example outputs: https://bit.ly/impgan_sup.

[2] https://github.com/zo7/painter-by-numbers/releases/tag/data-v1.0.

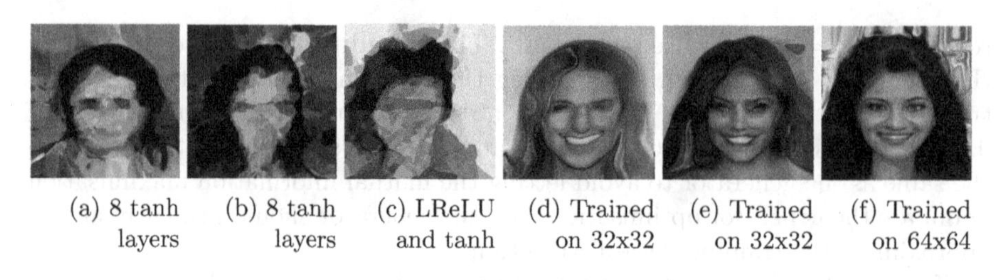

(a) 8 tanh layers (b) 8 tanh layers (c) LReLU and tanh (d) Trained on 32x32 (e) Trained on 32x32 (f) Trained on 64x64

Fig. 2. IMPGANs trained on the CelebA dataset

with experiments run for $\sim 250\,000$ training steps. The images in Fig. 2a–2c were generated by three different CPPNs with eight hidden layers of sizes $3*128$, $3*64$, and $2*32$. All network weights were initialised to large values and $\beta_1 = 0$ used in the Adam optimisers. The networks in Fig. 2a and 2b only use tanh *activation functions*, while Fig. 2c alternates between LReLU and tanh for each layer. Only Fig. 2a uses the distance term as an input, with layer normalisation applied in the auxiliary net instead of the default spectral normalisation. The networks in Fig. 2b and 2c use five latent code continuous variables instead of two. The two first networks obtain a high level of mutual information maximisation, i.e., varying the *latent code* predictably controls various features in the generated images. However, the third network fails to attribute meaning to the latent code, despite having a very similar architecture. The effects of including the *distance term* are visible in Fig. 2a, which has a distinctive circular pattern in the centre and an inherent bias towards symmetry, while Fig. 2c shows that using both LReLU and tanh functions gave more sharp lines and patterns.

To increase the visual quality of the generated images at high resolution, experiments on deeper architectures utilising *residual blocks* were conducted. Figure 2d shows an image with a mix of curved and sharp lines, generated by a network containing five residual blocks consisting of layers of sizes 64, 32 and 64, with tanh, tanh and LReLU activation functions, and small initial weights. The generator has clearly learned the concept of a face, but some finer details related to facial expression and shadowing are somewhat lacking. To improve the expressiveness of the generator, the network was extended to contain ten residual blocks with layers of sizes 128, 32 and 128. As shown in Fig. 2e, this gives a higher level of detail. The increased expressiveness manifests itself especially in the areas of the eyes, nose and lips. The low resolution of the training data naturally sets an upper bound on the amount of detail the generator will be able to produce, so the image in Fig. 2f was generated by a similar network trained on a 64×64 version of CelebA and containing seven residual blocks. The details of the facial features are finer and the increased amount of detail is especially apparent in the eyes, where this network is able to generate visible pupils.

CIFAR-10 is perhaps the most complex dataset used in the experiments, with a great diversity of objects and shapes. Training a CPPN with seven hidden layers on CIFAR-10 showed the network failing to properly capture the distribution underlying the dataset, albeit being able to approximate the colour and

Fig. 3. Grid of samples generated from CIFAR-10; each row with a different value of the categorical variable

shape distributions. To increase the CPPN's expressiveness, a deep network consisting of ten residual blocks was constructed, with each block containing three layers of size 128, 32 and 128, with tanh, tanh and LReLU activation functions. All network weights were initialised to small values. The network was able to generate samples approximating the original dataset to some degree, as indicated by the images depicting various shapes of cars, trucks and four-legged creatures (Fig. 3). The approximation to the training data is not perfect by any means, but the overall distribution of colours, objects and shapes implies that parts of the distribution have been learned. The *categorical variable* of the latent code corresponds to different object classes, e.g., cars, trucks and horses in this case, but some of the categories are harder to make out than others. The two *continuous variables* both have a large effect on the colour of the images, as well as various transformations of the main object in a given image.

Outputs from a network trained on the **Painter by Numbers** dataset are shown in Figs. 4 and 5. The training data was filtered to only include portrait paintings, resulting in ∼18 000 images. The network had small initial weights and consisted of one sin, two tanh and five LReLU hidden layers, with the first and the last four layers of sizes 128 and 64, respectively.

An **interactive application** was built using Vue.js and TensorFlow.js[3] to facilitate seamless exploration of the IMPGAN creative space. The cross-platform web application (Fig. 6) allows users to manipulate various pre-trained models, and offers three interaction modes:

[3] https://vuejs.org/ resp. https://www.tensorflow.org/js.

Fig. 4. A grid of random 32 × 32 px portrait samples

Fig. 5. Portrait generated at 1200 × 1200 px

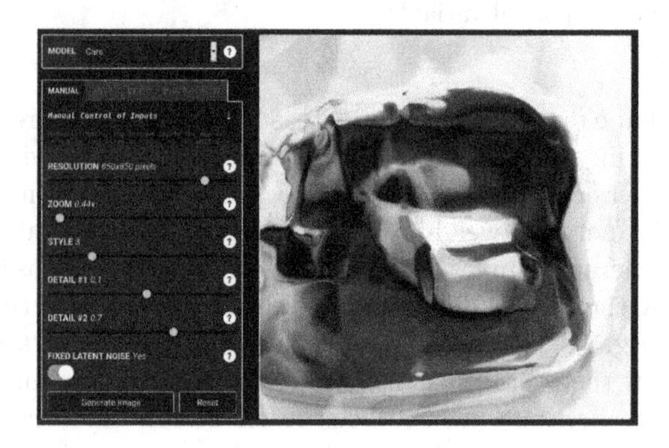

Fig. 6. Interactive web application, *manual* mode, with a generated CIFAR-10 image of a car

auto Users can select one or multiple images generated by the currently chosen CPPN model, and generate new variants of the image(s) in different resolutions and zoom levels, or as a mix of multiple other images.

manual Users can tweak all aspects of the CPPN model's inputs, generating an image using the new settings.

video Users can select multiple images generated by the current model, interpolating between them to create a video "morphing" one image into another.

5 Discussion

A diverse set of **activation functions** have been explored, and they have all contributed to distinct visual effects in the generated images. Using unconventional activation functions such as sin can have a detrimental effect on the convergence of a network, so the majority of the experiments utilise more conventional activation functions, such as LReLU and tanh, in an effort to increase the training stability. These functions contribute to distinct visual properties in the generated images, while performing well in an optimisation context. The LReLU leads to a bias towards sharp and straight lines in the high resolution images, while tanh gives rise to curved and smooth lines, and the periodic nature of sin leads to repeating patterns (e.g., Fig. 5). The strength of these features can to some extent be controlled by initialising a given layer to smaller or larger weights.

When varying the **network architectures**, increasing the depth of a given CPPN in general gives rise to more complex output images as the value of the final output is dependent on a higher number of functions, each with its own set of characteristics. When using an MLP, some of the deeper units might become saturated depending on the nature of the activation functions used. Saturated units in turn lead to sharper and more complex images, as the output pixel values become more extreme. Moreover, a residual architecture gives rise to some more complex patterns, as each block can adjust its weights to let the intermediate layers have a smaller or larger effect on its output. It is nonetheless interesting that training similarly configured IMPGANs on different datasets can net very different results: training a CPPN with residual blocks and small weights on the CelebA dataset results in sharper images compared to training a similar network on the CIFAR-10 dataset. This might be because CIFAR-10 contains a very wide range of objects, colours and shapes, and that the network output for any given pixel is pushed towards softer, more ambiguous values that can better accommodate the higher diversity of the dataset.

The degree to which the **distance term** affects the generated images varies based on the dataset used. For instance, since faces exhibit a certain degree of symmetry, the distance term might be of use in providing symmetry to images generated from CelebA: the CPPN does not need to learn the concept of symmetry from scratch, but can instead rely on it preexisting in the input. The CIFAR-10 dataset, in contrast, contains a wide range of shapes and objects, and there is no shared symmetry between most images in the dataset. Then the distance term might even hinder convergence, as the CPPN is biased towards a symmetry that might not actually exist in the dataset. Optimally, the networks could learn to simply ignore the distance term if it is of no use. However, in practice the networks utilise the included distance term during the early stages of training, when the fidelity of the generated images is low. Since the learning rate used for the generator is relatively small, it is then often difficult to reverse this dependency as training progresses.

The **scaling factor** s scales all of the input coordinates to lie in a specific interval. A relatively high s results in the coordinates having more influence over the output compared to the latent inputs, which destabilises the training process, as the inputs are of vastly different sizes. However, a small s leads to slower convergence when combined with large initial weights. After training, the scaling can freely be adjusted to control the zoom level of the generated images, and can provide insight into the representations the network has learned.

Most of the networks utilise a **latent code** of one categorical and two continuous variables. Changing the values of these variables usually results in meaningful changes in the generated output. In general, the categorical variable can be associated with visual properties that are more discontinuous in nature, such as a single digit class in the case of MNIST. The continuous variables, on the other hand, usually correspond to more continuous properties, such as rotation, size and colour. The two continuous variables are sometimes the inverse of each other: if the first variable rotates an object outwards, the second might rotate it inwards. In practice, however, there is often some additional property associated with these pair of inverses with each variable controling a unique set of features.

Most experiments pertaining to the sizes of the **initial weights** of CPPNs gave almost indistinguishable results. When merely initialising a CPPN with large weights (without training), the results are more dramatic and exhibit a sharper look due to the higher level of activation in the network. Similarly, when training a CPPN with large initial weights, the patterns and shapes unique to the activation functions are more apparent at the early stages of training, and then gradually lessen as the generator more closely approximates the distribution of the dataset. A network initialised to small weights exhibits a more stable training phase that converges quicker, but the generated high resolution images usually have a relatively soft and smooth appearance. The weight initialisation scheme used in a CPPN then reflects a trade-off between a high level of convergence and pronounced visual properties at high resolution. The images generated at the original, low resolution of the training data are usually quite similar regardless of the weight initialisation scheme used in the networks.

The infinite resolution premise offered by CPPNs combined with their novel visual properties, as well as controllable features from the disentangled representations, all lend themselves well to an **interactive system** where users can freely experiment with models to control the generated images. The web application created offers an easily accessible interactive system that is hosted on a public domain, runs fully client-side and works as well on phones as on laptops.

6 Evaluation

The overarching goal of the work was not to achieve state-of-the-art visual quality at original resolution, but to strike a balance between relatively good visual quality at original resolution, and distinct and novel visual properties at high resolution. Several evaluation metrics have been suggested to assess the visual quality of images generated by a GAN, such as Fréchet Inception Distance, FID [18] and Inception Score, IS [31]. The FID for Fig. 2e–2f was calculated using 50 000 random, low-resolution samples to be 18.11 and 41.34, respectively (a lower FID is better). Across the various datasets, the **visual quality at**

low resolution of the images generated by IMPGANs is comparable to that obtained by training similarly configured traditional GANs: a baseline FID and visual quality was established by training a convolutional generator on the 32 × 32 CelebA dataset, with a resulting FID of 12.55 using 50 000 samples.[4] Notably, in some cases CPPNs generate better images than similar MLPs or CNNs for the same amount of training steps, since CPPNs are activated and compute gradients $BatchSize \cdot ImgDim^2$ times during each step, while traditional generators do it only $BatchSize$ times.

Training time improvements are easier to quantify: training the model in Fig. 2f for 250 000 steps takes about four days on a Nvidia Tesla P100 GPU, while StyleGAN and StyleGAN2 need to be trained for some 41 and 70 days[5] (on 25k images), respectively, on similar hardware (Nvidia Tesla V100) to generate 1024 × 1024 images. The CPPN can be trained on the CelebA dataset downsampled to 64 × 64, and subsequently generate 2000 × 2000 images (although there obviously is less detail in 64 × 64 images than the 1024 × 1024 images that, e.g., StyleGAN is trained on). A CPPN-based network can also be configured to generate only one or a few pixels at a time, which enables it to generate large images on restricted hardware where other methods for high resolution synthesis, like StyleGAN, might be too computationally expensive.

Creativity is the ability to generate something which is novel, surprising and valuable [4], and the possible creative merits of an IMPGAN can be evaluated using those concepts. As the networks are only trained on low resolution images, the results obtained when generating images at higher resolutions are the product of a CPPN attempting to bridge the gap between the values it knows and values far outside what it is familiar with. The network thus generates images that are *novel* and of a different nature compared to anything it has seen and generated before. Moreover, the networks can elicit various forms of *surprise*, as when comparing the visual quality of the sample portraits in Fig. 4 to an image at a much larger resolution, Fig. 5. It is still a portrait of some kind of humanoid, but with visual properties the viewer would likely not have anticipated based on the smaller images. The notion of *value* is in general harder to quantify, but one option is to compare the quality to that of a set of target examples. Then the value of generated images can be described in terms of their similarity to some reference data, as well as their dissimilarity to results generated by a purely random process. The IMPGAN's value then lies in its ability to generate low-resolution images similar to the training data.

7 Conclusion and Future Work

The paper has shown that Compositional Pattern-Producing Networks can be trained using Generative Adversarial Networks to approximate the underlying

[4] See the supplement for visuals from these baseline experiments, further example outputs at different resolutions and hyper-parameter settings, and videos displaying super-resolution effects and the interactive application: https://bit.ly/impgan_sup.

[5] As given on http://github.com/NVlabs/ in `stylegan` resp. `stylegan2`.

distributions of a wide variety of datasets. An Information Maximising and Pattern-Producing architecture (IMPGAN) has been constructed using state-of-the-art methods and techniques for training GANs, and has demonstrated an ability to successfully train CPPNs of different configurations across a wide variety of datasets.

The novelty of the paper comes from the IMPGAN architecture which allows for training CPPNs on unlabelled low-resolution data and subsequently generating high resolution images with various emergent properties (super-resolution) based on network architecture, activation functions, initial weight sizes, etc. As for most image generating GANs, the output images are related to the training set images, as the CPPN tries to approximate the underlying distribution of the training data. The CPPNs, however, generate each pixel independent of the others, with the latent inputs $\mathbf{z}$ and $\mathbf{c}$ allowing for generation of different colours or intensities for the same pixels, so that the images at high resolution exhibit various novel visual properties and/or unexpected level of detail and realism. This sidesteps many of the requirements recent GAN methods have in terms of computational cost and high resolution datasets. Moreover, instead of using the labels, if any, of a given dataset to guide the training, the IMPGAN aims to map a part of the latent input to controllable, isolated features in the output images.

An IMPGAN has several input parameters that can be tweaked to affect the output: chiefly the latent noise and latent code, with unpredictable resp. relatively predictable effects on the output images, as well as the resolution and zoom level of the generated images. A range of experiments have been run varying certain setup aspects and input parameters, while keeping the others fixed (experiments with different activation functions, network depths, with and without distance term, with/without large initial weights, etc.). An interactive web application offers an easy-to-use interface to an assortment of pre-trained CPPN models, where a user is free to explore the creative space of the models using image interpolation and mixing, as well as through fine control over the inputs. This level of control over an image-generating GAN running entirely in a user's browser has not been seen before.

The goal has not been to achieve the highest degree of realism, but to explore how well CPPNs can generalise while also evaluating their creative properties. Thus the claim is not to outperform approaches such as StyleGAN2 [22] in terms of visual quality, but to offer a different architecture utilising recent techniques for improved GAN training, identify performant hyperparameters that work across many different and previously untested datasets, and present an in-depth evaluation of CPPN-GANs. StyleGAN and CPPN-GANs take two vastly different approaches to image generation. Training a StyleGAN to generate 1000 $\times$ 1000 images requires a dataset containing images of these dimensions and many weeks of training. A CPPN, however, can be trained on 64 $\times$ 64 images, and subsequently generate images, pixel-by-pixel instead of all at once, at arbitrarily high resolutions with interesting results. As such, these networks arguably have merits in terms of computational efficiency and creativity.

References

1. Anokhin, I., Demochkin, K., Khakhulin, T., Sterkin, G., Lempitsky, V., Korzhenkov, D.: Image generators with conditionally-independent pixel synthesis. CoRR abs/2011.13775, November 2020. https://arxiv.org/abs/2011.13775
2. Arjovsky, M., Chintala, S., Bottou, L.: Wasserstein GAN. CoRR abs/1701.07875, December 2017. https://arxiv.org/abs/1701.07875
3. Ba, J.L., Kiros, J.R., Hinton, G.E.: Layer normalization. CoRR abs/1607.06450, July 2016. https://arxiv.org/abs/1607.06450
4. Boden, M.A.: The Creative Mind: Myths and Mechanisms. 2 edn. Routledge (2003)
5. Brock, A., Donahue, J., Simonyan, K.: Large scale GAN training for high fidelity natural image synthesis. In: 7th International Conference on Learning Representations, New Orleans, LA, USA, May 2019. https://arxiv.org/abs/1809.11096
6. Chen, X., Duan, Y., Houthooft, R., Schulman, J., Sutskever, I., Abbeel, P.: InfoGAN: interpretable representation learning by information maximizing generative adversarial nets. In: Advances in Neural Information Processing Systems, vol. 30, pp. 2172–2180. NIPS, Barcelona, Spain, December 2016
7. Glorot, X., Bengio, Y.: Understanding the difficulty of training deep feedforward neural networks. In: Proceedings of the 13th International Conference on Artificial Intelligence and Statistics, pp. 249–256. PMLR, Chia Laguna, Italy, May 2010
8. Goodfellow, I., et al.: Generative adversarial networks. In: Advances in Neural Information Processing Systems, vol. 27, pp. 2672–2680. NIPS, Montréal, Canada, December 2014. https://arxiv.org/abs/1406.2661
9. Gulrajani, I., Ahmed, F., Arjovsky, M., Dumoulin, V., Courville, A.C.: Improved training of Wasserstein GANs. In: Advances in Neural Information Processing Systems, vol. 30, pp. 5767–5777. NIPS, Long Beach, CA, USA, December 2017
10. Ha, D.: Neural network generative art in Javascript, Jun 2015. http://blog.otoro.net/2015/06/19/neural-network-generative-art/
11. Ha, D.: Neurogram, July 2015. http://blog.otoro.net/2015/07/31/neurogram/
12. Ha, D.: The frog of CIFAR 10, April 2016. http://blog.otoro.net/2016/04/06/the-frog-of-cifar-10/
13. Ha, D.: Generating abstract patterns with TensorFlow, March 2016. http://blog.otoro.net/2016/03/25/generating-abstract-patterns-with-tensorflow/
14. Ha, D.: Generating large images from latent vectors, April 2016. http://blog.otoro.net/2016/04/01/generating-large-images-from-latent-vectors/
15. Ha, D.: Generating large images from latent vectors - part two, June 2016. http://blog.otoro.net/2016/06/02/generating-large-images-from-latent-vectors-part-two/
16. Ha, D.: Interactive abstract pattern generation Javascript demo, April 2016. http://blog.otoro.net/2016/04/24/interactive-abstract-pattern-generation-javascript-demo/
17. He, K., Zhang, X., Ren, S., Sun, J.: Deep residual learning for image recognition. In: 2016 IEEE Conference on Computer Vision and Pattern Recognition, pp. 770–778. IEEE, Las Vegas, NV, USA, June 2016
18. Heusel, M., Ramsauer, H., Unterthiner, T., Nessler, B., Hochreiter, S.: GANs trained by a two time-scale update rule converge to a local Nash equilibrium. In: Advances in Neural Information Processing Systems, vol. 31, pp. 6626–6637. NIPS, Long Beach, CA, USA, December 2017
19. Ioffe, S.: Batch renormalization: towards reducing minibatch dependence in batch-normalized models. In: Advances in Neural Information Processing Systems, vol. 31, pp. 1945–1953. NIPS, Long Beach, CA, USA, December 2017

20. Ioffe, S., Szegedy, C.: Batch normalization: accelerating deep network training by reducing internal covariate shift. In: Proceedings of the 32nd International Conference on Machine Learning, pp. 448–456. PMLR, Lille, France, June 2015
21. Karras, T., Laine, S., Aila, T.: A style-based generator architecture for generative adversarial networks. In: IEEE/CVF Conference on Computer Vision and Pattern Recognition, pp. 4396–4405. IEEE, Long Beach, June 2019
22. Karras, T., Laine, S., Aittala, M., Hellsten, J., Lehtinen, J., Aila, T.: Analyzing and improving the image quality of StyleGAN. CoRR abs/1912.04958, December 2019. https://arxiv.org/abs/1912.04958
23. Kingma, D.P., Ba, J.: Adam: a method for stochastic optimization. CoRR abs/1412.6980, January 2017. https://arxiv.org/abs/1412.6980
24. Krizhevsky, A.: Learning multiple layers of features from tiny images. Technical report, Department of Computer Science, University of Toronto, April 2009
25. Kurach, K., Lučić, M., Zhai, X., Michalski, M., Gelly, S.: A large-scale study on regularization and normalization in GANs. In: Proceedings of the 36th International Conference on Machine Learning, pp. 3581–3590. PMLR, Long Beach, CA, USA, June 2019
26. LeCun, Y., Cortes, C., Burges, C.J.: The MNIST database of handwritten digits (2010). http://yann.lecun.com/exdb/mnist
27. Liu, Z., Luo, P., Wang, X., Tang, X.: Deep learning face attributes in the wild. In: Proceedings of 2015 IEEE International Conference on Computer Vision, pp. 3730–3738. IEEE, Santiago, Chile, December 2015
28. Metz, L., Gulrajani, I.: Compositional pattern producing GAN. In: Workshop on Machine Learning for Creativity and Design, Long Beach, CA, USA, December 2017
29. Miyato, T., Kataoka, T., Koyama, M., Yoshida, Y.: Spectral normalization for generative adversarial networks. In: 6th International Conference on Learning Representations, Vancouver, Canada, May 2018. https://arxiv.org/abs/1802.05957
30. Odena, A.: Open questions about generative adversarial networks. Distill, April 2019. https://distill.pub/2019/gan-open-problems
31. Salimans, T., Goodfellow, I., Zaremba, W., Cheung, V., Radford, A., Chen, X.: Improved techniques for training GANs. In: Advances in Neural Information Processing Systems, vol. 30, pp. 2234–2242. NIPS, Barcelona, Spain, December 2016
32. Secretan, J., et al.: Picbreeder: a case study in collaborative evolutionary exploration of design space. Evol. Comput. **19**(3), 373–403 (2011)
33. Sims, K.: Artificial evolution for computer graphics. Comput. Graph. **25**(4), 319–328 (1991)
34. Stanley, K.O.: Compositional pattern producing networks: a novel abstraction of development. Genet. Program. Evolvable Mach. **8**(2), 131–162 (2007)
35. Stanley, K.O., Miikkulainen, R.: Evolving neural networks through augmenting topologies. Evol. Comput. **10**(2), 99–127 (2002)
36. The TensorFlow Team: Flowers, January 2019. http://download.tensorflow.org/example_images/flower_photos.tgz
37. Xiao, H., Rasul, K., Vollgraf, R.: Fashion-MNIST: a novel image dataset for benchmarking machine learning algorithms. CoRR abs/1708.07747, September 2017. https://arxiv.org/abs/1708.07747

Aesthetic Evaluation of Cellular Automata Configurations Using Spatial Complexity and Kolmogorov Complexity

Mohammad Ali Javaheri Javid(✉)

Business and Computing School, University of Chichester, West Sussex, UK
M.JavaheriJavid@chi.ac.uk

Abstract. This paper addresses the computational notion of aesthetics in the framework of multi-state two-dimensional cellular automata (2D CA). The measure of complexity is a core concept in computational approaches to aesthetics. Shannon's information theory provided an objective measure of complexity, which led to the emergence of various informational theories of aesthetics. However, entropy fails to take into account the spatial characteristics of 2D patterns; these characteristics are fundamental in addressing the aesthetic problem in general, and of CA-generated patterns in particular. We propose two empirically evaluated alternative measures of complexity, taking into account the spatial characteristics of 2D patterns along with experimental studies on human aesthetic perception in the visual domain. The first model, spatial complexity, is based on the probabilistic spatial distribution of neighbouring cells over the lattice of a multi-state 2D cellular automaton. The second model is based on algorithmic information theory (Kolmogorov complexity) which is extended to estimate the complexity of 2D patterns. The spatial complexity measure presents performance advantage over information-theoretic models enabling more accurate measurement of complexity in relation to aesthetic evaluations of 2D patterns. The results of experimentation demonstrate the presence of correlation between the models and aesthetic judgements of experimental 2D patterns.

Keywords: Computational aesthetics · Cellular automata · Information theory · Spatial complexity · Kolmogorov complexity

1 Introduction

The concept of cellular automaton, one of the early biologically inspired systems, has contributed to the creation of many forms of computer art. The popularity of Game of Life (GoL) cellular automaton by Conway [6] drew the attention of the wider community of digital artists and designers to the unexplored potential of CA in generating rich digital content from the iteration of simple deterministic rules. The machinery of CA is based on the local interaction of each automaton

J. Romero et al. (Eds.): EvoMUSART 2021, LNCS 12693, pp. 147–160, 2021.
https://doi.org/10.1007/978-3-030-72914-1_10

with its immediate neighbourhood automata according to a set of rules. The interaction of automata at a local level generates emergent behaviour, sometimes with attractive complexity, at the global level. The main characteristics of CA that make them particularly interesting to digital artists are their ability to generate visually appealing and complex patterns on the basis of simple rules. Also, the lattice of CA represents each automaton as picture element (pixel) interacting with neighbouring automata making them a suitable framework for generating computer graphics with unique characteristics [7].

Although Shannon's entropy (H) is dominant in computational notions of aesthetics, it failed to accurately discriminate structurally different patterns in two dimensions. The failing of entropy, as a measure of order and complexity, to reflect the structural characteristics of 2D patterns, is caused by its inclination to measure the distribution of symbols and not their arrangements. Therefore, it was possible to generate radically different patterns (structurally different 2D patterns) with the same entropy [9,10].

This fact was noted by Arnheim, stating that "entropy theory is indeed a first attempt to deal with global form; but it has not been dealing with structure. All it says is that a large sum of elements may have properties not found in a smaller sample of them" [2, p. 18]. Figure 1 illustrates the measurements of entropy for CA-generated 2D patterns with various structural characteristics. Figure 1(a) and 1(b) generated by a cellular automaton. Figure 1(a) is a fully symmetrical pattern, Fig. 1(b) is a pattern with local structures and Fig. 1(c) is a fairly structureless random pattern. The comparison of the structural characteristics of these patterns with their corresponding entropy values shows that despite their structural differences, all of the patterns have the same entropy value. This clearly demonstrates the failure of entropy to discriminate structurally different 2D patterns. In other words, entropy is invariant to the spatial arrangement of the composing elements of 2D patterns. This is in contrast to our intuitive perception of the complexity of patterns. Therefore, measuring the complexity of CA behaviour, particularly with multi-state structures, would be insufficient if only entropy is used as a measure.

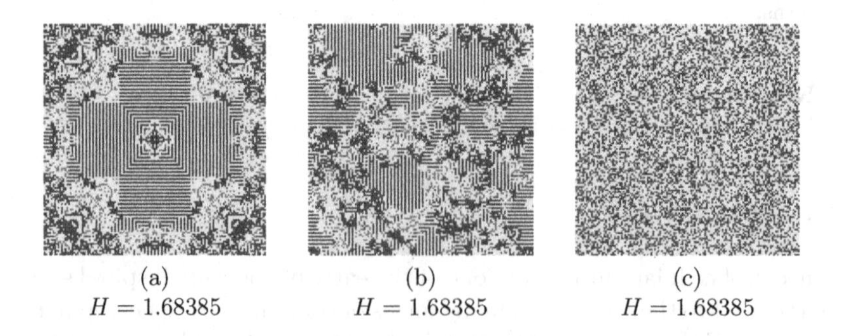

Fig. 1. The measurements of H for structurally different patterns.

2 Conceptual Model

Considering our intuitive perception of complexity and the structural characteristics of 2D patterns, a complexity measure must be bounded by two extreme points of complete order and disorder. It is reasonable to assume that *regular structures*, *irregular structures* and *structureless* patterns lie between these extremes, as illustrated in Fig. 2. A complete regular structure is a pattern of high symmetry, while an irregular structure is a pattern with some structure, though not as regular as a fully symmetrical pattern. Finally a structureless pattern is a random arrangement of elements [8].

order ⟵ regular structure | irregular structure | structureless ⟶ disorder

Fig. 2. The spectrum of spatial complexity.

3 Spatial Complexity Measure

Although Shannon further provided definitions of joint and conditional entropies in the framework of information theory [15, p. 52], its application in measuring structural complexity of dynamical systems remained unrecognised until studies [1,3,14,16] showed its merits. A measure introduced in [1,3,16], known as *information gain*, has been proposed as a means of characterising the complexity of dynamical systems and of 2D patterns. It measures the amount of information gained in bits when specifying the value, x, of a random variable X given knowledge of the value, y, of another random variable Y;

$$G_{x,y} = -\log_2 P(x|y), \tag{1}$$

where $P(x|y)$ is the conditional probability of a state x conditioned on the state y. The *mean information gain*, $\overline{G}_{X,Y}$, is the average amount of information gain from the description of all possible states of Y;

$$\overline{G}_{X,Y} = \sum_{x,y} P(x,y) G_{x,y} = -\sum_{x,y} P(x,y) \log_2 P(x|y), \tag{2}$$

where $P(x,y)$ is the joint probability, $\text{prob}(X = x, Y = y)$. $\overline{G}$ is also known as the conditional entropy, $H(X|Y)$ [5]. Conditional entropy is the reduction in uncertainty of the joint distribution of X and Y given knowledge of Y, $H(X|Y) = H(X,Y) - H(Y)$. The lower and upper bounds of $\overline{G}_{X,Y}$ are;

$$0 \leqslant \overline{G}_{X,Y} \leqslant \log_2 |\mathcal{X}|. \tag{3}$$

Definition 1. *A spatial complexity measure G, of a cellular automaton configuration is the sum of the mean information gains of cells having homogeneous/heterogeneous neighbouring cells over a lattice.*

For a cellular automaton configuration, $\overline{G}$ can be calculated by considering the distribution of cell states over pairs of cells r, s;

$$\overline{G}_{r,s} = -\sum_{s_r,s_s} P(s_r, s_s) \log_2 P(s_r, s_s) \tag{4}$$

where s_r, s_s are the states at r and s, respectively. Since $|S| = N$, $\overline{G}_{r,s}$ is a value in $[0, N]$.

The vertical, horizontal, primary diagonal ($\diagdown$) and secondary diagonal ($\diagup$) neighbouring pairs provide eight $\overline{G}s$; $\overline{G}_{(i,j),(i-1,j+1)}$, $\overline{G}_{(i,j),(i,j+1)}$, $\overline{G}_{(i,j),(i+1,j+1)}$, $\overline{G}_{(i,j),(i-1,j)}$, $\overline{G}_{(i,j),(i+1,j)}$, $\overline{G}_{(i,j),(i-1,j-1)}$, $\overline{G}_{(i,j),(i,j-1)}$ and $\overline{G}_{(i,j),(i+1,j-1)}$. The relative positions for non-edge cells are given by matrix M:

$$M = \begin{bmatrix} (i-1,j+1) & (i,j+1) & (i+1,j+1) \\ (i-1,j) & (i,j) & (i+1,j) \\ (i-1,j-1) & (i,j-1) & (i+1,j-1) \end{bmatrix}. \tag{5}$$

Correlations between cells on opposing lattice edges are not considered. The result of this edge condition is that $G_{i+1,j}$ is not necessarily equal to $\overline{G}_{i-1,j}$. In addition, the differences between the horizontal (vertical) and two diagonal mean information rates reveal left/right (up/down), primary and secondary orientation of 2D patterns. So the sequence of generated configurations by a multi-state 2D cellular automaton can be analysed using the differences between the vertical $(i, j \pm 1)$, horizontal $(i \pm 1, j)$, primary diagonal (P_d) and secondary diagonal (S_d) mean information gains by means of:

$$\Delta\overline{G}_{i,j\pm1}(\Delta\overline{G}_V) = |\overline{G}_{i,j+1} - \overline{G}_{i,j-1}|, \tag{6a}$$

$$\Delta\overline{G}_{i\pm1,j}(\Delta\overline{G}_H) = |\overline{G}_{i-1,j} - \overline{G}_{i+1,j}|, \tag{6b}$$

$$\Delta\overline{G}_{P_d} = |\overline{G}_{i-1,j+1} - \overline{G}_{i+1,j-1}|, \tag{6c}$$

$$\Delta\overline{G}_{S_d} = |\overline{G}_{i+1,j+1} - \overline{G}_{i-1,j-1}|. \tag{6d}$$

4 Kolmogorov Complexity of 2D Patterns

From an information theory perspective, the object X is a random variable drawn according to a probability mass function $P(x)$. If X is random, then the descriptive complexity of the event $X = x$ is $\log \frac{1}{P(x)}$, because $\lceil \log \frac{1}{P(x)} \rceil$ is the number of bits required to describe x. Thus, the descriptive complexity of an object depends on the probability distribution [5].

Kolmogorov attributed the algorithmic (descriptive) complexity of an object to the minimum length of a program such that a universal computer (universal Turing machine) can generate a specific sequence [11]. Thus, the Kolmogorov complexity of an object is independent of the probability distribution. Kolmogorov complexity is related to entropy, $H(X)$, in that the expected value of $K(x)$ for a random sequence is approximately the entropy of the source distribution for the process generating the sequence. However, Kolmogorov complexity differs from entropy in that it relates to the specific string being considered rather than the source distribution [5,12]. Kolmogorov complexity can be described as follows, where φ represents a universal computer, p represents a program, and x represents a string:

$$K_{\varphi}(x) = \left\{ \min_{\varphi(p)=x} \quad l(p) \right\} \tag{7}$$

Random strings have rather high Kolmogorov complexity - on the order of their length, as patterns cannot be discerned to reduce the size of a program generating such a string. On the other hand, strings with a high degree of structure have fairly low complexity. Universal computers can be equated through programs of constant length, thus a mapping can be made between universal computers of different types. The Kolmogorov complexity of a given string on two computers differs by known or determinable constants. The Kolmogorov complexity $K(y|x)$ of a string y, given string x as input is described by the equation below:

$$K_{\varphi}(y|x) = \left\{ \begin{array}{l} \min\limits_{\varphi(p,y)=y} \quad l(p) \\ \\ \infty, \text{if there is no } p \text{ such that } \varphi(p,x) = y \end{array} \right\} \tag{8}$$

where $l(p)$ represents program length p and φ is a particular universal computer under consideration. Thus, knowledge or input of a string, x may reduce the complexity or program size necessary to produce a new string, y. The major difficulty with Kolmogorov complexity is that it is uncomputable. Any program that produces a given string is an upper bound on the Kolmogorov complexity for this string, but it is not possible to compute the lower bound.

Lempel and Ziv defined a measure of complexity for finite sequences rooted in the ability to produce strings from simple copy operations [17]. This method, known as *LZ78* universal compression algorithm, harnesses this principle to yield a universal compression algorithm that can approach the entropy of an infinite sequence produced by an ergodic source. As such, *LZ78* compression has been used as an estimator for K. Kolmogorov complexity is the ultimate compression bound for a given finite string and thus, is a natural choice for the estimation of complexity in the class of universal compression techniques. In order to estimate the K value of 2D configurations generated by multi-state 2D CA, we generate linear strings of configurations by means of six different templates illustrated in Fig. 3.

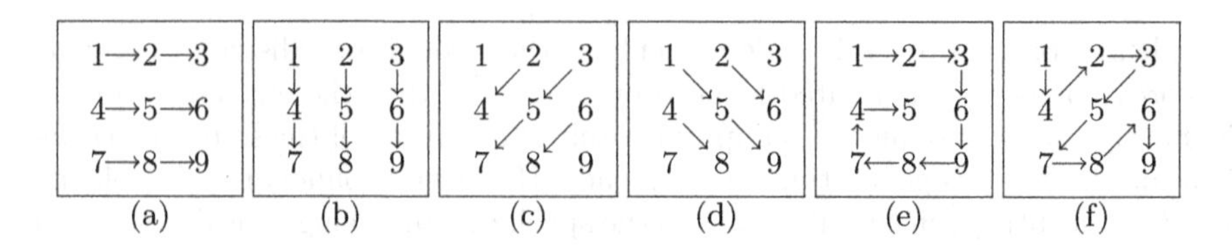

Fig. 3. Six different templates applied for the estimation of K in 2D plane.

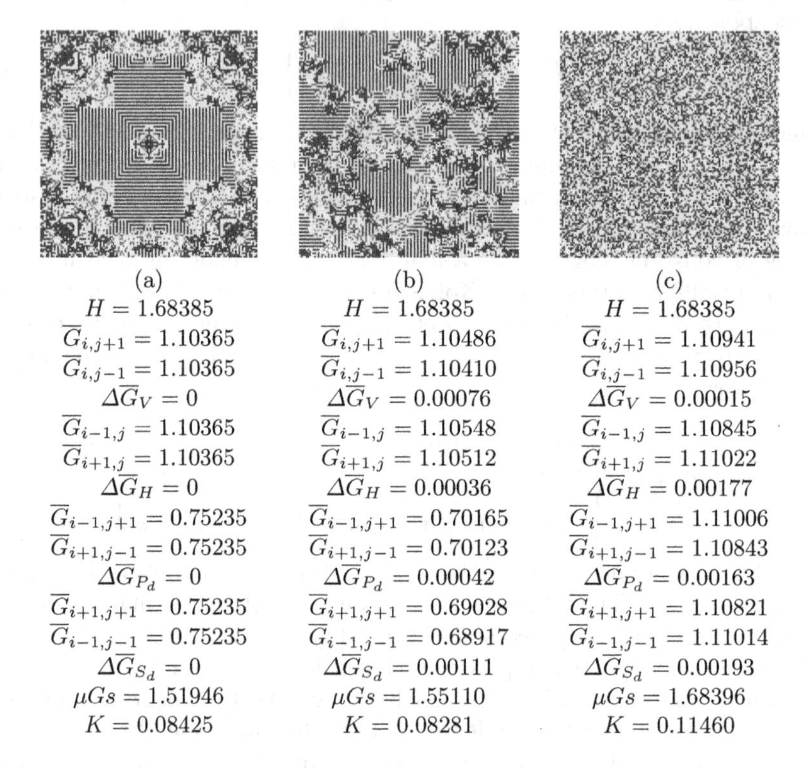

Fig. 4. The measurements of H, $\overline{G}s$, $\Delta\overline{G}s$ and K for structurally different patterns.

(a) Horizontal string: $S_h = \{1, 2, 3, 4, 5, 6, 7, 8, 9\}$
(b) Vertical string: $S_v = \{1, 4, 7, 2, 5, 8, 3, 6, 9\}$
(c) Diagonal string: $S_d = \{1, 2, 4, 3, 5, 7, 6, 8, 9\}$
(d) Reverse diagonal string: $S_{rd} = \{3, 2, 6, 1, 5, 9, 4, 8, 7\}$
(e) Spiral string: $S_s = \{1, 2, 3, 6, 9, 8, 7, 4, 5\}$
(f) Continuous spiral string: $S_{cs} = \{1, 4, 2, 3, 5, 7, 8, 6, 9\}$

Then, using the *LZ78* compression algorithm, the upper bound of K is estimated as the lowest value among the six different templates. The comparison of the measurements of H, $\overline{G}s$, $\Delta\overline{G}s$ and K for structurally different patterns is illustrated in Fig. 4. It is evident from the measurements, K is able to discriminate the complexity of patterns; however, it fails to discriminate the spatial orientations. Figure 4 demonstrates the merits of $\overline{G}$ in discriminating structurally different patterns for the sample patterns in Fig. 4. As can be observed, the measures of H are identical for structurally different patterns; however, $\overline{G}s$ and $\Delta\overline{G}s$ reflect both the complexity of patterns and the spatial distribution of their constituting elements $\mu Gs_{(a)} = 1.51946 > \mu Gs_{(b)} = 1.55110 > \mu Gs_{(c)} = 1.68396$.

5 Experiment and Results

This section details the experiment and its results on the correlation between three measures, namely spatial complexity, $\mu(G)s$, Kolmogorov complexity K and entropy, H, as well as human aesthetic judgement.

A set of 2D patterns with various structural properties were generated by seeding CA, then a survey was designed and used to compare their aesthetic values with the measurements of $\mu(G)s$, K and H.

5.1 Method

An online survey was conducted where users were paid for participating in the survey. A total of 100 participants aged between 18 and 60 in the USA participated in the survey.

5.2 Material

For this experiment, 10 patterns with various structural characteristics reflecting the spectrum of spatial complexity (Fig. 2) were generated by seeding 3-state CA with different ICs. The patterns fall in three categories namely:

- regular structures (Fig. 5(1), (2), (3), (4), (5)),
- irregular structures (Fig. 5(6), (7), (8), (9)) and,
- structureless patterns (Fig. 5(10)).

Figure 5 illustrates the generated experimental patterns. Grey scale colours were used for the colour assignments of the CA states to eliminate possible individual colour preferences in aesthetic judgements ($S = \{0, 1, 2\} \equiv \{ \ , ■, ■\}$). The size of lattice for all the patterns is $L = 65 \times 65$ (4225 cells). The spatial complexity measure, $\mu(G)s$, K and H, were then calculated for all the patterns. Figure 6 shows the measurements of $\mu(G)s$, K and H for the CA generated patterns, Fig. 7 and Fig. 8 show the details of these calculations.

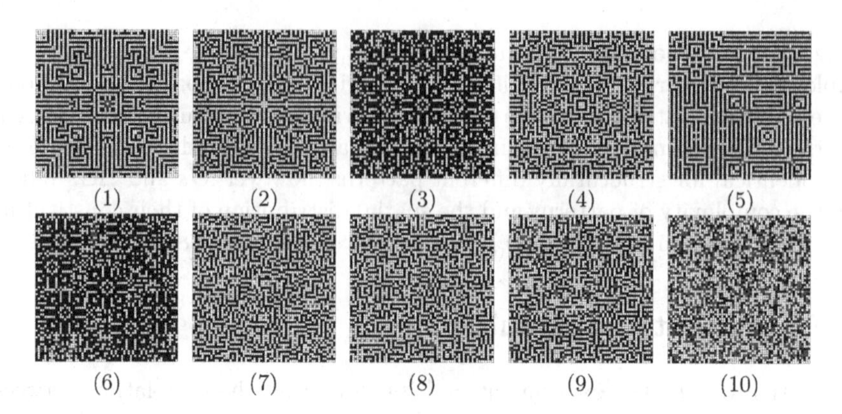

Fig. 5. Generated patterns with various structural characteristics reflecting the spectrum of spatial complexity (Fig. 2).

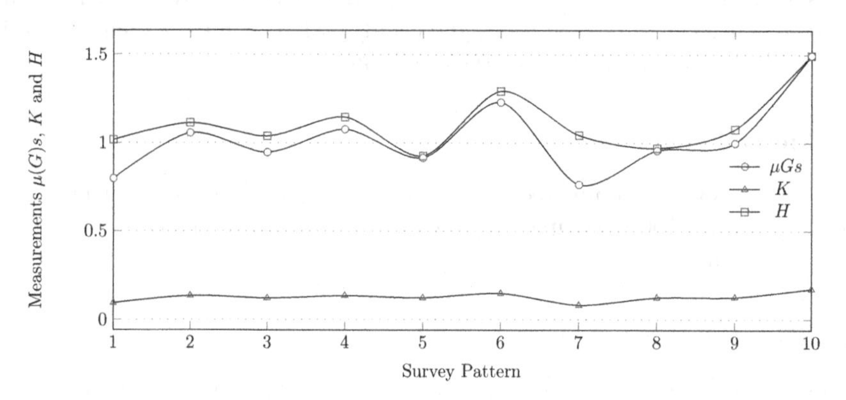

Fig. 6. The plot of $\mu(G)s$, K and H for the generated survey patterns.

5.3 Procedure

An online survey was designed with the 10 generated patterns ordered according to Fig. 9. A five-point Likert scale [13] was used to obtain quantitative measurements of the aesthetic judgements of respondents. The patterns were presented one at a time to the 100 participants with the following instructions:

The following 10 images have been generated by computers. Please rate them in terms of their aesthetic appeal.

The participants were asked to rate how well they agreed with the following statement: *This image is aesthetically pleasing (beautiful).* The five-point Likert scale consisted of "Strongly Disagree (SD)", "Disagree (D)", "Neutral (N)", "Agree (A)" and "Strongly Agree (SA)".

5.4 Results and Analysis

The online survey yielded 68 valid responses consisted of 25 males and 43 females with an age group distribution illustrated in Table 1 and Table 2 summarises the results of the survey.

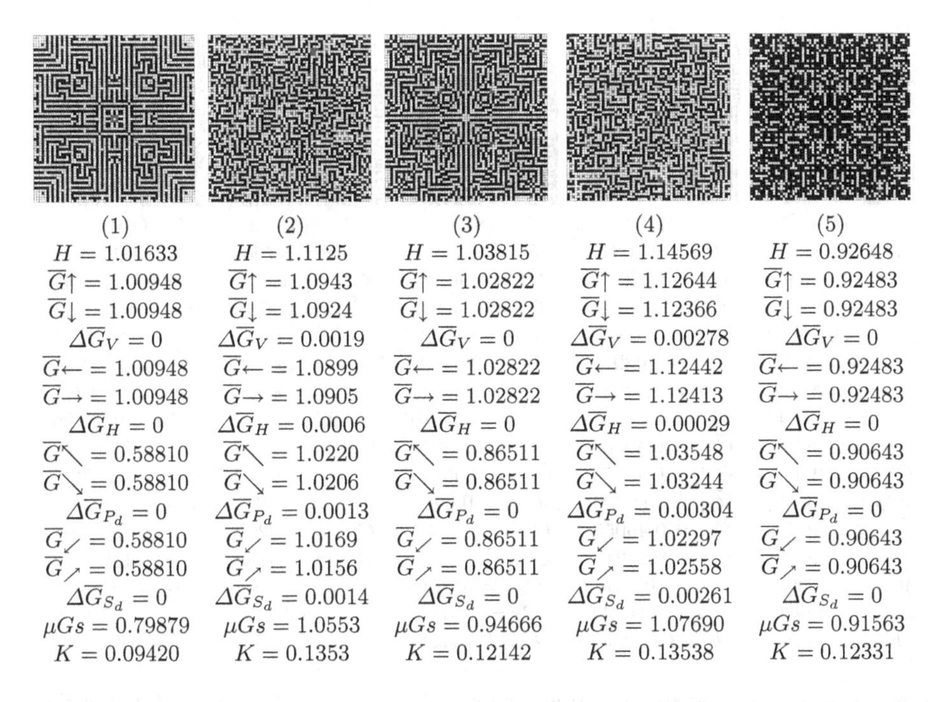

Fig. 7. The measurements of spatial complexity measure, $\mu(G)s$, K and H for the generated patterns with various structural characteristics.

Table 1. The distribution of age groups and genders for 68 collected valid responses.

Age group				Gender		Total
18–29	30–44	45–59	60	Male	Female	
17	16	26	9	25	43	68

Table 2. The results of survey.

Pattern	SD	D	N	A	SA	Total
1	7	21	17	17	6	68
2	23	35	10	0	0	68
3	4	14	19	26	5	68
4	22	37	8	1	0	68
5	19	5	20	20	4	68
6	25	33	8	2	0	68
7	5	24	20	13	6	68
8	7	30	19	11	1	68
9	4	9	20	26	9	68
10	30	22	13	3	0	68

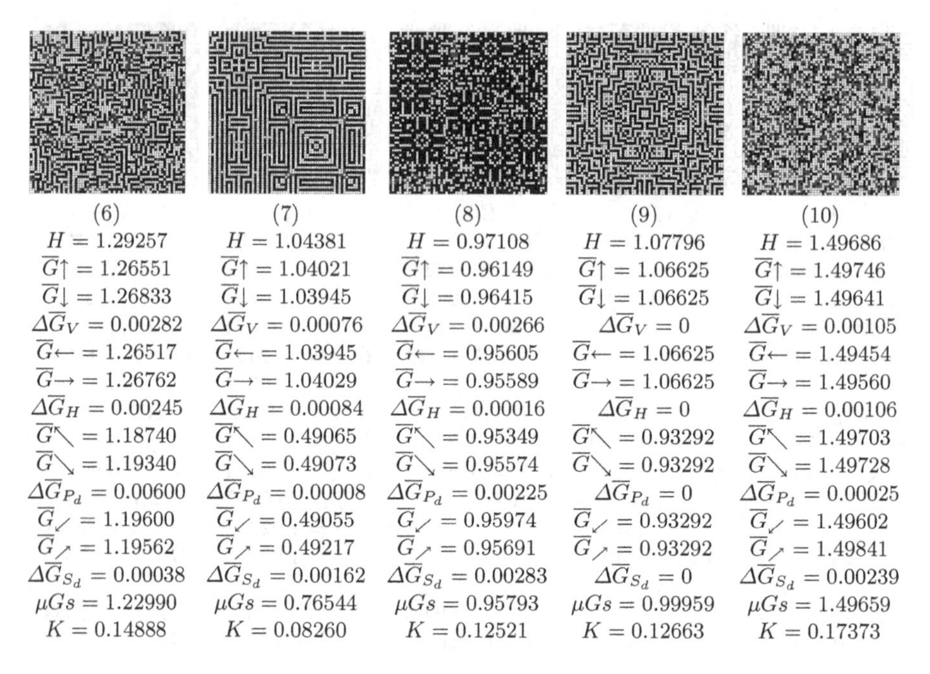

Fig. 8. The measurements of spatial complexity measure, $\mu(G)s$, K and H for the generated patterns with various structural characteristics.

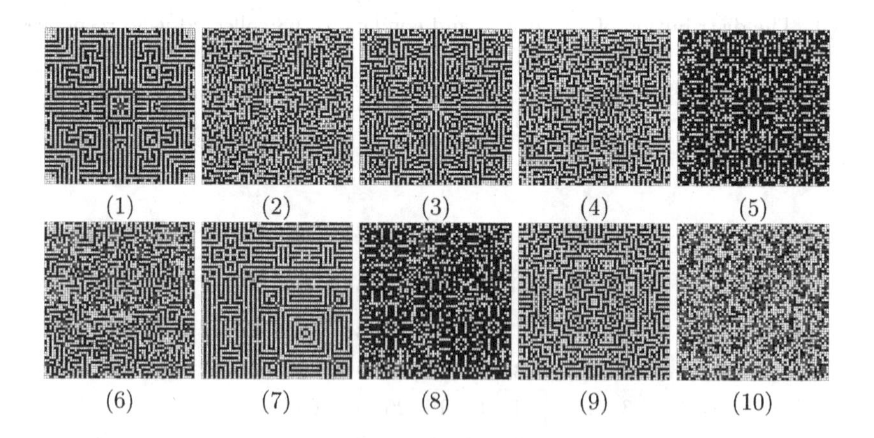

Fig. 9. The order of patterns used in the survey.

Table 3. Assigned weight to each scale.

Scale	SD	D	N	A	SA
Weight	1	2	3	4	5

Since Likert scale data are ordinal data (i.e. they only show that a rating is higher/lower than another and not the distance between the scales), the data must be aggregated across the collected data to ensure an accurate estimation of each scale's value for the patterns. In order to get an aggregated score (AScore) of the ratings, each of the scales were assigned a weight according to Table 3.

The following formula was used to calculate the aggregated score of the five-point Likert scale for the survey:

$$AScore = \frac{1}{5}\sum N.W, \tag{9}$$

where N is the total number of ratings for each scale and W is the assigned weight to each scale. For example, given the total number of participants' responses to the statement of the survey from Table 2, assigned weight to each scale from Table 3 and Eq. 9, the following calculation was performed to obtain the aggregated score for pattern one:

AScore for pattern one $= \frac{1}{5}\{(7\times 1)+(21\times 2)+(17\times 3)+(17\times 4)+(6\times 5)\} = 39.6$.

Figure 10 illustrates the total ratings and the plot of the aggregated scores for the patterns and Fig. 11 shows the individual aggregated scores for each of the 10 patterns. Figure 12 shows the survey pattern arranged based on aesthetic judgement from the most aesthetically appealing as shown in Fig. 12(1) to the least aesthetically appealing as shown in Fig. 12(10).

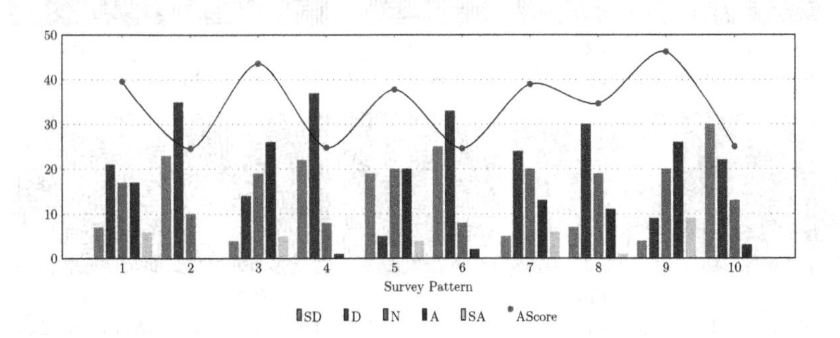

Fig. 10. The total ratings and the aggregated scores for the patterns.

Due to the ordinal nature of data, a non-parametric Spearman rank correlation, r_s was applied for the analysis of data with a significance level of $\alpha = 0.05$. Table 4 shows the results of the rank correlation test between the aggregated scores and the measurements of μGs, K and H.

Table 4. The results of Spearman rank correlation test (r_s) between μGs, K, H and the aggregated scores of survey patterns.

	μGs	K	H
AScore	−0.6383	−0.6859	−0.5288
p	0.04702	0.02852	0.11599

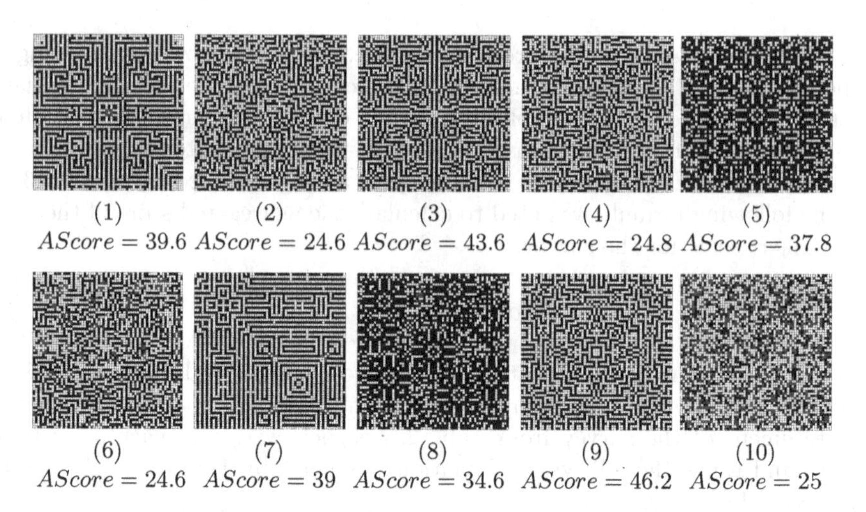

Fig. 11. The aggregated scores of survey patterns.

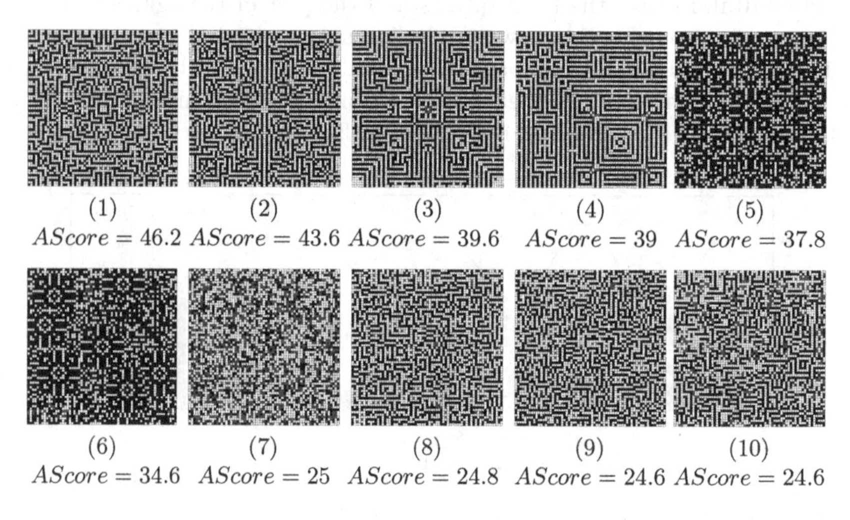

Fig. 12. The arrangement of patterns in descending order of their aggregated scores.

The value of r_s for μGs is -0.6383 and the 2-tailed $p = 0.04702 < 0.05$. Thus, there is *a negative linear correlation* between μGs and the aggregated scores and the association between the two variables is *statistically significant.* The value of r_s for K is -0.6859 and the 2-tailed value of $p = 0.02852 < 0.05$. Therefore, there is *a negative linear correlation* between K and the aggregated scores and the association between the two variables is *statistically significant.* The value of r_s for H is -0.5288 and the 2-tailed value of $p = 0.11599 > 0.05$. As such, the association between the two variables is *not statistically significant.* Considering the values of r_s for the Spearman rank correlation and regression analysis, the following conclusions can be drawn:

- There is a statistically significant relationship between the measurement of $\mu(G)s$ for CA-generated patterns and human aesthetic judgement and the direction of the relationship is negative.
- There is a statistically significant relationship between the estimation of K for CA-generated patterns and human aesthetic judgement and the direction of the relationship is negative.
- There is no statistically significant relationship between the measurement of H for CA-generated patterns and human aesthetic judgement.

6 Discussions

One of the major challenges in computational notions of aesthetics and generative art is the development of a quantitative model which conforms to human intuitive perceptions of aesthetic judgement. As discussed, informational theories of aesthetics based on the measurements of entropy have failed to discriminate structurally different patterns in a 2D plane. Consequently, spatial complexity, G and Kolmogorov complexity, K, were suggested for quantifying the spatial complexity of 2D patterns. An experimentation were conducted to examine the relationship between the measurements of G and K and human aesthetic judgement. Since entropy, H has emerged as a dominant measure of order and complexity in computational notions of aesthetics, the relationship with human aesthetic judgement was compared as well.

A set of CA-generated patterns with various structural properties reflecting the spectrum of spatial complexity was used to examine the relationship between the measurements of $\mu(G)s$, K and H and human aesthetic judgement. The result of this experiment highlighted the presence of a statistically significant negative linear relationship between $\mu(G)s$ and K for CA-generated pattern and human aesthetic judgement. The implication of these findings is that both spatial complexity and Kolmogorov complexity are conforming to the human aesthetic judgement but with an inverse direction. It confirms the validity of theories, which consider an *inverse* relationship between stimulus complexity and aesthetic preference (i.e. Birkhoff's aesthetic measure [4]).

These results could potentially provide researchers with a direction for future aesthetics analysis of CA-generated patterns and other types of imageries using $\mu(G)s$ and K, both of which exhibiting a relationship with human aesthetic judgement. The proposed models can have applications in the area of image processing, image aesthetic enhancement and computer art.

References

1. Andrienko, Y.A., Brilliantov, N.V., Kurths, J.: Complexity of two-dimensional patterns. Eur. Phys. J. B **15**(3), 539–546 (2000)
2. Arnheim, R.: Entropy and Art: An Essay on Disorder and Order. University of California Press (1974)

3. Bates, J.E., Shepard, H.K.: Measuring complexity using information fluctuation. Phys. Lett. A **172**(6), 416–425 (1993)
4. Birkhoff, G.: Aesthetic Measure. Harvard University Press, Cambridge (1933)
5. Cover, T.M., Thomas, J.A.: Elements of Information Theory (Wiley Series in Telecommunications and Signal Processing). Wiley-Interscience (2006)
6. Gardner, M.: Mathematical games - the fantastic combinations of John Conway's new solitaire game, life. Sci. Am. 120–123 (1970)
7. Javaheri Javid, M.A., Alghamdi, W., Ursyn, A., Zimmer, R., al Rifaie, M.M.: Swarmic approach for symmetry detection of cellular automata behaviour. Soft Comput. **21**(19), 5585–5599 (2017)
8. Javaheri Javid, M.A., Blackwell, T., Zimmer, R., Al-Rifaie, M.M.: Spatial complexity measure for characterising cellular automata generated 2D patterns. In: Pereira, F., Machado, P., Costa, E., Cardoso, A. (eds.) EPIA 2015. LNCS (LNAI), vol. 9273, pp. 201–212. Springer, Cham (2015). https://doi.org/10.1007/978-3-319-23485-4_21
9. Javaheri Javid, M.A., al Rifaie, M.M., Zimmer, R.: An informational model for cellular automata aesthetic measure. In: AISB 2015 Symposium on Computational Creativity. University of Kent, Canterbury (2015)
10. Javaheri Javid, M.A., Zimmer, R., Ursyn, A., al-Rifaie, M.M.: A quantitative approach for detecting symmetries and complexity in 2D plane. In: Dediu, A.-H., Magdalena, L., Martín-Vide, C. (eds.) TPNC 2015. LNCS, vol. 9477, pp. 150–160. Springer, Cham (2015). https://doi.org/10.1007/978-3-319-26841-5_12
11. Kolmogorov, A.N.: Three approaches to the quantitative definition of information. Probl. Inf. Transm. **1**(1), 1–7 (1965)
12. Li, M.: An Introduction to Kolmogorov Complexity and Its Applications. Springer, Heidelberg (1997)
13. Likert, R.: A technique for the measurement of attitudes. Arch. Psychol. **15**(140), 1–55 (1932)
14. Navratil, E., Zelinka, I., Senkerik, R.: Preliminary results of deterministic chaos control through complexity measures. In: 20th European Conference on Modelling and Simulation ECMS 2006: Modelling Methodologies and Simulation: Key Technologies in Academia and Industry. European Council for Modelling and Simulation (ECMS) (2006)
15. Shannon, C.: A mathematical theory of communication. Bell Syst. Tech. J. **27**, 379–423 & 623–656 (1948)
16. Wackerbauer, R., Witt, A., Atmanspacher, H., Kurths, J., Scheingraber, H.: A comparative classification of complexity measures. Chaos Solitons Fractals **4**(1), 133–173 (1994)
17. Ziv, J., Lempel, A.: Compression of individual sequences via variable-rate coding. IEEE Trans. Inf. Theory **24**(5), 530–536 (1978)

Auralization of Three-Dimensional Cellular Automata

Yuta Kariyado(✉), Camilo Arevalo, and Julián Villegas

University of Aizu, Aizu-Wakamatsu, Japan
{s1250192,m5231113,julian}@u-aizu.ac.jp
http://onkyo.u-aizu.ac.jp

Abstract. An auralization tool for exploring three-dimensional cellular automata is presented. This proof-of-concept allows the creation of a sound field comprising individual sound events associated with each cell in a three-dimensional grid. Each sound-event is spatialized depending on the orientation of the listener relative to the three-dimensional model. Users can listen to all cells simultaneously or in sequential slices at will. Conceived to be used as an immersive Virtual Reality (VR) scene, this software application also works as a desktop application for environments where the VR infrastructure is missing. Subjective evaluations indicate that the proposed sonification increases the perceived quality and immersability of the system with respect to a visualization-only system. No subjective differences between the sequential or simultaneous presentations were found.

Keywords: Cellular automata · Game of life · Auralization

1 Introduction

Cellular automata are one of the earliest computer-based models created to mimic biological systems [24]. They rapidly gained popularity among several communities including computer scientists, mathematicians, biologists, and enthusiast in the general audience. These automata create complex patterns that evolve with time and are usually based on simple rules. Traditionally, they have been first modeled in two-dimensional grids while three-dimensional extensions are considered a sophistication that in many cases have not been as successfully popularized as their two-dimensional counterparts. It is possible that the source of this unbalance is originated in the difficulty posed for their visualization. To begin with, 3D graphics techniques evolved long after 2D ones, and when successfully rendered in three dimensions, some cells are usually occluded depending on the perspective of the viewer even when the visual objects are semi-transparent.

Human vision and audition are more accurate in the front of a person [11] and gradually deteriorates away from that location, but audition let humans to perceive an environment in all directions, including the back. Thus, presenting spatially distributed sounds allows large representations of data [19].

J. Romero et al. (Eds.): EvoMUSART 2021, LNCS 12693, pp. 161–170, 2021.
https://doi.org/10.1007/978-3-030-72914-1_11

The purpose of this research is two-fold: To use sound to explore three-dimensional cellular automata, and to use these automata as a source of generative music. To that end, we focused on The Game of Life [18], arguably the most famous automaton, and the Abelian sand-pile model [2]. Associating sounds to each cell can be considered a case of sonification [11]. In addition to this cell–tone association, we also use audio spatialization to accurately represent the actual localization of cells. Adding spatialization to sonification can be considered as a kind of auralization [21]. However, the correct spatialization of audio sources is difficult since many of the filters used to that end are capable of accurately rendering horizontal and vertical angles (azimuth and elevation, respectively), but distance is usually barely approximated. Our sound spatializer is different in the sense that it allows the correct spatialization of audio sources including distance for near-field sources since is based on actual measurements of head-related transfer functions in the near-field [16].

Regarding the selection of automata, The Game of Life can be considered a simplification of the processes involved in the creation, evolution, and destruction of life. The fundamental components of this program are "cells" which were originally disposed in a two-dimensional grid. Each cell is either "alive" or "dead," and the transition between these two states depends on the states of its neighboring cells. Specifically, on the number of living neighboring cells. Originally, this research was made to celebrate the 50th anniversary of the Game of life; however, during the development stage, we learned of the tragic loss of Prof. Conway to the recent epidemic of COVID-19. This project is a humble homage to Prof. Conway's legacy and his important contribution to computer science.

2 Background

One of the earliest reports on three-dimensional versions of the game of life was presented by Bays [4]. In this study, he introduced a 3D version of the game of life implemented as an infinite number of parallel 2D-planes and several structures such as oscillators, collisions, etc. were presented.

More recently, other software implementations have been proposed, for example, CA3D [7] and Kaleidoscope of game of life [8]. The latter is as a Java applet, currently abandoned. Kaleidoscope allows users to define the game rules as well as a size of the grid. Additionally, users can select various known presets and store their own patterns. Our solution is inspired on this version. In CA3D, the initial conditions of the game (i.e., the location of the alive cells) are completely set at random with a 50% chance that a cell will be initially alive. Intersecting lines in the display are used in this program to mark the center of cells, and users can explore this 3D grid by using arrow keys to rotate it.

One of the most prominent software implementations of cellular automata among the current ones is Golly [6]. This is an open source, cross-platform application that includes features such as bounded and unbounded universes, fast generating algorithms, etc. Golly allows the creation of 3D models where users can define random initial conditions, or specify them to their own accord. The number of rendering steps can be adjusted, but at difference with CA3D, rendering speed cannot be adjusted in real-time.

3 Method

We developed our proof-of-concept application in Unity [22] mainly for two reasons: Unity is arguably the industry de facto three-dimensional environment creation tool, and also, it allows the deployment of the same code in several platforms including desktop-based and VR headset-based applications [3]. The source code, executable versions, and video demonstrations of this project are freely available from https://github.com/YKariyado/LG. In what follows we explain the main components of our solution.

3.1 Life

In the original game of life each cell has eight neighboring cells. The state of a cell is determined by the following rules [10]:

- Births: Each empty cell adjacent to exactly three neighbors is an alive cell in the next generation.
- Survivals: Every cell with two or three neighboring cells survives. I.e., it is an alive cell in the next generation.
- Deaths: Each alive cell with four or more neighbors dies.

In the case of three-dimensional game of life, each cell has 26 neighboring cells ($3 \times 3 \times 3$ minus itself). Therefore in our solution, the previous rules are defined with four arguments ($r1, r2, r3, r4$) in the range of 0 to 26 [8]:

- Births: A cell will be alive in the next generation if the number of neighbors n is $r1 \leq n \leq r2$.
- Survivals: A cell will be alive in the next generation if $n \geq r4$ and $n \leq r3$.
- Deaths: A cell will die if the number of neighbors is $n > r3$ or $n < r4$.

For reasons explained in the sonification section, we limited the size of the grid to 512 cells in a cube (i.e., 8 cells in each direction). But, the boundaries of this grid can be set to be "periodic," in which case cells at opposite faces of the cube become adjacent. Cells color was determined using the corresponding RGB coordinates of a given cell, as illustrated in the left panel of Fig. 1.

3.2 Abelian Sand Pile

Besides the Game of Life, we wanted to explore other cellular automata which could offer different sonification opportunities. To that end, we implemented the Abelian sand pile model (a.k.a. Bak-Tang-Wiesenfeld model) [2], a dynamic system showing self-organized criticality. Besides location, in the Abelian sand pile model cells also have a "capacity" (i.e., the number of "grains" it can pile

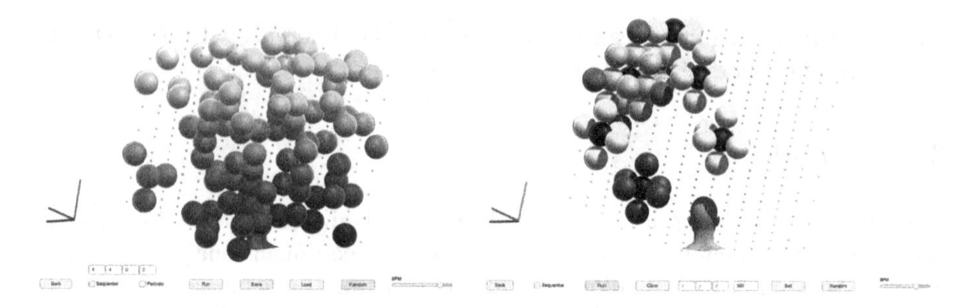

Fig. 1. Implemented game of life (left) and Abelian sand pile (right).

before toppling). A cell can receive grains from its neighbors, If the number of grains exceeds the cell's capacity, the pile topples and the cell transfers to each of its neighbors one grain yielding it empty if the number of grains matches exactly its capacity.

As in the case of the Game of Life, the sand pile model was originally designed in a two-dimensional grid and the capacity value of a cell was four (one for each edge). In the case of a three-dimensional sand pile model, cells have six neighbors (one for each face), so the capacity value in this case is six. In our implementation, the color of the cells represents the number of grains in a given cell: 0 (empty cell)—transparent, 1—red, 2—orange, 3—yellow, 4—green, 5—blue, 6—purple, > 6—black. As illustrated in the right panel of Fig. 1.

3.3 Navigation

To ease the exploration of these three-dimensional models, we used a combination of virtual perspectives [5]: The sound is presented from an endocentric point of view, i.e., all the sounds are in the near field and centered at the middle of the interaural axis. The graphic model is presented from an egocentric point of view, effectively tethering a camera behind of the head of the user. The Cartesian axes and a head model are displayed in our solution, as shown in Fig. 1. Whereas the exploration of the system is eased in the headset version by infrared trackers on the head of the user, in the desktop application, users can rotate the three-dimensional models around any axis using the mouse; the virtual head of the user can also be rotated around the y and x axes (i.e., yaw and pitch, respectively) by means of the arrow keys.

3.4 Sonification

For the sonification of these models we created a semantic auditory display [15] in which sounds were related to the location and content of each cell. Our symbolic representation offers the possibility of displaying sounds in sequence (along the z-axis) for a melodic and rhythmic representation, or simultaneously displaying all alive/non-empty cells in a given iteration. In this case, the representation is more indexical in the sense that the location of a cell is directly mapped to a sound feature such a fundamental frequency, phase, amplitude, etc.

Individual WAV files for each cell were created in Matlab [14] sampled at 16 bit/48 kHz. Each tone was 250 ms long, with a triangular amplitude envelope of 62.5 ms fade-in. The tones (fundamental frequencies) of a minor scale (0, 200, 300, 500, 700, 900, 1100, 1200 cents) starting on A_3 (220 Hz) were directly mapped to the x-axis values. We mapped harmonics of these tones whose amplitudes were determined by the harmonic series to the y-axis. Finally, the phase of these tones changed according to their location in the z-axis from 0° in increments of 40°.

For the sand pile model, we replaced the harmonic amplitudes with an amplitude modulation depending on the number of grains within a non-empty cell. Modulation frequencies were set to be a fraction of the corresponding fundamental frequency of the cell, inversely proportional to the number of grain in the cell. Modulation indices were linearly distributed from 0 to 100 in six parts, depending on the number of the grains in the cell. Finally, it was not clear what the number of sound sources could be simultaneously spatialized in Unity. Unity has a default of 512 audio sources that can be specified in a given scene, as determined by the programming interface. We tested how many of these could be actually simultaneously rendered (i.e., made audible) using the Unity profiler [23] and found that only half (255) register a valid SPL (i.e., $> -\infty$ dB). Nevertheless, we decided to use 512 virtual sources, so the grid in our solution is a cube of $8 \times 8 \times 8$.

3.5 Audio Spatialization

Each alive cell in our system is a sound source whose location relative to the user is accordingly simulated. Specifically, there may be 512 active cells and 255 audio rendered cells at a given moment. As previously mentioned, this top is imposed by Unity. We created a spatializer based on the reconstruction of HRTFs by Eigen decomposition, as described in [1]. This spatializer express the location of a sound source relative to the listener in the virtual scene and uses these endocentric coordinates to reconstruct its corresponding HRTF. Since not all possible locations are included in the spatializer HRTF database, the closest HRTF is used instead. The discretization of the HRTFs locations varies depending on the spherical coordinates: for elevation, $-40° \leq \phi \leq 90°$, every 2° steps; for azimuth θ, the discretization depends additionally on the elevation $\theta_\phi = 359° \cos(\phi) + 1$; for distance, the original intervals at which the HRTFs were captured in [16] (i.e., $[20, 30]$, $[30, 40]$, etc.) were sub-divided in ten equal parts. With these changes, the HRTF database used for our spatializer comprises $1,212,680$ locations. The spatialization is performed on a dedicated DSP thread on the game engine by doing the convolution (via Fourier transformation) between HRTF retrieved from the database and the audio signal associated with the alive cells (no interpolation of HRTFs is needed in our solution at this stage). The audio is processed in blocks of 256 samples at a sampling frequency of 48 kHZ (default in Unity). (Fig. 3)

4 Results

We run the Game of Life model simulation searching for interesting patterns that feature alive cells after 100 iterations. Through this process we found oscillators such as those reported previously in [8]. We also found some patterns such as blinkers and oscillators [18], with rules [4, 4, 0, 0]. These patterns are available via a "load" button in the application. Likewise, users can save their own patterns using a "save" button.

One of the patterns we found is an oscillator with a period of 14 generations. This oscillator, dubbed by us "Rocket," starts with four cells at the top of the first generation, and grow into a pattern that resembles the shape of a rocket in the fifth and sixth generation, as shown in Fig. 2.

Furthermore, the proposed system was evaluated to assess 1) how the sonification system performed in terms of audio dropouts, and 2) its perceived quality and immersability, as detailed in the next sections.

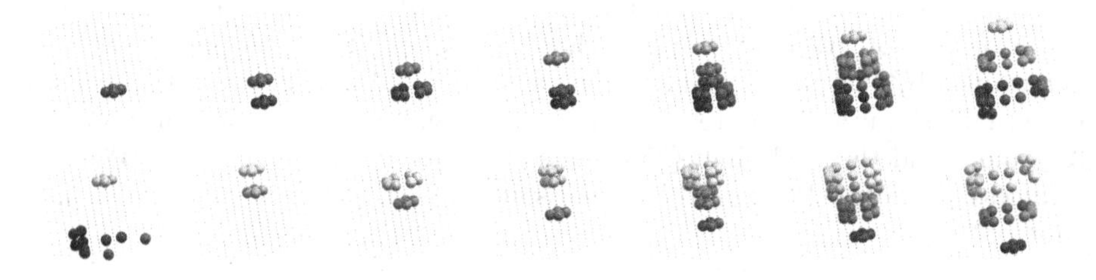

Fig. 2. "Rocket" oscillator in the game of life with [4, 4, 0, 0] rules. Different stages of this oscillator are presented starting from the top row, left to right.

Fig. 3. "Blinker" oscillators in the game of life with [4, 4, 0, 0] rules. Some of the most frequent oscillators in this game. These oscillators are characterized by repeating a very short cycle.

5 Objective Evaluation

A performance test was done on a MacBook Pro 2019 (processor Intel core *I9-9880H* with 8 cores at 2.3 GHz and DDR4 SDRAM memory with 32 GB at 2,666 MHz) on Unity 2020.1.6f1. In this simple test, audio files associated with cells in the Game of Life (in increasing coordinate order) were sequentially added every 5 s. Taking into account that the DSP process is done every 256 samples at 48 kHz, the DSP thread must perform the spatialization in < 5.3 ms. Using the Unity profiler [23], the CPU usage was tracked to determine the time

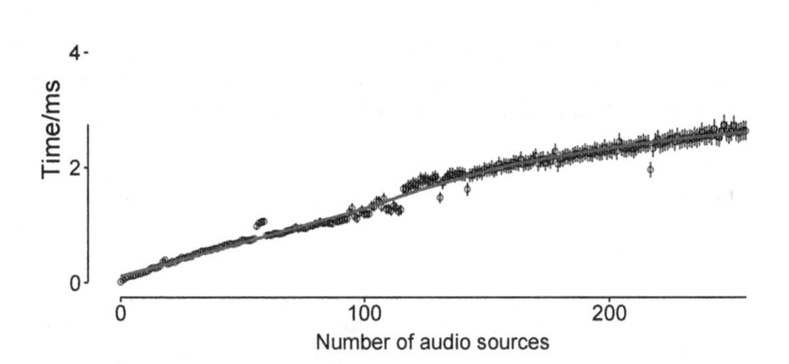

Fig. 4. CPU time used for the spatializer for different number of audio sources. Horizontal black dashed line shows the block processing time. Error bars correspond to the 95% CI, blue line shows the smoothed mean trend. (Color figure online)

taken by the DSP thread in the spatialization. Spatializer requires on average 13.362μs (with $\sigma = 14.461\,\mu$s) to spatialize an audio source, as shown in Fig. 4. These results indicate that the default amount of audio sources (255) can be spatialized without audio drop-outs with our current hardware set-up.

6 Subjective Evaluation

To investigate the effect of the auralization on the immersability and perceived quality of the model, we conducted a subjective experiment.

6.1 Participants

21 students from the University of Aizu (five females) volunteered for this experiment. The subjects were on average 21.81 years of age ($\sigma = 1.26$) with no apparent hearing issues (self-reported). Permission for performing this experiment was obtained following the University of Aizu ethics guidelines.

6.2 Procedure

Participants were requested to wear masks and Nitrile gloves during the experiment, as a measure against the spreading of COVID-19. Likewise, keyboard, headphones, mouse, etc., were cleaned after each participant with alcohol. Only one participant was allowed at a time in a ventilated room for this experiment.

Participants were subjected to three blocks of the 3D Game of Life: no sound (control condition), sequential, and simultaneous audio conditions. These blocks were randomly permuted for each participant. They were instructed to explore the model, rotating it using the arrow keys and the mouse. After each block (lasting about two minutes), participants gave their opinions on the immersion and overall quality of the block, rating it in a discrete scale from 1 (worst) to 7 (best). After the three blocks were presented, participants also expressed their preferred block. The experiment lasted less than 20 min in all cases.

6.3 Apparatus

A pair of Sennheiser HD 380 Pro headphones connected to an iMac 27-in. Late 2012 (processor Intel core *i7-3770* with 4 cores at 3.4 GHz and DDR3 SDRAM memory with 8 GB at 1,600 MHz) running Unity 2020.1.6f1 were used.

6.4 Results

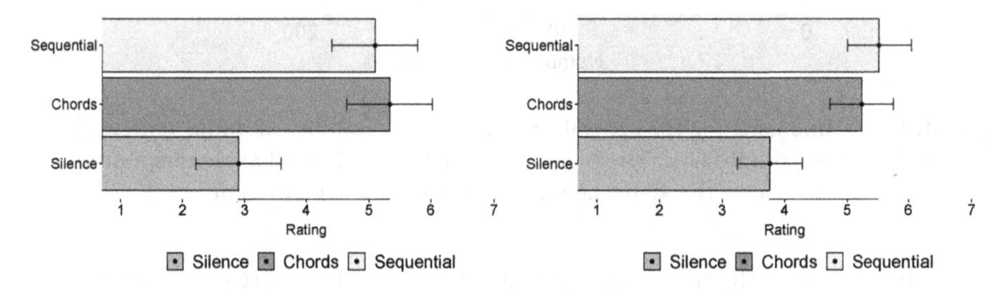

Fig. 5. Results of the subjective experiment. Immersion mean ratings on the left, perceived quality on the right. Error bars correspond to Fisher's least significant difference. Disjoint error bars indicate significant differences.

Statistical analysis of these data was done in R [17]. Immersability and quality judgements were analyzed with repeated-measures ANOVAs as implemented in the library ez [13]. Mode (Silence, Chords, or Sequential presentation) was the sole within-subject factor considered.

According to these analyses, Mode has a significant effect on Immersion ratings $[F(2,40) = 27, p < .001, \eta_G^2 = .393]$, as shown in the left panel of Fig. 5. No significant differences were found between the Chord and Sequential presentation, but the control condition (silence) was rated significantly worse than the other two conditions, as visually confirmed with a post-hoc analysis based on Fisher's Least Significant Difference (FLSD).

Mode also has a significant effect on the perceived quality. Sphericity assumptions were violated in this case (as confirmed with a Mauchly's test, $p = .039$), so degrees of freedom and p-values were corrected using Greenhouse-Geisser ϵ estimates of sphericity $[F(1.58, 31.52) = 3.57, p < .001, \eta_G^2 = .284]$. These results are summarized in the right panel of Fig. 5. The FLSD analysis in this cases, revealed a similar situation as before: No significant differences between Chord and Sequential presentation, and silence rated worse than them.

Finally, none of the participants preferred the silent presentation. 13 (out of the 21 participants) preferred the sequential presentation, but the difference with the chord presentation is not significant, as determined with a 1-sample proportion test with continuity correction $[\chi^2(1) = 0.8, p = .200]$.

7 Discussion and Concluding Remarks

We have limited the implementation of our solution to two cellular automata: Life, and sand pile models. The results obtained from these auralizations indicate that adding sound to traditional visualizations are beneficial for the exploration of patterns within these models, in terms of immersion and perceived quality. Auralization in our case is limited by the number of audio sources that Unity can render simultaneously. Our results show that the implemented spatializer was able to handle the limited number of audio sources without audio drop-outs. In complex models, such as a sand pile model initialized with a great number of grains in one or several cells, the sequential display prevent possible dropouts from happening. Compared to traditional Game of Life models featuring hundreds or thousands of cells, our proposal seems rather limited. However, we found our small grid sufficient for generative music, with interesting rhythmical possibilities. We are now working on offering greater freedom in the mapping of acoustic features to cell properties by means of an editor. This editor should also offer some templates (e.g., using simple-, AM-, FM-tones, etc.) as starting points for novice users.

We are also interested in including other cellular automata and extending them to three-dimensions when required. Such is the case of Langton's ants [12], Brian's Brain [20], Physarum machines [9], etc. In the case of Langton's ant, the use of the periodic mode will ensure that the tool zooms into the most recent region and keeping the size of the simulation manageable.

References

1. Arévalo, C., Villegas, J.: Compressing head-related transfer function databases by Eigen decomposition. In: IEEE International Workshop on Multimedia Signal Processing (MMSP), Tampere, Finland, September 2020
2. Bak, P., Tang, C., Wiesenfeld, K.: Self-organized criticality: an explanation of the 1/f noise. Phys. Rev. Lett. **59**, 381–384 (1987)
3. Barczak, A., Woźniak, H.: Comparative study on game engines. Studia Informatica. Syst. Inf. Technol. **1**(23), 5–24 (2019)
4. Bays, C.: Candidates for the game of life in three dimensions. Complex Syst. **1**(3), 373–400 (1987)
5. Cohen, M., Villegas, J.: Applications of audio augmented reality. Wearware, everyware, anyware, and awareware (chap. 13). In: Fundamentals of Wearable Computers and Augmented Reality , 2nd edn., pp. 309–329. CRC Press (2015)
6. Delahaye, J.P.: Le royaume du jeu de la vie (the kingdom of the game of life). Pour la Science **378**, 86–91 (2009). (in French)
7. Demidov, E.: CA3D (2000). https://grelf.net/gr3d/ca3d.html. Accessed 11 Mar 2021
8. Demidov, E.: Kaleidoscope of 3D Life (2000). https://www.ibiblio.org/e-notes/Life/Game.htm. Accessed 11 Mar 2021
9. Evangelidis, V., Jones, J., Dourvas, N., Tsompanas, M.A., Sirakoulis, G.C., Adamatzky, A.: Physarum machines imitating a roman road network: the 3D approach. Sci. Rep. **7**(1), 7010 (2017)

10. Gardner, M.: The fantastic combinations of John Conway's new solitaire game "life". Sci. Am. **223**, 120–123 (1970)
11. Hermann, T., Hunt, A., Neuhoff, J.G.: The Sonification Handbook. Logos Verlag Berlin, Berlin (2011)
12. Langton, C.G.: Studying artificial life with cellular automata. Physica D Nonlinear Phenom. **22**(1–3), 120–149 (1986)
13. Lawrence, M.A.: ez: Easy Analysis and Visualization of Factorial Experiments (2016). https://CRAN.R-project.org/package=ez, r package version 4.4-0
14. Mathworks: Matlab. Software (2020). www.mathworks.com. Accessed 11 Mar 2021
15. Polotti, P.: Sound to Sense, Sense to Sound: A State of the Art in Sound and Music Computing. Logos Verlag Berlin GmbH, Berlin (2008)
16. Qu, T., Xiao, Z., Gong, M., Huang, Y., Li, X., Wu, X.: Distance-dependent head-related transfer functions measured with high spatial resolution using a spark gap. IEEE Trans. Audio Speech Lang. Process. **17**(6), 1124–1132 (2009)
17. R Core Team: R: A Language and Environment for Statistical Computing. R Foundation for Statistical Computing, Vienna, Austria, version 4.0.3 (2021). http://www.R-project.org/. Accessed 11 Mar 2021
18. Rendell, P.: Turing Machine Universality of the Game of Life. ECC, vol. 18. Springer, Cham (2016). https://doi.org/10.1007/978-3-319-19842-2
19. Roginska, A., Childs, E., Johnson, M.K.: Monitoring real-time data: a sonification approach. In: International Conference on Auditory Display (ICAD). Georgia Institute of Technology (2006)
20. Rucker, R.: The Lifebox, the Seashell, and the Soul: What Gnarly Computation Taught Me About Ultimate Reality, the Meaning of Life, and How to Be Happy, 2nd edn. Transreal Books (2016)
21. Summers, J.E.: What exactly is meant by the term "auralization?". J. Acoust. Soc. Am. **124**(2), 697–697 (2008)
22. Unity Technologies: Unity, v. 2020.1.6f1. https://unity.com. Accessed 11 Mar 2021
23. Unity Technologies: Unity - manual: Profiler overview. https://docs.unity3d.com/2020.1/Documentation/Manual/Profiler.html. Accessed 19 Nov 2020
24. Von Neumann, J., et al.: The general and logical theory of automata, pp. 1–41 (1951)

Chord Embeddings: Analyzing What They Capture and Their Role for Next Chord Prediction and Artist Attribute Prediction

Allison Lahnala[1(✉)], Gauri Kambhatla[1], Jiajun Peng[1], Matthew Whitehead[2], Gillian Minnehan[1], Eric Guldan[1], Jonathan K. Kummerfeld[1], Anıl Çamcı[3], and Rada Mihalcea[1]

[1] Department of Computer Science and Engineering, University of Michigan, Ann Arbor, MI 48109, USA
{alcllahn,gkambhat,pjiajun,gminn, eguldan,jkummerf, mihalcea}@umich.edu

[2] School of Information, University of Michigan, Ann Arbor, MI 48109, USA
mwwhite@umich.edu

[3] Department of Performing Arts Technology, University of Michigan, Ann Arbor, MI 48109, USA
acamci@umich.edu

Abstract. Natural language processing methods have been applied in a variety of music studies, drawing the connection between music and language. In this paper, we expand those approaches by investigating *chord embeddings*, which we apply in two case studies to address two key questions: (1) what musical information do chord embeddings capture?; and (2) how might musical applications benefit from them? In our analysis, we show that they capture similarities between chords that adhere to important relationships described in music theory. In the first case study, we demonstrate that using chord embeddings in a next chord prediction task yields predictions that more closely match those by experienced musicians. In the second case study, we show the potential benefits of using the representations in tasks related to musical stylometrics.

Keywords: Chord embeddings · Representation learning · Musical artificial intelligence

1 Introduction

Natural language processing (NLP) methods such as classification, parsing, or generation models have been used in many studies on music, drawing the connection that music is often argued to be a form of language. However, while word embeddings are an important piece of almost all modern NLP applications, embeddings over musical notations have not been extensively explored. In this paper, we explore the use of *chord embeddings* and argue that it is yet

J. Romero et al. (Eds.): EvoMUSART 2021, LNCS 12693, pp. 171–186, 2021.
https://doi.org/10.1007/978-3-030-72914-1_12

another NLP methodology that can benefit the analysis of music as a form of language. Our objectives are (1) to probe embeddings to understand what musical information they capture, and (2) to demonstrate the value of embeddings in two example applications.

Using `word2vec` [18] to create embeddings over chord progressions, we first perform qualitative analyses of chord similarities captured by the embeddings using Principal Component Analysis (PCA). We then present two case studies on chord embedding applications, first in next-chord prediction, and second in artist attribute prediction. To show their value, we compare models that use chord embeddings with ones using other forms of chord representations.

In the next chord prediction study, we provide a short chord sequence and ask what the next chord should be. We collected human annotations for the task as a point of reference. By comparing model predictions with the human annotations, we observe that models using chord embeddings yield chords that are more similar to the predictions of more experienced musicians. We also measure the system's performance on a larger set drawn from real songs. This task demonstrates a use case for chord embeddings that involves human perception, interaction, and composition. For the artist attribute prediction study, we perform binary classification tasks on artist type (solo performer or group), artist gender (when applicable), and primary country of the artist. Results on these tasks demonstrate that chord embeddings could be used in studies of musical style variations, including numerous studies in musicology.

This paper contributes analyses of the musical semantics captured in *chord2vec* embeddings and of their benefits to two different computational music applications. We find that the embeddings encode musical relationships that are important in music theory such as the *circle-of-fifths* and relative major and minor chords. The case studies provide insight into how musical applications may benefit from using chord embeddings in addition to NLP methods that have previously been employed.

2 Related Work

Methods for learning word embeddings [4,18,22,23] have been useful for domains outside of language (e.g., network analysis [7]). Recent work has explored embeddings for chords, including an adaptation of *word2vec* [15], their use in a chord progression prediction module of a music generation application [3], and for aiding analysis and visualization of musical concepts in Bach chorales [24]. However, understanding the musical information captured as latent features in the embeddings has been limited by the decision to ground evaluation in language modeling metrics (e.g., perplexity) rather than analyses of their behavior in downstream tasks. In this work, our first case study shows that language models with no remarkable differences in performance by perplexity exhibit remarkable relationships in their predictions to the experience of musicians, and furthermore we provide insights into what is captured by the embeddings.

NLP methods have benefited computational musicology topics such as authorship attribution [30], lyric analysis [6], and music classification tasks using audio and lyrics [16]. In the task of composer identification, many approaches draw inspiration from NLP, applying musical stylometry features and melodic n-grams [2,8,31]. One study used language modeling methods on musical n-grams to perform composer recognition [10] and another used a neural network over encodings of the pitches of musical pieces [12]. Similar methods have been used to study stylistic characteristics of eight jazz composers [1,19] over chord sequences similar to ours. While these studies operated on small datasets (on the order of hundreds of samples) to identify and analyze music of a small set of musicians, we use a large dataset (on the order of tens of thousands) and predict attributes of artists based on the music.

Our attribute prediction tasks are related to NLP work in authorship attribution and are motivated by studies on the connection between language and music in psychology [11,21], and the intersection of society, music, and language, or sociomusicology [5,28]. For instance, Sergeant & Himonides studied the perception of gender in musical composition, and found no significant match between the listener's guess of a composer's gender and their actual gender [27].

3 Dataset

We compile a dataset of 92,000 crowdsourced chord charts with lyrics from the Ultimate Guitar website.[1] We identify and remove duplicate songs[2] in our data using Jaccard similarity, then extract the chord progressions from our data representation to learn chord embeddings.

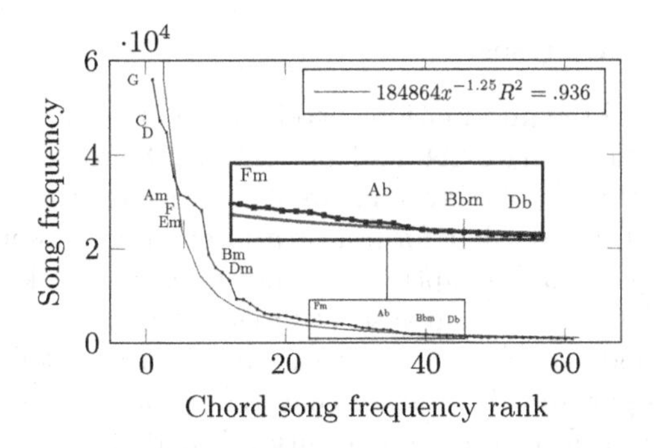

Fig. 1. Number of chords in the dataset and a fitted trendline with the parameters given in the figure for the 61 most common chords, showing a power law distribution. A sample of chords are labeled that also appear in the embedding visualization in Fig. 2.

We remove songs with fewer than six chords, leaving us with a final set of 88,874 songs and 4,913 unique chords. The chords' song frequencies are distributed according to a power law distribution, much like Zipf's word frequency

[1] https://www.ultimate-guitar.com/.

[2] Sometimes multiple users submit chord charts for a song.

law exhibited in natural language. Figure 1 shows the song frequency and the power law trend line fitted to the top 61 chords for demonstration, though the trend is even stronger over all chords. Beyond there existing many possible chord notations, we observe a relation between chord frequency and physical difficulty of playing the chord on the guitar or other instruments (e.g., `G`, `C`, and `D`), and variations in notation.

4 Chord Embeddings

Word embeddings have been used extensively to represent the meaning of words based on their use in a large collection of documents. We consider a similar process for chords, forming representations based on their use in a large collection of songs.

We create chord embeddings for chords that appear in at least 0.1% of our songs (237 chords) using the continuous bag of words model ($\texttt{CE}_{cbow}$) and the *skip-gram* language model ($\texttt{CE}_{sglm}$) from **word2vec** [18], a widely used method for creating word embeddings. For $\texttt{CE}_{cbow}$, a target chord is predicted based on context chords, while $\texttt{CE}_{sglm}$ is the reverse: context chords are predicted given a target chord. In both cases, this has the effect of learning representations that are more similar for chords that appear in similar contexts. We tested context sizes of two, five, and seven, and varied the vector dimensions between 50, 100, 200, and 300, but observed only minor differences across different models, and chose to use a context window of five and a vector dimension of 200.

4.1 Qualitative Analyses

To better understand the information encoded in chord embeddings, we perform a qualitative analysis using PCA and present a 2D projection for the $\texttt{CE}_{sglm}$ model in Fig. 2 (our main observations are consistent for the case of $\texttt{CE}_{cbow}$).

We observe that chords that form a fifth interval are closer together, which suggests that the embeddings capture an important concept known as the *circle of fifths*. Fifth-interval relationships serve broad purposes in tonal music (music that adheres to a model of relationships to a central tone). Saker [26] encapsulates their structural importance by stating that "circle of fifths relationships dominate all structural levels of tonal compositions" and that "the strongest, most copious harmonic progressions to be found in tonal music are fifth related." The *circle of fifths* relationship is observed in our chord embeddings over different chord *qualities*,[3] specifically over major chords (highlighted in Fig. 2b), minor chords (highlighted in Fig. 2c), major-minor 7 chords, and minor 7 chords. For both chord qualities, the layout captured by the chord embeddings is similar to the ideal/theoretical circle of fifths, illustrated in Fig. 2a. This pattern is particularly interesting as it does not follow the style of word analogy patterns observed

[3] *Qualities* refers to sound properties that are consistent across chords with different roots but equidistant constituent pitches. The interaction of intervals between pitches determines the quality.

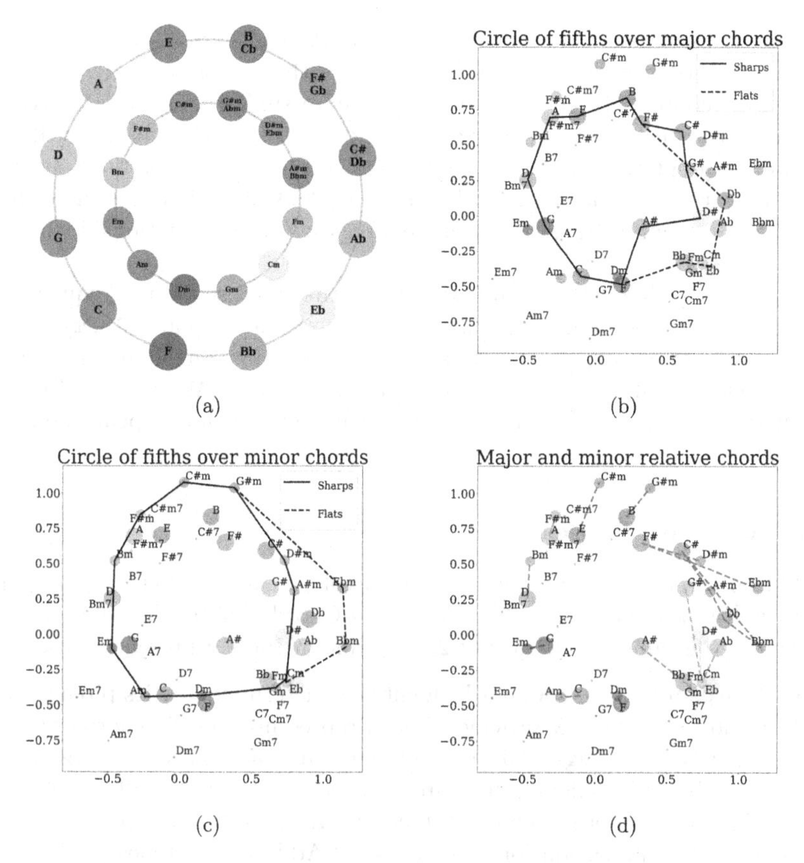

Fig. 2. In (a), we show the circle of fifths with the same colors as in (b), (c) and (d), which show the same 2-dimensional PCA projection of the chord embedding space with lines denoting the circle of fifths over major chords (b) and minor chords (c), and lines denoting major-minor relatives (d).

in language. This makes sense, as the "is-a-fifth" relation forms a circle in chords, whereas word analogies connect pairs of words without forming a circle.

Additionally, we observe that relative major and minor chords[4] appear relatively close together in the embedding space, as shown by their proximity in the PCA plots (highlighted in Fig. 2d). We also observe that enharmonics, notes with different names but the same pitch, are often close together. Not only that, but there is a consistent pattern in the positioning of enharmonics, with sharps to the left and flats to the right.

These observations suggest that chord embeddings are capable of representing musical relationships that are important to music theory. Transitions between

[4] *Relative* refers to the relation between the chords' roots, in which the scale beginning on the minor chord's root shares the same notes as the scale beginning on the major chord's root, but the ordering of the notes give different qualities to the scales.

the tonic (I), dominant (V), and subdominant (IV) chords of a scale are prescriptive components in musical cadences [25]. Since these chords frequently appear in the same context, their embeddings are more similar. A common *deceptive cadence* is a transition between the fifth and sixth root chords of a scale [20]. An example progression with a deceptive cadence in C major is C-F-G-Am; these chords appear in a similar neighborhood in the PCA plots. Because these chords frequently co-occur in music, the embeddings capture a relationship between them.

We also note that relationships for chords that are used more frequently are more strongly represented. The major and minor relative pairs (G, Em), (C, Am), and (D, Bm), are among the top ten chords ranked by song frequency (Fig. 1) and have clear relations in Fig. 2. In contrast, the pairs (Ab, Fm) and (Db, Bbm) are ranked lower and their minor-major relative relationship appears weaker by their distance.

4.2 Alternative Representations

In addition to chord embeddings, we also explore two other chord representations: Pitch Representations (PR) and Bag-of-Chords (BOC). For a fair comparison, we use the same vocabulary of 237 chords for these representations.

Pitch Representations. A chord's identity is determined by its pitches, so we test if the individual pitches provide a better representation of a chord as a whole than our chord embeddings. This method represents each chord by encoding each of its pitches by their index in the chromatic scale $\{\texttt{C} = 1, \texttt{C\#} = 2, \cdots, \texttt{B} = 12\}$. The pitches are in order of the triad, followed by an additional pitch if marked, and by one extra dimension for special cases.[5] Additional pitches are indicated by the interval relative to the root of the pitch that is being added. We also represent chord inversions, e.g., the chord G/B which is an inversion of G such that the bottom pitch is B.

Bag-of-Chords. In this representation, each chord is represented as a "one-hot" vector, where the vectors have length equal to the vocabulary size. We consider two ways of determining the value for a chord. For $\texttt{BOC}_{count}$, we use the frequency of a chord in a song, divided by the number of chords in the song. For $\texttt{BOC}_{tfidf}$, we use the TF-IDF of each chord (term-frequency, inverse document frequency).

5 Case Study One: Next Chord Prediction

In this section, we present our first case study, which investigates if there is a relationship between chord embedding representations and the ways humans perceive and interact with chords. We test the use of chord representations for

[5] Special cases include: the "*" marking on a chord, which is a special marker specific to the ultimate-guitar.com site; "UNK" which we use to replace chords that do not meet the 0.1% document frequency threshold; and "H" and "Hm" which indicates "hammer-ons" in the notation on ultimate-guitar.com.

predicting the most likely next chord following a given sequence of chords, and then compare these to human-annotated responses.

We train a next chord prediction model using a *long short-term memory* model [9] (LSTM).[6] We follow standard practice and do not freeze the embeddings, meaning the chord representations undergo updates throughout training, adjusting to capture the musical features most important for the task. Our main model uses the pre-trained chord embeddings to initialize the chord prediction architecture. We test both $\texttt{CE}_{cbow}$ and $\texttt{CE}_{sglm}$ embeddings, and will refer to these models by these acronyms. We also define a baseline model, where the encoder is randomly initialized (denoted `NI`, for no initialization). Finally, we also evaluate a model where we initialize the encoder with the pitch representations introduced in Sect. 4.2 (denoted `PR`).

We divide our data into three sets, with 69,985 songs (80%) for training, 8,748 songs (10%) for validation, and 8,748 (10%) for testing. We train using a single GPU with parameters: epochs = 40, sequence length = 35, batch size = 20, dropout rate = 0.2, hidden layers = 2, learning rate = 20, and gradient clipping = 0.25.

5.1 Human Annotations

To evaluate the next chord prediction models, we collect data with a human annotation task in which annotators are asked to add a new chord at the end of a chord progression. For example, given the progression "`A`, `D`, `E`," they must pick a chord that would immediately follow `E`. They are also asked to pick one or two alternatives to this selection. Continuing the example, if an annotator provides `E7` and `A`, then the chord progressions they have specified are "`A`, `D`, `E`, `E7`," and "`A`, `D`, `E`, `A`." The annotators are given a tool to play 48 different chords (all major, minor, major-minor 7, and minor 7 chords) so they can hear how different options would sound as part of the progression.[7] They were given a total of 39 samples shown in the same order to all annotators. The samples were chosen randomly from our entire dataset of songs, permitting only one sequence to come from a single song, and requiring they contain the same 48 chords. We presented sequences of length three and six as we expect that patterns in the given sequence affect the responses.

Participants. The annotators were first asked to estimate their expertise in music theory on a scale from 0–100, where 0 indicates no knowledge of music theory, 25–75 indicates some level of knowledge from pre-university training though self-teaching, private lessons/tutoring, or classroom settings, and 75–100 indicates substantial expertise gained by formal university studies, performing and/or composing experience. They were given the option to provide comments about how they estimated their expertise. We collected this information because we expected that the annotations provided by a participant may vary depending

[6] We use an open-source repository of neural language models https://github.com/pytorch/examples/blob/master/word_language_model/model.py.

[7] We did not limit our next chord prediction models to these 48 chords.

on their background education in music theory. It also allows us to perform comparisons of our system with sub-groups defined by self-reported knowledge. Nine participants provided complete responses, with expertise ratings of 0, 0, 10, 10, 19, 25, 25, 50 and 73. For the following analyses, we define a *beginner* set containing annotators who provided 0 for their self-rating, an *intermediate* set containing annotators with ratings >0 and <50, and an *expert* set containing annotators whose ratings are at least 50.

Inter-annotator Agreement. For pairwise agreement, we compute the proportion of chord progressions in which a pair of annotators provided the same chord, averaged over all pairs of annotators. The pairwise agreement across all annotators is 22.51, it is 23.08 for the beginner set, 25.38 for the intermediate set, and 17.95 for the expert set.

To account for responses of similar but not identical chords (discussed in Sect. 5.2), we measure pairwise agreement on response pitches. We compute the fraction of matching pitches for a pair of annotators' responses for a given chord progression, averaged over all pairs of annotators. The pairwise pitch agreement score for all annotators is 38.00, it is 37.01 for the beginner set, 40.90 for the intermediate set, and 33.59 for the expert set. The average number of unique chords used by each annotator is 30.2, it is 35.5 for the beginner set, 27.4 for intermediate set, and 32.0 for the expert set.

5.2 Evaluation Metrics

The main objective of this case study is to investigate whether chord similarities captured by our embeddings reflect human-perceived similarities. We use the chord-prediction systems to perform the same task given to the annotators. Each model provides a probability distribution over the full set of chords, therefore we treat the chords with highest probabilities as each model's selection.

We evaluate the predictions with the following metrics, which are inspired by the metrics employed by the Lexical Substitution task [17] but modified for our setup, which weight more frequent responses higher:

$Match_{best}$: For each example we calculate the fraction of people who included the model's top prediction in their answer. These values are then averaged over all examples.

$Match_{oo4}$: This adds together values for the previous metric across the model's top four predictions.

$Mode_{best}$: The fraction of cases in which the top model prediction is the same as the most common annotator response, when there is a single most common response.

$Mode_{oo4}$: The fraction of cases in which one of the top four model predictions is the same as the most common annotator response.

Note that only 25 out of the 39 examples had a unique most common response. Of these, 20 had a chord chosen by three or four annotators, and five had a chord chosen by five to seven annotators. The rest of the examples are not considered in the *Mode* metrics.

Pitch Matches: The metrics above penalise all differences in predictions equally, even though some chord pairs are more different than others, e.g., A7 and A differ only in the addition of a pitch, whereas B and A share no pitches. To address this, we use a metric that is the total number of pitches that match for each question between the model's top response and the annotator's first response. We calculate this separately per-annotator, and then average across annotators (PM_{ave}).

Loss and *Perplexity*(*PPL*): These are two measures from the language modeling literature that we apply to see how well the models do on the true content of songs. Note that this evaluation is over a different set: 8,748 randomly chosen songs that are not included in model training.

Table 1. Results for all models when compared with different sets of annotators based on their expertise.

	Match		*Mode*		*PM*
	best	*oo4*	*best*	*oo4*	*ave*
All					
NI	7.52	30.10	**32.00**	64.00	48.33
PR	7.49	29.64	24.00	**72.00**	**51.11**
CE_{cbow}	7.45	29.82	16.00	64.00	47.78
CE_{sglm}	**7.60**	**30.22**	12.00	68.00	49.00
Intermediate					
NI	8.17	32.68	**33.33**	**71.43**	48.80
PR	8.19	32.42	23.81	**71.43**	**53.60**
CE_{cbow}	8.04	32.14	14.29	66.67	47.20
CE_{sglm}	**8.34**	**33.19**	14.29	**71.43**	50.00

	Match		*Mode*		*PM*
	best	*oo4*	*best*	*oo4*	*ave*
Beginner					
NI	**5.61**	**22.44**	0.00	**44.44**	40.50
PR	**5.61**	**22.44**	**11.11**	33.33	**43.00**
CE_{cbow}	4.97	19.87	0.00	22.22	39.50
CE_{sglm}	5.29	21.15	0.00	33.33	42.00
Expert					
NI	7.92	31.67	28.57	**85.71**	55.00
PR	7.72	30.26	28.57	**85.71**	53.00
CE_{cbow}	**8.56**	**34.23**	**42.86**	**85.71**	**57.50**
CE_{sglm}	8.15	32.18	14.29	57.14	53.50

5.3 Results

Match and Mode. Table 1 reports metrics for each system when compared with the beginner set, intermediate set, expert set, and full set of annotators. When evaluating against all annotators, CE_{sglm} is best in $Match_{best}$ and $Match_{oo4}$, NI is best in $Mode_{best}$, and PR is best in $Mode_{oo4}$.

In the results for subsets of annotators, all systems tend to match experts better than the beginner or intermediate groups. In particular, CE_{cbow} obtains the lowest scores when evaluated against the beginner group, but the highest when evaluated against the expert group. To investigate this pattern, we compute $Match_{best}$ and $Match_{oo4}$ for each individual annotator and then the Spearman Correlation Coefficient r_s between these metrics and their expertise. In Table 2, we observe strong significant correlation between the $\mathtt{CE}_{cbow}$ and expertise, and no significant correlation for the other models. This can be explained by the fact that the models are trained on a large collection of songs composed by experts, and the chord embeddings seem to capture the chord use style of the experts.

Pitch Matches. We report PM_{ave} for the models and expertise groups in Table 1. Similarly to the *Match* and *Mode* metrics, we observe that all models perform better when compared to annotators with higher expertise, and the differences between groups is most extreme with $\mathtt{CE}_{cbow}$. Table 2 shows the results of correlation analysis for pitch matches (Pearson Correlation Coefficient r_p), with a significant linear correlation for only $\mathtt{CE}_{cbow}$. This trend is visually shown in Fig. 3.

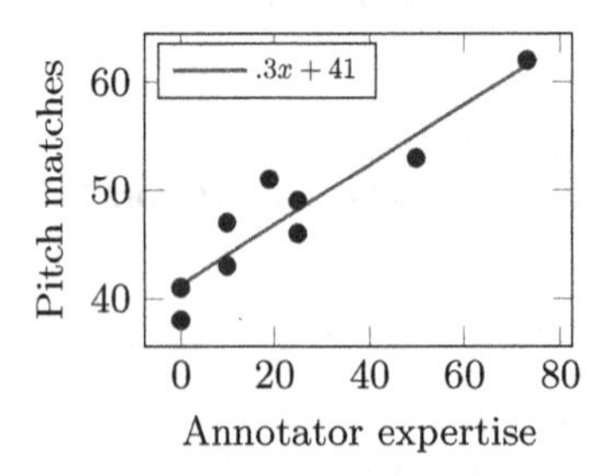

Fig. 3. Total pitch matches between annotator and output of $\mathtt{CE}_{cbow}$ model, plotted by the annotator's expertise. The red line is the line of best fit computed by linear regression. (Color figure online)

Table 2. Correlation coefficients for expertise and Match and Pitch Match metrics.

	best r_s	p-val	*oo4* r_s	p-val	PM r_p	p-val
NI	0.44	0.24	0.44	0.24	0.71	0.03
PR	0.64	0.06	0.57	0.11	0.43	0.25
$\mathtt{CE}_{cbow}$	**0.72**	**0.03**	**0.72**	**0.03**	**0.94**	**2e−4**
$\mathtt{CE}_{sglm}$	0.41	0.28	0.41	0.28	0.42	0.27

Table 3. Loss and perplexity metrics for the chord prediction models on a held-out test set.

	Test loss	Test PPL
NI	1.42	4.16
PR	1.44	4.20
$\mathtt{CE}_{cbow}$	1.44	4.22
$\mathtt{CE}_{sglm}$	1.42	4.15

Loss and Perplexity. Table 3 shows the results on our 8,748 song test set. All models perform similarly in this setting.

5.4 Discussion

We observe that automatic models produce predictions that resemble human responses. Comparing against annotators grouped by expertise, $\mathtt{CE}_{cbow}$ compares best to the high expertise group for all metrics, while the best model varies among the other groups. $\texttt{CE}_{cbow}$'s predictions also correlate significantly with pitch matches and annotator expertise. NI and PR achieve the highest $Mode_{best}$ and $Mode_{oo4}$ scores, however, fewer samples are considered because only twenty-five had a unique *mode* chord across the annotators and only five of these samples had more than half the annotators agree. Additionally, the chord symbol based metrics are strict, requiring an exact match on chords, and had lower interannotator agreement than the pitch-based metrics.

While $\texttt{CE}_{cbow}$'s predictions exhibit a strong pitch-match correlation, $\texttt{CE}_{sglm}$'s predictions exhibit no significant correlate at all. However, differences between the $\texttt{CE}_{cbow}$ and $\texttt{CE}_{sglm}$ embeddings may not be as apparent in other downstream applications; in fact, by the perplexity and test loss metrics shown in Table 3, there is barely a difference between these two, or any, models. Investigating key differences between these embedding models in musical contexts is a direction for future work.

6 Case Study Two: Artist Attribute Prediction

Our second case study introduces the task of performing artist attribute prediction, demonstrating that these chord representations could be used more broadly in tasks involving musical stylometry and musical author profiling. With binary classifiers using our chord representations, we predict three attributes as separate tasks: gender (male or female), performing country (U.S. or U.K.), and type of artist (group or solo artist).

Data. For these experiments, we augment the dataset with information obtained with the MusicBrainz API,[8] which includes the song artist's location, gender, lifespan, tags that convey musical genres, and other available information for 35,351 English songs (identified using Google's Language Detection project [29]). From this extracted information, we choose artist type, performing country, and gender because of the sufficient quantity of data available with these attributes enabling the tasks; we note that tasks dedicated to genre or time period are of interest for future investigations, and our preliminary experiments using the artists' lifespan and tags as proxies for time period and genre indicated these tasks are promising use cases for chord embeddings.

We use the top two most frequent classes of each attribute, and balance the data to have the same number of examples for each class. For artist type, there are 20,000 songs per class (group and solo). For performing country, there are 8,000 songs per class (U.S. and U.K.). For gender, there are 6,000 songs per class (male and female). The number of samples varies because of differences in the raw class counts and because not all songs have a label for each property.

Experimental Setup. We build two binary classifiers and compare their performance with each chord representation. The first uses logistic regression (LR) over a single vector for each song by aggregating chord representations. We

[8] https://musicbrainz.org/.

also experimented with an SVM classifier, but LR was more efficient with minimal performance trade-offs. The BOC methods are defined in Sect. 4.2. The PR method aggregates the chords with a many-hot encoding vector counting each chord pitch, normalized by the total number of pitches. The CE methods aggregate chord embeddings by max-pooling, taking the most extreme absolute value in each dimension across all chords.

The second classifier is a Convolutional Neural Network (CNN) that considers the chords in sequences. We experimented with an LSTM, and found that the CNN functions better for these tasks. We use the CNN model for sentence classification by Kim [13][9] over the chord progressions for each song, using the same NI, PR, CE_{sglm}, and CE_{cbow} representations from the first case study.

For our model parameters, chosen in preliminary experiments on a subset of the data, we use L2 regularization for the LR classifier, and the CNN model uses filter window sizes 3, 4, 5 with 30 feature maps, drop-out rate 0.5 and Adam [14] optimization. The sequence limit is 60 chords, cutting off extra chords and padding when there are fewer.

Table 4. Accuracy scores from 10-fold cross validation in artist gender, country, and type prediction tasks. The significance tests are performed among the logistic regression models and CNN models separately.

Model	Gender	Country	Artist Type
	Logistic regression		
BOC_{count}	*†‡**57.53**	†55.71	*†**57.04**
BOC_{tfidf}	†55.94	†55.17	†56.06
PR	52.32	53.13	53.75
CE_{cbow}	*†56.90	†55.37	*†56.92
CE_{sglm}	†56.37	†**55.85**	*†56.88

Model	Gender	Country	Artist Type
		CNN	
NI	†58.93	†56.79	†59.20
PR	57.98	55.31	57.54
CE_{cbow}	58.67	†57.30	†58.92
CE_{sglm}	†**58.95**	†**57.54**	†**59.29**

$^*p < 0.05$ over BOC_{tfidf}, $\dagger p < 0.05$ over PR, $\ddagger p < 0.05$ over CE_{sglm}, $\S p < 0.05$ over NI

6.1 Results and Analyses

Table 4 shows the models' accuracy scores from experiments using 10-fold cross validation. CE_{sglm} CNN is the top performer for all tasks, significantly outperforming CNN PR and all LR models for all tasks, and NI for country.[10] All models outperform a random baseline (50%) significantly in each task. In each task, CNN models significantly outperform their LR counterparts.

For insight into the models' performance, we analyze the gender prediction task, the only attribute where the LR CE_{cbow} and LR BOC_{count} predictions differed significantly. First, we compare the rate of use of each chord between genders. To show the differences in Fig. 4, we divide the higher rate by the lower rate, subtract

[9] CNN model is built on https://github.com/Shawn1993/cnn-text-classification-pytorch.

[10] By a paired t-test with $p < .05$.

one to set equal use to zero, and flip the sign when female use is higher. We observe greater variations among chords with lower song frequency. For instance, `C/G`, `F#7`, `Bbm`, and `Ab` are twice as salient for one gender than the other. The highest variation among the top 20 chords reaches 1.5 times more salient, and for the top 10, 1.2 times more salient for one gender.

To investigate the impacts of the musical relationships captured in embeddings (Sect. 4) to the `CE` models, we also compared use of five *chord qualities.* Figure 5 shows higher relative frequency of suspended and diminished chords among the songs of male artists, augmented and minor chords among the female artist songs, and fairly similar use of major chords.

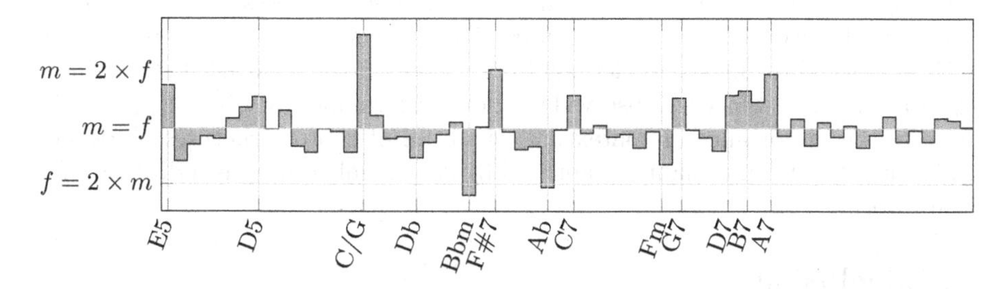

Fig. 4. Chord variations by gender, ordered from least to greatest by song frequency. The labeled chords are at least 1.5 more salient for one gender than the other.

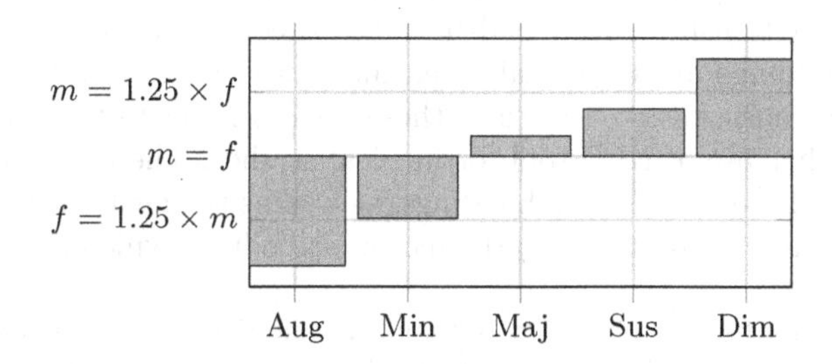

Fig. 5. Variation in chord quality usage by gender by ratio of use percentage.

6.2 Discussion

To our knowledge, this is the first time that an association between chord representations and author attributes has been explored. Each model for each attribute showed significant improvement over a random baseline of 50%, indicating there are quantifiable differences in our music data between the genders, countries, and artist types. In addition to the tasks we presented, we also observed improvement when using the system to predict the life-period of the artist and their associated genres. As life-period is a proxy for the music's time period, chord embeddings could benefit future work in musical historical analysis.

For gender, the variation of rare chords may contribute to $\texttt{BOC}_{count}$ outperforming $\texttt{CE}_{cbow}$. However, $\texttt{CE}_{cbow}$ significantly outperforms $\texttt{BOC}_{tfidf}$ which gives more weight to rare chords by their inverse-document frequency. This suggests that chord rarity is not the only critical feature. The variations of chord quality use may contribute to the performance of the CE models as the embeddings capture musical relationships.

LR PR consistently underperforms all other models, which may indicate the importance chord structures. Different chords with the same pitches (e.g., G and G/B) have the same PR vector. Chords with overlapping pitches have similar PR vectors. However, CNN PR, which performs closer to the others, encodes pitch orderings (Sect. 4.2) and BOC methods encode chord symbols which indicate structure. CE representations are learned from chord symbols, likely capturing contextual functions of chord structures. These functions would matter for the CNN models which make predictions from chord sequences rather than a single aggregated vector. Since we observed the best performance by CNN $\texttt{CE}_{sglm}$, there suggests the importance of contextual semantics of chord structures. A deeper study into structural semantics captured by chord embeddings is a direction for future work.

7 Conclusion

In this paper, we presented an analysis of the information captured by chord embeddings and explored how they can be applied in two case studies. We found that chord embeddings capture chord similarities that are consistent with important musical relationships described in music theory. Our case studies showed that the embeddings are beneficial when integrated in models for downstream computational music tasks. Together, these results indicate that chord embeddings are another useful NLP tool for musical studies. The code to train chord embeddings and the resulting embeddings, as well as the next-chord annotations are publicly available from https://lit.eecs.umich.edu/downloads.html.

Acknowledgements. We would like to thank the anonymous reviewers and the members of the Language and Information Technologies lab at Michigan for their helpful suggestions. We are grateful to MeiXing Dong and Charles Welch for helping with the design and interface of the next-chord annotation task. This material is based in part upon work supported by the Michigan Institute for Data Science, and by Girls Encoded and Google for sponsoring Jiajun Peng through the Explore Computer Science Research program. Any opinions, findings, and conclusions or recommendations expressed in this material are those of the authors and do not necessarily reflect the views of the Michigan Institute for Data Science, Girls Encoded or Google.

References

1. Absolu, B., Li, T., Ogihara, M.: Analysis of chord progression data. In: Raś, Z.W., Wieczorkowska, A.A. (eds.) Advances in Music Information Retrieval. Studies in Computational Intelligence, vol. 274, pp. 165–184. Springer, Berlin (2010). https://doi.org/10.1007/978-3-642-11674-2_8

2. Brinkman, A., Shanahan, D., Sapp, C.: Musical stylometry, machine learning and attribution studies: a semi-supervised approach to the works of Josquin. In: Proceedings of the Biennial International Conference on Music Perception and Cognition, pp. 91–97 (2016)
3. Brunner, G., Wang, Y., Wattenhofer, R., Wiesendanger, J.: JamBot: music theory aware chord based generation of polyphonic music with LSTMs. In: Proceedings of ICTAI (2017)
4. Devlin, J., Chang, M.W., Lee, K., Toutanova, K.: BERT: pre-training of deep bidirectional transformers for language understanding. In: Proceedings of NAACL (2019)
5. Feld, S., Fox, A.A.: Music and language. Annu. Rev. Anthropol. **23**(1), 25–53 (1994)
6. Fell, M., Sporleder, C.: Lyrics-based analysis and classification of music. In: Proceedings of COLING (2014)
7. Grover, A., Leskovec, J.: node2vec: scalable feature learning for networks. In: Proceedings of KDD (2016)
8. Hillewaere, R., Manderick, B., Conklin, D.: Melodic models for polyphonic music classification. In: Second International Workshop on Machine Learning and Music (2009)
9. Hochreiter, S., Schmidhuber, J.: Long short-term memory. Neural Comput. **9**(8), 1735–1780 (1997)
10. Hontanilla, M., Pérez-Sancho, C., Iñesta, J.M.: Modeling musical style with language models for composer recognition. In: Sanches, J.M., Micó, L., Cardoso, J.S. (eds.) IbPRIA 2013. LNCS, vol. 7887, pp. 740–748. Springer, Heidelberg (2013). https://doi.org/10.1007/978-3-642-38628-2_88
11. Jäncke, L.: The relationship between music and language. Front. Psychol. **3**, 123 (2012)
12. Kaliakatsos-Papakostas, M.A., Epitropakis, M.G., Vrahatis, M.N.: Musical composer identification through probabilistic and feedforward neural networks. In: Di Chio, C., et al. (eds.) EvoApplications 2010. LNCS, vol. 6025, pp. 411–420. Springer, Heidelberg (2010). https://doi.org/10.1007/978-3-642-12242-2_42
13. Kim, Y.: Convolutional neural networks for sentence classification. In: Proceedings of the 2014 Conference on Empirical Methods in Natural Language Processing (EMNLP), pp. 1746–1751 (2014)
14. Kingma, D.P., Ba, J.: Adam: a method for stochastic optimization. In: ICLR 2015 (2014)
15. Madjiheurem, S., Qu, L., Walder, C.: Chord2vec: learning musical chord embeddings. In: Proceedings of the Constructive Machine Learning Workshop (2016)
16. Mayer, R., Rauber, A.: Musical genre classification by ensembles of audio and lyrics features. In: Proceedings of ISMIR (2011)
17. McCarthy, D., Navigli, R.: The English lexical substitution task. Lang. Resour. Eval. **43**(2), 139–159 (2009). https://doi.org/10.1007/s10579-009-9084-1
18. Mikolov, T., Chen, K., Corrado, G., Dean, J.: Efficient estimation of word representations in vector space. In: ICLR (2013)
19. Ogihara, M., Li, T.: N-gram chord profiles for composer style representation. In: ISMIR, pp. 671–676 (2008)
20. Owen, H.: Music Theory Resource Book. Oxford University Press, USA (2000)
21. Patel, A.D.: Language, music, syntax and the brain. Nat. Neurosci. **6**(7), 674 (2003)
22. Pennington, J., Socher, R., Manning, C.D.: Glove: global vectors for word representation. In: Proceedings of EMNLP (2014)

23. Peters, M., et al.: Deep contextualized word representations. In: Proceedings of NAACL (2018)
24. Phon-Amnuaisuk, S.: Exploring Music21 and Gensim for music data analysis and visualization. In: Tan, Y., Shi, Y. (eds.) DMBD 2019. CCIS, vol. 1071, pp. 3–12. Springer, Singapore (2019). https://doi.org/10.1007/978-981-32-9563-6_1
25. Randel, D.M.: The Harvard Concise Dictionary of Music and Musicians. Harvard University Press, Cambridge (1999)
26. Saker, M.N.: A theory of circle of fifths progressions and their application in the four Ballades by Frederic Chopin. Ph.D. thesis, University of Wisconsin-Madison (1992)
27. Sergeant, D.C., Himonides, E.: Gender and music composition: a study of music, and the gendering of meanings. Front. Psychol. **7**, 411 (2016). https://doi.org/10.3389/fpsyg.2016.00411, https://www.frontiersin.org/article/10.3389/fpsyg.2016.00411
28. Shepherd, J.: A theoretical model for the sociomusicological analysis of popular musics. Popular Music **2**, 145–177 (1982)
29. Shuyo, N.: Language detection library for java (2010). http://code.google.com/p/language-detection/
30. Stamatatos, E.: A survey of modern authorship attribution methods. J. Am. Soc. Inf. Sci. Technol. **60**(3), 538–556 (2009). https://doi.org/10.1002/asi.21001, https://onlinelibrary.wiley.com/doi/abs/10.1002/asi.21001
31. Wołkowicz, J., Kulka, Z., Kešelj, V.: N-gram-based approach to composer recognition. Arch. Acoust. **33**(1), 43–55 (2008)

Convolutional Generative Adversarial Network, via Transfer Learning, for Traditional Scottish Music Generation

Francesco Marchetti(✉), Callum Wilson, Cheyenne Powell, Edmondo Minisci, and Annalisa Riccardi

Intelligent Computational Engineering Laboratory (ICE Lab),
Department of Mechanical and Aerospace Engineering,
University of Strathclyde, Glasgow, UK
{francesco.marchetti,callum.j.wilson,
cheyenne.powell,edmondo.minisci,annalisa.riccardi}@strath.ac.uk

Abstract. The concept of a Binary Multi-track Sequential Generative Adversarial Network (BinaryMuseGAN) used for the generation of music has been applied and tested for various types of music. However, the concept is yet to be tested on more specific genres of music such as traditional Scottish music, for which extensive collections are not readily available. Hence exploring the capabilities of a Transfer Learning (TL) approach on these types of music is an interesting challenge for the methodology. The curated set of MIDI Scottish melodies was preprocessed in order to obtain the same number of tracks used in the BinaryMuseGAN model; converted into pianoroll format and then used as a training set to fine tune a pretrained model, generated from the Lakh MIDI dataset. The results obtained have been compared with the results obtained by training the same GAN model from scratch on the sole Scottish music dataset. Results are presented in terms of variation and average performances achieved at different epochs for five performance metrics, three adopted from the Lakh dataset (qualified note rate, polyphonicity, tonal distance) and two custom defined to highlight Scottish music characteristics (dotted rhythm and pentatonic note). From these results, the TL method shows to be more effective, with lower number of epochs, to converge stably and closely to the original dataset reference metrics values.

Keywords: Generative Adversarial Network · Transfer Learning · Convolutional Neural Network · Scottish music

1 Introduction

The ability for Artificial Intelligence (AI) to generate music is a challenge that has been taken on by many for various reasons. Music is known to assist humans with emotional comfort and needs and thus the ability for a machine to create such melodies to accomplish that is being looked into with the advancement of technology [18].

J. Romero et al. (Eds.): EvoMUSART 2021, LNCS 12693, pp. 187–202, 2021.
https://doi.org/10.1007/978-3-030-72914-1_13

The first examples dated in 1957 and 1958 are known as the Illiac Suite, which used a process of music composition through sequential record of experiments performed by a computer [21]. The next occurred in 1960, when R. Kh. Zaripov explored the use of a computer for creating short monophonic melodies, with Zaripov later identifying the creation of melodies as being one of the most important aspects of the process [1].

Progress continued throughout the years of 1975 and 1980, as Mark Steedman investigated machine perception of musical rhythms [22] and David Cope's experiments in musical intelligence focused on imitating his own musical style [6]. This paved the way for major companies like Google and Sony to begin development in their own laboratories which have produced a number of computer-generated music compositions. The Google Brain team, Magenta, took two approaches to understand the progress of audio modelling. The first used the WaveNet style auto-encoder that is used on temporal codes by the conditioning of an auto-regressive decoder to learn from the raw audio waveforms. The second was the introduction of a data set consisting of musical notes in the order of a larger scale of similar public data-sets known as NSynth [9]. Sony's technologies on the other hand were used in various ways to create different types of music. The first of these is titled "Daddy's Car", which was created with the inspiration of the Beatles and another by SKYGGE to generate an album that consisted of an AI-human collaboration titled "Hello World" [2].

Generative Adversarial Networks (GANs) are one of the methods used for AI music generation. A GAN is a framework proposed by Goodfellow et al. [11] where two Neural Networks (NNs) are trained simultaneously in an adversarial manner. One network is called the *Generator* and the other is called the *Discriminator*. The Generator creates samples starting from uniformly distributed random data and the Discriminator receives as input either the sample generated by the Generator network or real data. The goal of the Discriminator is to learn whether the data that it receives as input are samples generated by the Generator or are real data. The goal of the Generator is to learn how to "trick" the Discriminator into making it believe that the samples generated by the Generator are real data. In this way, the Generator learns how to produce samples which resemble real data but are completely artificial [14].

GANs are used for several tasks, resulting in different types of GAN specific to each task. A few of these are: cGAN, a type of conditional GAN used to aid computer diagnosis systems in the localisation and detection of prostate tissue on MRI scans [12]. CycleGAN, an image-to-image translation approach where image mapping is learned using image pairs that are aligned [25], in addition to being used for images they have also been applied to the concept of music in the form of a symbolic music representation in MIDI format [4]. FusionGAN, also used for generating music, is a type of fusion framework and an optional use of dual learning that can be implemented on the styles of the provided domains [5]. There is also a style-based GAN architecture (StyleGAN) for image modelling [15] and an Image Super-Resolution GAN (SRGAN) which uses a low-resolution image to estimate a corresponding high-resolution image [16]. Overall, various types of GANs are used for art generation, whether it is a picture, a video, or music. Some examples of their art generation capabilities involve the generation of photographs of human faces [14], the generation of pictures from a text description [24], the prediction of video frames up to a second [23], and music generation [17].

A Binary Multi-track Sequential GAN (BinaryMuseGAN) is used for generating multi-track polyphonic music consisting of multi-track inter-dependency, temporal, and harmonic and rhythmic structures. Two different scenarios were taken to integrate the temporal model; one of these scenarios incorporates the learning of a temporal structure from a human made track, while the other has the ability to make its own music without any human intervention. With the use of these principles, private Generators are used for individual tracks; another is used with a combination of private generators for each track and their inputs that are shared among tracks; and last is another approach with all tracks executed at once with one Generator. When results are combined to view bars, Convolutional Neural Networks (CNNs) are used to translate patterns [7]. In this paper, BinaryMuseGANs will be used throughout for the unsupervised generation of Scottish traditional music patterns.

The paper is divided in three more sections, excluding introduction, to highlight the main contribution of the paper

- In Sect. 2 the curated Scottish music dataset is presented together with the proposed instruments mapping used to associate Scottish traditional instruments into respective BinaryMuseGAN tracks list. Data curation and preparation of Scottish traditional music for evolutionary approaches has never been completed or studied before, hence it represents the first contribution of our work.
- In Sect. 3 the training methodology used is presented. The main contribution of our paper is the utilisation of a TL scheme where a pre-learned model, trained with BinaryMuseGAN on a dataset of non specialist music pianoroll tracks (Lakh MIDI dataset), is fine tuned on the Scottish music dataset. This has a double advantage: the model can be fine tuned effectively on smaller datasets, the training can achieve greater performances because the core features of music generation has already been learnt in the pre-learning phase. TL approaches for music generation is a largely unexplored area, to the best of our knowledge no previous approach has attempted anything similar.
  - In Subsect. 3.1 two new metrics, dotted rhythm and pentatonic note, are defined, as deemed to be representatives of the key features of Scottish music.
- Finally in Sect. 4, the training of the TL model is compared with the model trained from scratch. The results show the superiority of a TL approach in terms of robustness and metrics performances.

2 Data Gathering and Preprocessing

The training data consists of 137 midi files sequenced by Barry Taylor which are traditional or contemporary Scottish tunes [3]. Here the process used to clean and preprocess the data is described. This is a non-trivial task as the data have many differing characteristics. Starting with the time signatures, these are used to narrow down the dataset to suitable files. Tracks in the raw dataset are then assigned to new tracks in the processed dataset depending on their instrumentation and certain other features as will be described. Finally, the shaping of the data in a format which can be used to train the GAN is described.

2.1 Time Signatures

First the time signature of each file is considered, which is stated in the midi file metadata. In the original BinaryMuseGAN dataset [8], the authors only use pieces with a time signature of 4/4. Traditional Scottish music often features compound times such as 6/8 or 9/8, which would therefore be unsuitable to allow TL. The proportions of the time signatures in the Scottish music dataset are shown in Fig. 1. Since 4/4 time signatures only make up 69.2% of the total dataset, it was decided also to include 2/4 and 2/2 as these have a very similar feel to 4/4. This reduces the valid dataset size to 78 files.

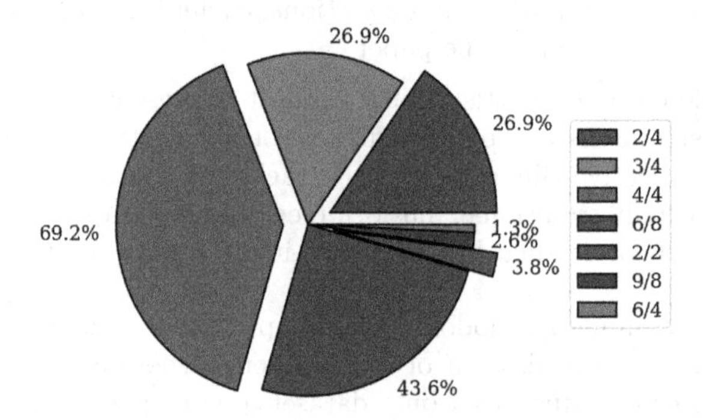

Fig. 1. Proportion of each time signature present in the Scottish music dataset - exploded segments indicate the time signatures included in the processed dataset

2.2 Tracks and Instrumentation

Now the instrumentation of each file is considered. Each track in a file has an associated General MIDI program change number - equivalent to an instrument [20]. BinaryMuseGAN generates 8 tracks with the following midi instruments: Drums, Piano, Guitar, Bass, Ensemble, Reed, Synth Lead, and Synth Pad. Effective TL requires the same number of tracks to be generated, however the exact instrumentation used in their implementation is not suitable for a Scottish ensemble. In our dataset, certain files contain multiple tracks of the same instrument; for example, a file with 4 tracks could have all tracks with program change number 1, which is an Acoustic Grand Piano. Most files have fewer than 8 tracks which is highlighted in Fig. 2. As a result, the preprocessed dataset is sparse with respect to which of the 8 tracks are not empty. The instrumentation was adjusted to the following: Drums, Piano, Guitar, Bass, Fiddle, Wind, Accordion, and Clarsach. These are listed in Table 1 along with their corresponding General MIDI program change number. The table also shows which program change numbers (i.e. instruments) from the raw dataset are included in each new track.

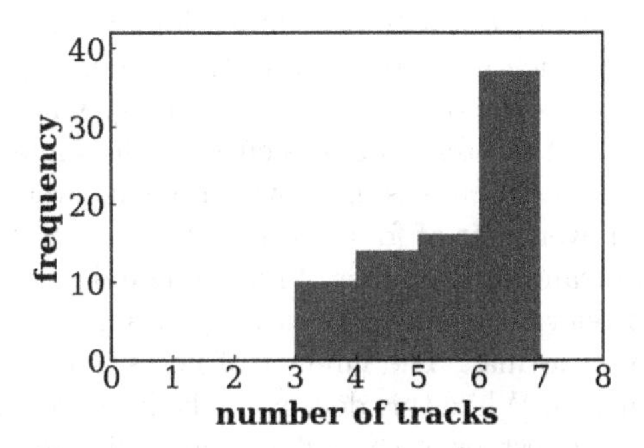

Fig. 2. Histogram of the number of tracks in each file in the Scottish music dataset.

Table 1. Instrument changes for scottish music generation

Track no	Their instrument	Our instrument	Included instruments
0	Drums	Drums	Drums
1	Piano	Piano (Acoustic Grand Piano - 1)	1, 2, 3, 5, 6, 8
2	Guitar	Guitar (Acoustic Guitar (steel) - 26)	25, 26, 31, 106
3	Bass	Bass (Electric Bass (finger) - 34)	33, 34, 36, 37, 43
4	Ensemble	Fiddle (111)	41, 42, 49, 51, 111
5	Reed	Wind (Recorder - 75)	57, 58, 59, 62, 66, 67, 2, 68, 69, 71, 72, 73, 74, 75, 76, 77, 80, 110, 11
6	Synth Lead	Accordion (22)	22, 23, 24
7	Synth Pad	Clarsach (Orchestral Harp - 47)	11, 15, 16, 89, 95, 47

We refer to the 8 tracks in the preprocessed dataset as "new tracks" and tracks from files in the raw dataset as "old tracks". Any new track for a given file which contains one or more old tracks is referred to as a non-empty track, whereas empty tracks do not contain any old tracks. Only using the transformations shown in Table 1 results in very uneven distributions of old tracks among new tracks which leaves many new tracks empty as shown in Fig. 4. To avoid having large numbers of new tracks in the dataset which do not contain any notes, the old tracks should be spread more evenly across the new tracks. To achieve this balancing, heuristics are employed to change some of the tracks based on the following features of each track: mean note, number of notes, and polyphonic ratio.

Every note in any track is represented by a number from 0 to 127 which defines the note's pitch as per the General MIDI standard [20]. The processed format of the dataset is a pianoroll, which is an array with a temporal dimension and pitch dimension. Along the temporal dimension, every column contains 128 binary-valued pitches which indicate whether that pitch is being played at that time or not. This requires the definition of a frequency parameter that controls the effective resolution of the temporal dimension. The same frequency of 24 as the original MuseGAN research was used, which means every quarter-note beat has 24 timesteps and every bar of four beats contains 96 timesteps.

To calculate the mean note, number of notes (here denoted n_{note}), and polyphonic ratio (herein referred to as "poly ratio") for a given track it is first converted to the pianoroll format. The number of notes is the number of nonzero locations in the pianoroll. While this does not take into account notes which are sustained for multiple timesteps, it can still give an indication of how many notes are played in that track. The mean note is the average location of nonzero elements along the pitch dimension. Finally, the poly ratio is the ratio between the number of locations along the temporal dimension with more than one nonzero location to the number of locations along the temporal dimension with any nonzero location. This indicates whether the track is playing mostly chords of multiple notes or mostly single lines of notes.

To balance the tracks, we derived heuristics based on these metrics for moving old tracks from non-empty tracks which contain three or more old tracks to other, "target" new tracks. These heuristics are designed to move old tracks to target new tracks with suitable instrumentation, e.g., tracks which play low notes go to bass, tracks which play more chords go to piano. The heuristics used are summarised in Algorithm 1.

Algorithm 1. Psuedocode of the heuristic used to balance the tracks

1: **if** mean note < 50 **then**
2:     move to bass
3: **end if**
4: **if** mean note > 60 and $n_{note} > 100$ **then**
5:     move to fiddle or wind
6: **end if**
7: **if** poly ratio > 0 **then**
8:     move to piano
9: **end if**
10: **if** poly ratio < 0.1 **then**
11:     move to any empty track except drum or bass
12: **end if**

The final heuristic in Algorithm 1 (lines 10–12) is used in case not enough suitable tracks are found using the other rules. The decision to move an old track to a different new track using the heuristics depends on the number of old tracks in the current new track and sometimes on the number of old tracks already in the target new track, as described here. Let n_{ne} denote the number of non-empty new tracks, n^i_{new} denote the number of old tracks in new track i, and n^t_{new} denote the number of old tracks in a target track t to which an old track could be moved. To balance the new tracks, the aim is to have all $n^i_{new} <= 2$ and $n_{ne} >= 3$. First new tracks with $n^i_{new} > 2$ are found and the rules above are applied to those tracks until all $n^i_{new} <= 2$. Then if n_{ne} is still less than 3, the tracks where $n^i_{new} = 2$ are considered and the rules are applied again. This is shown schematically in Fig. 3 which describes which tracks are moved and the conditions necessary for moving.

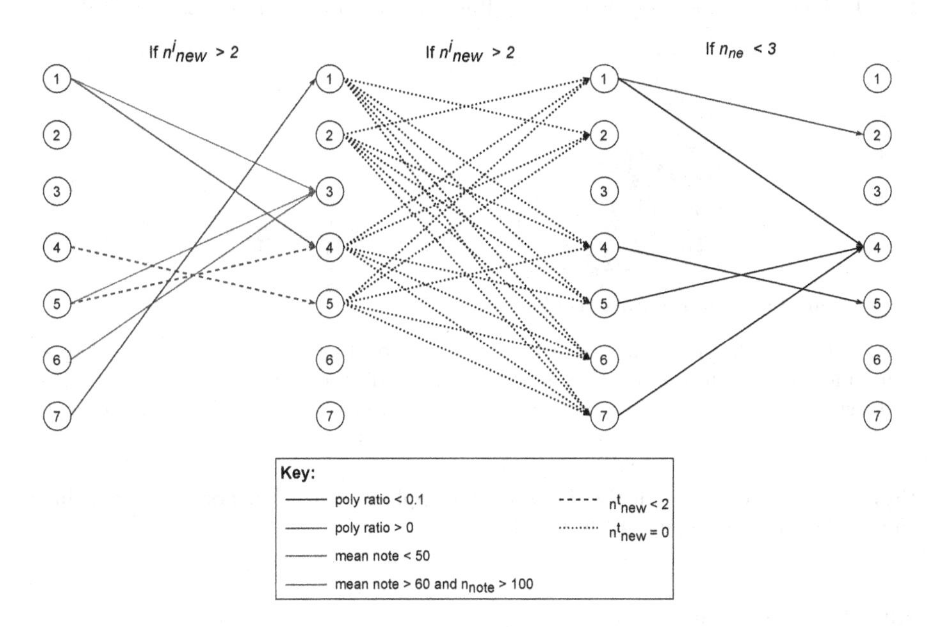

Fig. 3. Graph showing the heuristics used to balance the new tracks. Line colour indicates the condition applied to the old track to be moved. Dashed or dotted lines indicate the condition on the number of old tracks in the target track. Track 0 is drums and is not shown here since no tracks are transferred to or from this track.

Figure 5 shows the number of non-empty tracks and the distribution of old tracks after applying these balancing heuristics. Compared to Fig. 4, the number of non-empty tracks for each file is greater on average with none having fewer than 3 non-empty tracks. Moreover, considering the number of old tracks in each new track, the maximum is now 2 for all new tracks except drums and the average is more evenly distributed across the new tracks.

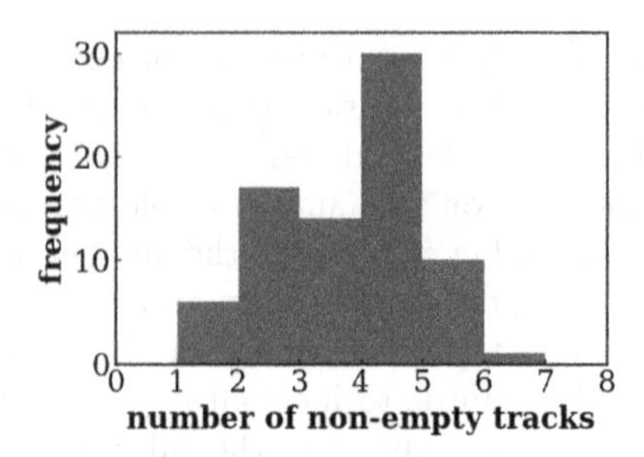

(a) Histogram showing numbers of new tracks for each file in the Scottish music dataset

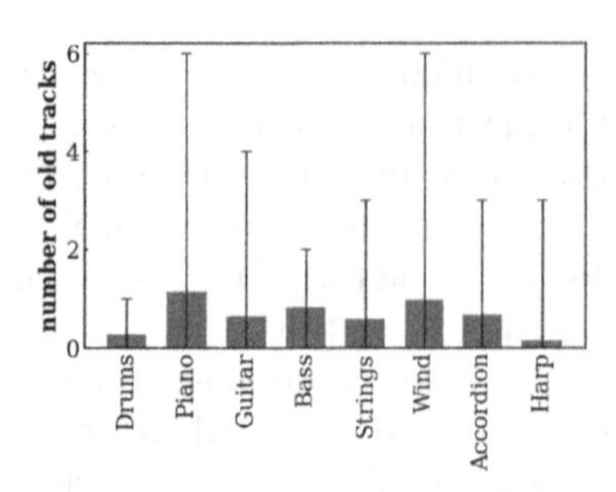

(b) Bar chart of average number of old tracks in each new track - errorbars indicate minimum and maximum

Fig. 4. Balance of tracks in the dataset after applying track conversions from Table 1.

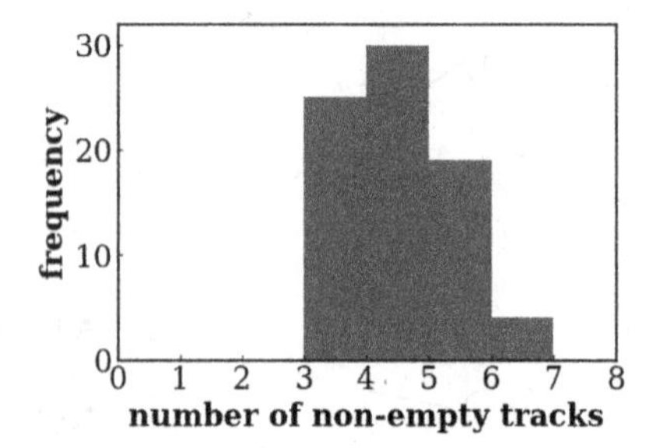

(a) Histogram showing numbers of new tracks for each file in the Scottish music dataset

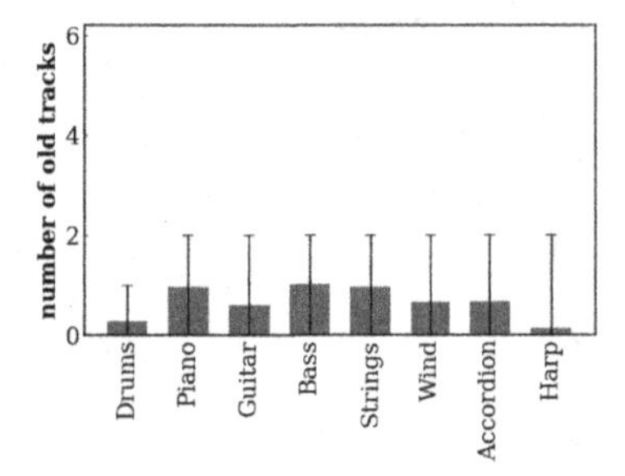

(b) Bar chart of average number of old tracks in each new track - errorbars indicate minimum and maximum

Fig. 5. Balance of tracks in the dataset after applying track conversions from Table 1 and further track balancing from Fig. 3.

2.3 Data Shaping

The final step in preprocessing the data is shaping the pianoroll array. As discussed previously, the pianoroll format has a temporal and pitch dimension. This new dataset should also be the same shape as the original BinaryMuseGAN dataset to effectively apply TL. In the original work, the array is cropped in the pitch dimension to 84 notes between 24 (C1) and 107 (B7) and the same was done here. Since none of the notes in the Scottish music dataset are outwith this range, this does not cause any notes to be lost.

Along the temporal dimension, data are split into bars of 96 timesteps. Each point in the dataset contains 4 bars. The final dimensions of the array are then $[\mathit{number\ of\ data\ points} \times \mathit{4(bars)} \times \mathit{96(timesteps)} \times \mathit{84(note\ pitches)} \times \mathit{8(tracks)}]$. For the original BinaryMuseGAN dataset, each piece has 6 4 bar phrases randomly sampled and added to the dataset. Due to the limited number of files in the Scottish music dataset, all of the 78 valid pieces are instead divided into 4 bar phrases of 96 timesteps and all of these are included in the dataset. This gives 1047 data points.

3 GAN Model

The technique chosen in this work to train a network able to learn from the curated dataset of Scottish music, is an adaptation of the GAN model developed by Dong et al. [8]. They developed a deep convolutional GAN that employs binary output neurons to generate music in the pianoroll format described in Sect. 2. The model developed in [8] is depicted in Fig. 6 and is composed by:

- a Generator network shared among all the tracks, G_s in Fig. 6, which is responsible of generating a high-level representation of the output music shared by all the tracks. The shared Generator is composed by an input dense layer with 1536 neurons and five transposed convolutional layers.
- A private Generator network for each track, G_p in Fig. 6, which convert the high-level music output provided by the shared generator into the final pianoroll output for the corresponding track. Each private Generator network is composed by three transposed convolutional layers.
- A Refiner network for each track, which refines the real-valued output of the Generators into binary ones. In this network, the tensor size remains unaltered.
- A private Discriminator for each track, D_p in Fig. 6, which extracts low-level features from the corresponding track. Each private Discriminator network is composed by three convolutional layers.
- A Discriminator network shared among all the tracks, D_s in Fig. 6, which extracts a high-level abstraction. The shared discriminator is composed by two convolutional layers.
- An onset/offset stream Discriminator, D_o in Fig. 6, formed by three convolutional layers.
- A chroma stream Discriminator, D_c in Fig. 6, formed by two convolutional layers.
- A final Discriminator, D_m in Fig. 6, which takes as input the outputs of D_s, D_o and D_c. This last Discriminator is composed by one convolutional layer and two dense layers respectively with 1536 and 1 neurons.

The output of the Generator group (shared plus private ones) has the shape $\mathbb{R}^{4\times 96\times 84\times 8}$, which is the same of the input for the Discriminator group. The output of the Discriminator group has shape $\mathbb{R}^1$. The total number of parameters for the BinaryMuseGAN model is 3735737, 1580440 in the Generator group and 2155297 in the Discriminator one. For more detailed informations on each network topology and on the onset/offset and chroma streams, please see [8].

Such GAN model can be either trained from scratch or via a Transfer Learning framework as shown in Fig. 7 by using a pretrained model. Transfer Learning

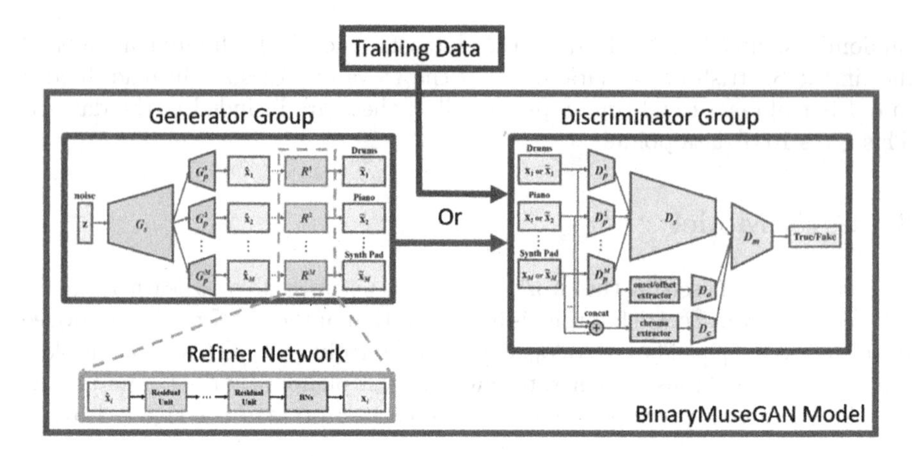

Fig. 6. High level depiction of the BinaryMuseGAN model. The separate images were taken from [8]. This image depicts the high level interaction between the three type of networks. The Generator group receives as input random data and it produces a music sample by first passing through the Refiner network which produces a binary output. The Discriminator group takes as input either training data or the samples produced by the generator and try to determine if the input was either real or fake data.

is a Machine Learning (ML) technique where a model trained on a set of data is used as a starting point for training on a new set of data. It can be used to enhance the generalisation capabilities of the pretrained model, by training it again on different kinds of data; or it can be used to speed up the training process on a set of training data similar to the one used for the pretraining phase; or it can be used to train complex model where large datasets are not available. In this last case, the model is pretrained on a larger alike dataset and then fine tuned on the dataset of interest. In this work the aforementioned GAN model was tested both by training it from scratch on the dataset presented in Sect. 2 and via transfer learning with the pretrained model provided by Dong et al. [8]. The dataset used to pretrain the model is the piano-roll version of the Lakh dataset, which is a collection of 176,581 unique MIDI files from the Million Song Dataset. This dataset was proposed by Raffel [19] and was converted into piano-roll format by Dong et al. [7].

3.1 Performance Metrics

To evaluate the results obtained, the same metrics employed in [8] were used, plus two new metrics here which are features typically observed in Scottish music [10], the Dotted Rhythms and Pentatonic Notes. The full set of metrics used are:

- Qualified note rate: evaluates the ratio of the number of qualified notes to the total number of notes. A qualified note is a note no shorter than three time steps. Hence a low qualified note rate value means that the produced track is overly-fragmented.

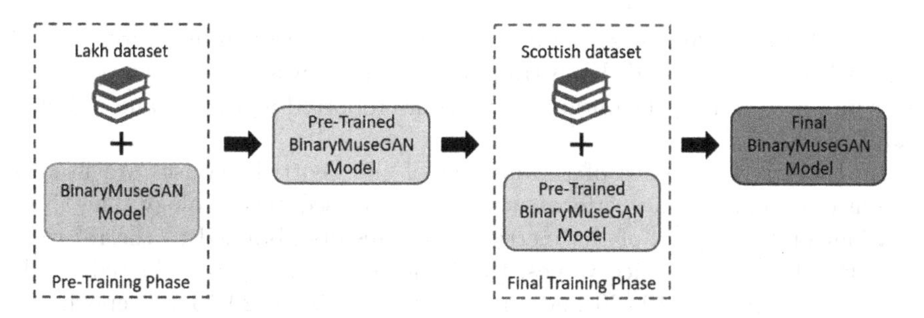

Fig. 7. Schematic of the Transfer Learning framework employed in this work.

- Polyphonicity: is the ratio of the number of time steps where more than two pitches are played simultaneously to the total number of time steps.
- Tonal distance: is the distance between the chroma features (one for each beat) of a pair of tracks in the tonal space [13]. As for what was done is [8], also here the tonal distance was measured between the piano and guitar tracks. A larger tonal distance implies weaker inter-track harmonic relations.
- Dotted rhythms: these rhythms feature regularly in Scottish music which are a dotted quaver (18 timesteps) followed by a semiquaver (6 timesteps) or vice versa. In particular, the rhythm of a semiquaver followed by a dotted quaver is often referred to as a "scotch snap" due to its prevalence in traditional Scottish music. This metric assesses what proportion of beats (i.e. 24 timesteps) in a section of a piece contain a dotted rhythm.
- Pentatonic Notes: the pentatonic scale is commonly used in Scottish music, as well as many other styles of music due to its versatility. Since an indication of key signature for the pieces in the Scottish music dataset is not provided, the pentatonic scale to which compare the notes to must be inferred, based on the notes in the piece. To do this, each samples section of a piece was compared to all possible pentatonic scale (starting on each of the 12 possible semitones) and see which gives the highest proportion of pentatonic notes.

The Qualified note rate and the Poliphonicity are defined as intra-track metrics, since they capture the features of the different track separately, while the tonal distance is defined as an inter-track metric since it captures the relation between different tracks. For more informations on these metrics, please see [7,8].

4 Results

As explained in the previous sections, the aim of this work is to use a GAN model to create original music samples from a training database consisting of traditional Scottish music. The used code and the training data can be found at https://github.com/strath-ace/HAGGIS, while the produced results are available from the University of Strathclyde KnowledgeBase at https://doi.org/10.15129/4ae2eb7e-678d-4644-90ad-1cf2a953287f. To assess the effectiveness of this approach, the training process of the model described in Sect. 3 was repeated a total of 40 times: 20 times where the model was trained from scratch and 20 times where TL was used. Each training simulation was run for 100 epochs to asses its convergence. The obtained results are reported in Table 2 and in Figs. 8 and 9.

In Table 2, the values of the metrics of the Scottish dataset are used as reference to quantify the performance of the proposed methodology. The aim is to achieve closer values of the metrics of the results obtained to the reference ones. Besides the reference values, the values of the metrics evaluated on the produced samples at 10, 20 and their average between 20 to 100 epochs are listed. These results are expressed in terms of median and standard deviation, except for the values from 20 to 100 epochs, which represent an average of the median and standard deviation values obtained in the considered epochs range. Data are not available for the dotted rhythm and pentatonic notes metrics prior to 20 epochs because the new metrics were evaluated on MIDI samples output from the training, which by default were produced starting from epoch 20.

Table 2. Median and standard deviation metrics values at 10, 20 and between 20 and 100 Epochs. Scratch refers to the model trained from scratch while TL to the one trained using Transfer Learning. Training refers to the corresponding metric values of the training data. The highlighted values are those with a median value closer to the Training values for each case.

	Training	10 Epochs		20 Epochs		20-100 Epochs	
		Scratch	TL	Scratch	TL	Scratch	TL
Qualified note rate	0.987	0.205 $\pm$ 0.091	**0.583 $\pm$ 0.055**	0.450 $\pm$ 0.162	**0.583 $\pm$ 0.053**	**0.715 $\pm$ 0.083**	0.582 $\pm$ 0.040
Poliphonicity	0.393	0.077 $\pm$ 0.061	**0.094 $\pm$ 0.019**	0.053 $\pm$ 0.015	**0.097 $\pm$ 0.012**	0.077 $\pm$ 0.022	**0.126 $\pm$ 0.018**
Tonal Distance	1.394	0.728 $\pm$ 0.283	**1.381 $\pm$ 0.156**	**1.374 $\pm$ 0.394**	1.334 $\pm$ 0.201	1.335 $\pm$ 0.301	**1.342 $\pm$ 0.143**
Dotted Rhythm	0.063	–	–	**0.023 $\pm$ 0.029**	0.115 $\pm$ 0.032	**0.085 $\pm$ 0.050**	0.120 $\pm$ 0.039
Pentatonic Notes	0.594	–	–	**0.542 $\pm$ 0.121**	0.466 $\pm$ 0.066	**0.487 $\pm$ 0.080**	0.437 $\pm$ 0.061

From these results, it can be observed that in the three reported cases (10, 20, 20-100 epochs), the TL approach achieves better results on the majority of the considered conventional metrics in terms of median and standard deviation values. About the two cases that go against this trend, namely the Tonal Distance at 20 Epochs and the Qualified Note Rate at 20-100 Epochs, it can be observed that, in both cases, the model trained from scratch achieves a slightly better median value despite having a standard deviation about twice than the one achieved by the model trained with TL. Regarding the standard deviation, it is clear that for all cases the value measured on the model trained with TL is always smaller than that of the model trained from scratch. Hence more robust results are obtained using the TL approach. This can also be observed by looking at Figs. 8a–8f. These plots represent the evolution of the metrics during the training

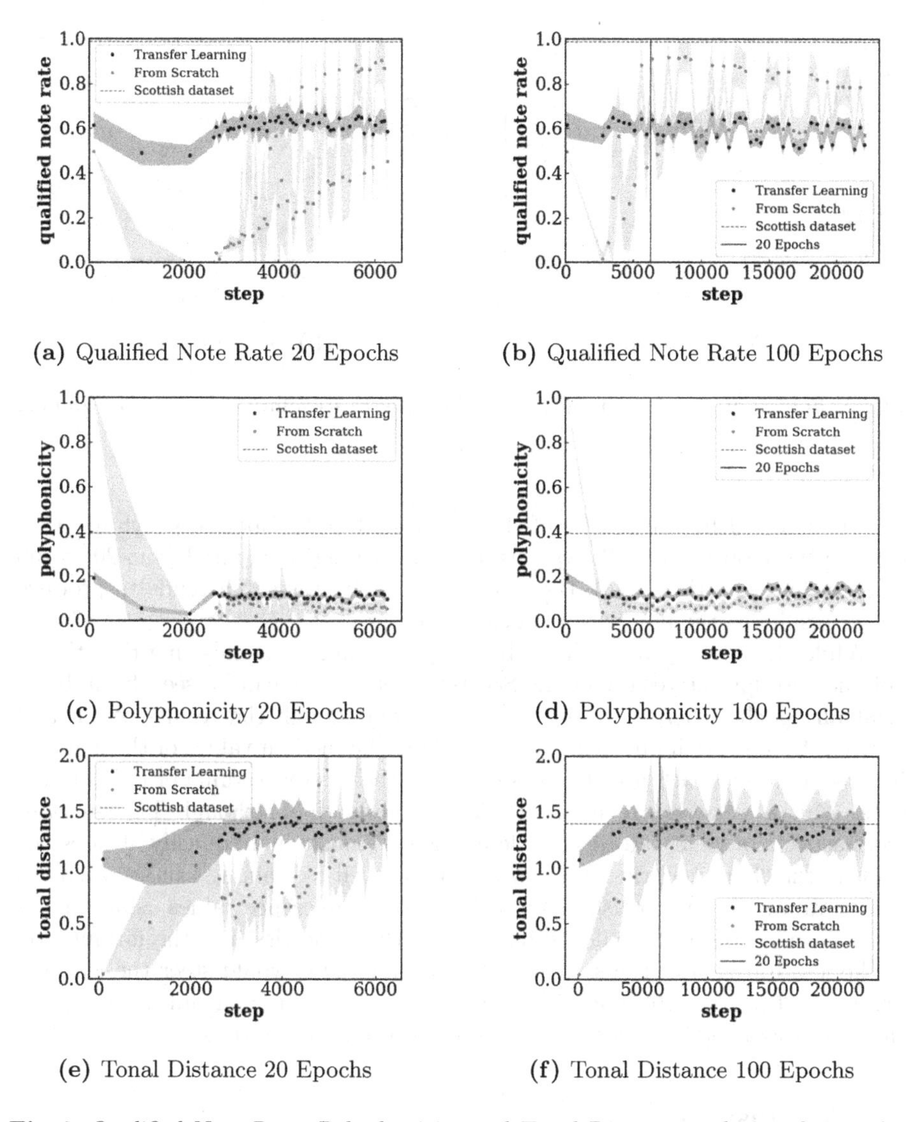

(a) Qualified Note Rate 20 Epochs

(b) Qualified Note Rate 100 Epochs

(c) Polyphonicity 20 Epochs

(d) Polyphonicity 100 Epochs

(e) Tonal Distance 20 Epochs

(f) Tonal Distance 100 Epochs

Fig. 8. Qualified Note Rate, Polyphonicity and Tonal Distance evolution during the training process. The plots on the left are up to 20 Epochs, while those on the right are up to 100. On the right plots, the vertical line represents where 20 epochs is.

process. In these six figures, the plots on the left are up to 20 epochs to give greater resolution over the initial epochs, while those on the right are up to 100 epochs. It is clear that the use of TL obtains more robust results with respect to these conventional metrics, i.e. with a lower standard deviation, and also with a less oscillating behaviour.

As briefly discussed in Sect. 3, one motivation for using TL is to obtain superior results with fewer training epochs than the model trained from scratch. This

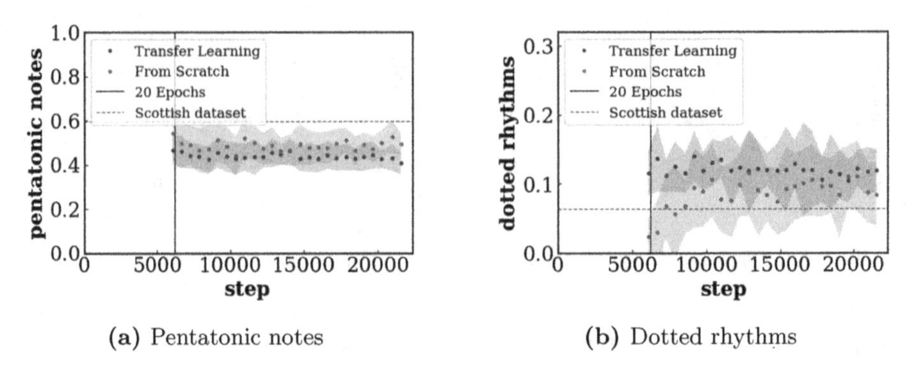

(a) Pentatonic notes

(b) Dotted rhythms

Fig. 9. Pentatonic notes and Dotted rhythms metrics evolution during the training process.

is demonstrated by the results of the TL case after 10 Epochs, which are very close to those obtained at 20 Epochs or to the average computed from 20 to 100 Epochs. On the other hand, the model trained from scratch tends to produce better results as the number of epochs increases.

While the TL approach shows better performance for most metrics, this is not the case for the results of the Scottish metrics. As can be seen from Table 2, at all epochs where data are available, even though the results of TL and learning from scratch are close to each others, the median values of the learning from scratch approach are the ones closer to the reference value of the training data. This is also shown in Fig. 9. This suggests that although TL generates most aspects of what can be considered "good" music more quickly, it does not capture the characteristics of a new dataset as well as training from scratch. To obtain values of the Scottish metrics closer to the reference values using the TL strategy, one approach could be integrating these metrics into the formulation of the loss function of the second stage training. This would steer the learning process to the characteristics of Scottish music, while taking full advantage of the robustness and performance recorded for the other metrics.

5 Conclusions

The paper presents an application of the Binary Multi-track Sequential Generative Adversarial Network (BinaryMuseGAN) to generate original Scottish music from a reduced dataset of songs by exploiting a TL approach. The GAN model is first trained on a larger and diverse music collection, to learn representative features of music in general, and then fine tuned on the smaller dataset of traditional Scottish music. The proposed approach demonstrates that more robust and performing results can be obtained via a TL method than by training the same network from scratch with the sole smaller dataset. The results are evaluated with three standard metrics, taken from the literature and two novel metrics defined here to highlight Scottish music characteristics. The results obtained for these last two metrics show that, despite the TL approach achieving superior training performance according to standard metrics, the most relevant features of Scottish music are lost in the process if compared to a learning from scratch approach. This is mainly due to imbalance between the two datasets used and no adaptation of the loss function formulation during the fine tuning training step. While the proposed approach is an initial step in generating original music from reduced datasets of specific music kind, some limitations are highlighted for future studies. While for the TL approach the BinaryMuseGAN network topology needs to be invariate from the one used in the pretraining phase, for the learning from scratch approach, the topology of the network can be further optimised to improve the results. In addition, unsupervised analysis techniques such as clustering or principal components analysis could provide further insights into differences between datasets and outputs when using TL, as opposed to using human generated metrics.

References

1. Arutyunov, V., Averkin, A.: Genetic algorithms for music variation on genom platform. Procedia Comput. Sci. **120**, 317–324 (2017). https://doi.org/10.1016/j.procs.2017.11.245
2. Avdeeff, M.: Artificial intelligence & popular music: SKYGGE, flow machines, and the audio uncanny valley. Arts **8**(4), 130 (2019). https://doi.org/10.3390/arts8040130
3. Barry, T.: Traditional Scottish Tunes in Midi Format. http://www.whitestick.co.uk/midi.html
4. Brunner, G., Wang, Y., Wattenhofer, R., Zhao, S.: Symbolic music genre transfer with CycleGAN. In: Proceedings - International Conference on Tools with Artificial Intelligence, ICTAI, November 2018, pp. 786–793 (2018). https://doi.org/10.1109/ICTAI.2018.00123
5. Chen, Z., Wu, C.W., Lu, Y.C., Lerch, A., Lu, C.T.: Learning to fuse music genres with generative adversarial dual learning. In: Proceedings - IEEE International Conference on Data Mining, ICDM, November 2017, pp. 817–822 (2017). https://doi.org/10.1109/ICDM.2017.98
6. Cope, D.H.: Experiments in music intelligence (EMI). In: Proceedings of the 1987 International Computer Music Conference (1987)
7. Dong, H.W., Hsiao, W.Y., Yang, L.C., Yang, Y.H.: MuseGAN: multi-track sequential generative adversarial networks for symbolic music generation and accompaniment, September 2017. http://arxiv.org/abs/1709.06298
8. Dong, H.W., Yang, Y.H.: Convolutional generative adversarial networks with binary neurons for polyphonic music generation. In: Proceedings of the 19th International Society for Music Information Retrieval Conference, ISMIR 2018, pp. 190–196, April 2018. https://doi.org/10.5281/zenodo.1492377
9. Engel, J., et al.: Neural audio synthesis of musical notes with WaveNet autoencoders. In: 34th International Conference on Machine Learning, ICML 2017, no. 3, pp. 1771–1780 (2017)
10. Finnerty, A.: BrightRED Study Guide National 5 Music (2017)
11. Goodfellow, I.J., et al.: Generative adversarial nets. Adv. Neural Inf. Process. Syst. **3**(January), 2672–2680 (2014)

12. Grall, A., Hamidinekoo, A., Malcolm, P., Zwiggelaar, R.: Using a conditional generative adversarial network (cGAN) for prostate segmentation. In: Zheng, Y., Williams, B.M., Chen, K. (eds.) MIUA 2019. CCIS, vol. 1065, pp. 15–25. Springer, Cham (2020). https://doi.org/10.1007/978-3-030-39343-4_2
13. Harte, C., Sandler, M., Gasser, M.: Detecting harmonic change in musical audio. In: Proceedings of the ACM International Multimedia Conference and Exhibition, pp. 21–26 (2006). https://doi.org/10.1145/1178723.1178727
14. Karras, T., Aila, T., Laine, S., Lehtinen, J.: Progressive growing of GANs for improved quality, stability, and variation. In: 6th International Conference on Learning Representations, ICLR 2018 - Conference Track Proceedings, pp. 1–26 (2018)
15. Karras, T., Laine, S., Aila, T.: A style-based generator architecture for generative adversarial networks. In: Proceedings of the IEEE Computer Society Conference on Computer Vision and Pattern Recognition 2019-June, pp. 4396–4405 (2019). https://doi.org/10.1109/CVPR.2019.00453
16. Ledig, C., et al.: Photo-realistic single image super-resolution using a generative adversarial network. In: Proceedings - 30th IEEE Conference on Computer Vision and Pattern Recognition, CVPR 2017, January 2017, pp. 105–114 (2017). https://doi.org/10.1109/CVPR.2017.19
17. Lee, S.G., Hwang, U., Min, S., Yoon, S.: Polyphonic music generation with sequence generative adversarial networks (2017). http://arxiv.org/abs/1710.11418
18. Loughran, R., O'Neill, M.: Generative music evaluation: why do we limit to 'human'? In: Proceedings of the 1st Conference on Computer Simulation of Musical Creativity (Ml), pp. 1–16 (2016)
19. Raffel, C.: Learning-based methods for comparing sequences, with applications to audio-to-MIDI alignment and matching. Ph.D. thesis (2016)
20. Rothstein, J.: Midi Comprehensive Introduction, 7th edn. AR Editions Inc., Middleton (1995)
21. Sandred, Ö., Laurson, M., Kuuskankare, M.: Revisiting the Illiac Suite–a rule-based approach to stochastic processes. Sonic Ideas/Ideas Sonicas, pp. 1–8 (2009)
22. Steedman, M.J.: The perception of musical rhythm and metre. Perception **6**(5), 555–569 (1977). https://doi.org/10.1068/p060555
23. Vondrick, C., Pirsiavash, H., Torralba, A.: Generating videos with scene dynamics. In: Advances in Neural Information Processing Systems (Nips), pp. 613–621 (2016). https://doi.org/10.13016/m26gih-tnyz
24. Zhang, H., et al.: StackGAN: realistic image synthesis with stacked generative adversarial networks. IEEE Trans. Pattern Anal. Mach. Intell. **41**(8), 1947–1962 (2019). https://doi.org/10.1109/TPAMI.2018.2856256
25. Zhu, J.Y., Park, T., Isola, P., Efros, A.A.: Unpaired image-to-image translation using cycle-consistent adversarial networks. In: Proceedings of the IEEE International Conference on Computer Vision, October 2017, pp. 2242–2251 (2017). https://doi.org/10.1109/ICCV.2017.244

The Enigma of Complexity

Jon McCormack[1(✉)], Camilo Cruz Gambardella[1], and Andy Lomas[2]

[1] SensiLab, Monash University, Melbourne, Australia
{Jon.McCormack,Camilo.CruzGambardella}@monash.edu
[2] Goldsmiths, University of London, London, UK
https://sensilab.monash.edu/
https://andylomas.com/

Abstract. In this paper we examine the concept of complexity as it applies to generative art and design. Complexity has many different, discipline specific definitions, such as complexity in physical systems (entropy), algorithmic measures of information complexity and the field of "complex systems". We apply a series of different complexity measures to three different generative art datasets and look at the correlations between complexity and individual aesthetic judgement by the artist (in the case of two datasets) or the physically measured complexity of 3D forms. Our results show that the degree of correlation is different for each set and measure, indicating that there is no overall "better" measure. However, specific measures do perform well on individual datasets, indicating that careful choice can increase the value of using such measures. We conclude by discussing the value of direct measures in generative and evolutionary art, reinforcing recent findings from neuroimaging and psychology which suggest human aesthetic judgement is informed by many extrinsic factors beyond the measurable properties of the object being judged.

Keywords: Complexity · Aesthetic measure · Generative art · Generative design · Evolutionary art · Fitness measure

1 Introduction

"The number of all the atoms that compose the world is immense but finite, and as such only capable of a finite (though also immense) number of permutations. In an infinite stretch of time, the number of possible permutations must be run through, and the universe has to repeat itself. Once again you will be born from a belly, once again your skeleton will grow, once again this same page will reach your identical hands, once again you will follow the course of all the hours of your life until that of your incredible death."

—Jorge Luis Borges, *The doctrine of cycles*, 1936

J. Romero et al. (Eds.): EvoMUSART 2021, LNCS 12693, pp. 203–217, 2021.
https://doi.org/10.1007/978-3-030-72914-1_14

Complexity is a topic of endless fascination in both art and science. For hundreds of years scholars, philosophers and artists have sought to understand what it means for something to be "complex" and why we are drawn to complex phenomena and things. Today, we have many different understandings of complexity, from information theory, physics, and aesthetics [6,8,32,37].

In this paper we again revisit the concept of complexity, with a view to understanding if it can be useful for the generative or evolutionary artist. The application of complexity measures and their relation to aesthetics in generative and evolutionary art are numerous (see e.g. [14] for an overview). A number of researchers have tested complexity measures as candidates for fitness measures in evolutionary art systems for example. Here we are interested in the value of complexity to the individual artist or designer, not the system (or though that may benefit too). Put another way, we are asking what complexity can tell us about an individual artist's personal aesthetic taste or judgement, rather than the value of such measures in general.

A long held intuition is that visual aesthetics are related to an artefact's order and complexity [2,15,22]. From a human perspective, complexity is often regarded as the amount of "processing effort" required to make sense of an artefact. Too complex and the form becomes unreadable, too ordered and one quickly looses interest. Birkhoff [4] famously formalised an aesthetic measure $M = O/C$, the ratio of order to complexity [4], and similar approaches have built on this idea. To mention some examples, Berlyne and colleagues, defined visual complexity as "irregularities in the spatial elements" that compose a form [30], which lead to the formalisation of the relationship between pleasantness and complexity as an "inverted-U" [2]. That is, by increasing the complexity of an artefact beyond the "optimum" value for aesthetic preference, its appeal starts to decline [35]. Another example is Biderman's theory of "geons", which proposes that human understanding of spatial objects depends on how discernible its basic geometric components are [3,30] Thus, the harder an object is to decompose into primary elements, the more complex we perceive it is. This is the basis for some image compression techniques, which are also used as a measure of visual complexity [16].

More recent surveys and analysis of computational aesthetics trace the history [9,12] and current state of research in this area [14]. Other approaches introduce features such as symmetry as a counterbalance to complexity, situating aesthetic appeal somewhere within the range spanning between these two properties [30]. The most recent approaches combine measures of algorithmic complexity with different forms of filtering or processing to eliminate noise but retain overall detail [16,38].

Multiple attempts to craft automated methods for the aesthetic judgement of images have made use of complexity measures. Moreover, some of these show encouraging results. In this paper we test a selection of these methods on three different image datasets produced using generative art systems. All of the images in these datasets have their own "aesthetic" score as a basis for understanding the aesthetic judgements of the system's creator.

2 Complexity and Aesthetic Measure

Computational methods used to calculate image complexity are based on the definitions of complexity described the previous section (Sect. 1): the amount of "effort" required to reproduce the contents of the image, as well as the way in which the patterns found in an image can be decomposed. Some methods have been proposed as useful measures of aesthetic appeal, or for predicting a viewer's preference for specific kinds of images. In this section we outline the ones relevant for our research.

In the late 1990s Machado and Cardoso proposed a method to determine aesthetic value of images derived from their interpretation of the process that humans follow when experiencing an aesthetic artefact [20]. In their method the authors use a ratio of *Image Complexity* – a proxy for the complexity of the art itself – to *Processing Complexity* – a proxy of the process humans use to make sense of an image – as a representation of how humans perceive images.

In 2010, den Heijer and Eiben compared four different aesthetic measures on a simple evolutionary art system [11], including Machado and Cardoso's *Image Complexity*/*Processing Complexity* ratio, Ross & Ralph's colour gradient bell curve, and the fractal dimension of the image. Their experiments demonstrated that, when used as fitness functions, different metrics yielded stylistically different results, indicating that each assessment method biases the particular image features or properties being evaluated. Interestingly, when interchanged – when the results evolved with one metric are evaluated with another – metrics showed different affinities, suggesting that regardless of the specificity of each individual measure, there are some commonalities between them.

3 Experiments

To try and answer our question about the role and value of complexity measures in developing generative or evolutionary art systems, we compared a variety of complexity measures on three different generative art datasets, evaluating them for correlation with human or physical measures of aesthetics and complexity.

3.1 Complexity Measures

We tested a number of different complexity measures described in the literature to see how they correlated with individual evaluations of aesthetics. We first briefly introduce each measure here and will go into more detail on specific measures later in the paper.

Entropy (S): the image data entropy measured using the luminance histogram (base e).

Energy (E): the data energy of the image.

Contours (T): the number of lines required to describe component boundaries detected in the image. The image first undergoes a morphological binarisation (reduction to a binary image that differentiates component boundaries) before detecting the boundaries.

Euler (γ): the morphological Euler number of the image (effectively a count of the number of connected regions minus the number of holes). As with the T measure, the image is first transformed using a morphological binarisation.

Algorithmic Complexity (C_a): measure of the algorithmic complexity of the image using the method described in [16]. Effectively the compression ratio of the image using Lempil-Ziv-Welch lossless compression.

Structural Complexity (C_s): measure of the structural complexity, or "noiseless entropy" of an image using the method described in [16].

Machardo-Cardoso Complexity (C_{mc}): a complexity measure used in [21], without edge detection pre-processing.

Machardo-Cardoso Complexity with edge detection (C_{mc}^E): the C_{mc} measure with pre-processing of the image using a Sobel edge detection filter.

Fractal Dimension (D): fractal dimension of the image calculated using the box-counting method [7].

Fractal Aesthetic (D_a): aesthetic measure similar to that used in [10], based on the fractal dimension of the image fitted to a Gaussian curve with peak at 1.35. This value is chosen based on an empirical study of aesthetic preference for fractal dimension.

While each of these measures is in some sense concerned with measuring image complexity, the basis of the measure for each is different. *Entropy* (S) and *Energy*(E) measures are based on information theoretic understandings of complexity but concern only the distribution of intensity, while *Contours* (T) and *Euler* (γ) try to directly count the number of lines or features in the image, somewhat in line with perceptual notions of complexity. Lakhal et al.'s *Algorithmic Complexity* (C_a) and *Machardo & Cardoso's Complexity* (C_{mc}) measures use algorithmic or Kolmogrov-like understandings of complexity, relying on image compression algorithms to proxy for visual complexity. Lakhal et al. also define a *Structural Complexity* measure (C_s) designed to address the limitations of algorithmic complexity measures in relation to high frequency noise or many fine details. This is achieved by a series of "course-graining" operations, effectively low-pass filtering the image to remove high frequency detail in both the spatial and intensity domains. Finally, the fractal methods recognise self-similar features as proxies for complexity. They are based on past analysis of art images that demonstrated relationships between fractal dimension and aesthetics [7,31,36].

3.2 Datasets

For the experiments described in this paper, we worked with three different generative art datasets (Fig. 1). As the goal of this work was to understand the effectiveness of complexity measures in actual generative art applications, we wanted to work with artistic systems of demonstrated success, rather than invented or "toy" systems often used in this research. This allows us to understand the ecological validity [5] of any system or technique developed. Ecological validity requires the assessment of creative systems in the typical environments and contexts under which they are actually developed and used, as opposed to laboratory or artificially constructed settings. It is considered an important methodology for validating research in the creative and performing arts [13]. Additionally, all the datasets are open access, allowing others to validate new methods on the same data.

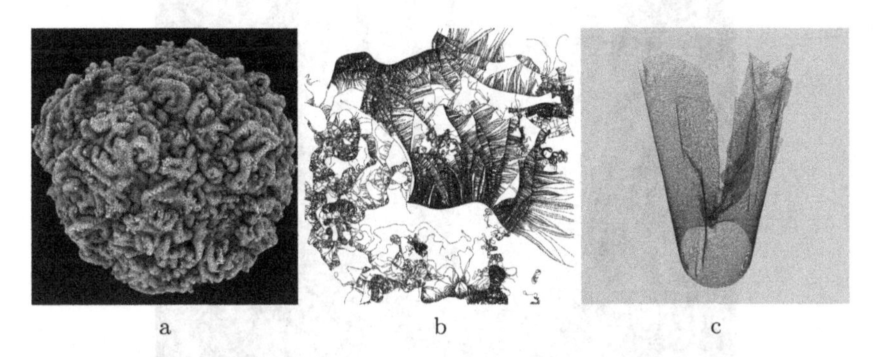

Fig. 1. Example images from the Lomas (a), Line drawing (b) and 3D DLA forms (c) datasets.

Dataset 1: Andy Lomas' Morphogenetic Forms. This dataset [28] consists of 1,774 images generated using a 3D morphogenetic form generation system, developed by computer artist Andy Lomas [18,19]. Each image is a two-dimensional rendering (512 × 512 pixels) of a three-dimensional form that has been algorithmically "grown" from 12 numeric parameters. The images were evolved using an *Interactive Genetic Algorithm* (IGA)-like approach with the software *Species Explorer* [18,19]. As the 2D images, not the raw 3D models are evaluated by the artist, we perform our analysis similarly.

The dataset contains an integer numeric aesthetic rating score for each form (ranging from 0 to 10, with 1 the lowest and 10 the highest, 0 meaning a failure case where the generative system terminated without generating a form or the result was not rated). These ratings were all performed by Lomas, so represent his personal aesthetic preferences. Additionally, each form is categorised by Lomas into one of eight distinct categories (these were not used in the experiments described in this paper).

Dataset 2: DLA 3D Prints. This dataset [27] consists of 2,500 3D forms created using a Differential Line Algorithm (DLA) based method [1]. Multiple closed 2D line segments develop over time. At each time-step the geometry is captured and forms a sequential z-layer in a 3D form. After several hundred time-steps, the final 3D form is generated, suitable for 3D printing (Fig. 2). Each image is 600 × 600 pixels resolution. Images in this set are 3D line renderings of the final form, from a perspective projection and orthographic projection in the xy plane. In the experiments described here we tested both the top-down

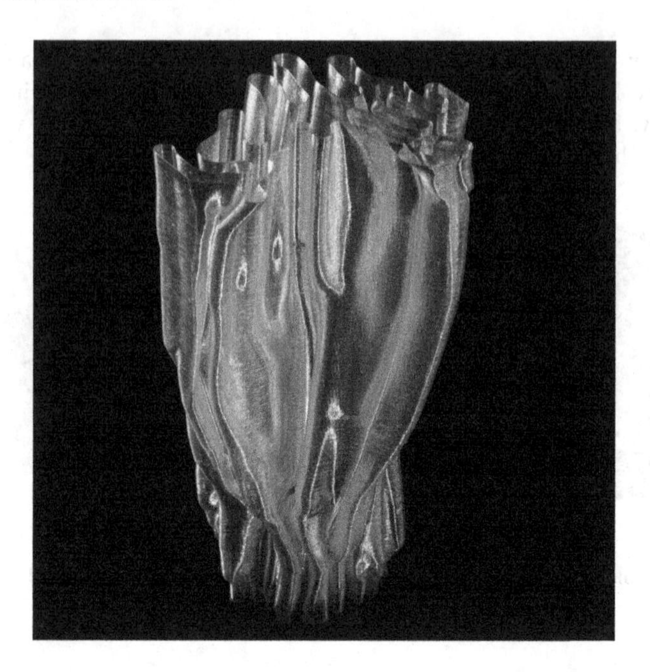

Fig. 2. Example 3D printed from the DLA 3D Prints dataset.

orthographic images and perspective images, finding the perspective images gave better results and so are the ones reported here.

Rather than human-designated aesthetic measures, this dataset has a physically computed complexity measure. This measure is based on two geometric aspects of the 3D form: *convexity* (how much each layer deviates from its convex hull) and the quartile *coefficient of dispersion* of angles between consecutive edges that make up each layer in the 3D form. These measures are calculated for each layer (weighted equally) and the final measure is the mean of all the layers in the form. This physical complexity measure appears to be a reasonable proxy to the visual complexity of the forms generated by the system.

Dataset 3: Line Drawings. A set of 53 line drawings generated using an agent-based method based on the biological principles of niche construction [24,26]. Each image is 1024×1024 pixels resolution. The dataset [25] also contains artist assigned aesthetic scores normalised to the range $[0, 1]$.

3.3 Settings and Measure Details

Our preliminary investigations showed that some measures are sensitive to parameter settings. The structural complexity measure (C_s) has two parameters: r_{cg}, a course-grain filter radius (in pixels), $\delta \in [0, 0.5]$ a threshold for determining the black to white pixel ratio, $\eta \in [0, 1]$ (white if $\eta \leq \delta$, grey if

$\delta < \eta \leq 1-\delta$, black for $\eta > 1-\delta$). In the original study, the authors [16] used values $(r_{cg}, \delta) = (7, 0.23)$ for one set of test images (abstract textures generated by Fourier synthesis) and $(13, 0.12)$ for the second set (abstract random boxes placed using an inverse of the fractal box counting method) for 256×256 resolution images. For the experiments described her we used $(r_{cg}, \delta) = (5, 0.23)$ as our image sizes were larger and the images contain significant high frequency detail.

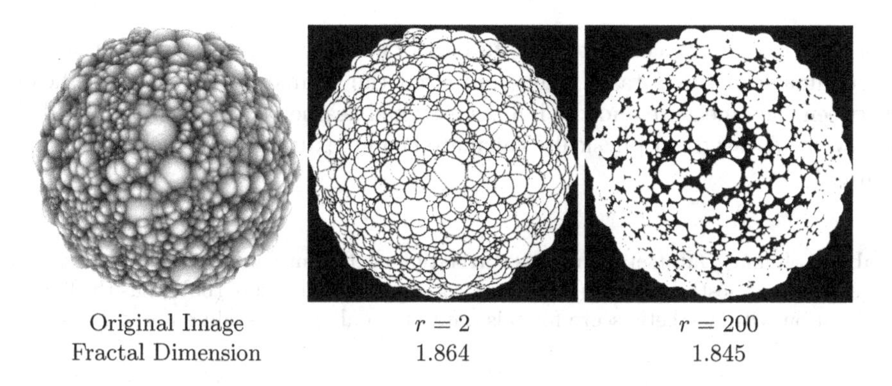

Fig. 3. The effect of different adaptive binarisation radii on an image from the Lomas dataset

For the fractal dimension measurements (D, D_a), images are pre-processed using a local adaptive binarisation process to convert the input image to a binary image (typically used to segment the foreground and background). A radius, r, is used to compute the local mean and standard deviation over $(2r+1) \times (2r+1)$ blocks centered on each pixel. Values above the mean of the r-range neighbourhood are replaced by 1, others by 0. Figure 3 shows a sample image from the Lomas dataset (left) with binary versions for $r = 2$ (middle) and $r = 200$ (right). Higher values of r tend to reduce high frequency detail and result in a lower fractal dimension measurement. For the DLA 3D prints and Line Drawing datasets, which are already largely comprised of lines, the value of r has negligible effect on the measurement.

Our Fractal Aesthetic Measure (D_a) is defined as:

$$D_a(i) = exp(\frac{-(D(i)-p)^2}{2\sigma^2}), \tag{1}$$

where p is the peek preference value for fractal dimension and σ the width of the preference curve. D_a returns a normalised aesthetic measure $\in [0,1]$. For the results reported here we used $(p, \sigma) = (1.35, 0.2)$, based on prior findings for this preference [34].

The Machardo-Cardoso Complexity measure (C_{mc})) is defined as:

$$C_{mc}(i) = RMS(i, f(i)) \times \frac{s(f(i))}{s(i)}, \tag{2}$$

where i is the input image, RMS a function that returns the root mean squared error between it's two arguments, f a lossy encoding scheme for i and s a function that returns the size in bytes of its argument.[1] For the lossy encoding scheme we used the standard JPEG image compression scheme with a compression level of 0.75 (0 is maximum compression).

4 Results

For each dataset we computed the full set of complexity measures (Sect. 3.1) on every image in the dataset, then computed the Pearson correlation coefficient between each measure and the human assigned aesthetic score (Lomas and Line Drawings datasets) or physically calculated complexity measure (DLA 3D Prints dataset).

Table 1. Lomas Datatset: Pearson's correlation coefficient values between image measurements and aesthetic score (Sc). The C_{mc} complexity measure (bold) has the highest correlation with aesthetic score for this dataset. In all cases p-values are $< 1 \times 10^{-3}$

	S	E	T	γ	C_a	C_s	C_{mc}	C_{mc}^E	D	D_a	Sc
S	1										
E	−0.989	1									
T	0.425	−0.375	1								
γ	−0.423	0.373	−0.999	1							
C_a	0.974	−0.945	0.496	−0.495	1						
C_s	0.922	−0.874	0.660	−0.659	0.940	1					
C_{mc}	0.793	−0.732	0.590	−0.589	0.907	0.860	1				
C_{mc}^E	0.779	−0.699	0.603	−0.602	0.869	0.907	0.930	1			
D	−0.352	0.452	0.294	−0.295	−0.164	−0.052	0.223	0.257	1		
D_a	0.105	−0.211	−0.318	0.319	−0.064	−0.165	−0.393	−0.442	−0.931	1	
Sc	0.634	−0.590	0.537	−0.536	0.757	0.685	**0.873**	0.774	0.284	−0.389	1

The results are shown for each dataset in Tables 1 (Lomas), 2 (DLA 3D Prints) and 3 (Line Drawings) with the highest correlation measure shown in bold.

As the tables show, a different complexity measure performed best for each dataset. For the **Lomas dataset** there is a strong correlation (0.873) between the artist assigned aesthetic score and the C_{mc} complexity measure, and that all the algorithmic and structural complexity measures are highly correlated. This is to be expected since they all involve image compression ratios. It is further highlighted in Fig. 4, which shows a plot of aesthetic score vs C_{mc} (a) and C_s vs

[1] We adopted this measure as it specifically deals with complexity as defined in [22]. Machardo & Cardoso also define an aesthetic measure as the ratio of image complexity to processing complexity [20], as used by den Heijer & Eiben in their comparison of aesthetic measures [10].

Table 2. DLA 3D Prints datatset: Pearson's correlation coefficient values between image measurements and physically computed complexity score (Sc). The C_s structural complexity measure (bold) has the highest correlation with aesthetic score for this dataset.

	S	E	T	γ	C_a	C_s	C_{mc}	C_{mc}^E	D	D_a	Sc
S	1										
E	−0.995	1									
T	0.857	−0.880	1								
γ	0.107	−0.083	-0.363	1							
C_a	0.953	−0.956	0.936	−0.106	1						
C_s	0.882	−0.892	0.925	−0.204	0.942	1					
C_{mc}	0.915	−0.935	0.968	−0.197	0.969	0.950	1				
C_{mc}^E	0.914	−0.935	0.961	−0.188	0.965	0.954	0.999	1			
D	0.928	−0.949	0.869	0.012	0.896	0.801	0.898	0.895	1		
D_a	−0.870	−0.888	−0.761	−0.112	−0.798	−0.678	−0.779	−0.774	−0.972	1	
Sc	0.760	−0.726	0.652	−0.066	0.756	**0.774**	0.704	0.706	0.550	−0.434	1

Table 3. Line Drawing datatset: Pearson's correlation coefficient values between image measurements and aesthetic score (Sc). The Contours T measure (bold) has the highest correlation with aesthetic score for this dataset.

	S	E	T	γ	C_a	C_s	C_{mc}	C_{mc}^E	D	D_a	Sc
S	1										
E	−0.910	1									
T	0.558	−0.677	1								
γ	−0.559	0.677	−1.000	1							
C_a	0.994	−0.934	0.541	−0.541	1						
C_s	0.576	−0.717	0.474	−0.474	0.618	1					
C_{mc}	0.515	−0.690	0.233	−0.233	0.592	0.761	1				
C_{mc}^E	0.648	−0.811	0.312	−0.312	0.712	0.822	0.927	1			
D	0.580	−0.807	0.431	−0.431	0.640	0.835	0.867	0.914	1		
D_a	−0.434	0.641	−0.323	0.323	−0.686	−0.771	−0.725	−0.770	−0.942	1	
Sc	0.209	−0.407	**0.565**	−0.564	0.218	0.364	0.267	0.199	0.456	−0.457	1

C_{mc} (b). The banding in Fig. 4a is due to the aesthetic scores being integers. A clear non-linear relationship between the complexity measures C_s and C_{mc} can be seen in Fig. 4b.

Also of note is that fractal measures performed the worst of the measures tested. This seems to be confirmed visually: while certainly the images are complex (many are composed of 1 million or more cells) and have patterns at different scales, the patterns are not self-similar.

For the **DLA 3D Prints** the most highly correlated measure was structural complexity (C_s) with a correlation of 0.774. The structural complexity aims to give a "noiseless entropy" measure by filtering high frequency spatial and intensity details. Given that the images are composed of many hundreds of thin lines stacked on top of each other, there is a significant amount of high frequency

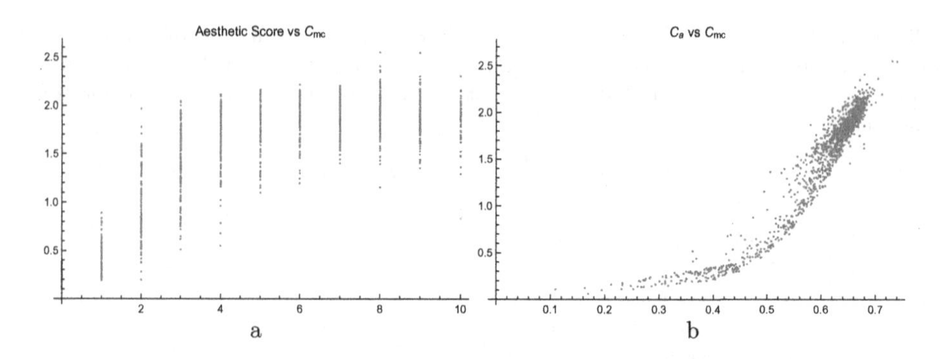

Fig. 4. Plots for the Lomas dataset showing the relationship between aesthetic score and C_{mc} (a) and C_a vs C_{mc} (b).

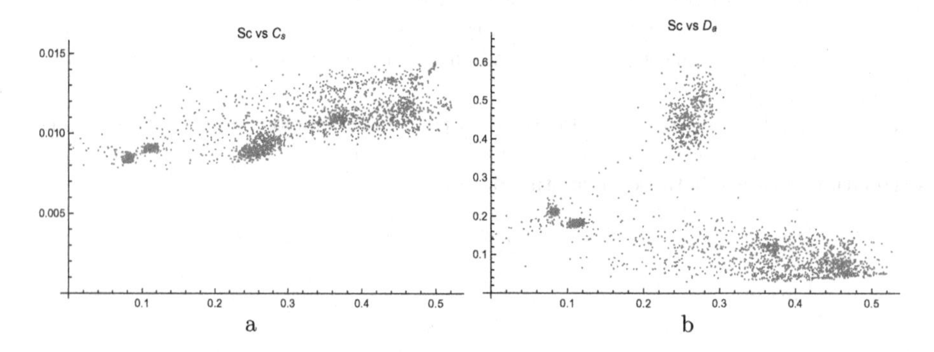

Fig. 5. Plots for the DLA 3D Prints dataset showing the relationship between physical complexity score (Sc) and C_s (a) and D_a (b).

information, hence filtering is likely to give a better measure of real geometric details in each form. As can be seen in Fig. 5 a clear correlation can be seen between the physical complexity (Sc) and Structural[2] Complexity measure (C_s). Again we note that the fractal measures (D, D_a) had the lowest correlation and that all the algorithmic complexity measures are highly correlated. As shown in Fig. 5b however, there appears to be a kind of bifurcation and clustering in the relationship between Sc and D_a, indicating a more complex relationship between fractal dimension and complexity.

The **Line Drawing dataset** exhibited quite different results over the previous two. Here the Contours (T) measure had the highest, but only moderate, correlation with artist-assigned aesthetic scores (0.565). Given the nature of the drawings, measures designed to capture morphological structure seem most appropriate for this dataset. It is also interesting to note that the algorithmic

[2] Readers should not draw any direct relation between the terms "structural" and "physical" in relation to complexity used here. Structural refers to image structures, whereas physical refers to characteristics of the 3D form's line segments.

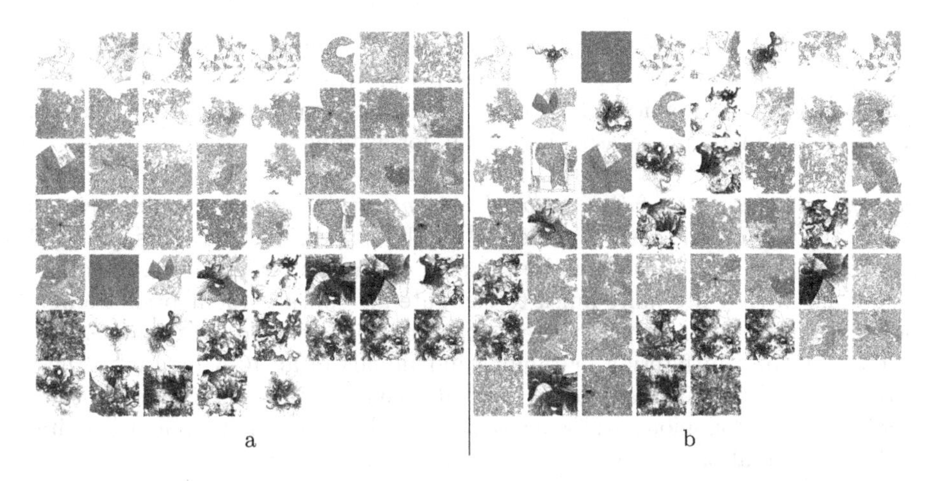

Fig. 6. Thumbnail grid of the entire Line Drawing dataset, ordered with increasing aesthetic score (lowest top left, highest bottom right) (a) and ordering by structural complexity (C_s). As the size of the dataset is relatively small in comparison with the others, the images can be shown in the figure.

complexity measures perform relatively poorly in this case. The original basis for the drawings came from the use of niche construction as a way to generate density variation in the images. The dataset contains images both with and without the use of niche construction, and generally those with niche construction are more highly ranked than those without. Figure 6 shows the entire dataset ordered in terms of artist-assigned aesthetic score (a) and structural complexity (b). The drawings with niche construction are easy to see as they are more highly ranked than those without. The structural complexity measure has greater difficulty in differentiating them (b).

With this in mind, we ran an additional image measure on this dataset that looks at asymmetry in intensity distribution (*Skew*). Since the niche construction process results in contrasting areas of high and low density it was hypothesised that this measure might be able to better capture the differences. This measure had a correlation of 0.583 ($p = 4.5 \times 10^{-6}$), so better than any of the other measures, but still only mildly correlated.

5 Discussion

Our results show that there appears to be no single measure that is best to quantify image complexity in the the context of generative art. Hence it seems wise to select a measure most appropriate to the style or class of imagery or form being generated.

It is also important to point out that, in general, computer synthesised imagery and in particular images generated by algorithmic methods, have important characteristics that differ from other images, such as photographs or paintings. Apart from any semantic differences or differentiation between figurative and abstract, intensity and spatial distributions in computer synthesised images differ from real world images. This is one reason why we selected datasets that are specific to the application of these measures (generative art and design), rather than human art datasets in general, for example.

The rational for this research was to further the question: *how can complexity measures be usefully employed in generative and evolutionary art and design?* Based on the results presented in this paper, our answer is that – if chosen appropriately – they can be valuable aids in course-level discrimination. Additionally, they are quite quick to compute and work without prior training or exposure to large numbers of examples or training sets. So, for example, they could be helpful in filtering or ranking individuals in an IGA or used to help classify or select individuals for further enhancement. However they are insufficient as fully autonomous fitness measures – the human designer remains a vital and fundamental part of any aesthetic evaluation.

5.1 Aesthetic Judgement

In Sect. 1 we discussed possible relationships between complexity measures and aesthetics. It is worth reflecting further here on this relationship and the long-held "open problem" for evolutionary and generative art of quantifying aesthetic fitness [23].

In the last decade or so, the biggest advances in the understanding of computational and human aesthetic judgements have come from (i) large, open access datasets of imagery with associated human aesthetic rankings and (ii) psychological and neuroscience discoveries on the mechanisms of forming an aesthetic judgement and what constitutes aesthetic experience.

In a recent paper, Skov summarised aesthetic appreciation from the perspective of neuroimaging [33]. Some of the key findings included neuroscientific evidence suggesting that "aesthetic appreciation is not a distinct neurobiological process assessing certain objects, but a general system, centered on the mesolimbic reward circuit, for assessing the hedonic value of any sensory object" [33]. Another important finding was that hedonic values are not solely determined by object properties. They are subject to numerous extrinsic factors outside the object itself. Similar claims have come from psychological models [17]. These findings suggest that any algorithmic measure of aesthetics which only considers an object's visual appearance ignores many other extrinsic factors that humans use to form an aesthetic judgement. Hence they are unlikely to correlate strongly with human judgements.

Our results appear to tally with these findings. Complexity measures, carefully chosen for specific styles or types of generative art can capture some broad aspects of personal aesthetic judgement, but they are insufficient alone to fully replace human judgement and discretion. Using other techniques, such as deep learning, *may* result in slightly better correlation to individual human judgement [29], however such systems require training on large datsets which can be tedious and time-consuming for the artist and still do not do as well as the trained artist's eye in resolving aesthetic decisions.

6 Conclusion

Making and appreciating art is a shared human experience. Computers can expand and grow the creative possibilities available to artists and audiences. The fact that humans artists are successfully able to create and communicate artefacts of shared aesthetic value indicates some shared concept of this value between people and cultures. Could machines ever share such concepts? This remains an open question, but evidence suggests that achieving such a unity would require consideration of factors beyond the quantifiable properties of objects themselves.

In this paper we have examined the relationship between complexity measures and personal or specific understandings of aesthetics. Our results suggest that some measures can serve as crude proxies for personal visual aesthetic judgement but the measure itself needs to be carefully selected. Complexity remains an enigmatic and contested player in the long-term game of computational aesthetics.

References

1. Barlow, P., Brain, P., Adam, J.: Differential growth and plant tropisms: a study assisted by computer simulation. In: Differential Growth in Plants, pp. 71–83. Elsevier (1989)
2. Berlyne, D.E.: Aesthetics and Psychobiology. Appleton-Century-Crofts, New York (1971)
3. Biederman, I.: Geon theory as an account of shape recognition in mind and brain. Irish J. Psychol. **14**(3), 314–327 (1993)
4. Birkhoff, G.D.: Aesthetic Measure. Harvard University Press, Cambridge (1933)
5. Brunswik, E.: Perception and the Representative Design of Psychological Experiments, 2nd edn. University of California Press, Berkley and Los Angeles (1956)
6. Crutchfield, J.P.: Complexity: metaphors, models, and reality. In: Is Anything Ever New?: Considering Emergence, vol. XIX, pp. 479–497. Addison-Wesley, Redwood City (1994)
7. Forsythe, A., Nadal, M., Sheehy, N., Cela-Conde, C.J., Sawey, M.: Predicting beauty: fractal dimension and visual complexity in art. Br. J. Psychol. **102**(1), 49–70 (2011)
8. Gell-Mann, M.: What is complexity? Complexity **1**(1), 16–19 (1995)
9. Greenfield, G.: On the origins of the term computational aesthetics. In: Neumann, L., Sbert, M., Gooch, B., Purgathofer, W. (eds.) Computational Aesthetics in Graphics, Visualization and Imaging, pp. 9–12. The Eurographics Association (2005). https://doi.org/10.2312/COMPAESTH/COMPAESTH05/009-012
10. den Heijer, E., Eiben, A.E.: Comparing aesthetic measures for evolutionary art. In: Applications of Evolutionary Computation, pp. 311–320. Springer, Heidelberg (2010). https://doi.org/10.1007/978-3-642-12242-2_32
11. den Heijer, E., Eiben, A.E.: Comparing aesthetic measures for evolutionary art. In: European Conference on the Applications of Evolutionary Computation, pp. 311–320. Springer, Heidelberg (2010)

12. Hoenig, F.: Defining computational aesthetics. In: Neumann, L., Sbert, M., Gooch, B., Purgathofer, W. (eds.) Computational Aesthetics in Graphics, Visualization and Imaging. The Eurographics Association (2005). https://doi.org/10.2312/COMPAESTH/COMPAESTH05/013-018
13. Jausovec, N., Jausovec, K.: Brain, creativity and education. Open Educ. J. **4**, 50–57 (2011)
14. Johnson, C.G., McCormack, J., Santos, I., Romero, J.: Understanding aesthetics and fitness measures in evolutionary art systems. Complexity 2019 (Article ID 3495962), 14 pages (2019). https://doi.org/10.1155/2019/3495962
15. Klinger, A., Salingaros, N.A.: A pattern measure. Environ. Plan. B: Plan. Design **27**(4), 537–547 (2000)
16. Lakhal, S., Darmon, A., Bouchaud, J.P., Benzaquen, M.: Beauty and structural complexity. Phys. Rev. Research **2**(2), 022058 (2020). https://doi.org/10.1103/PhysRevResearch.2.022058
17. Leder, H., Nadal, M.: Ten years of a model of aesthetic appreciation and aesthetic judgments: the aesthetic episode - developments and challenges in empirical aesthetics. Br. J. Psychol. **105**, 443–464 (2014)
18. Lomas, A.: Species explorer: an interface for artistic exploration of multi-dimensional parameter spaces. In: Bowen, J., Lambert, N., Diprose, G. (eds.) Electronic Visualisation and the Arts (EVA 2016). Electronic Workshops in Computing (eWiC), BCS Learning and Development Ltd., London, 12th–14th July 2016
19. Lomas, A.: On hybrid creativity. Arts **7**(3), 25 (2018). https://doi.org/10.3390/arts7030025
20. Machado, P., Cardoso, A.: Computing aesthetics. In: de Oliveira, F.M. (ed.) SBIA 1998. LNCS (LNAI), vol. 1515, pp. 219–228. Springer, Heidelberg (1998). https://doi.org/10.1007/10692710_23
21. Machado, P., Romero, J., Nadal, M., Santos, A., Correia, J., Carballa, A.: Computerized measures of visual complexity. Acta Psychol. **160**, 43–57 (2015). https://doi.org/10.1016/j.actpsy.2015.06.005
22. Machado, P., Romero, J., Nadal, M., Santos, A., Correia, J., Carballal, A.: Computerized measures of visual complexity. Acta psychol. **160**, 43–57 (2015)
23. McCormack, J.: Open problems in evolutionary music and art. In: Rothlauf, F., et al. (eds.) EvoWorkshops 2005. LNCS, vol. 3449, pp. 428–436. Springer, Heidelberg (2005). https://doi.org/10.1007/978-3-540-32003-6_43
24. McCormack, J.: Enhancing creativity with niche construction. In: Fellerman, H., et al. (eds.) Artificial Life XII, pp. 525–532. MIT Press, Cambridge (2010)
25. McCormack, J.: Niche Constructions Generative Art Dataset, January 2021. https://bridges.monash.edu/articles/dataset/Niche_Constructions_Generative_Art_Dataset/13662383
26. McCormack, J., Bown, O.: Life's what you make: Niche construction and evolutionary art. In: Giacobini, M., et al. (eds.) EvoWorkshops 2009. LNCS, vol. 5484, pp. 528–537. Springer, Heidelberg (2009). https://doi.org/10.1007/978-3-642-01129-0_59
27. McCormack, J., Gambardella, C.C.: DLA Form Generation dataset, January 2021. https://doi.org/10.26180/13663400.v1. https://bridges.monash.edu/articles/dataset/DLA_Form_Generation_dataset/13663400
28. McCormack, J., Lomas, A.: Andy Lomas generative art dataset. https://doi.org/10.5281/zenodo.4047222
29. McCormack, J., Lomas, A.: Deep learning of individual aesthetics. Neural Comput. Appl. **33**(1), 3–17 (2020). https://doi.org/10.1007/s00521-020-05376-7

30. Papadimitriou, F.: Spatial complexity, visual complexity and aesthetics. Spatial Complexity, pp. 243–261. Springer, Cham (2020). https://doi.org/10.1007/978-3-030-59671-2_16
31. Peitgen, H.O., Richter, P.H.: The Beauty of Fractals: Images of Complex Dynamical Systems. Springer, Berlin (1986). https://doi.org/10.1007/978-3-642-61717-1
32. Prigogine, I.: From Being to Becoming: Time and Complexity in the Physical Sciences. W. H. Freeman, New York (1980)
33. Skov, M.: Aesthetic appreciation: the view from neuroimaging. Empirical Stud. Arts **37**(2), 220–248 (2019). https://doi.org/10.1177/0276237419839257
34. Spehar, B., Clifford, C.W.G., Newell, B.R., Taylor, R.P.: Universal aesthetic of fractals. Comput. Graph. **27**(5), 813–820 (2003)
35. Sun, L., Yamasaki, T., Aizawa, K.: Relationship between visual complexity and aesthetics: application to beauty prediction of photos. In: Agapito, L., Bronstein, M.M., Rother, C. (eds.) ECCV 2014. LNCS, vol. 8925, pp. 20–34. Springer, Cham (2015). https://doi.org/10.1007/978-3-319-16178-5_2
36. Taylor, R.P., Micolich, A.P., Jonas, D.: Fractal analysis of Pollock's drip paintings. Nature **399**, 422 (1999)
37. Wolfram, S.: A New Kind of Science. Wolfram Media, Champaign (2002)
38. Zanette, D.H.: Quantifying the complexity of black-and-white images. PLoS ONE **13**(11), e0207879 (2018). https://doi.org/10.1371/journal.pone.0207879

SerumRNN: Step by Step Audio VST Effect Programming

Christopher Mitcheltree[1,2](✉) and Hideki Koike[1]

[1] Tokyo Institute of Technology, Tokyo 152-8550, Japan
christhetree@gmail.com, koike@c.titech.ac.jp
[2] Qosmo Inc., Tokyo 153-0051, Japan
https://www.vogue.cs.titech.ac.jp/
https://qosmo.jp

Abstract. Learning to program an audio production VST synthesizer is a time consuming process, usually obtained through inefficient trial and error and only mastered after years of experience. As an educational and creative tool for sound designers, we propose *SerumRNN*: a system that provides step-by-step instructions for applying audio effects to change a user's input audio towards a desired sound. We apply our system to Xfer Records Serum: currently one of the most popular and complex VST synthesizers used by the audio production community. Our results indicate that *SerumRNN* is consistently able to provide useful feedback for a variety of different audio effects and synthesizer presets. We demonstrate the benefits of using an iterative system and show that *SerumRNN* learns to prioritize effects and can discover more efficient effect order sequences than a variety of baselines.

Keywords: Synthesizer programming · Audio effects · VST · Sound design · Educational machine learning · Ensemble modeling · Recurrent neural networks · Convolutional neural networks

1 Introduction and Background

Sound design is the process of using a synthesizer and audio effects to create a desired output sound, typically by leveraging virtual studio technology (VST) on a computer. Often, the audio effects applied to the synthesizer play the biggest role in producing a desired sound. Sound design for the music industry is a very difficult task typically done by professionals with years of experience. Educational tools are limited and beginners are usually forced to learn via trial and error or from online resources created by others who typically also learned in a similar way. This makes the learning curve for sound design very steep.

1.1 Serum

Serum is a powerful VST synthesizer made by Xfer Records [6] that can apply up to 10 audio effects to the audio it generates. Serum is currently one of the most

J. Romero et al. (Eds.): EvoMUSART 2021, LNCS 12693, pp. 218–234, 2021.
https://doi.org/10.1007/978-3-030-72914-1_15

popular VST synthesizers in the audio production community and is routinely used by hobbyists and professionals alike. We chose Serum because we wanted to apply our research to a relevant, widely adopted, fully-featured synthesizer that is challenging for humans to master and will therefore maximize the practicality of our work.

1.2 Related Work

While applying AI to sound design is a relatively niche research area, there has been some prior work on leveraging AI to program audio VSTs. K-means clustering+tree-search [3] and evolutionary algorithms such as genetic algorithms [17,19] and genetic programming [12] have been applied to this problem with varying levels of success. Genetic algorithms have also been used to model audio effects directly [11]. However, these systems suffer from one or more of the following problems:

- They are applied to toy VSTs with little practical use.
- They are incompatible with existing VST plugins.
- Their inference time is prohibitively long.
- They are black-boxes with uninterpretable results.

With the recent rise in deep learning and neural networks, there has also been some related work using deep convolutional neural networks (CNNs) to program audio VSTs [2,19]. *InverSynth* [2] is probably most similar to *SerumRNN* since it also makes use of CNNs to program synthesizer parameters. However, as mentioned previously, these neural systems approach the problem with a one-shot, black-box process and focus more on synthesis rather than applying effects. This end-to-end approach replaces a user more than it augments them and results in fewer opportunities to learn. Most users typically know how to begin programming their desired sound via oscillator, attack, decay, sustain, and release parameters, but have difficulty programming the relevant effects due to the sheer number of combinatorial possibilities.

Finally, there has also been research on using deep learning to model audio effects and/or applying it directly to raw audio [5,7,13,14]. These systems typically use interesting signal processing techniques and neural network architectures from which we draw inspiration. However, they typically cannot be applied to existing VST synthesizers making their usefulness for our goals limited.

Overall, we believe that when using an AI assisted system, the user's sense of ownership over their work should be preserved. As a result, *SerumRNN* is inspired by step-by-step white-box automatic image post-processing systems [9] and collaborative production tools [15,18] that can educate and augment a user rather than aiming to replace them.

1.3 Contributions

In this paper we propose a system (*SerumRNN*) that iteratively changes an input audio towards the same timbre of a desired target audio by sequentially applying

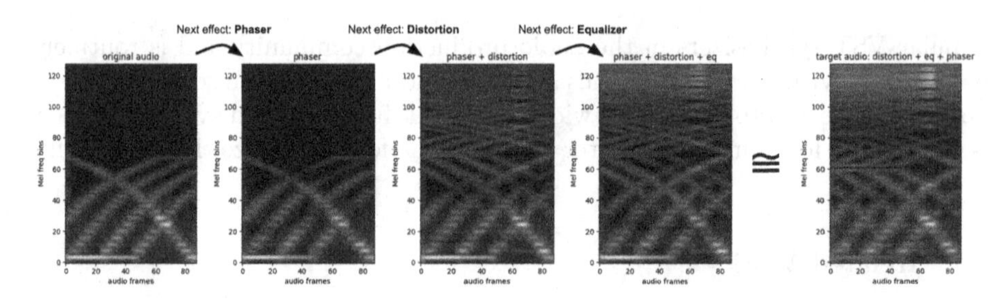

Fig. 1. Mel spectrogram progression of our system applying three effects to a user's input audio.

audio effects via the Serum VST synthesizer. As a result, the system provides a sequence of interpretable intermediate steps. It uses, to the best of our knowledge, a novel approach consisting of an ensemble of models working together: an *effect selection model* to determine which effect to apply next to the input audio and then a collection of *effect parameter models*, one per supported effect, to program the selected effect parameters. We demonstrate through extensive evaluation that *SerumRNN*:

- Significantly reduces the error between the input and target audio.
- Benefits from applying effects iteratively in a specific order.
- Learns which effects are most important.
- Provides interpretable and valuable intermediate steps.
- Can discover more efficient effect order sequences than a variety of baselines.

An example of *SerumRNN* applying three steps to some input audio can be seen in Fig. 1. Audio examples can be listened to at https://bit.ly/serum_rnn.

2 Data Collection

Data collection and processing systems represent a significant portion of the software engineering required for our system. Training data for all models is generated by rendering audio samples from Serum and then converting them into spectrograms and cepstra. Due to the complexity of Serum and its effects, virtually infinite amounts of training data can be generated.

2.1 Audio Rendering

Five commonly used and predominantly timbre altering effects are chosen for training data collection: multi-band compression, distortion, equalizer (EQ), phaser, and hall reverb. Table 1 summarizes which Serum synthesizer parameters are sampled for each supported effect. Continuous parameters (knobs on the Serum synthesizer) can be represented as floating-point numbers between zero and one inclusively. Categorical parameters can be represented as one-hot

Table 1. Parameters sampled from the Serum VST synthesizer.

Effect	Parameter name	Type	Sampled values
Compressor	Low-band compression	Continuous	[0.0, 1.0]
Compressor	Mid-band compression	Continuous	[0.0, 1.0]
Compressor	High-band compression	Continuous	[0.0, 1.0]
Distortion	Mode	Categorical	12 classes
Distortion	Drive	Continuous	[0.3, 1.0]
Equalizer	High frequency cutoff	Continuous	[0.50, 0.95]
Equalizer	High frequency resonance	Continuous	[0.0, 1.0]
Equalizer	High frequency gain	Continuous	[0.0, 0.4] and [0.6, 1.0]
Phaser	LFO depth	Continuous	[0.0, 1.0]
Phaser	Frequency	Continuous	[0.0, 1.0]
Phaser	Feedback	Continuous	[0.0, 1.0]
Hall reverb	Mix	Continuous	[0.3, 0.7]
Hall reverb	Low frequency cutoff	Continuous	[0.0, 1.0]
Hall reverb	High frequency cutoff	Continuous	[0.0, 1.0]

vectors of length C where C is the number of classes. Continuous parameter sampling value ranges are occasionally limited to lie within practical, everyday use regions.

Furthermore, in order to apply the system to a variety of different sounds, we collect data from 12 different synthesizer presets split into three groups (in increasing order of complexity): *Basic Shapes*, *Advanced Shapes*, and *Advanced Modulating Shapes*. The *Basic Shapes* preset group consists of the single oscillator sine, triangle, saw, and square wave default Serum presets. Next, the *Advanced Shapes* preset group consists of the dry (no effects) dual oscillator `"LD Power 5ths"`, `"SY Mtron Saw"`, `"SY Shot Dirt Stab"`, and `"SY Vintage Bells"` default Serum presets. Finally, the *Advanced Modulating Shapes* preset group consists of the dry dual oscillator `"LD Iheardulike5ths"`, `"LD Postmodern Talking"`, `"SQ Busy Lines"`, and `"SY Runtheharm"` default Serum presets. These final four presets also use intense modulations on top of their use of dual oscillators.

Since five different effects are supported by the system, there are 32 different combinations possible when applying a minimum of zero and a maximum of five effects to an input audio signal. For each of these combinations, a modified automated VST rendering tool [8] is used to render and save 4000 mono audio clips. Parameters are sampled randomly and uniquely from the permissible values shown in Table 1 and effect order when multiple effects are used is randomized. All audio samples are played and rendered for one second using a MIDI pitch of C4, maximum note velocity, and a sampling rate 44100 Hz. This process is repeated for each of the 12 presets resulting in 128k audio clips per preset, 384k audio clips per preset group, and 1.536M audio clips total. We only sample a single C4 pitch due to the inclusion of high quality timbre-preserving pitch shifting algorithms in all commonly used digital audio workstations, thus allowing users to easily warp a desired sound they would like to use as input to the system to C4. This preprocessing step could also be done by the system automatically.

2.2 Audio Processing

Audio clips are converted to Mel spectrograms using a Short-time Fourier transform (STFT) with a hop length of 512 samples, a fast Fourier transform (FFT) window length of 2048, a minimum and maximum frequency 20 Hz and 16 kHz respectively, and a Mel filter-bank of size 128. For each spectrogram the first 30 Mel-frequency cepstral coefficients (MFCCs), representing the smoothed spectral envelope, are also computed and saved. Lastly, each spectrogram is converted to decibels (dB). We use both spectrograms and cepstra in order to provide the subsequent neural networks as much volume, pitch, timbre, and temporal information as possible. Early prototypes of the system using only spectral or cepstral features benefited from including both.

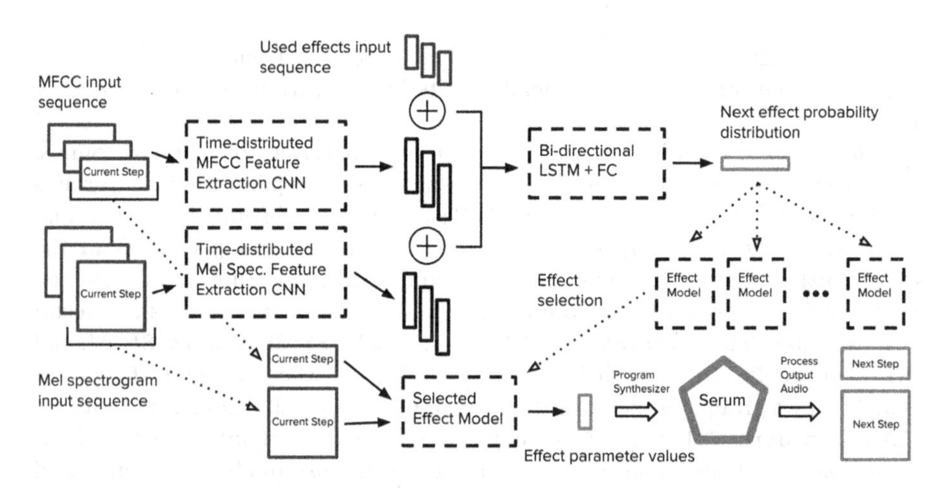

Fig. 2. System diagram: model components are dashed rectangles, inputs are blue, outputs are red, and the Serum synthesizer is green. Dotted arrows represent the conceptual flow of the system and solid arrows represent the flow of tensors.

3 Modeling

Our system consists of an ensemble of models that work together to take iterative steps to transform some input audio towards the timbre of some target audio. First, an *effect selection model* determines which effect to apply next and then a collection of five *effect parameter models*, one per supported effect, program the synthesizer accordingly. Figure 2 depicts a diagram of the entire system performing one step.

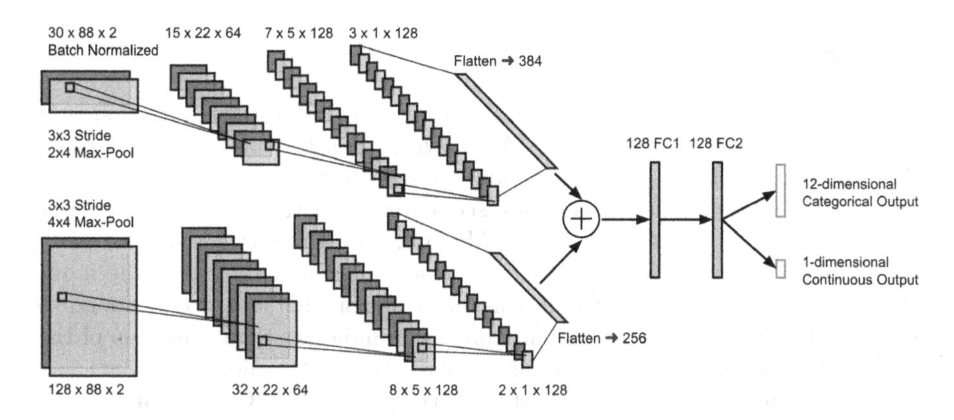

Fig. 3. *Distortion* effect parameter model architecture (not to scale).

3.1 Effect Parameter Models

The output values of effect parameter models are used to program the parameters of an effect. They take two tensors as input: a Mel spectrogram and the corresponding first 30 MFCCs. Both of these tensors consist of two channels, the first representing the input audio and the second representing the target audio. The MFCC tensor is normalized by feeding it through a batch normalization layer that has learnable shift and scale parameters. Two almost identical convolutional neural networks (CNNs) are used to extract features from each input. The Mel spectrogram CNN consists of three convolutional layers each with a 3×3 stride, 4×4 max-pool layer, and 64, 128, and 128 filters respectively. The MFCCs CNN is identical except, due to the smaller X input dimension, it uses 2×4 max-pool layers. The final max-pool layers of the two CNNs are flattened and concatenated together. This is then followed by two 128-dimensional fully connected (FC) layers with a dropout rate [16] of 50% each. All layers use ELU activations [4].

The output layers and loss functions of an *effect parameter model* vary depending on what effect is being modeled. Continuous parameters of an effect, where N is the number of continuous parameters, are represented as a single N-dimensional fully connected layer with a linear activation. Categorical parameters of an effect, where C is the number of classes, are each represented as a C-dimensional fully connected layer with a softmax activation. All output layers are connected to the second fully connected layer.

As a result, the effect parameter models for the *compressor*, *EQ*, *phaser*, and *hall reverb* effects have a single 3-dimensional continuous output layer and the *distortion* model has two output layers: a 12-dimensional categorical output layer and a 1-dimensional continuous output layer. The architecture of the *distortion* effect parameter model is shown in Fig. 3.

3.2 Effect Selection Model

The effect selection model outputs a probability distribution of what effect should be applied next to the input audio. As shown in Fig. 2, it takes as input three sequences of the same length corresponding to the number of steps that have been taken by the system. The first two are sequences consisting of effect parameter model inputs: Mel spectrograms and MFCCs. The third is a sequence of 6-dimensional one-hot vectors representing which effect was applied to the input audio at each step. The first five dimensions of the one-hot vector correspond to the five supported effects and the last dimension indicates the initial step of the system when no effect has been applied yet to the input audio.

Features are extracted from the Mel spectrogram and MFCC sequences using time distributed CNNs identical in architecture to the ones in the effect parameter models, except each convolutional layer uses only half as many filters and there is only one fully connected layer which is the output. The extracted feature vectors are then concatenated with the one-hot vector effect sequence and are then fed through a 128-dimensional bi-directional long short-term memory (LSTM) layer. This is followed by a 128-dimensional fully connected layer with an ELU activation and 50% dropout applied to it, and lastly a 5-dimensional fully connected output layer with a softmax activation.

3.3 Training

One effect parameter model is trained per effect for each preset group (therefore 15 in total) on the Cartesian product of all possible input audio and target audio Mel spectrogram and MFCC pairs. This results in each model being trained on approximately 1.2M data points. Categorical output layers are trained with categorical cross-entropy loss and continuous output layers are trained with mean squared error loss. Multiple losses due to multiple output layers, such as for the *distortion* effect parameter model, are weighted equally and a batch size of 128 is used.

One effect selection model is trained for each preset group (therefore three in total) on unique sequences consisting of one to five steps (applied effects) generated randomly from the available audio clip Mel spectrograms and MFCCs. To encourage efficiency, each effect can only be applied once to the input audio. Consequently, the shortest generated sequence is of length two since this represents the original input audio and one effect being applied and similarly, the longest generated sequence is of length six which represents the original input audio and the five supported effects being applied. Approximately 500k data points are generated for each preset group effect selection model. A categorical cross-entropy loss is used for training along with a batch size of 32.

All models are trained for 100 epochs with early stopping (8 epochs of patience) and a validation and test split of 0.10 and 0.05 respectively. The Adam optimizer [10] is used with a learning rate of 0.001.

4 Evaluation

We evaluate our system at three levels of granularity: the effect parameter models, the effect selection models, and the entire system/ensemble of models as a whole. Each of the three preset groups are evaluated separately and yield very similar and consistent results. We display detailed results for the most challenging preset group, *Advanced Modulating Shapes*, and place the figures and tables for the other two preset groups in the Appendix. Since understanding how perceptually close two audio samples are from metrics alone is difficult, audio examples for *SerumRNN* can be found at https://bit.ly/serum_rnn.

4.1 Audio Similarity Metrics

In order to effectively evaluate our system, the distance between two audio clips needs to be computed; specifically the similarity between their timbres. We make use of a variety of six different metrics to evaluate our models vigorously.

The first three metrics are quite basic: the mean squared error (MSE), mean absolute error (MAE), and log-spectral distance (LSD) between two magnitude spectrograms. LSD is defined as the average root mean square difference between each frame of two log power spectrograms.

$$MSE = \frac{1}{mn} \sum_{i=1}^{m} \sum_{j=1}^{n} (S_{ij} - \hat{S}_{ij})^2 \tag{1}$$

$$MAE = \frac{1}{mn} \sum_{i=1}^{m} \sum_{j=1}^{n} \mid S_{ij} - \hat{S}_{ij} \mid \tag{2}$$

$$LSD = \frac{1}{n} \sum_{i=1}^{n} \sqrt{\frac{1}{m} \sum_{j=1}^{m} (S_{ij} - \hat{S}_{ij})^2} \tag{3}$$

S is an output decibel Mel spectrogram matrix, $\hat{S}$ is the target decibel Mel spectrogram matrix, m is the number of Mel filters, and n is the number of frames.

The fourth metric (MFCCD) is the mean Euclidean distance between the first 30 MFCCs which represent the smoothed spectral envelope of an audio clip.

$$MFCCD = \frac{1}{n} \sum_{i=1}^{n} \|C_i - \hat{C}_i\|_2 \tag{4}$$

C is a matrix of the first 30 output audio MFCCs for each frame, $\hat{C}$ is a matrix of the first 30 target audio MFCCs for each frame, and n is the number of frames.

The Pearson Correlation Coefficient (PCC) can also be used to measure audio reconstruction quality [2] and is defined as

$$PCC = \frac{\text{cov}(x, y)}{\sigma_x \sigma_y} \tag{5}$$

where x and y are 1-dimensional vectors, not matrices. We apply this metric to decibel Mel spectrogram matrices S and $\hat{S}$ by flattening each one via row concatenation.

Our last and most robust metric is based off a multi-scale spectral loss [7] that we have modified and we call it multi-scale spectral mean absolute error (MSSMAE). Given an output audio clip and a target audio clip, magnitude spectrograms (S_i and $\hat{S}_i$ respectively) are calculated for several different FFT sizes i. We then also compute $\log S_i$ and $\log \hat{S}_i$ and define the metric as

$$MSSMAE = \frac{1}{w} \sum_i (MAE(S_i, \hat{S}_i) + \alpha MAE(\log S_i, \log \hat{S}_i)) \tag{6}$$

where w is the number of different FFT sizes, α is a constant scaling term, and MAE is defined in Eq. 2 as the mean squared error between two spectrograms. We set α to 0.1 in our experiments to make the two MAE values roughly equal since the log values are generally around one order of magnitude larger. We use FFT sizes of (64, 128, 256, 512, 1024, 2048) making $w = 6$ and the neighboring frames in the resulting spectrograms have 75% overlap. By calculating multiple regular and logarithmic spectrograms with different FFT sizes, two audio clips can be compared along different spatial-temporal resolutions.

For the MSE, MAE, LSD, MFCCD, and MSSMAE metrics, lower values indicate a higher similarity between two audio clips whereas the opposite is true for PCC values. MSE, MAE, PCC, and MSSMAE values are multiplied by 100 for better readability.

4.2 Effect Parameter Models

Effect parameter models are evaluated using the six metrics defined in Sect. 4.1 on 1000 input and target audio pairs from their corresponding test data sets. We observe from Table 2 that for almost every metric and effect, the models substantially decrease the distance between the input audio clip and the target audio clip.

*It is worth noting that this trend appears to be less dramatic for the *Phaser* effect, but can be explained by the fact that spectrograms and cepstra do not represent phase information well [7]. As a result, our error metrics might be less representative of the *Phaser* effect when compared to the other effects.

Table 2. Effect parameter models eval. metrics (*Adv. Mod. Shapes* preset group).

Effect	Metric	Mean error against target audio			
		Input audio	Output audio	Δ	Δ%
Compressor	MSE	1.86	0.82	−1.04	−56.06
	MAE	10.50	6.43	−4.07	−38.75
	LSD	10.04	6.52	−3.53	−35.10
	MFCCD	108.53	64.48	−44.05	−40.59
	PCC	82.01	86.85	4.84	5.90
	MSSMAE	71.64	45.57	−26.07	−36.39
Distortion	MSE	4.49	0.78	−3.71	−82.66
	MAE	14.57	6.21	−8.36	−57.40
	LSD	13.90	6.41	−7.50	−53.92
	MFCCD	146.04	59.76	−86.29	−59.08
	PCC	74.67	80.42	5.75	7.70
	MSSMAE	94.33	65.35	−28.97	−30.72
Equalizer	MSE	1.48	0.80	−0.68	−45.92
	MAE	8.76	6.29	−2.47	−28.24
	LSD	8.64	6.49	−2.14	−24.82
	MFCCD	92.52	64.17	−28.35	−30.64
	PCC	84.90	86.45	1.55	1.82
	MSSMAE	64.05	64.55	0.50	0.78
*Phaser**	MSE	0.86	0.77	−0.09	−10.15
	MAE	6.89	6.31	−0.58	−8.44
	LSD	7.06	6.59	−0.47	−6.67
	MFCCD	72.50	64.42	−8.08	−11.14
	PCC	86.05	85.58	−0.47	−0.55
	MSSMAE	51.53	47.33	−4.20	−8.16
Hall reverb	MSE	1.28	0.48	−0.80	−62.60
	MAE	8.32	4.89	−3.44	−41.28
	LSD	8.32	5.04	−3.28	−39.43
	MFCCD	80.49	46.21	−34.28	−42.59
	PCC	83.87	89.47	5.60	6.68
	MSSMAE	60.87	38.56	−22.31	−36.65

Table 3. Effect selection models prediction accuracy (*Adv. Mod. Shapes* preset group).

Effect selection model	Step number (number of effects for *SerumCNN*)					
	1	2	3	4	5	All
SerumRNN	**0.994**	**0.984**	**0.983**	**0.979**	0.999	**0.985**
SerumRNN NI	0.992	0.974	0.961	0.962	**1.000**	0.971
SerumCNN	0.993	0.975	0.946	0.916	0.889	0.955

4.3 Effect Selection Model

We compare the effect selection model, which we call *SerumRNN*, against two baselines to showcase the advantage of 1: using an iterative system and 2: considering the order of effects.

The first baseline is a non-iterative version of the effect selection model (called *SerumRNN NI*) that uses an identical architecture and training process except that only the first step (the original input and target audio) of the Mel spectrogram and MFCC input sequences is used. As a result, the sequence of effects to be applied to the input audio can be predicted all at once using *SerumRNN NI* without iteratively modifying the audio in between steps using the effect parameter models.

The second baseline is the effect parameter model CNN with a 5-dimensional fully connected output layer and a sigmoid activation. It is trained with a binary cross-entropy loss to predict all effects at once without considering order. We call this baseline *SerumCNN*.

The accuracy of all three models for effect sequences of length one to five is measured using their test data sets of approximately 25k data points and is summarized in Table 3. We observe that all three models perform remarkably well, with *SerumRNN* outperforming the other two for virtually all cases. Since the RNN models predict effects one by one, we believe accuracy is highest for the first and last step due to the first step representing the most dramatic difference between input and target audio clips and the last step being deducible given the previous effects since repeated effects are not supported. The behavior of *SerumCNN* is also expected: accuracy decreases the longer the effect sequence is since more effects need to be predicted correctly.

4.4 Ensemble/Entire System

Similarly to the effect parameter models, we evaluate the entire ensemble of models (*SerumRNN*) using the six metrics defined in Sect. 4.1 on 1000 input and target audio pairs. These pairs are sampled from the unseen test sets and the target audio is the result of a sequence of one to five effects (200 audio pairs per sequence length) being applied to the input audio. Inference time for the entire system (audio processing, effect selection, parameter programming, and audio rendering) is approximately 300 ms per effect applied to the input audio.

Baselines. We compare *SerumRNN* against five baselines. The first two are *SerumRNN NI* and *SerumCNN* from Sect. 4.3 which allow us to test for any gains from using an iterative system and considering the order of effects, respectively. The next baseline is a perfect effect selection model (*Oracle*) that always selects the "correct" next effect as is defined by the effect sequence that created the target audio. This allows us to compare our systems to what should be the technically "correct" solution. In order to test the benefit of using an effect selection model in the first place, we also have a *Random* baseline which simply chooses a random effect each time it is called. Finally, to test the benefit of applying the effect parameter models iteratively, we include a baseline called *SerumCNN 1S*. This baseline is identical to the *SerumCNN* baseline except that the effect parameter models are applied all at once to the input audio (after all effects have been predicted at once by *SerumCNN*). This results in a one-shot system with the fastest inference time possible, but no intermediate steps. This baseline is also the most similar to existing neural network synthesizer programming architectures of prior works in related literature [2,19].

All six systems use the same five effect parameter models for each preset group. Since repeated effects are not supported, if an effect selection model chooses an effect that has already been applied in a previous step, the next most probable unused effect is selected. As a result, all systems terminate at the latest after applying the maximum of five supported effects.

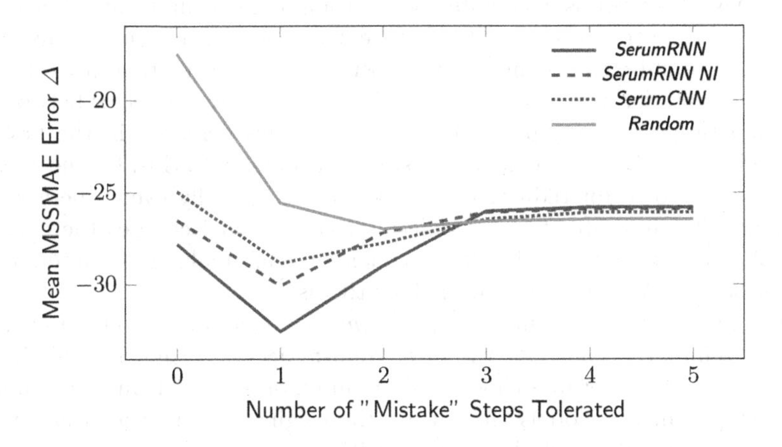

Fig. 4. Relationship between the tolerated number of mistake steps vs. the overall error Δ against the target audio (*Adv. Mod. Shapes* preset group).

Stopping Condition. Before the systems can be evaluated, a stopping condition must be defined for *SerumRNN*, *SerumRNN NI*, *SerumCNN*, and *Random*. Otherwise five effects will always be applied to the input audio, even when the target audio was created using only one effect. We believe a reasonable stopping condition is terminating the system once it makes a mistake, i.e. when the distance between the input and target audio increases. However, there may be situations where making a "mistake" is advantageous since the ideal sequence of steps to the target audio can be non-monotonic. Therefore, we investigate in Fig. 4 how the number of tolerated bad steps affects the resulting overall error reduction of each system. From the plotted results we can conclude that the ideal stopping condition is tolerating one step in the wrong direction before terminating.

Discussion. Table 4 summarizes the evaluation results of *SerumRNN* compared to the five baselines and Table 5 gives more granular evaluation details for *SerumRNN* and the *Advanced Modulating Shapes* preset group.

All systems significantly reduce the error between the input and target audio which indicates that the effect parameter models individually are robust and highly effective, even when applied in a random order. We believe using two channel Mel spectrogram and MFCC inputs and training on the Cartesian product of possible input and target audio combinations (as explained in Sect. 3.1) helped achieve this robustness.

Perhaps unsurprisingly, *SerumRNN* consistently outperforms *SerumRNN NI*, *SerumCNN*, and *SerumCNN 1S* for all metrics. This highlights the importance of considering the order of effects and using an iterative approach for both effect selection and effect parameter programming. Performance of the one-shot *SerumCNN 1S* system is also quite good considering it is up to five times faster than any of the other systems. If inference speed is critically important, a reasonable compromise can be made by choosing to use this system instead.

We also note for *SerumRNN* that all five steps reduce the error between the input and target audio clips, with the largest gains achieved in the first two steps and the smallest in the final two steps. In contrast to this, for the *Random* baseline (Table 6) error reduction is distributed essentially evenly between the five steps. This indicates that the effect selection model chooses the effects in order of importance for multi-effect sequences, thus making the intermediate steps of *SerumRNN* more educational for the user.

The most surprising result is that *SerumRNN* consistently outperforms the Oracle baseline, especially for the most robust MSSMAE metric. This implies that *SerumRNN* sometimes chooses effects in an order that is more optimal for the effect parameter models than using the original effect order that created the target audio. We find this result exciting because it suggests our neural architecture and ensemble of models is able to extrapolate and discover solutions outside of the scope of what would be considered "correct" by the training data.

From an auditory point of view, the reduction in error for each step of *SerumRNN* can be heard very clearly with the final output audio often being indistinguishable from the target audio. We highly recommend listening to the audio examples at https://bit.ly/serum_rnn.

Table 4. Comparison of overall mean error reduction for all systems.

Preset group	System	Mean error Δ against target audio					
		MSE	MAE	LSD	MFCCD	PCC	MSSMAE
Advanced modulating shapes	Oracle	−4.02	**−9.45**	**−8.48**	**−97.14**	14.80	−28.67
	SerumRNN	**−4.04**	−9.41	−8.36	−96.17	**15.53**	**−32.53**
	SerumRNN NI	−4.02	−9.20	−8.25	−94.55	15.26	−30.06
	SerumCNN	−3.97	−9.21	−8.18	−93.72	14.71	−28.84
	Random	−3.33	−7.45	−6.52	−74.40	11.96	−25.55
	SerumCNN 1S	−3.96	−9.12	−8.10	−92.68	14.51	−28.50
Advanced shapes	Oracle	−2.96	**−8.09**	**−7.21**	−83.01	15.90	−37.74
	SerumRNN	**−2.97**	−8.08	−7.17	**−83.34**	**16.87**	**−41.24**
	SerumRNN NI	−2.96	−7.98	−7.07	−81.80	16.79	−39.14
	SerumCNN	−2.92	−7.84	−7.05	−81.21	15.84	−38.89
	Random	−2.48	−6.39	−5.71	−65.95	14.20	−35.18
	SerumCNN 1S	−2.88	−7.65	−6.83	−78.40	15.55	−37.80
Basic shapes	Oracle	−4.00	−9.49	−8.19	−92.12	18.78	−41.16
	SerumRNN	**−4.08**	**−9.62**	**−8.35**	**−94.08**	**20.03**	**−42.32**
	SerumRNN NI	−3.79	−9.02	−7.75	−86.36	19.23	−40.67
	SerumCNN	−3.45	−8.50	−7.25	−81.01	19.32	−41.95
	Random	−3.14	−7.20	−6.21	−69.70	15.14	−35.30
	SerumCNN 1S	−3.56	−8.52	−7.36	−82.60	18.93	−40.06

Table 5. Eval. metrics for *SerumRNN* (*Adv. Mod. Shapes* preset group).

Metric	Mean error against target				Mean error Δ per step				
	Init.	Final	Δ	Δ%	1	2	3	4	5
MSE	4.97	0.94	−4.04	−81.19	−2.64	−0.86	−0.38	−0.16	−0.10
MAE	16.47	7.06	−9.41	−57.11	−5.49	−2.30	−1.13	−0.48	−0.33
LSD	15.59	7.23	−8.36	−53.63	−5.04	−1.95	−0.97	−0.41	−0.27
MFCCD	167.52	71.35	−96.17	−57.41	−56.37	−23.26	−11.41	−5.12	−3.19
PCC	68.48	84.01	15.53	−22.68	5.81	6.23	2.37	1.33	1.53
MSSMAE	83.38	50.85	−32.53	−39.02	−3.70	−15.63	−9.47	−3.81	−2.23

Table 6. Evaluation metrics for the random baseline (*Adv. Mod. Shapes* preset group).

Metric	Mean error against target				Mean error Δ per step				
	Init.	Final	Δ	Δ%	1	2	3	4	5
MSE	4.99	1.66	−3.33	−66.71	−0.84	−0.95	−0.85	−0.81	−0.69
MAE	16.49	9.05	−7.45	−45.15	−1.70	−2.00	−1.98	−1.99	−1.89
LSD	15.62	9.10	−6.52	−41.72	−1.47	−1.74	−1.78	−1.70	−1.69
MFCCD	167.81	93.41	−74.40	−44.34	−16.85	−19.98	−19.84	−19.26	−19.53
PCC	68.24	80.20	11.96	17.52	1.73	3.72	3.85	4.00	4.15
MSSMAE	83.16	57.60	−25.55	−30.73	−4.04	−6.34	−7.63	−5.91	−8.74

5 Conclusion

Overall, *SerumRNN* is consistently able to produce intermediate steps that bring the input audio significantly closer to the target audio while using a fully featured, industry leading synthesizer VST plugin. *SerumRNN* also provides near real-time, quantitative feedback about which effects are the most important. In addition to this, it can also find effect sequence orders that are potentially more optimal than what the target audio was originally created with. With *SerumRNN* the user can pick and choose which intermediate steps they would like to use and can feed tweaked versions back into the system for additional fine-tuning or to learn more. We also noticed fun creative applications when our system produced unexpected results or was given significantly out of domain target audio.

Currently, the training data for *SerumRNN* is sampled randomly and thus many of the generated sounds are not useful musically. As future work we would like to improve the efficiency of this using adversarial techniques. We would also like to evaluate the system's usefulness and perceived audio similarity with a user study and plan on packaging *SerumRNN* into a Max for Live [1] plugin to make it easily accessible to anyone learning sound design.

We believe our research is just the tip of the iceberg for building AI powered sound design tools. We can imagine a future where tools like ours might be able to find more efficient and simpler methods of creating sounds, thus educating students more effectively and democratizing sound design.

A Appendix

See Tables 7, 8 and 9.

Table 7. Effect selection models prediction accuracy (*Advanced Shapes* and *Basic Shapes* preset groups).

Preset group	Effect sel. model	Step number (no. of effects for *SerumCNN*)					
		1	2	3	4	5	All
Advanced shapes	*SerumRNN*	**0.993**	**0.983**	**0.981**	**0.986**	0.994	**0.985**
	SerumRNN NI	0.990	0.968	0.959	0.966	**1.000**	0.969
	SerumCNN	0.992	0.968	0.941	0.921	0.909	0.952
Basic shapes	*SerumRNN*	**0.992**	**0.983**	**0.983**	**0.974**	**1.000**	**0.983**
	SerumRNN NI	0.991	0.970	0.954	0.971	**1.000**	0.969
	SerumCNN	0.990	0.963	0.936	0.913	0.884	0.947

Table 8. Evaluation metrics for *SerumRNN* (*Advanced Shapes* preset group).

Metric	Mean error against target				Mean error Δ per step				
	Init.	Final	Δ	Δ%	1	2	3	4	5
MSE	3.92	0.95	−2.97	−75.80	−1.74	−0.76	−0.35	−0.14	−0.06
MAE	15.28	7.20	−8.08	−52.87	−4.48	−2.01	−1.13	−0.51	−0.20
LSD	14.44	7.27	−7.17	−49.64	−4.12	−1.68	−0.96	−0.45	−0.17
MFCCD	153.24	69.90	−83.34	−54.38	−46.05	−20.51	−11.57	−5.89	−2.34
PCC	64.26	81.12	16.87	26.25	7.59	5.63	2.32	1.75	1.48
MSSMAE	95.76	54.52	−41.24	−43.07	−19.55	−11.00	−5.82	−4.16	−2.29

Table 9. Evaluation metrics for *SerumRNN* (*Basic Shapes* preset group).

Metric	Mean error against target				Mean error Δ per step				
	Init.	Final	Δ	Δ%	1	2	3	4	5
MSE	5.25	1.16	−4.08	−77.83	−2.58	−0.93	−0.39	−0.21	−0.08
MAE	16.82	7.20	−9.62	−57.19	−5.80	−2.23	−1.05	−0.66	−0.16
LSD	15.69	7.34	−8.35	−53.25	−5.09	−1.86	−0.92	−0.57	−0.18
MFCCD	166.76	72.68	−94.08	−56.42	−55.78	−22.05	−10.78	−6.57	−2.03
PCC	65.97	86.00	20.03	30.37	13.02	3.33	2.06	1.76	1.19
MSSMAE	87.18	44.85	−42.32	−48.55	−21.67	−13.33	−3.46	−3.36	−1.95

References

1. Ableton: Max for Live—Ableton 2020. Accessed 19 Nov 2020. https://www.ableton.com/en/live/max-for-live/
2. Barkan, O., Tsiris, D., Katz, O., Koenigstein, N.: Inversynth: deep estimation of synthesizer parameter configurations from audio signals. IEEE/ACM Trans. Audio Speech Lang. Process. **27**(12), 2385–2396 (2019). https://doi.org/10.1109/TASLP.2019.2944568
3. Cáceres, J.P.: Sound design learning for frequency modulation synthesis parameters (2007)
4. Clevert, D.A., Unterthiner, T., Hochreiter, S.: Fast and accurate deep network learning by exponential linear units (elus). CoRR abs/1511.07289 (2016)
5. Damskägg, E.P., Juvela, L., Thuillier, E., Välimäki, V.: Deep learning for tube amplifier emulation. In: ICASSP 2019–2019 IEEE International Conference on Acoustics, Speech and Signal Processing (ICASSP), pp. 471–475 (2019)
6. Duda, S.: Serum: Advanced Wavetable Synthesizer - Xfer Records 2020. 19 Accessed Nov 2020. https://xferrecords.com/products/serum
7. Engel, J., Hantrakul, L.H., Gu, C., Roberts, A.: Ddsp: differentiable digital signal processing. In: International Conference on Learning Representations (2020). https://openreview.net/forum?id=B1x1ma4tDr
8. Fedden, L.: RenderMan 2018. Accessed 19 Nov 2020. https://github.com/fedden/RenderMan

9. Hu, Y., He, H., Xu, C., Wang, B., Lin, S.: Exposure: a white-box photo post-processing framework. ACM Trans. Graph. **37**(2) (2018). https://doi.org/10.1145/3181974
10. Kingma, D.P., Ba, J.: Adam: a method for stochastic optimization. CoRR abs/1412.6980 (2015)
11. Ly, E., Villegas, J.: Genetic reverb: synthesizing artificial reverberant fields via genetic algorithms. In: EvoMUSART (2020)
12. Macret, M., Pasquier, P.: Automatic design of sound synthesizers as pure data patches using coevolutionary mixed-typed cartesian genetic programming. In: Proceedings of the 2014 Annual Conference on Genetic and Evolutionary Computation, GECCO 2014, pp. 309–316 (2014). https://doi.org/10.1145/2576768.2598303
13. Ramírez, M.A.M., Reiss, J.: End-to-end equalization with convolutional neural networks. In: Proceedings of the 21st International Conference on Digital Audio Effects (DAFx-18) (2018)
14. Sheng, D., Fazekas, G.: A feature learning siamese model for intelligent control of the dynamic range compressor. In: 2019 International Joint Conference on Neural Networks (IJCNN), pp. 1–8 (2019)
15. Sommer, N., Ralescu, A.: Developing a machine learning approach to controlling musical synthesizer parameters in real-time live performance. In: MAICS (2014)
16. Srivastava, N., Hinton, G.E., Krizhevsky, A., Sutskever, I., Salakhutdinov, R.: Dropout: a simple way to prevent neural networks from overfitting. J. Mach. Learn. Res. **15**, 1929–1958 (2014)
17. Tatar, K., Macret, M., Pasquier, P.: Automatic synthesizer preset generation with presetgen. J. New Music Res. **45**, 124–144 (2016)
18. Thio, V., Donahue, C.: Neural loops. In: NeurIPS Workshop on Machine Learning for Creativity and Design (2019)
19. Yee-King, M.J., Fedden, L., d'Inverno, M.: Automatic programming of VST sound synthesizers using deep networks and other techniques. IEEE Trans. Emerg. Top. Comput. Intell. **2**(2), 150–159 (2018). https://doi.org/10.1109/TETCI.2017.2783885

Parameter Tuning for Wavelet-Based Sound Event Detection Using Neural Networks

Pallav Raval[1](✉) and Jabez Christopher[2]

[1] Department of Electrical and Electronics Engineering, BITS Pilani Hyderabad Campus, Hyderabad, Telangana, India
f20170615@hyderabad.bits-pilani.ac.in

[2] Department of Computer Science and Information Systems, BITS Pilani Hyderabad Campus, Hyderabad, Telangana, India

Abstract. Wavelet-based audio processing is used for sound event detection. The low-level audio features (timbral or temporal features) are found to be effective to differentiate between different sound events and that is why frequency processing algorithms have become popular in recent times. Wavelet based sound event detection is found effective to detect sudden onsets in audio signals because it offers unique advantages compared to traditional frequency-based sound event detection using machine learning approaches. In this work, wavelet transform is applied to the audio to extract audio features which can predict the occurrence of a sound event using a classical feedforward neural network. Additionally, this work attempts to identify the optimal wavelet parameters to enhance classification performance. 3 window sizes, 6 wavelet families, 4 wavelet levels, 3 decomposition levels and 2 classifier models are used for experimental analysis. The UrbanSound8k data is used and a classification accuracy up to 97% is obtained. Some major observations with regard to parameter-estimation are as follows: wavelet level and wavelet decomposition level should be low; it is desirable to have a large window; however, the window size is limited by the duration of the sound event. A window size greater than the duration of the sound event will decrease classification performance. Most of the wavelet families can classify the sound events; however, using Symlet, Daubechies, Reverse biorthogonal and Biorthogonal families will save computational resources (lesser epochs) because they yield better accuracy compared to Fejér-Korovkin and Coiflets. This work conveys that wavelet-based sound event detection seems promising, and can be extended to detect most of the common sounds and sudden events occurring at various environments.

Keywords: Sound event detection · Audio processing · Wavelet transform · Artificial neural network

1 Introduction

A sound event is a sudden onset of a trigger event indicating caution. It can be found everywhere around us such as the siren of an ambulance on the road, glass-break or dog-bark inside the house and many more. Sound event detection (SED) is an active field

J. Romero et al. (Eds.): EvoMUSART 2021, LNCS 12693, pp. 235–247, 2021.
https://doi.org/10.1007/978-3-030-72914-1_16

of research because of its vast application in real word. SED has been applied to detect baby-cry sounds in the home environment [1], ringtones, sirens and smoke-alarms in a workspace environment [2, 3], gunshots and screams in road surveillance applications [4]. Video and image-based signal processing approaches are also supportive of SED [5]. SED has been implemented using multiple models with various network architectures in the deep learning framework. The convolutional neural network (CNN) model is a popular deep learning framework; it is useful for sound classification applications because it can detect shift patterns in the time and frequency domains [6]. The recurrent neural network (RNN) model can classify sound events by extracting features using long temporal-context information. With gated recurrent units (GRU) [7] or long short-term memory (LSTM) units [8], the RNN can be appropriately trained for SED. Models combining CNN and RNN (called recurrent convolutional neural network RCNN or CRNN) have also been implemented, in which the former exploits the frequency-shift features, and the latter models the temporal structures in classification [9, 10]. Some other methods on polyphonic mixtures [11] have also been proposed. In one of the pioneering works, of Wan et al., they reported the significant advantages of wavelet-based processing over traditional time and frequency methods [12].

Frequency transform-based sound event detection has become popular recently because of similar frequency information associated with a sound event. A sudden sound event is characterized by an instantaneous change in the signal. In the time domain, it is difficult to resolve these peaks arising due to the sudden sound event. The short time fourier transform (STFT) is popular method for frequency domain processing, however it is limited by the tradeoff between time and frequency resolution because of the uncertainty principle. Hence the temporal resolution for sudden sound events is not flexible in the STFT. Wavelet transform is also better suited than Fourier transform in our application because of its advantage compared to Fourier transform in analyzing non stationary or real-world signals [13].

To overcome this problem, this work adapts the use of wavelets. Wavelets are extremely localized in nature and therefore can be used to detect sharp transitions or contrasts in a signal. Because of this unique feature, wavelets have already found many applications in audio denoising, fingerprinting and compression [14]. Wavelet transform is done in two ways: discrete wavelet transforms (DWT) and continuous wavelet transform (CWT). Both these transform extract features based on scaling and translation values to represent the audio signal without losing any information. The proposed approach applies DWT over CWT as CWT yields redundant data, processing of which will be computationally expensive.

2 Methodology

The focus and novelty of the proposed approach lies in leveraging the advantage of wavelet processing approaches over traditional time frequency processing methods.

Relevant audio data containing sound events are integrated and standardized in order to make it uniform. Additionally, data cleaning approaches are also used to remove any mislabeled or outlier data in the audio files. The data obtained after normalization is very large, and classifying events based only on the basis of audio amplitude information is less intuitive and moreover computationally expensive. To extract meaningful time frequency coefficients, wavelet transforms are applied on a window, by which the same data can be represented without any loss of information. Features are extracted from these DWT coefficients (data) to ensure that the information obtained from a window is optimally represented. The data obtained is used for training and testing of the supervised machine learning approaches and for further analysis. The architecture of the proposed SED approach along with all the modules and the components is presented in Fig. 1.

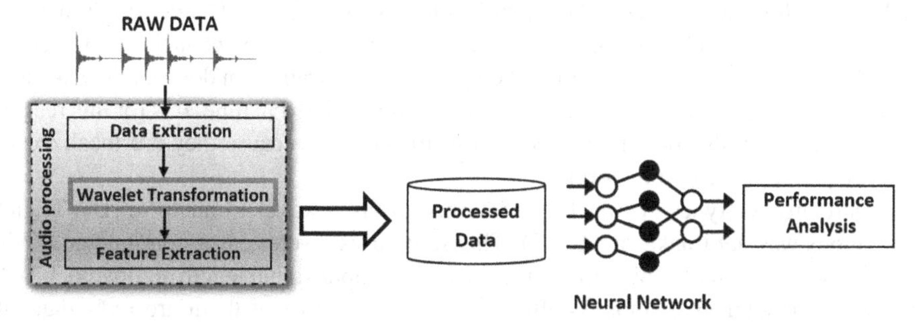

Fig. 1. SED architecture

2.1 Data Extraction

The main application for sound event is in road and home surveillance. The sound event is often a sudden onset in the environment, that is why it is extremely important to get relevant data. Our project's aim is to detect sound events at every particular time frame or window. The data that is required for this needs to have distinct events for short periods; not like one event occurring in a 30 s time frame. The UrbanSound8k data [15] is used; it contains sound events limited to a time frame (a few seconds) which is appropriate for experimental and analytic purposes of this work. From the urban sound 8k dataset, two sound events corresponding to road surveillance in the form of Siren and Car horn are taken. For the default situation in which there are no sound events, a combination of urban environmental sounds (park or airport scene) from the DCASE dataset [16] is considered. Once the data is extracted standard normalization is done over the audio files. The proposed approach has only 3 classes as it intends to identify the parameters for wavelet-based SED and focus less on every event detection; however, from the observations it can be proved that this method can be used to detect most commonly occurring sound events.

2.2 Wavelet Transformation

The wavelet transform (WVT) and short time Fourier transform (STFT) are two popular transformations in signal processing. The STFT coefficients represents time and frequency domain information simultaneously using a window. However, because of the uncertainty principle, the resolution in one domain affects the resolution in the alternate domain. WVT is better than STFT because it solves the above problem. WVT uses wavelets that can be translated and scaled to represent a signal. Higher frequencies can be represented using small time domain window size whereas low frequencies can be represented using bigger window size. This way a good frequency resolution is achieved for low frequency signals and good time resolution is achieved for high frequency signals.

Real world signals are often slowly changing oscillation combined with high amplitude changes. The sudden changes in these signals carry the most unique features (perceptually and informatively) related to identification of sound event. The Fourier transform is an effective method to process this data. However, the sudden changes aren't represented effectively. The reason for that being the Fourier transform denotes the data as a combination of sine and cosine waves, which are unlocalized in time. It is for this reason that wavelets are also preferred over Fourier transforms because they are localized in time frequency Fourier transform.

There are two types of wavelet transforms: discrete wavelet transform (DWT) and continuous wavelet transform (CWT). This study uses discrete over continuous. DWT gives almost the same number of coefficients as the input signal length compared. CWT on the other hand gives more coefficients, but since most of them are redundant, it becomes computationally expensive to process those coefficients. Discrete wavelet transforms are used in representing a signal efficiently. In simple understanding, DWT can be thought of as an encryption method because signals can easily be reconstructed using wavelet coefficients obtained after applying DWT. It is applied to a signal using filter banks: a signal of length N is decomposed using hid low pass filters to obtain two individual streams: approximate and detailed coefficients (also known as high pass and low pass coefficients). To maintain the original length, the detailed and approximate coefficients are down sampled by half. For every further decomposition level, the low pass stream is further split into approximate and detailed coefficients. The approximate and detailed coefficients can be merged to reconstruct the original signal.

2.3 Feature Extraction

The wavelet coefficients are extracted by translating and scaling of the wavelet over the audio signal. If wavelet transform is applied to a sinusoidal waveform, it will look like the figure shown above. Figure 2 shows how DWT coefficients represents sinusoidal signal into a combination of high frequency and low frequency.

After applying wavelet transforms, the following statistical features are extracted from the coefficients obtained:

1. Mean: Mean is a measure of the average of the sequence.
2. Median: The middle value in a sequence when arranged in an ascending/descending order is the median.
3. Mode: The most frequent value in a sequence is called the mode.
4. Kurtosis: Kurtosis is the feature that measures the "tailness" of a sequence.
5. Skewness: Skewness is a statistical feature indicating the variation of any given distribution from the standard bell curve.

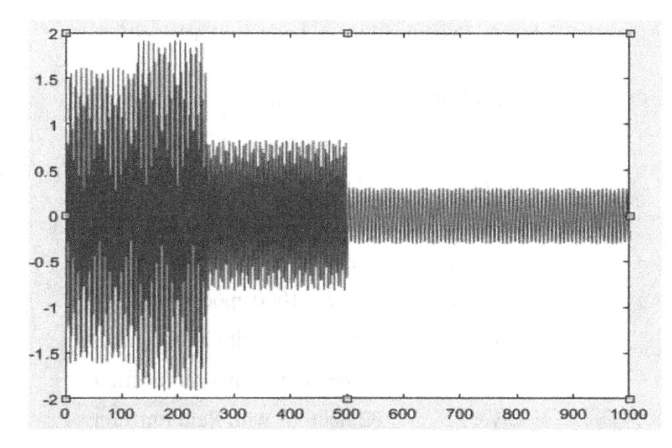

Fig. 2. DWT Coefficients extracted from a constant frequency sinusoidal signal

6. Standard Deviation (STD): It measures the dispersion of the sequence from its mean value using mean square distance.
7. Root mean square (RMS): Square root of mean square value obtained in a sequence is called the root mean square.
8. Mean Absolute Deviation (MAD): This feature is also a measure of the dispersion of the data from its mean value, but unlike standard deviation this is calculated using absolute distance.
9. Variance: The square of standard deviation is called variance.
10. Minimum (MIN): minimum coefficient in a sequence.
11. Maximum (MAX): maximum coefficient in a sequence.
12. Range: This indicates the spread of a sequence.
13. Average rate of change (CHG): This feature measures the amplitude difference between two consecutive data points (rate of change) and averages it out for the whole sample.
14. Absolute average rate of change (MCHG): This feature measures the absolute amplitude difference between two consecutive data points (rate of change) and averages it out for the whole sample.

The processed data is now ready to be trained using supervised machine learning approaches.

3 Classification

After obtaining the feature notation for a window, the simple feedforward artificial neural networks classifier trained using a gradient descent-based backpropagation algorithm is used for classifying the sound events. The specifications of the two multi-layered neural network architectures are shown in Table 1:

The output layer in both models identifies the 3 different sound events considered in this work: Siren, car horn and default situation (No sudden event).

Table 1. Specifications of the neural network

	Specification	Description
Model 1	Layer 1	100 neurons with TanH function
	Layer 2	50 neurons with TanH function
	Layer 3	18 neurons with TanH function
	Output layer	3 neurons with Softmax function
	Learning rate	0.01 for 1000 epochs
Model 2	Layer 1	256 neurons with Relu function
	Layer 2	64 neurons with Relu function
	Layer 3	32 neurons with Relu function
	Output layer	3 neurons with Softmax function
	Learning rate	0.02 for 500 epochs

4 Results and Discussions

Simple classification using traditional machine learning approaches proves that wavelet method is an appropriate method for sound event detection. However, this research work intends to identify the optimal wavelet parameters to achieve best accuracy. The parameters considered in this study are as follows:

- *Window size*: Window size is the time frame parameter. Every second of sound roughly contains 44,000 samples based on the Nyquist Shannon sampling theorem. A 10,000-window size, therefore corresponds to 1/4th of a second. In this report 3 different window sizes of 5000, 10000 and 20000 are considered.
- *Wavelet level*: Most common wavelets belong to a family of wavelets like Daubechies, Symlet, etc. We are defining the order of the wavelet as a wavelet level. 3 different wavelet levels are considered in this work.
- *Wavelet decomposition level*: The extent of decomposition of discrete wavelet transform coefficients is called the wavelet decomposition level. For example, if the total sample length is 1000 and the wavelet decomposition level is 2 then the wavelet transform coefficients would be split up into lengths of 250, 250 and 500 but if the level was 3 then it would be split into 125,125,250 and 500. 4 wavelet decomposition levels up to level 5 are considered.
- *Wavelet family*: The aim of this work is to determine the best performing wavelets for SED application. Wavelets can be grouped as a family when they are derived from a base wavelet as depicted in Fig. 3. A family of wavelets share similar properties with each other. The following wavelet families are considered because they have scaling functions unlike other wavelets like morlet and Mexican hat which do not have scaling properties. These particular families of wavelets have also been applied in other wavelet-based research [17].

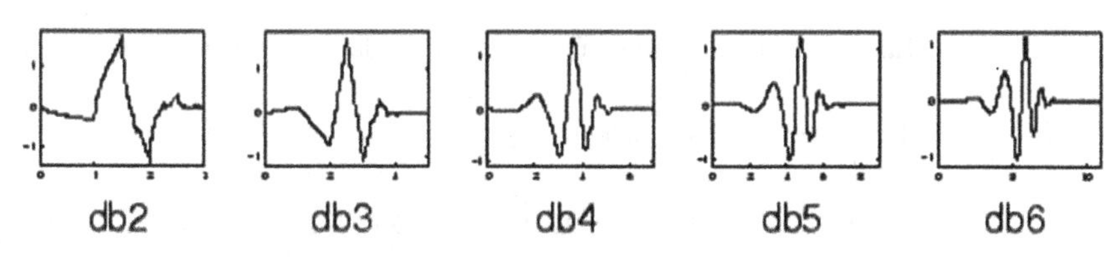

Fig. 3. Db family of wavelets

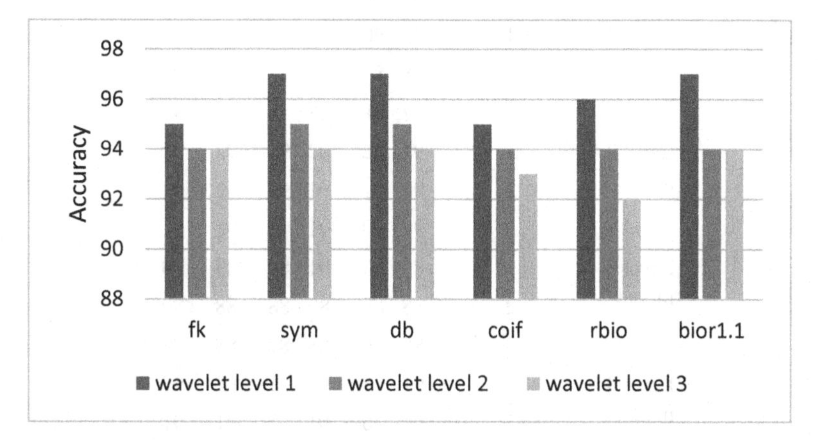

Fig. 4. Performance analysis by varying wavelet level

The following 6 families of wavelets are considered:

- 'db' Daubechies
- 'sym' Symlet
- 'coif' Coiflets
- 'bior' Biorthogonal
- 'fk' Fejér-Korovkin
- 'rbio' Reverse biorthogonal

Figure 3 denotes the decomposition in db family of wavelets. Most wavelets decompose in a similar manner as db, db2, db3 etc., however some families decompose in the manner fk4, fk6, fk8… or rbio1.1, rbio1.3, rbio1.5 and so on.

On the whole, this work considers 3 window sizes, 6 wavelet families, 4 wavelet levels, 3 decompositions and 2 classifier models. Classification of sound events is performed considering all permutations of these parameters. Tables 2, 3 and 4 present the testing accuracy (percentage) obtained for all various wavelet parameter combinations considered. The complete working code and other specifications are available at https://github.com/Palladium27/SED.

4.1 Wavelet Level

From Tables 2, 3 and 4, keeping Window size fixed at 20000 and wavelet decomposition level to 2, Classification performance is analyzed by varying the wavelet level for all families of wavelets. The results of the analysis are shown in Fig. 5. It can be observed that lower the wavelet level, better the classification accuracy.

Table 2. Classification performance for window size 5000

Decomposition level	Wavelet level	fk	sym	db	coif	rbio	bior
Level 2 model 1	1	91	92	92	90	92	92
	2	90	90	92	89	91	91
	3	90	90	90	89	91	91
Level 2 model 2	1	89	90	90	90	90	91
	2	88	89	89	88	89	90
	3	87	88	88	88	89	89
Level 3 model 1	1	91	92	92	91	92	92
	2	90	91	91	90	91	91
	3	90	90	91	90	91	91
Level 3 model 2	1	89	90	91	89	89	90
	2	88	89	89	88	88	89
	3	88	89	88	87	88	88
Level 4 model 1	1	91	92	92	91	92	93
	2	90	90	91	90	91	91
	3	89	90	91	89	91	91
Level 4 model 2	1	89	90	90	89	90	90
	2	88	89	89	89	89	89
	3	88	88	89	88	89	88
Level 5 model 1	1	91	92	92	91	91	91
	2	90	90	91	90	90	90
	3	90	91	90	88	90	89
Level 5 model 2	1	88	89	90	89	90	90
	2	88	88	89	88	89	89
	3	87	88	88	87	88	88

4.2 Window Size

From Tables 2, 3 and 4, keeping wavelet level to 2 and wavelet decomposition level to 2, classification is done by varying the window sizes for all families of wavelets. The performance analysis is shown in Fig. 6.

It can be observed that greater window size, leads to greater accuracy. To put these window sizes in perspective: a 1000 window size corresponds to 250 ms. This however doesn't mean that any large windows size can be taken. An observation was made in our experiment that whenever window size exceeded the time duration of the event, a drop in accuracy was noted. Window size therefore needs to calibrated based on the time duration of the combination of events considered in a classification.

Table 3. Classification performance for window size 10000

Decomposition level	Wavelet level	fk	sym	db	coif	rbio	bior
Level 2 model 1	1	93	95	94	93	95	95
	2	92	92	92	92	92	93
	3	91	91	91	91	92	92
Level 2 model 2	1	90	92	92	90	92	92
	2	89	90	89	89	90	91
	3	88	89	89	88	90	91
Level 3 model 1	1	93	94	94	93	94	94
	2	92	92	92	92	92	93
	3	92	91	92	92	92	92
Level 3 model 2	1	89	90	91	90	91	91
	2	89	88	89	89	89	90
	3	88	87	88	89	88	89
Level 4 model 1	1	93	94	93	93	94	94
	2	92	93	93	93	93	93
	3	92	92	92	92	93	92
Level 4 model 2	1	89	90	90	90	90	90
	2	88	89	89	89	88	89
	3	88	88	87	88	88	88
Level 5 model 1	1	93	94	94	94	95	94
	2	93	93	93	93	94	93
	3	92	92	93	92	93	92
Level 5 model 2	1	89	91	90	89	91	90
	2	88	88	89	88	90	90
	3	86	87	88	88	89	89

4.3 Wavelet Saturation

In machine learning, increasing the number of epochs, may enable the network to improve accuracy until it saturates after a certain number of epochs. The saturation accuracy of these wavelets are as follows; (all other parameters are constant).

From this analysis, the observation is that when the model is allowed to train for 1500 epochs all wavelets achieve 98% accuracies which intuitively means that all wavelets are suitable for classification. However, from Table 4 it can be noticed that sim, db, rbio and bior were already achieving an accuracy of 97%. This conveys that any wavelet could be used, but using certain wavelets will save computational resources. A clear

Table 4. Classification performance for window size 20000

Decomposition level	Wavelet level	fk	sym	db	coif	rbio	bior
Level 2 model 1	1	95	97	97	95	96	97
	2	94	95	95	94	94	94
	3	94	94	94	93	92	94
Level 2 model 2	1	91	95	93	93	92	92
	2	89	92	90	92	90	91
	3	89	92	91	90	89	90
Level 3 model 1	1	94	95	95	94	95	95
	2	94	94	94	93	94	94
	3	94	94	94	93	95	94
Level 3 model 2	1	88	92	89	88	89	91
	2	89	90	87	90	88	90
	3	88	90	90	86	87	89
Level 4 model 1	1	94	95	95	95	96	95
	2	94	94	94	94	95	94
	3	94	94	94	94	95	94
Level 4 model 2	1	91	91	91	90	91	92
	2	90	90	90	90	90	90
	3	89	89	90	87	91	89
Level 5 model 1	1	94	95	95	95	95	95
	2	94	94	94	94	94	94
	3	93	94	94	93	94	94
Level 5 model 2	1	90	91	91	90	90	90
	2	88	90	90	89	89	90
	3	86	87	88	87	86	89

observation is also seen how maximum accuracy decreases with increasing wavelet level decomposition in sync with results highlighted in Sect. 4.1.

4.4 Wavelet

From the results presented in Tables 2, 3 and 4, it can be said that the sym1, db1, rbio1.1 and bior1.1 all are yielding the best possible results across parameters with very less difference between the accuracy outputs amongst them. Fk4 and coif1 are comparatively less accurate than those wavelets.

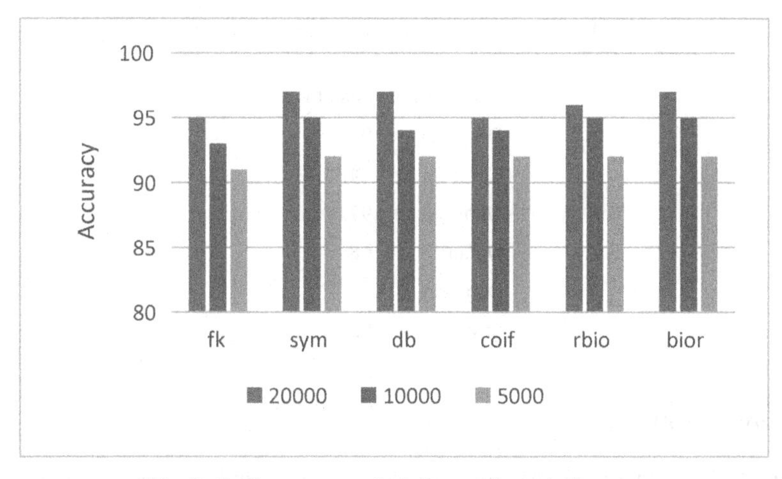

Fig. 5. Performance analysis by varying window size

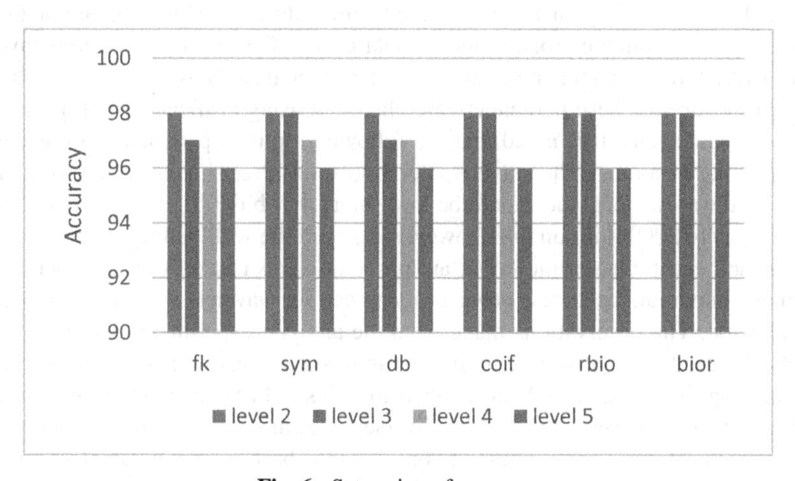

Fig. 6. Saturation of accuracy

4.5 Real World Sound Event Detection

Through our research, the best parameters suitable for an optimal accuracy is identified. To prove that this method can be used for multiple event detection, other sound events from the urban sound 8k dataset are extracted for classification using a simple one vs all classification approach using the Neural Network. The results are presented in Table 5.

Table 5. Sound event detection performance

Sound event	Accuracy
Dog bark	94
Car horn	94
Gunshot	97
Default	98
Baby cry	91
Siren	92

5 Conclusion

This work shows the importance of wavelet parameters and its importance and necessity to enhance classification performance for sound event detection. The input data is generated using wavelet transform combined with feature extraction applied on the raw audio files taken from the urban sound 8k dataset. It is the various parameters involved in this wavelet transform that motivates the experimental analysis in this work. The data obtained by varying these parameters are classified using artificial neural networks to chart out conclusions. It is noted that the following optimal parameters are necessary for improving accuracy: The wavelet decomposition level should be low (preferably level 2) and wavelet level too should be low, (bior1.1 > bior1.5). A larger window size (approximately 20000) is suitable; however, it should be well within the time duration of the sound event. One of the important observations is that all wavelets can be used to achieve maximum accuracy, however using certain wavelets will save computation resources too. The observations made from the analytic experimentation of this work can also be used to identify or classify common sound events. This method seems to be promising; it can be scaled and applied to industrial applications by incorporating larger real-world datasets representing diverse environments and sound events. Moreover, the focus of this work is limited to only find the best wavelet parameters for DWT based SED. It can be further extended for more sound events tried with diverse machine learning strategies.

References

1. Lavner, Y., Cohen, R., Ruinskiy, D., IJzerman, H.: Baby cry detection in domestic environment using deep learning. In: 2016 IEEE International Conference on the Science of Electrical Engineering (ICSEE), pp. 1–5. IEEE (2016)
2. Lilja, A.P., Raboshchuk, G., Nadeu, C.: A neural network approach for automatic detection of acoustic alarms. In: BIOSIGNALS, pp. 84–91 (2017)
3. Surampudi, N., Srirangan, M., Christopher, J.: Enhanced feature extraction approaches for detection of sound events. In: 2019 IEEE 9th International Conference on Advanced Computing (IACC), pp. 223–229 (2019)
4. Upadhyay, S.G., Bo-Hao, S., Lee, C.-C.: Attentive convolutional recurrent neural network using phoneme-level acoustic representation for rare sound event detection. Proc. Interspeech **2020**, 3102–3106 (2020)

5. Xu, Z., Yang, Y., Hauptmann, A.G.: A discriminative CNN video representation for event detection. In: Proceedings of the IEEE Conference on Computer Vision and Pattern Recognition, pp. 1798–1807 (2015)
6. Salamon, J., Bello, J.P.: Deep convolutional neural networks and data augmentation for environmental sound classification. IEEE Signal Process. Lett. **24**(3), 279–283 (2017)
7. Cho, K., Van Merriënboer, B., Bahdanau, D., Bengio, Y.: On the properties of neural machine translation: Encoder-decoder approaches. arXiv preprint arXiv:1409.1259 (2014)
8. Hayashi, T., Watanabe, S., Toda, T., Hori, T., Le Roux, J., Takeda, K.: Duration-controlled LSTM for polyphonic sound event detection. IEEE/ACM Trans. Audio Speech Lang. Process. **25**(11), 2059–2070 (2017)
9. Cakır, E., Parascandolo, G., Heittola, T., Huttunen, H., Virtanen, T.: Convolutional recurrent neural networks for polyphonic sound event detection. IEEE/ACM Trans. Audio Speech Lang. Process. **25**(6), 1291–1303 (2017)
10. Choi, K., Fazekas, G., Sandler, M., Cho, K.: Convolutional recurrent neural networks for music classification. In: 2017 IEEE International Conference on Acoustics, Speech and Signal Processing (ICASSP), pp. 2392–2396. IEEE (2017)
11. Martín-Morató, I., Mesaros, A., Heittola, T., Virtanen, T., Cobos, M., Ferri, F.J.: Sound event envelope estimation in polyphonic mixtures. In: ICASSP 2019-2019 IEEE International Conference on Acoustics, Speech and Signal Processing (ICASSP), pp. 935–939. IEEE (2019)
12. Wan, Y., et al.: Precise temporal localization of sudden onsets in audio signals using the wavelet approach. In: Audio Engineering Society Convention 147. Audio Engineering Society (2019)
13. Sifuzzaman, M., Rafiq Islam, M., Ali, M.Z.: Application of wavelet transform and its advantages compared to Fourier transform (2009)
14. Yu, G., Bacry, E., Mallat, S.: Audio signal denoising with complex wavelets and adaptive block attenuation. In: 2007 IEEE International Conference on Acoustics, Speech and Signal Processing-ICASSP 2007, vol. 3, pp. III-869. IEEE (2007)
15. Salamon, J., Christopher, J., Bello, J.P.: A dataset and taxonomy for urban sound research. In: Proceedings of the 22nd ACM international conference on Multimedia, pp. 1041–1044 (2014)
16. Heittola, T., Mesaros, A., Virtanen, T.: TUT urban acoustic scenes 2018. Development dataset [Data set]. Zenodo (2018). http://doi.org/10.5281/zenodo.1228142
17. Xu, R., Yun, T., Cao, L., Liu, Y.: Compression and recovery of 3D broad-leaved tree point clouds based on compressed sensing. Forests **11**(3), 257 (2020)

Raga Recognition in Indian Classical Music Using Deep Learning

Devansh P. Shah(✉), Nikhil M. Jagtap, Prathmesh T. Talekar, and Kiran Gawande

Department of Computer Engineering, Sardar Patel Institute of Technology, Mumbai, India
{devansh.shah,nikhil.jagtap,prathmesh.talekar,kiran_gawande}@spit.ac.in

Abstract. Raga is central to Indian Classical Music in both Carnatic Music as well as Hindustani Music. The benefits of identifying raga from audio are related but not limited to the fields of Music Information Retrieval, content-based filtering, teaching-learning and so on. A deep learning and signal processing based approach is presented in this work in order to recognise the raga from the audio source using raw spectrograms. The proposed preprocessing steps and models achieved 98.98% testing accuracy on a subset of 10 ragas in CompMusic dataset. A thorough study on the effect of various hyperparameters, sound source separation and silent part removal is carried out and is listed with results and findings of the same. A discussion on the predictions and behavior of deep learning models on audio apart from the dataset is also carried out. This new approach yields promising results and gives real time predictions from raw audio or recordings.

Keywords: Raga recognition · Indian Classical Music · Music Information Retrieval · Deep learning

1 Introduction

Raga is a very important aspect of Indian Classical Music (ICM). Formally, raga is a melodic framework for improvising melodic mode in music. Compositions and performances of ICM heavily rely on these ragas. In short, ragas can be termed as the grammar of Indian Classical Music. The concept of ragas is fundamental in both Hindustani as well as Carnatic Classical Music. A raga gets its identity from the *svaras* used in it. The seven svaras themselves have their own variations. The set of these svaras in unique sequence form a raga's unique property. However, some ragas can have the same set of svaras and still be different ragas depending on prominent *svaras* and the overall mood created by the characteristic phrases of the raga. For example, raga Jaunpuri and raga Darbari have the same set of notes. But raga Darbari can be distinguished from raga Jaunpuri by the characteristic oscillation around the *komal Ga* note. This oscillation is not present in raga Jaunpuri. ICM practitioners introduce their own

J. Romero et al. (Eds.): EvoMUSART 2021, LNCS 12693, pp. 248–263, 2021.
https://doi.org/10.1007/978-3-030-72914-1_17

variations in a raga to enhance melodic performance. This makes identifying raga not only interesting but a challenging task. Moreover, in Indian Classical Music, svaras are relative to each other and don't have fixed pitch value which is the case in Western Music. A performer chooses his/her base pitch of svara *Sa* and all the other svaras are relative to that. This pitch is called the Tonic Pitch of the performer. The most significant svara in a raga is termed as vadi and the second most significant svara is termed as samvadi. In a raga, svaras ascend and descend which are called arohana and avrohana respectively.

There have been various approaches to recognise the raga from music. Early approaches were based on the simple rule base heuristic, pitch tracking and hidden Markov models. The CompMusic dataset which is used in this work as well was published by Gulati et al. [6,7]. It made the data driven approach possible in this domain. These works are detailed further in the section of Literature Survey.

The intense variation of relative pitches, variation in tempos, arohana and avrohana, the tonic pitch of the performer along with his/her own style of performing particular raga makes raga identification an even more complex task. Recognising raga of a recording or performance has its own benefits. It can be used for Music Information Retrieval (MIR) of ICM, music recommendations, learning and practicing ragas, composing and many more. Moreover, it can further enhance the music experience of appreciators of music. The real time feedback from model can be used for teaching-learning, compositions and performances.

The proposed work is based on the thorough understanding of the ragas of the author Devansh Shah and his decade long experience in the field of Indian Classical Music. The need of Raga Recognition, knowledge of Ragas and the understanding of deep learning frameworks have motivated this research and experimentation. A deep learning model of hybrid architecture and preprocessing pipeline is proposed in this work which can identify raga in real time. Through this work, the hypotheses of authors that the models will perform better with sound source separation of audio, models will have less confident predictions in the faster tempo parts of audio and removal of the silent part of audio will improve performance are also validated.

The work presented here has following contributions. 1) A deep learning model based on hybrid architecture is proposed. 2) Use of source separation technique to get vocals from audio. 3) Removing the silent parts which accumulates as a noise. 4) Using raw spectrograms to predict the raga from audio. 5) Overlapping windows in generation of spectrograms and sequences.

The rest of the document is structured as follows. The next section of Literature Survey briefs the previous work done in the similar problem domain. In the section Proposed Methodology, the novel approach is presented with Data Preprocessing and hybrid model. The Experimentation section details the experiments conducted during this research work. The results and findings of experiments are presented in the Results and Discussion section. The final section of Conclusion concludes the output of the work. The bibliography is listed at the end.

2 Literature Survey

In a conventional way, the task of *raga* recognition has been addressed during both learning-teaching phase and musical performances. Pupils of Indian Classical Music are taught various ragas by scholars in traditional *guru-shishya* way. Listeners of Indian Classical Music recognise the ragas from a musical phrase or a set of musical phrases called *pakad*. This *pakad* contains melodic theme of a raga encapsualtes the essence of that raga. One can identify raga by listening to the *pakad*. However, for beginners, it is very difficult to identify the raga from *pakad* and requires very high attention of the listener.

Being a very key element of Indian Classical Music, multiple attempts have been made to recognise the raga from the audio. As mentioned earlier, the ragas are ordered sets of *svaras*. Also, svaras are computationally very basic feature to extract from the audio. In several works including [1], authors extracted the svaras from the audio and then matched them with the already known svaras of the ragas. This method works on very basic features i.e. the svaras. Though it can capture the melodic details of music till some extent, it does not consider the temporal details of the same. Further, some ragas in the Indian Classical Music have same set of svaras but they are different ragas. In the works [2–5], authors used ascending and descending progressions of svaras called *arohana* and *avrohana*. Using arohana and avrohana, authors captured the temporal aspects of the music. They generated discrete representation of the melody using quantized pitch vectors. Even though it is a very simple and fast approach in raga recognition, it is not very accurate because of the discrete representation.

The work of Sarkar et al. [14] also used *vadi*, *samvadi*, *arohana* and *avrohana* of raga to classify them. The svara profile based on pitch is formed. The audio was then split into frequency bands and a magnitude spectrogram was generated. The audio was then represented as 168 dimensional vector and used with SVM classifier. Another novel approach was by Padmasundari et al. [15] in their work to identify the raga using Locality Sensitivity Hashing. By extracting pitch vector and doing locality search in its neighbourhood from the known database. The approach is highly scalable and can identify up to 3000 ragas. However, the accuracy is affected and the similar ragas cannot be identified very well.

In the work of Gulati et al. [6], authors used vector space model to represent the audio signals. Similar to vector space models in the field of Natural Language Processing, they built a predictive model to classify the ragas. They obtained 70% accuracy in classifying 40 ragas and 92% accuracy in classifying a subset of 10 ragas. Another work of Gulati et al. [7] showed even better results by using their Time-Delayed Melody Surface (TDMS). Authors performed pre-processing steps including Predominant Melody Estimation and Tonic Normalization followed by surface generation to generate TDMS. The works [6,7] were the first to make their CompMusic dataset publicly available and open the opportunities of research and application of data-driven approaches in Music Information Retrieval for Indian Classical Music.

The applications of artificial neural networks is vast and Music Information Retrieval is no exception. Convolutional Neural Networks (CNN) are proven to be effective in capturing patterns in images and time series data. In [8], author used CNN to recognise the ragas on the CompMusic dataset [6,7]. Using the grayscale plots of pitch vectors as inputs to a CNN model, author achieved accuracy of 96.7% on the set of 5 distinct ragas and 85.6% on the set of 11 distinct ragas. However, the accuracy of the model was around 50% when tested on allied ragas, the ragas which are similar to each other.

In another work titled PhonoNet [9], author used a multi-stage deep neural network model. In PhonoNet, audio clips are converted into mel-Chromagrams. Short Time Fourier Transform was used because of *taans* in Indian Classical Music. The first stage, CNN was trained to detect the ragas on CompMusic dataset fed in smaller chunks. In the second stage, the fully connected layers at the end of CNN was replaced by LSTM cells Layer. The motivation behind this CNN-LSTM hybrid model is that the CNN when trained first will learn the specific filters to identify the nuances of the raga. The second stage LSTM trained on larger chunks will learn the temporal melodic details of the audio and will be able to identify the raga. After training both stages along with data augmentation, PhonoNet achieved state of the art accuracy.

Similar to the Word Embeddings in Natural Language Processing, J. Ross et al. [10] proposed identification of similar ragas from embeddings which they termed as *note-embeddings*. These embeddings were learnt from the notations. Authors used the *bandish* notation of the compositions to train the recurrent network and learn the note-embeddings. Even though these embeddings cannot directly identify a raga from an audio recording, the embeddings can be used to detect the similarity between two recordings in terms of raga. The cosine similarity between two vectors can be a useful parameter in MIR.

The latest SOTA approach to the raga recognition problem on the CompMusic dataset is given by [11]. This was the only paper where the audio was preprocessed using audio source separation techniques. The authors separated vocals from the mixed audio using Mad Twinnet [11]. After performing tonic normalization and pitch tracking, the pitch values were fed to an LSTM recurrent neural network with Attention layers. The model named "SRGM1" achieved state of the art results on the CompMusic dataset with the accuracy of 97% on 10 Ragas subset. The authors also created another model "SRGM2" which solved the problem of sequence ranking.

3 Proposed Methodology

The recognition of raga from an audio is complex problem and can be tackled by signal processing and deep learning methods. Using the appropriate preprocessing, the audio files can be molded into sequences of some form interpretable by deep learning models. Thus, the problem of raga recognition can be transformed into a problem of a sequence classification. The proposed method takes an input audio and preprocesses it using the Data Preprocessing pipeline explained next.

It then feeds the output of the preprocessing to the deep learning model. The output of the model is a prediction of the raga for the input audio.

3.1 Data Preprocessing

Data Preprocessing can play a huge part in the performance of the whole system. The main aim of the preprocessing steps is how to best represent the audio data so that the deep learning models can extract features easily. A data preprocessing pipeline has been proposed which takes in the raw audio as its input and outputs spectrogram images. Figure 1 shows the detailed block diagram of the proposed data preprocessing pipeline. The various components in the preprocessing pipeline are explained below.

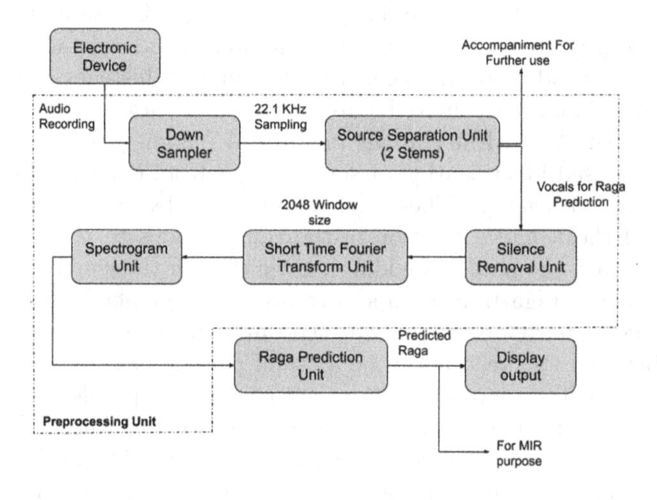

Fig. 1. Data preprocessing pipeline

Down Sampler. Audio files are recorded at some standard frequency, usually 44.1 KHz or 22.1 KHz. In order to have all the audio files at same frequency, the down sampler samples the input audio to 22.1 KHz. This down sampler is not required if the audio files are already at 22.1 KHz.

Source Separation. Audio source separation is a very active area of research. Source Separation unit separates out vocal track of a performance from the entire audio track. By performing source separation, the output of the data preprocessing unit will be more filtered. Another point to consider while performing source separation is that in an Indian Classical Music performance, the vocals of the singer are sufficient to determine the raga of the audio. While there is a loss of useful data of accompanying instruments like harmonium and *sarangi*, the data is more clear and the model should be able to extract features much easily.

Figures 2 and 3 demonstrate the difference by performing source separation. Figure 3 is more clear with the disappearance of the *tanpura* in the audio.

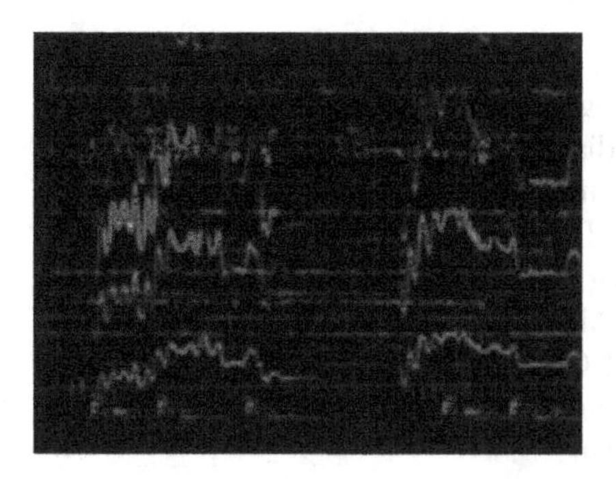

Fig. 2. Spectrogram image without source separation

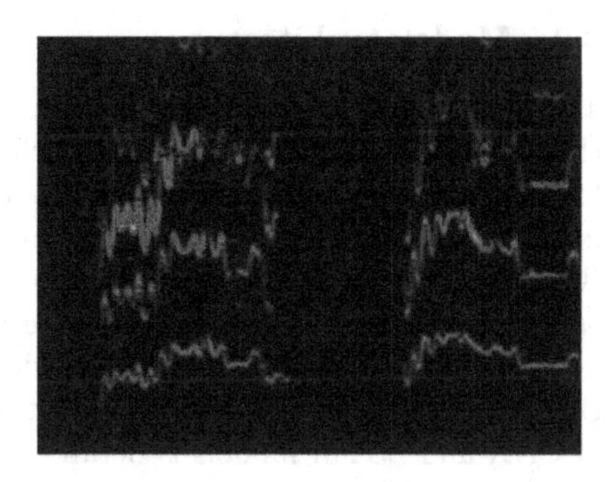

Fig. 3. Spectrogram image with source separation

Silence Removal. Although silence plays an important role in Indian Classical Music, an empty spectrogram is not much of use to predict the raga. This unit removes the silent parts of the audio. There are two methods used in this unit out of which at a time only one is used. The first method is to use InaSpeech Segmenter [12] which uses deep learning to predict silent parts of audio along with other things. The second method is to use simple thresholding. Any part of the audio where the intensity drops below 60 dB is automatically considered as silence.

Short Time Fourier Transform. The audio received from the previous block is fed to the Short Time Fourier Transform Unit (STFT Unit) which performs the Fourier Transform on the audio signal. This translates the audio from time domain to the frequency domain. It is here that one has to strike a balance between resolution in time and frequency. The window size and the overlapping percentage are some of the hyperparameters experimented on in this work. For example, an experiment can be done with *n fft* 2048 and 50% overlapping.

Spectrogram Unit. The output of the STFT unit is the input to the Spectrogram unit, which generates a spectrogram image of the audio. This unit plays an important role in deciding the representation of audio data as images. Changing the color scales, the y-axis scales, the y-axis limit can all change how deep learning models can easily extract features from the images.

All these units together form the Preprocessing unit. All units can be dropped except the STFT unit and the Spectrogram unit depending on the experiment.

Further experiments are carried out using this preprocessing pipeline in different modes.

3.2 Model Architecture

The data processed using the preprocessing pipeline is fed to the model to get the prediction about the raga. The earlier discussion showed the need of the deep learning model which can extract the features from the spectrogram and then identify the sequences of these features. The Convolutional Neural Networks (CNN) are exceptionally good at extracting the features and can represent them as a single vector. The Long Short Term Memory (LSTM) Cells can learn and identify the sequences. Thus, an hybrid of CNN and LSTM is very suitable for this problem statement. The architecture of the proposed model which consists of CNN and LSTM models is detailed further.

Convolutional Neural Network. The last step of the Preprocessing pipeline is the Spectrogram Unit which converts the audio into raw spectrogram images using the output of Short Time Fourier Transform Unit. These images are 324 pixels in width, 216 pixels in height and RGB in colors. The amount of audio data which is represented in one such image can be tuned easily, say 5 s or 10 s. The Fig. 4 shows the architecture of the CNN model with the dimensions for each layer labeled.

The input image in the form of a tensor of shape <3, 216, 324>(Channels, Height, Width) is fed to the CNN model. Convolution layer with kernel size 3, Batch Normalization followed by ReLu activation, Dropout and Max-Pooling layers are applied. The Max Pooling layer in the initial layer has kernel size and stride of 3 each while for next layers it is 2. The same sequence is applied 3 more times to get the final tensor of shape <64, 9, 13>. This tensor is then flattened into a single vector of shape 7488. This vector represents the features of an input and hence called as a feature vector. A fully connected hidden layer of 512 neurons is used after this. Then, a ReLU activation and dropout is used. This fully connected layer is then finally connected to the output layer of size equal to the number of ragas present in the dataset. The Log-Softmax activation is used at the end.

The same architecture is used in all the experiments in the Experimentation section. The model is trained separately from LSTM on dataset and then tested for accuracy. Further details on training procedure and hyperparameters is attached with each experiment.

Long Short Term Memory. The CNN model described above gives an output as a prediction of the raga. This prediction is based on only one spectrogram and thus has no information about notes played before that spectrogram. To capture this temporal information, a sequence of spectrograms or their feature vectors is required. In order to get the feature vector from the spectrogram, the last fully connected layers are dropped from the trained CNN model. Thus, a

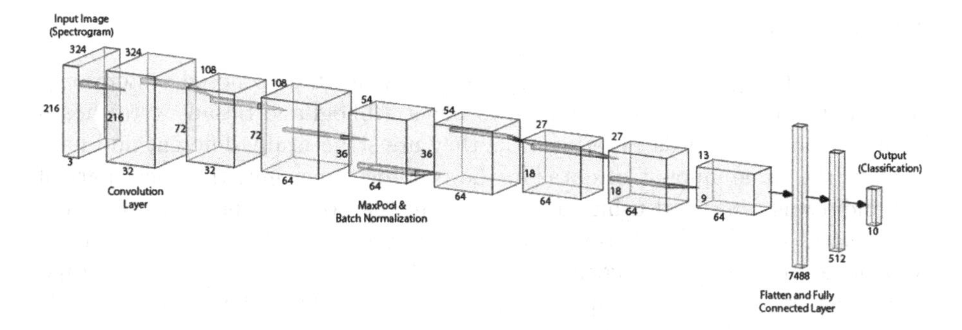

Fig. 4. The architecture of the proposed CNN model

flattened vector of size 7488 is obtained for each spectrogram input fed to the CNN. Using the consecutive spectrograms from the audio and converting them into feature vectors using CNN model, a sequence of feature vectors is obtained.

The LSTM model is trained on these sequences with the weights of the CNN model frozen and backpropagation done only on the LSTM model. The LSTM model with one layer and hidden size of 256 is created. The sequences of three or five feature vectors are used to train this LSTM model with suitable hyperparameters for each experiment.

The data preprocessing and the proposed models are used for all the experiments discussed in the next section. Some of the units in preprocessing pipelines are tuned or dropped or replaced in order to investigate the effects on model predictions. The model architecture stays the same throughout the implementation with slight changes in the hyperparameters.

4 Experimentation

In order to achieve the best possible results, best preprocessing pipeline, optimised training strategy for both CNN and LSTM models and validate our hypotheses, certain experiments were conducted. Each of these experiments used preprocessing and models from the previous section and is detailed further. The results and findings of all the experiments is presented in the following section of Results and Discussion.

The dataset used in this paper is the Hindustani Music Raga Recognition Dataset [7], which consists of around 116.2 h of audio data of 30 commonly heard ragas. There are 10 recordings for each raga in this dataset, though the length of those recording may vary.

It is important to understand the problem definition and train-test splits. For CNN, the problem is framed as identifying raga of n-seconds spectrogram. For LSTM, it is to identify raga of a sequence of spectrograms. The train-validation-test split is done on spectrograms (in case of CNN) and on sequences of spectrograms (in case of LSTM). Another small set of audios which is out of the CompMusic Dataset is used to validate our hypotheses.

4.1 Experiment 1

The goal of this very first experiment was to try out the proposed system on a very small subset of the data. Thus, two ragas, Abhogi and Basant were chosen from the dataset for this experiment as they are structurally different and have roughly the same amount of data. In the preprocessing unit, the vocal part of the audio was separated using Spleeter [13] in the Source Separation Unit and silent parts were removed using InaSpeech Segmenter [12]. The spectrograms were generated by processing the audio file in a stream using the librosa library. The parameters for taking input are frame length of 2048, hop length of 1024 and block size of 256. While performing STFT on the data, the number of output bins *nfft* was 4096, and the hop size was 64. A total of 1194 images were generated in this way. These images were then further split into train, test and validation datasets with a ratio of 80:10:10. A CNN model was trained in this experiment with the batch size of 8 for 500 epochs with the learning rate of 0.00001. The CNN model was able to achieve 96.64% accuracy on the test dataset in classifying these two ragas just by looking at spectrograms.

4.2 Experiment 2

The good results in experiment 1 were indication that CNNs are effective. The question was how scalable they will be with respect to more classes. To test that, we kept all the other parameters exactly the same as the experiment 1 except the number of ragas. The subset of CompMusic dataset was taken with the five ragas, Abhogi, Basant, Alhaiya Bilawal, Ahir Bhairav, Bairagi. A total of 3233 images were generated from these 5 ragas. In order to avoid overfitting, the model was trained for 200 epochs. After training for 200 epochs the model achieved 90% testing accuracy on the test dataset which had 324 images.

4.3 Experiment 3

In this experiment we study the effect of changing hyperparameters by tweaking the *n fft* parameter in the STFT unit to 2048. The intuition behind changing the *nfft size* parameters is that different representation of the same data might affect the way a model learns the distribution of the data. A similar model was trained with this data and we achieved the similar results in 200 epochs. The testing accuracy was 96.6%.

4.4 Experiment 4

As Experiment 3 is similar to Experiment 1 with the *nfft size* cut down to half, this experiment is similar to the Experiment 2 with *nfft size* 2048. There was not any significant improvement in the testing accuracy which was 90.625% compared to 90% of experiment 2.

4.5 Experiment 5

After noticing that the accuracy of CNN dangling around 90% for the subset of 5 ragas, the generation of spectrograms was tuned heavily. With no adjustment in the STFT, the spectograms were generated using the window length of 10 s and the hop length of 5 s. It means, there was 10 s long data in the spectrogram and the consecutive spectrograms had the overlap of 50% data. Due to the overlap, there was a significant increase in the number of images generated for the same amount of ragas, i.e. 7540 images. After training for just 100 epochs, batch size of 8 and the learning rate of 0.001, the CNN model was able to achieve the testing accuracy of 96.81%. This clearly outperforms the models in experiment 2 and 4.

4.6 Experiment 6

To expand more on the complexity of the dataset, five more ragas (Bhup, Kedar, Bilaskhani Todi, Puriya Dhanashree, Rageshri) were added to the subset. The choice of ragas is very interesting here. Raga Basant and Raga Puriya Dhanashree have the same notes and are melodically very close to each other. All the selected ragas had roughly the same amount of data to avoid class imbalance. The audio files are vocal separated and silent parts are removed using InaSpeech Segmenter [12]. Spectrograms are generated with a window size of 10 s and a hop size of 5 s. The parameters for the STFT algorithm are *nfft* 4096 with hop of 64 frames. The total number of images generated in this experiment is 15854 which were split into 80:10:10 ratio for training, validation and testing. Since this experiment was a scaled out version of experiment 5, we increased the number of epochs back to 200 with rest of the hyperparameters same. The trained model achieved the testing accuracy of 96.01%.

4.7 Experiment 7

This experiment was same to previous one, the only difference being in the silent separation unit. The previously mentioned method classified some data as noise and silent parts due to the varying quality of background noises in the audio. Simple threshold based silence removal is used in this experiment. The CNN model was set to train for 100 epochs but the best validation score was achieved before 100 epochs. The best model saved had achieved a validation accuracy of 94.8%. On the test dataset, the best model achieved the testing accuracy of 94.5%. In this experiment we used the best performing CNN model and trained the LSTM model to classify sequences of feature vectors. Sequences of length 5 were generated from all images. These sequences have to be consecutive in nature, i.e. the audio represented by the images must be in one consecutive sequence. Also, the sequences cannot span over multiple audio files, as that would not be a valid input. A total of 18,661 sequences were created in such a manner. The batch size for training sequences was 16 and a learning rate of 0.0001. LSTM model was trained for 20 epochs with a best validation accuracy of 91.4%. The best performing model achieved a testing accuracy of 98.06%.

4.8 Experiment 8

The aim of this experiment was to analyse the effect of vocal separation on the accuracy of the deep learning models. Therefore the vocal separation unit in the data preprocessing pipeline was skipped, and the models were trained on images directly generated from the raw audio with other hyperparameters same. Batch size was selected to be 8 and the learning rate was 0.0001. The model after 100 epochs achieved a testing accuracy of 99.78%. The Fig. 5 shows the loss vs epoch curve for training of the CNN in this experiment. Similar to the previous

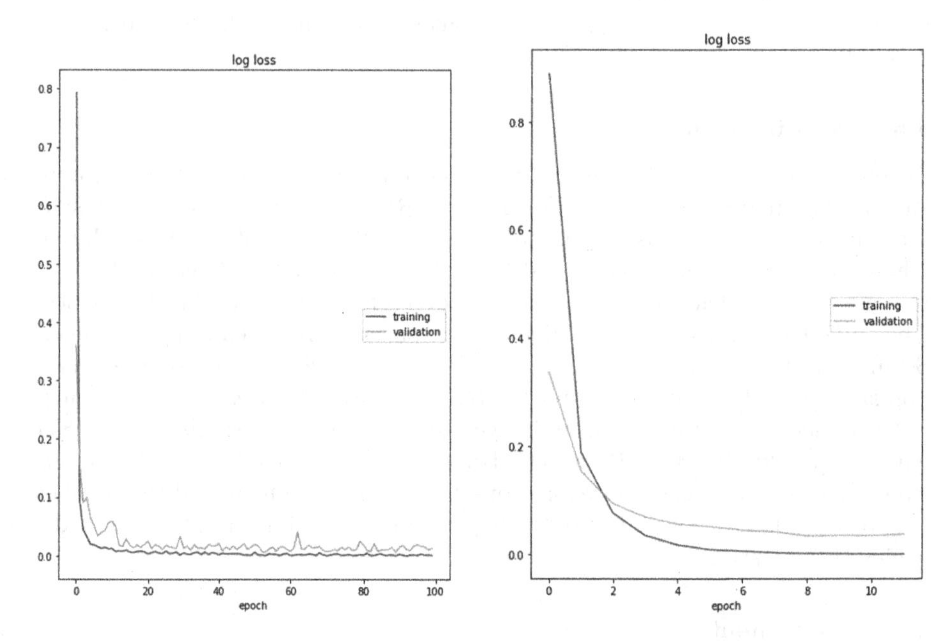

Fig. 5. CNN training in Experiment 8

Fig. 6. LSTM training in Experiment 8

experiment, a total of 22,646 sequences were generated. The batch size was 16 and the learning rate was 0.0001. The model was set to train for 20 epochs, but the best performing model was at epoch number 10. Using the early stopping method to prevent overfitting, after the validation loss did not decrease for 3 consecutive epochs, the model was stopped for training. The best performing model had a validation accuracy of 98.7%. It also achieved a testing accuracy of 98.98% on the test dataset. The loss vs epoch curve for LSTM training in this experiment is shown in the Fig. 6.

4.9 Experiment 9

This experiment was conducted to test the scalability of the proposed solution, whether it would scale up to 30 ragas or not. A CNN model was trained on data of all the 30 ragas. Some of the ragas in those 30 ragas were closely related to each other. Some examples being Raga Basant and Raga Puriya Dhanashree, Raga Bihag and Raga Maru Bihag. There were 82868 images generated from around 116.2 h of audio data. The batch size was increased to 16 from 8 and the learning rate was 0.0001. The CNN model was trained for 20 epochs in which it achieved a best validation accuracy of 99.4%. The model had achieved the testing accuracy of 99.51%. Similar to the previous experiments, sequences were generated of sequence length 5 from the available data. A total of 81368 sequences were generated. The LSTM model was trained with a batch size of 16 and a learning rate of 0.0001. The LSTM model after 5 epochs achieved a testing accuracy of 98.4% by correctly classifying 8007 sequences out of 8137 sequences.

Table 1. Table of comparisons of preprocessing steps of the experiments performed.

Expt. no.	Source separation	Silence removal	Spectrogram parameters (*nfft size*, hop size)	Total images	Testing accuracy
1	Yes	Using Ina model	4096, 64	1194	96.64
2	Yes	Using Ina model	4096, 64	3233	90.0
3	Yes	Using Ina model	2048, 64	1192	96.6
4	Yes	Using Ina model	2048, 64	3213	90.625
5	Yes	Using Ina model	4096, 64	7540	96.81
6	Yes	Using Ina model	4096, 64	15854	96.01
7	Yes	Simple threshold	4096, 64	19158	98.06
8	No	Simple threshold	4096, 64	23143	98.98
9	No	Simple threshold	4096, 64	82868	99.51

The table 1 shows the summary of the experiments. The experiments listed here are easily replicable and all the code used by authors is openly available. The plots for each experiment, the experiment notebooks and the inference code is available at this repository https://github.com/dev1911/raga_plus_plus.

5 Results and Discussion

This section briefs the results of the experiments and the findings. The models and preprocessing strategy proposed here work very well on the CompMusic dataset. The experiments conducted show these testing accuracies peaking more than 98%. The results from various hyperparameters can be drawn from these experiments as well. The *nfft size* for STFT Unit was taken 4096 after trying out various combinations. From experiments number 1, 2, 3 and 4 it is very clear that the *nfft size* of 2048 does not improve the model by a very large margin. Even lower sizes will generate distorted spectrograms with more noise. The confusion matrix for the CNN model which was trained with data of 30 ragas is shown in Fig. 7.

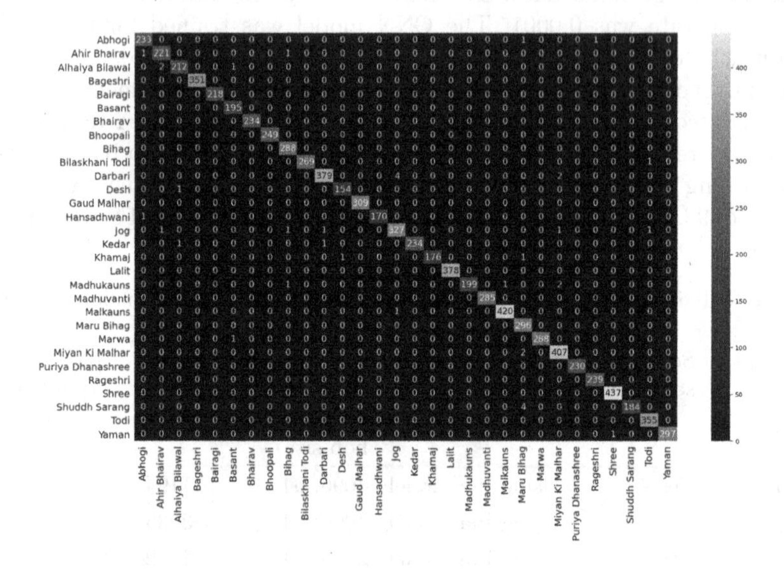

Fig. 7. Confusion matrix

In order to validate the applicability of this approach, some audio files which are completely out of the dataset where taken to test the approach. One of the songs was raga *Bhoopali* sung by Gaansaraswati Kishori Amonkar https://youtu.be/ipauyMfVYso on which both CNN and Hybrid models were tested. The Fig. 8 shows the predictions of the raga by CNN and the Fig. 9 shows the same by Hybrid model. With the better sequence length, the LSTM model will be able to give even better results.

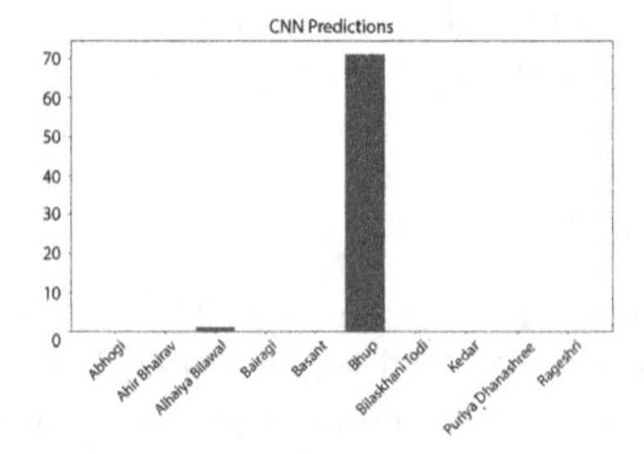

Fig. 8. Predictions by CNN for Raga Bhoopali

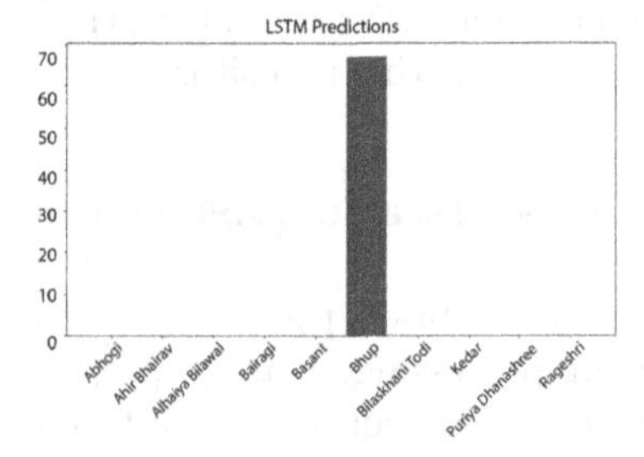

Fig. 9. Predictions by LSTM for Raga Bhoopali

One of our hypotheses is that the models will be able to predict the raga in the slower parts of performances with higher confidence and as the speed of the performance increases, the model might suffer from less confident predictions. To validate that, an audio of raga *Alhaiya Bilawal* by Ustad Rashid Khan https://youtu.be/mGobQ9ztOnY was taken. The models were able to predict the correct raga however the confidence level of the models is very interesting throughout the performance. Figures 10 and 11 shows the confidences of the predictions of the CNN model and LSTM model respectively with respect to time. It can be clearly noticed that when the faster tempo parts of performance start, there is a drastic fall in the confidences of models. As discussed earlier, this is because the features of audio during the fast tempos are not very distinctly clear in spectrograms. Accuracy of these models can be improved by training a different model which is specialised in extracting features from the spectrograms of fast tempo audio.

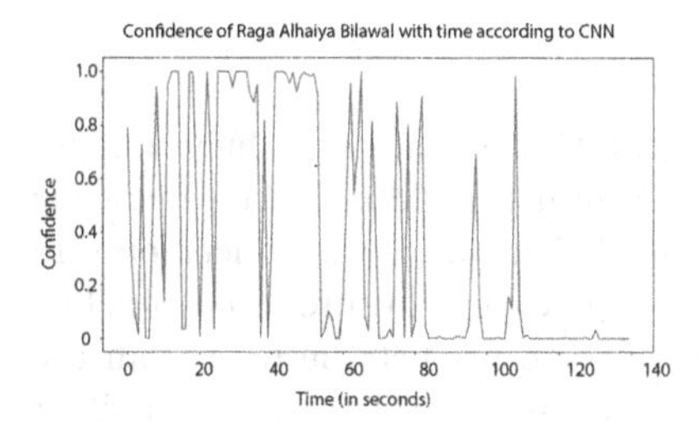

Fig. 10. Confidence of Alhaiya Bilawal vs time - CNN

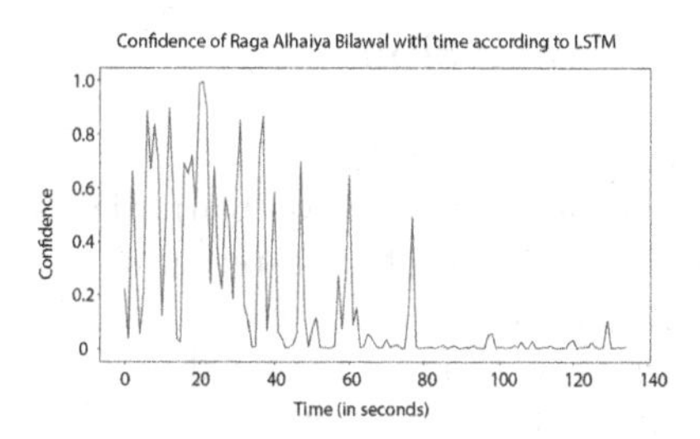

Fig. 11. Confidence of Alhaiya Bilawal vs time - LSTM

Another hypothesis to be validated through these experiments was to study the effect of Sound Source Separation on the performance of the models. Overall, the sound source separation results in better and consistent predictions by model. This was again tested using various songs including the ones which are out of the dataset. As an example, the audio of raga *Puriya Dhanashree* by Pandit Bhimsen Joshi https://youtu.be/AXHMckpzX40 was taken and fed to the same models as it is and also with sound source separation using Spleeter [13]. The bar graph in the Fig. 12 shows the predictions of CNN when the original song was used as the input. The one in the Fig. 13 shows the predictions of CNN when the source separated audio was used. The source separation improves the performance of model by a very large margin.

However, there are cases where the original audio performed better than the source separated counterpart of the audio. Mostly, this is because the models might make predictions by looking at frequency lines of instrumentals like Harmonium or *tabla*. Thus, this hypothesis is not entirely correct at this point. It needs more investigation because the fact that the deep learning models are just black boxes at one end. The separation unit uses Spleeter [13] which is not specifically developed for Indian Classical Instruments yet very decent at it. This

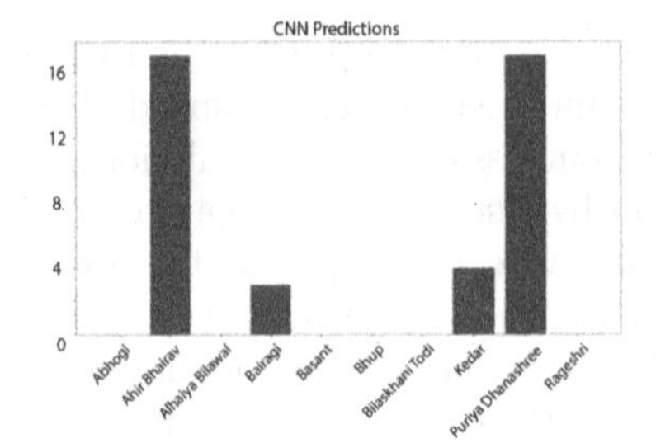

Fig. 12. CNN predictions on original audio of Puriya Dhanashree

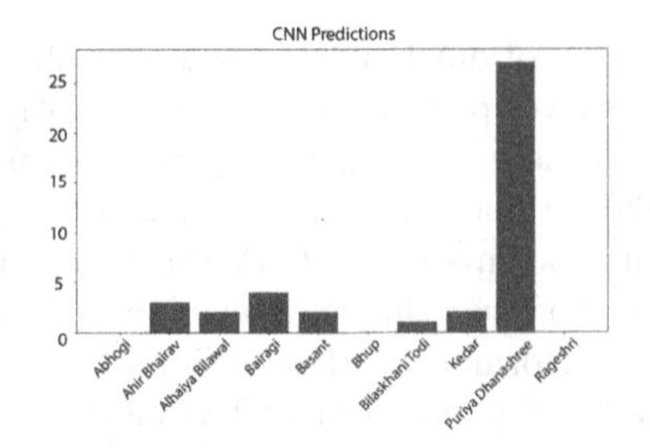

Fig. 13. CNN predictions on source separated audio of Puriya Dhanashree

hypothesis can fully evaluated by using a source separation model trained truly for ICM instruments.

6 Conclusion

Raga being an essential part of Indian Classical Music has a very huge impact on both forms of ICM, Carnatic as well as Hindustani Music. The intricacy of the problem of raga recognition can be tackled using the advancements in Signal Processing and the field of Deep Learning. Using the proposed pipeline of preprocessing and trained models, it is possible to predict the raga from the audio and its raw spectrograms. The use of sound source separation and silent part removal can be very well used along with the Short Time Fourier Transform for the preprocessing. The hybrid architecture of Convolutional Neural Networks and Long Short Term Memory Cells is used to achieve very accurate models. The models can classify the raga from the set of ten ragas with close to 99% accuracy. These models can also identify the allied ragas with better accuracy than most of the previous works.

The proposed hypotheses were validated through the experiments and their results. The accuracy of the model surely decreases during the faster temp parts of the performance. The sound source separation does affect the accuracy of the model. The source separated audio gives better results in most of the cases however the original audio is marginally better in some cases. The use of raw spectrograms works very well in predicting the ragas along with the proposed preprocessing steps.

We would like to express our sincere gratitude to author Devansh Shah's guru Pt. Pradeep Dhond and Mr. Sudhir Parekh.

References

1. Chakraborty, S., De, D.: Object oriented classification and pattern recognition of Indian classical ragas, pp. 505–510 (2012)
2. Pandey, G., Mishra, C., Ipe, P.: TANSEN: a system for automatic raga identification. In: IICAI (2003)

3. Chordia, P., Rae, A.: Raag recognition using pitch-class and pitch-class dyad distributions, pp. 431–436 (2007)
4. Dighe, P., Agrawal, P., Karnick, H., Thota, S., Raj, B.: Scale independent raga identification using chromagram patterns and swara based features. In: 2013 IEEE International Conference on Multimedia and Expo Workshops (ICMEW), pp. 1–4 (2013)
5. Kumar, V., Pandya, H., Jawahar, C.V.: Identifying ragas in Indian music. In: 2014 22nd International Conference on Pattern Recognition, pp. 767–772 (2014)
6. Gulati, S., Serra, J., Ishwar, V., Sentürk, S., Serra, X.: Phrase-based raga recognition using vector space modeling. In: IEEE International Conference on Acoustics, Speech and Signal Processing (ICASSP), Shanghai, China, pp. 66–70 (2016)
7. Gulati, S., Serra, J., Ganguli, K.K., Sentürk, S., Serra, X.: Time-delayed melody surfaces for raga recognition. In: International Society for Music Information Retrieval Conference (ISMIR), New York, USA, pp. 751–757 (2016)
8. Anand, A.: Raga identification using convolutional neural network. In: 2019 Second International Conference on Advanced Computational and Communication Paradigms (ICACCP), pp. 1–6 (2019)
9. Chowdhuri, S.: PhonoNet: multi-stage deep neural networks for raga identification in Hindustani classical music, pp. 197–201 (2019)
10. Ross, J.C., Mishra, A., Ganguli, K.K., Bhattacharyya, P., Rao, P.: Identifying raga similarity through embeddings learned from compositions' notation. In: ISMIR, pp. 515–522 (2017)
11. Madhusudhan, S.T., Chowdhary, G.: DeepSRGM - sequence classification and ranking in indian classical music via deep learning. In: Proceedings of the 20th International Society for Music Information Retrieval Conference, pp. 533–540. ISMIR, Delft, The Netherlands (2019)
12. Doukhan, D., Carrive, J., Vallet, F., Larcher, A., Meignier, S.: An open-source speaker gender detection framework for monitoring gender equality. In: 2018 IEEE International Conference on Acoustics, Speech and Signal Processing (ICASSP) (2018)
13. Hennequin, R., Khlif, A., Voituret, F., Moussallam, M.: "Spleeter: a fast and efficient music source separation tool with pre-trained models. J. Open Source Softw. **5**, 2154 (2020)
14. Sarkar, R., Naskar, S., Saha, S.: Raga identification from Hindustani classical music signal using compositional properties (2017)
15. Padmasundari, G., Murthy, H.A.: Raga identification using locality sensitive hashing. In: 2017 Twenty-Third National Conference on Communications (NCC), pp. 1–6 (2017)

The Simulated Emergence of Chord Function

Yui Uehara(✉) and Satoshi Tojo(✉)

Japan Advanced Institute of Science and Technology,
1-1 Asahidai, Nomi, Ishikawa 923-1292, Japan
{yuehara,tojo}@jaist.ac.jp

Abstract. In this paper, we propose an autonomous, unsupervised learning of chord classification, based on the Neural Hidden Markov Model (HMM), and extend it to the Semi-Makov Model (HSMM) to integrate such additional contexts as the pitch-class histogram, the beat positions, and the preceding chord sequences. We train our model on a minimally pre-processed dataset in a mixture of major/minor pieces, expecting the models to learn the chord clusters in accordance with the contexts without assignment of tonality. Thereafter, we evaluate their performance by perplexity, and show that the added contexts would considerably improve the efficiency. In addition, we show that the proposed model reflects the context of major and minor in its state transitions, even though trained in mixed tonality.

Keywords: Unsupervised learning · Chord classification · Hidden Markov Model · Hidden Semi-Markov Model · Neural network

1 Introduction

Chord functions, such as tonic, dominant, and subdominant, govern the harmonic progression in music, and thus, the analysis of them has been widely applied to structure recognition [7,20], composition [1,16], and melody harmonization [8,25]. Although textbook theories on chord functions [18,19] have been employed in these, the theories are somewhat restrictive to western classical music and thus seem not adequate to wider genres of music. This research aims at an autonomous acquisition of chord categories in an unsupervised learning, to obtain data-oriented, statistically stable classification of chords, following preceding efforts [11,21,26–28]. These works, however, required the chord labeling or grouping as a pre-process. This process includes notes segmentation [21], key finding [11,28], and Berklee chord annotation [26], however, is often hindered by overlapping phrases, metrical changes, local modulations, and so on, and is not easy to obtain consistent Berklee annotations [13] or key assignments [23].

In this work, we present a model that segments surface notes to be classified into chords in an unsupervised manner where contexts of chord progressions

This work is supported by JSPS Kaken 16H01744.

J. Romero et al. (Eds.): EvoMUSART 2021, LNCS 12693, pp. 264–280, 2021.
https://doi.org/10.1007/978-3-030-72914-1_18

are considered. Technically, the task is close to the unsupervised part-of-speech (POS) tag induction in the natural language processing (NLP). We employ the unsupervised Neural HMM that is one of the prominent model proposed for the POS tag induction [24]. The most advantageous feature of the Neural HMM is that it can easily embed additional contexts. We customize the model and its additional contexts to be suitable for the music. Furthermore, to make the model be in accordance with metrical structures, we introduce an extended version of the Neural HMM, *i.e.*, the Neural Hidden Semi-Markov Model (HSMM)[1].

We evaluate our model on the J.S. Bach's four-part chorales dataset. Although the dataset has been consulted by several previous works [11,21,27,28], we use it with minimum pre-processings, *i.e.*, without pre-defined chord symbols and key-assignments. Unlike previous works, we only adopt an automated normalization so that pieces do not have any key signature, and treat major pieces/phrases and minor ones in mixed condition. We use perplexity as the automatic evaluation metric in accordance with previous works [26,27]. Experiments show that our model appropriately segments and classifies surface pitch-classes especially by the model of smallest perplexity. Additional contexts with neural network modeling are shown to be effective to improve the perplexity.

This paper is organized as follows. In Sect. 2, we review related studies. We introduce proposed models in Sect. 3. We describe experimental setups in Sect. 4 and show results in Sect. 5. And then we summarize our contributions in Sect. 6.

2 Related Work

Learning Models of Chord Progression. Unsupervised learning of harmony or chord progressions has been studied through clustering methods [11,21], HMMs [26–28], and Probabilistic Context-free Grammars (PCFG) [26]. These studies aimed at acquiring the progression rules, not based on possibly subjective human annotation, but relying on data-driven analysis.

They have found similarities between statistically induced clusters and chord functions in textbook theories, when compared models with the same number of clusters as the textbook [11,26,28]. Especially in HMM based models, not only obtained clusters but also their state transition property was found to have similarity to the known chord functions [27,28]. However, selecting an optimal number of clusters would not be trivial since the larger number of clusters would help the model to gain the more accuracy but lose simplicity or interpretability. Therefore, Jacoby et al. investigated the balance of the model by introducing the optimal complexity-accuracy curve [11]. Similarly, Tsushima et al. evaluated the generative models by the perplexity and found that more number of states leads a better score [26]. On the other hand, White and Quinn proposed a methodology to find the best number of states for HMMs [28]. They adopted the *k-medoids* clustering over hundreds of HMMs with different initial parameterizations. And then, they assumed that the number of state N that showed higher *silhouette width* (*i.e.*, clearer cluster boundary) than $N-1$ and $N+1$

[1] The code is available at https://github.com/yui-u/emerge-chord-function.

was appropriate. Uehara et al. proposed another tandem approach to find the number of states [27].

Previous works on Bach's chorales experimented chord clustering, being major/minor pieces separated [21,27] possibly with modulation segmentation [11,28], and thus resulted in such clusters that coincided with known chord functions. Conversely, Hugo Riemann discussed that a role change of a chord in a context implied a modulation [19]. According to this, we verify not that a chord possesses a common feature independent of a key, but that a chord bears a different role dependent on a key. Therefore, we will not consider tonalities nor local modulations.

Although unsupervised models mentioned so far focused on harmonic structure solely, a recent work proposed a supervised learning of the combined model of harmony and rhythm, and reported the efficacy of introducing rhythmical information [9]. We also incorporate metrical information to our unsupervised learning by a Semi-Markov Model.

Neural Hidden Markov Models. Tran et al. proposed the unsupervised Neural HMM and applied it to part of speech induction [24]. An important strength of the Neural HMM is the seamless integration with additional contexts. They introduced the two additional contexts: infinite context with the Long-Short Term Memory (LSTM) and morphological information via character Convolutional Neural Networks. Despite its simple framework, the Neural HMM outperformed even the highly polished models such as the Bayesian mixture model [3] or the hierarchical Pitman-Yor Process HMM [2].

Unlike the Hybrid DNN (Deep Neural Network)-HMM model [29] that converts a pre-trained GMM (Gaussian Mixture Model)-HMM to a DNN-HMM or the Tandem model that combines features by a supervised DNN [10] to a HMM, the Neural HMM is seamless and fully unsupervised.

Our extension, *i.e.*, the unsupervised Neural HSMM differs from the Recurrent HSMM that makes use of the bi-LSTM for the purpose of reducing the error in the variational approximation [6]. Additional contexts in the Neural HMM are also can be used for the unsupervised Neural HSMM in the same manner.

3 Unsupervised Neural HMM and HSMM for Chord Classification

3.1 Neural Hidden Markov Model

A graphical representation of a HMM is shown in Fig. 1. Though unsupervised, we expect that the learned hidden states represent clusters of chords. We regard every change of surface pitch-classes vector, *e.g.*, $(0, 4, 9)^2$, as the observations

[2] Each pitch name in an octave: C, C#, D, ..., B has a corresponding pitch-class value: $0, 1, 2, \ldots, 11$. We adopt binary pitch class vector representation, *e.g.*, $(0, 4, 9) =$
$\begin{matrix} 0 & 1 & 2 & 3 & 4 & 5 & 6 & 7 & 8 & 9 & 10 & 11 \\ (1 & 0 & 0 & 0 & 1 & 0 & 0 & 0 & 0 & 1 & 0 & 0) \end{matrix}$.

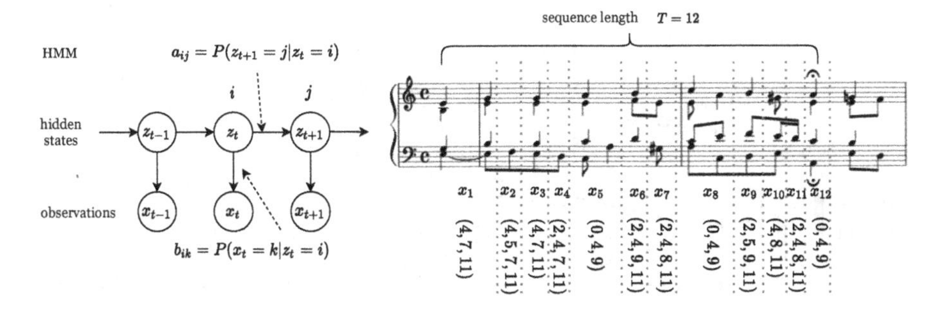

Fig. 1. Graphical representations of the Hidden Markov Model (HMM).

for HMMs (Fig. 1). We employ discrete HMMs in this study, therefore, each observation is associated with an index $k \in V$, *e.g.*, $(0, 4, 9) = (k = 2)$, where V is the number of unique pitch-class values in the dataset, *i.e.*, the vocabulary size.

Unlike the conventional HMM that has categorical parameters just as matrices of parameters, making it "neural" means that we prepare neural network components as functions for calculating initial, transition, and emission categorical parameters. Thus the same graphical representation (Fig. 1) applies to Neural HMMs, but a_{ij} and b_{ik} are obtained as outputs of neural networks that can employ additional musical contexts for the calculation. We show these networks and additional contexts in the following paragraphs.

Initial State Probability. The initial state probability ρ_i is calculated as follows[3], which is the only component remained almost the same as the conventional model except that it is normalized by softmax function.

$$\rho_i = \text{softmax}_i(\boldsymbol{\pi})$$

where i is a state index, $\boldsymbol{\pi} \in \mathbb{R}^S$ is the learnable weight vector, and S is the number of states.

State Transition Probability. The state transition probability a_{ij} is calculated as follows.

$$a_{ij} = P(z_{t+1} = j|z_t = i) = \text{softmax}_j(\text{MLP}_3([\boldsymbol{s}_i; \boldsymbol{c}_t])) \tag{1}$$

where MLP_3 is a 3-layer Multi Layer Perceptron that has one input layer, two hidden layers and one output layer with a hyperbolic tangent (tanh) activation function after each hidden layer[4]. $\boldsymbol{s}_i$ is a state embedding, which is a learnable

[3] ρ_i is equivalent to $P(z_1 = i)$ for HMMs and $P(z_0 = i)$ for HSMMs. We set an external initial state z_0 for HSMMs to meet the initial boundary condition as described in Sect. 3.3.

[4] Similarly, MLP_2 denotes a MLP with one hidden layer, and we use the same notation in following sections as well.

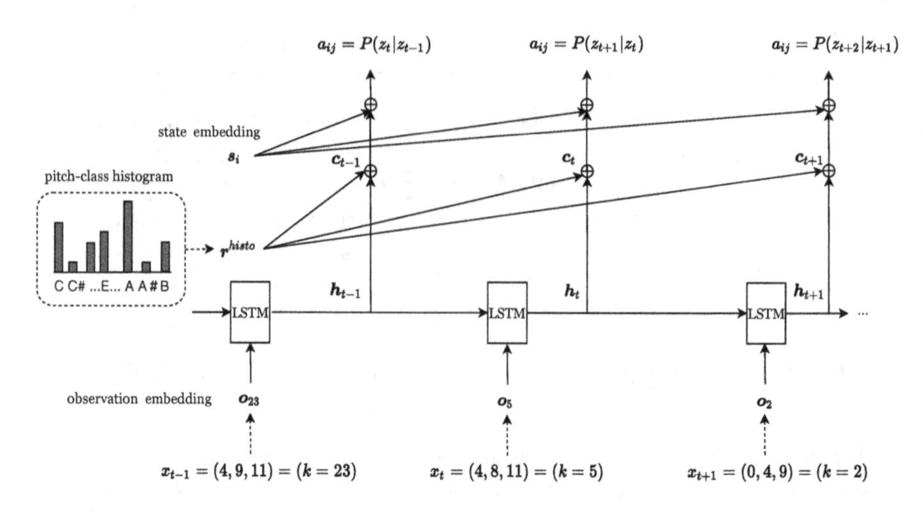

Fig. 2. The network architecture for calculating transition probabilities a_{ij}.

vector associated to a state index i. Taking the advantage of Neural HMMs, we introduce two types of additional contexts, *i.e.*, pitch-class histograms and infinite contexts from preceding observations, and feed a concatenation of the two contexts $\boldsymbol{c}_t$ to calculate the state transition probability.

The first additional context, pitch-class histogram, is a kind of context for tonality. We split a piece into chord sequences at each *fermata* (point d'orgue) that works as a full-stop marker of a lyric, and then accumulate the duration of each pitch-class to obtain a pitch-class histogram of a sequence. Even though the main tonality of a piece would be easily distinguished by the key signature[5], there would be local modulations and thus we need an extra effort to employ such a histogram as soft information of local tonality. We obtain the feature of a pitch-class histogram of a chord sequence $\boldsymbol{r}^{histo}$ as follows.

$$\boldsymbol{r}^{histo} = \text{MLP}_2(\boldsymbol{v}^{histo}) \tag{2}$$

where $\boldsymbol{v}^{histo} \in \mathbb{R}^{12}$ is the raw pitch-class histogram of a sequence.

The second additional context is the infinite context from preceding observations. This idea was originally proposed by Tran et al. [24], which can contextualize the infinite sequence of the preceding observation by employing the Long-Short Term Memory (LSTM). We feed an observation embeddings $\boldsymbol{o}_k$ to it at every time step, and exploit the hidden state as a context to the transition probability calculation. Unlike the state embeddings that are simple

[5] However, the main tonalities of pieces that are written in church modes (*e.g.*, *dorian*) would not be corresponds to the key signature system of modern tonality. In addition, major/minor is not distinguished by the key signature solely. Furthermore, some minor pieces retain the feature of church modes [22] and frequently proceed relative major keys [21,27].

learnable values, inputs of observation embeddings are binary pitch-class vectors. An observation embedding is obtained as follows.

$$\boldsymbol{o}_k = \tanh(\mathrm{MLP}_2(\boldsymbol{v}_k^{pitch})) \tag{3}$$

We expect that binary pitch-class vectors give more direct information of observations. As shown in Fig. 2, the context at time step t is obtained as follows.

$$\boldsymbol{h}_t = \mathrm{LSTM}(\boldsymbol{o}_k, \boldsymbol{h}_{t-1}) \tag{4}$$

Finally, we concatenate the two contexts and obtain the context vector as (5), where [;] denotes vector concatenation. This context vector is applied to the transition probability network (1).

$$\boldsymbol{c}_t = [\boldsymbol{r}^{histo}; \boldsymbol{h}_t] \tag{5}$$

Emission Probability. The emission probability b_{ik} is calculated as follows, based on Tran et al. [24].

$$b_{ik} = P(x_t = k|z_t = i) = \mathrm{softmax}_k(\boldsymbol{s}_i^{\mathsf{T}}\, \boldsymbol{o}_k + l_k) = \frac{\exp\left(\boldsymbol{s}_i^{\mathsf{T}}\, \boldsymbol{o}_k + l_k\right)}{\sum_{k'} \exp\left(\boldsymbol{s}_i^{\mathsf{T}}\, \boldsymbol{o}_{k'} + l_{k'}\right)} \tag{6}$$

Instead of employing weight matrices or MLPs, the emission probability is learned as the dot product of a state embedding and an observation embedding. We use the same observation embedding as one used in the input to LSTM (4). The bias value $l_k \in \mathbb{R}$ would help to consider the frequency of each chord.

3.2 Extension to Neural Hidden Semi-Markov Model

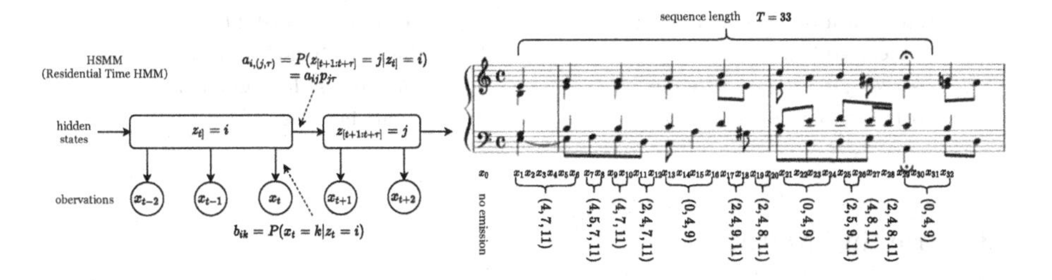

Fig. 3. Graphical representations of the Hidden Semi-Markov Model (HSMM).

In the Neural HMM, we have regarded every change of pitch-classes as an output token to avoid self state transition, and thus, the duration of it has been ignored. To remedy this problem, we introduce the Neural Hidden Semi-Markov Model (HSMM) by extending the Neural HMM. An advantage of HSMMs is its ability to predict duration of states [30]. Here we segment a sequence by the duration of the sixteenth note, and assign an output token, *i.e.*, a pitch-class vector at each time step, as shown in Fig. 3.

Among variants of HSMMs, we choose the Residential-Time HMM that does not allow self state transitions. We can use the same transition (a_{ij}) and emission (b_{ik}) networks in the Neural HMM (Sect. 3.1), except for that the self transition a_{ii} should be set to zero. The only component we need to add is the one for the duration probability. We define it as follows.

Duration Probability

$$p_{i\tau} = \mathrm{MLP}_3([\boldsymbol{s}_i; \boldsymbol{r}_t^{beat}]) \quad (7)$$

$$\boldsymbol{r}_t^{beat} = \mathrm{MLP}_2([v^{timesig}; v_t^{beat}]) \quad (8)$$

where $\boldsymbol{r}_t^{beat}$ is an additional context, which is expected to give a metrical information. $[v^{timesig}; v_t^{beat}]$ is a 2-dimension vector consists of a beat position and a numerator of the time signature. In this study, $v^{timesig}$ is constant (=4) since we only contain four-four time (4/4) pieces, while v_t^{beat} varies among $(1.0, 1.25, \cdots, 4.5, 4.75)$.

We summarize the notations for the Neural HMM and HSMM in Table 1.

Table 1. Notations in the neural HMM and HSMM.

N	: number of sequences in a mini-batch	$n \in N$	: a sequence index in a mini-batch
S	: number of states	V	: vocabulary size
T	: maximum time step of a sequence	$t \in T$	: time step
z_t	: hidden state at t	$i, j \in S$	: state indices
D	: maximum state duration	$\tau \in D$	: duration index
x_t	: index of observed pitch-class content at t	$k \in V$	: observed symbol index
$\mathbf{x}_{1:T}$	: an observed sequence	$p_{i\tau}$	: duration probability
ρ_i	: initial state probability	π	: unnormalized initial state probability
a_{ij}	: state transition probability	b_{ik}	: emission probability
$\boldsymbol{s}_i$	: state embedding	$\boldsymbol{v}_k^{pitch}$	: binary pitch-class vector
$\boldsymbol{v}^{histo}$	: observed pitch-class histogram	$\boldsymbol{r}^{histo}$	: pitch-class histogram encoding
$\boldsymbol{o}_k$	: observation encoding	l_k	: emission bias

3.3 Training Method

There is no analytical solution to obtain an optimal parameterization of a HMM that maximizes the likelihood of the observed sequence [17]. Although the widely used Baum-Welch re-estimation algorithm (that is a kind of the expectation–maximization algorithm) is able to obtain a locally optimal parameterization, it would be stuck in bad local optima when the likelihood surface is complex [17]. On the other hand, we can utilize gradient based methods to maximize the likelihood or the marginal probability $\sum_{n=1}^{N} \ln P(\mathbf{x}_{1:T}^{(n)})$ where $\mathbf{x}_{1:T}^{(n)}$ is an observed sequence since it is an optimization problem [12,14,17]. The known dynamic programming named the forward algorithm for HMMs can be used to calculate the marginal probability [17], and similar one for HSMMs [30]. Neural networks models are generally trained by gradient based optimizers as well as the back propagation for the gradient computing. Recently, along with the success of neural networks, efficient optimizers also have been proposed. By implementing the H(S)MM as a neural network, we can naturally utilize the gradient based optimization [12][6]. In this study, we use one of the latest optimizer RAdam [15], which adjusts the learning rate automatically.

Forward Algorithm for Hidden Semi-Markov Models. To conduct the gradient based optimization, we calculate the marginal probability of an observed sequence $\ln P(\mathbf{x}_{1:T})$ by the forward algorithm. Although we have omitted the details of the forward algorithm for HMMs since it is well known [17], we give one for HSMMs in the following.

To explain the duration of states, we borrow the notation used by Yu et al. [30].

$t_1 : t_2]$: a state lasts at latest from t_1 and ends at time t_2.
$[t_1 : t_2]$: a state starts at t_1 and ends at t_2 (with duration $t_2 - t_1 + 1$).

Among variants of HSMMs, the Residential-Time HMM does not allow the self state transition, and a transition to another state is required when "remaining" duration $\tau = 1$. On the other hand, when $\tau > 1$, the duration just decrements at each time step.

The $\alpha_t(j, \tau)$ represents the forward probability that the state at t is j and the remaining duration of which is τ, and output tokens from $t = 1$ to t, *i.e.*, $\mathbf{x}_{1:t}$ are observed. It is decomposed as follows.

$$\begin{aligned}
\alpha_t(j, \tau) &= P(z_{t:t+\tau-1]} = j, \mathbf{x}_{1:t}) \\
&= \alpha_{t-1}(j, \tau+1) P(x_t = k | z_t = j) \\
&\quad + P(\tau | z_t = j) P(x_t = k | z_t = j) \sum_{i \backslash j} \alpha_{t-1}(i, 1) P(z_t = j | z_{t-1} = i) \\
&= \alpha_{t-1}(j, \tau+1) b_{jk} + p_{j\tau} b_{jk} \sum_{i \backslash j} \alpha_{t-1}(i, 1) a_{ij}
\end{aligned}$$

where $P(\tau | z_t = i) = p_{i\tau}$ is a duration probability[7], $P(x_t = k | z_t = j) = b_{jk}$ is an emission probability, and $P(z_t = j | z_{t-1} = i) = a_{ij}$ is a transition probability. Note that when the state at t is j, there are two possibilities: (i) the state at time $t-1$ is also j with the remaining duration at there is $\tau + 1$, (ii) the state at time $t-1$ is i ($i \neq j$) and the remaining duration is 1 then is transferred another state j at the time t.

---

[6] Sequences need not to be concatenated as in other (*e.g.*, Baum-Welch algorithm and Gibbs sampling) optimization approaches.

[7] $\tau \in D$ is a discrete value of remaining time steps, and D is the maximum duration.

We apply scaling by replacing $\alpha_t(i, \tau)$ to the conditional probability, similar to HMM [30], and obtain a modified forward algorithm.

$$\hat{\alpha}_t(j,\tau) = P(z_{t:t+\tau-1]} = j|\mathbf{x}_{1:t}) = \frac{\alpha_t(j,\tau)}{P(x_1,\ldots,x_t)}$$
$$C_t = P(x_t|x_1,\ldots,x_{t-1})$$
$$P(x_1,\ldots,x_t) = \prod_{t'}^{t} C_{t'} \quad (9)$$
$$C_t\hat{\alpha}_t(j,\tau) = \frac{\alpha_t(j,\tau)}{P(x_1,\ldots x_{t-1})}$$
$$= \hat{\alpha}_{t-1}(j,\tau+1)b_{jk} + p_{i\tau}b_{jk}\sum_{i\backslash j}\hat{\alpha}_{t-1}(i,1)a_{ij} \quad (10)$$

As can be seen from (9), the marginal probability is obtained by $P(\mathbf{x}_{1:T}) = P(x_1,\ldots,x_T) = \prod_{t'}^{T} C_{t'}$, where T is the length of an entire observation sequence, which means we can train the model only by executing the forward algorithm, without the backward algorithm usually used in EM algorithm. Note that C_t is obtained by summing up (10) about all j and τ, since $\sum_j \sum_\tau \hat{\alpha}_t(j,\tau) = 1$. Unlike HMMs, we set initial probability to yield the initial hidden state z_0 that does not yield observation by considering the initial boundary condition: $\alpha_0(i,1) = P(z_0 = i)$ and $\alpha_0(i,\tau) = 0$ for $\tau > 1$ [30]. Therefore, $\alpha_0 = \hat{\alpha}_0$ and $C_0 = 1.0$ at $t = 0$.

4 Experiments

4.1 Dataset

Table 2. The statistics of the dataset.

	#pieces	(maj.)	(min.)	(dor.)	(other)	#phrases	#obs. (HMM)	#obs. (HSMM)
Train	185	(89)	(70)	(23)	(3)	1155	15922	38452
Eval.	51	(21)	(23)	(6)	(1)	314	4184	10408
Test	54	(28)	(20)	(5)	(1)	350	5009	11868

Resource. We use J.S. Bach's four-part chorales by the Riemenschneider numbering system (1–371) from the Music21 Corpus [4] as our dataset resource. At first, we have removed duplicated pieces according to the analysis by Dahn [5]. We only contain four-four time (4/4) pieces[8], thus the number of pieces is 290. We regard each phrase as an independent sequence, however, when we provide these phrases to training, evaluation, and testing set, we give randomness over pieces (not over phrases), since one piece may contain similar phrases.

[8] We have also excluded 27 pieces that are not four-part voices or have some problems such as a collapsed format.

Pre-processing. We try to adopt the least pre-processing as possible. We transpose pieces to have no key signature, however, there can be local modulations, as well as a few pieces of church modes (*e.g., dorian*). We leave these local modulations untransposed, to be represented by altered chords in some relative keys [22].

We do not divide major and minor pieces/phrases since the chorales retain the feature of church modes and seamless transformation to relative keys have been observed especially in minor pieces [21,27]. We build vocabulary of observations from every combination of the four-part pitch-classes. In practice, we assign indices (k) for pitch-classes the accumulated pitch-duration of them in the dataset over 95%, and then, remained chords are merged to "Others". The obtained vocabulary size was 79 including the "Rest".

For HMMs, we regard every change of pitch-classes as an output token as shown in Fig. 1. This process would be efficient for HMMs since it reduces undesirably large amount of self state transitions. However, it would have a demerit of ignoring the duration of each output token. On the other hand, we assign output tokens at every time step of sixteenth note lengths[9] for HSMMs as shown in Fig. 3. The statistics of the dataset are shown in Table 2. Although we do not split major and minor pieces, we show the number of major, minor, *dorian*, and other pieces for reference, which is based on the key analysis from [5]. The difference between the number of observations for HMMs and for HSMMs comes from the method for the output token creation as described above.

4.2 Model Settings

Table 3. The size of layers and related equations.

Layer	Size	Eq.
Hidden layer size of the transition MLP_2	16	(1)
State embedding size ($\boldsymbol{s}_i$)	16	(1), (6)
Hidden and output layer sizes of the observation encoder MLP_2	16	(3)
Observation encoding size ($\boldsymbol{o}_k$)	16	(3), (6)
Hidden and output layer sizes of the histogram encoder MLP_2	8	(2)
LSTM hidden size	16	(4)
Hidden layer size of the duration MLP_2	16	(7)
Hidden and output layer size of the beat encoder MLP_2	8	(8)

We set the size of each layer of the networks as shown in Table 3. The output layer size of the transition MLP_2 in (1) should be the number of states S for HMMs, and $S-1$ for HSMMs (since it does not allow self state transitions). The output layer size of the duration MLP_2 in (7) is 16 that is a maximum duration length correspond to a whole note.

[9] The sixteen note is the minimum duration of the corpus, except quite few cases of 32th ones.

We use RAdam [15] optimizer with the learning rate 0.001. The mini-batch size is set to 8. We train models up to 1000 epochs for HMM and 500 epochs for HSMM, however, the process is stopped when the lowest loss on the evaluation set is not updated over 20 epochs. We apply the dropout rate of 12.5% to each hidden layer of MLPs.

4.3 Evaluation

Following previous works [26,27], we use perplexity as an automatic evaluation metric, defined by the following equation.

$$\mathcal{P} = \exp\left(-\frac{1}{T}\ln P(x_1, \ldots x_T; \boldsymbol{\theta})\right) \tag{11}$$

where $(x_1, \ldots x_T)$ is a sequence of output tokens and $\boldsymbol{\theta}$ is a set of model parameters. Smaller perplexity for test data means model's better generalization performance since perplexity corresponds to log-average of inverse probability, and thus it is commonly used for evaluating probabilistic models. We examine the performance on multiple number of states: 3 to 16. We conduct the experiments with three different random seed of $\{0, 1, 2\}$ for each number of hidden states, and report the averaged scores in Sect. 5.1.

4.4 Baseline Model

In addition to the Neural H(S)MM, we implement a baseline model that represents the initial, transition, and emission probabilities as simply learnable weight vector or matrices with softmax output layers. The baseline model is almost "Not-Neural" but a H(S)MM tuned by the same gradient-based optimizer. We compare the proposed model with this baseline to see how efficient the elaborated neural network components works.

5 Results and Discussion

5.1 Automatic Evaluation by Perplexity

Table 4. Ablation studies for Neural HMMs and Neural HSMMs. Averaged perplexities by three trials with random seeds of $\{0, 1, 2\}$ on the testing sets. The bold numbers are best score in the same number of hidden state.

#states	4	8	16	32	#states	4	8	16	32
BASE	32.32	27.74	23.55	21.18	**BASE**	16.47	11.42	7.96	5.64
NHMM	**26.42**	**22.04**	**18.58**	**17.25**	**NHSMM**	**14.43**	**9.36**	**6.14**	**4.74**
−LSTM	27.80	23.50	19.95	17.64	**−LSTM**	14.94	9.67	6.47	5.07
−PITCH	31.14	24.99	21.50	19.11	**−PITCH**	16.88	10.64	7.23	5.38
−HISTO	28.71	24.27	22.15	19.84	**−HISTO**	15.18	9.78	6.51	4.92
					−BEAT	14.85	9.65	6.29	4.77

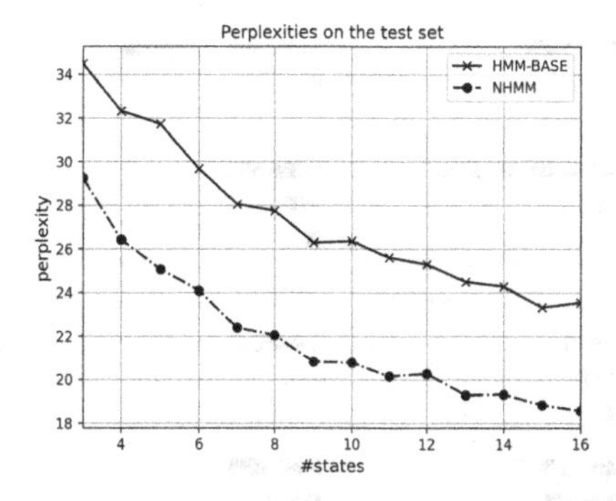

Fig. 4. Averaged perplexities by three trials with random seeds of $\{0, 1, 2\}$ on the testing sets for HMMs.

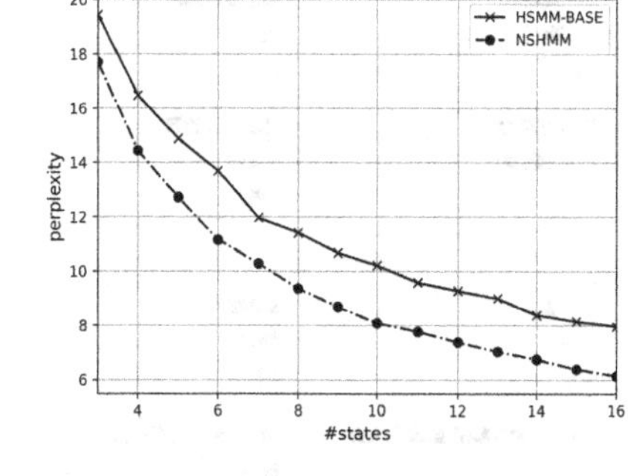

Fig. 5. Averaged perplexities by three trials with random seeds of $\{0, 1, 2\}$ on the testing sets for HSMMs.

As mentioned in Sect. 3.3, the BASE model is a baseline that represents the initial, transition, emission, and duration probabilities as weight matrices (like conventional HMMs or HSMMs) with softmax output layers. In both of HMMs (Fig. 4) and HSMMs (Fig. 5), the elaborated neural models (the Neural HMM and the Neural HSMM) considerably outperformed the baselines. Note that perplexities between HMM and HSMM cannot be compared since they use differently processed dataset, however, the setting of HSMM is more faithful to the original score.

The ablation study also shows efficacy of additional contexts as shown in Table 4. Removing any additional contexts, *i.e.*, LSTM (-LSTM), the pitch histograms (-HISTO), the beat positions (-BEAT), and the observation representation by pitch-classes (-PITCH) degraded the perplexity. Among them, removing[10] the observation representation by pitch-classes (-PITCH) led a significant drop. Since we do not employ MLPs for the architecture of calculation of emission probability but force a model to learn relationships between states and vocabularies directly, the information of raw pitch-classes would help the learning.

5.2 Induced Clusters and Model's Perpexities

From this section, we give qualitative analysis and discussion. We focus on the eight-state models since we have transposed pieces not to have key signatures and have assumed that each cluster would roughly correspond to a triad on the diatonic scale, or otherwise to the rests. Although we should not estimate appropriateness of a model only by perplexity, we can find clearer clusters in the higher scored model in obtained examples. We show the emission probabilities of the best Neural HSMM, the worst Neural HSMM, and the best baseline model in Fig. 6. Note that we have executed three trials of training for each model with

[10] In this case, we used a observation embedding of learnable weights in the same way of the state embedding instead of the pitch-class one.

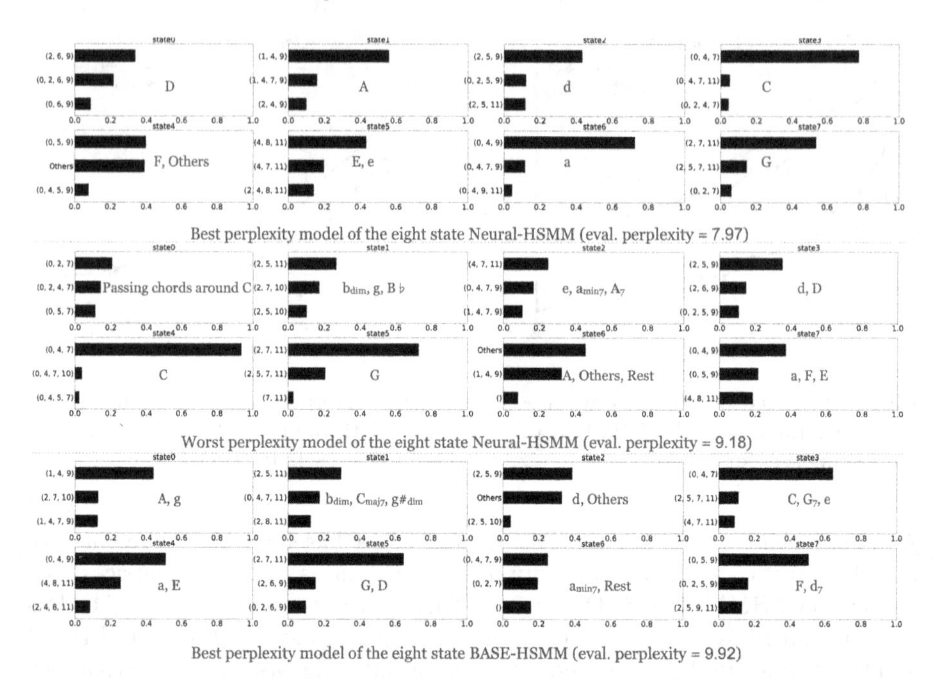

Fig. 6. Comparison between (Top) Neural HSMM of the best evaluation perplexity among the three random seeds, (Middle) Neural HSMM of the worst evaluation perplexity, and (Bottom) Base HSMM of the best evaluation perplexity. The bar charts shows the top three emissions per each hidden state.

a different random seed among $\{0, 1, 2\}$. Here the "best" model means that the evaluation perplexity of which is the smallest among the three trials.

We observe the best Neural HSMM consists of clusters that can be clearly interpreted as: s_3 : {C}, s_2 : {d}, s_5 : {E, e}, s_4 : {F, Others}, s_7 : {G}, s_6 : {a}, s_1 : {A}. In other words, the clusters mainly consist of the chords on diatonic scales (C major and a minor), and its commonly-used borrowed chords such as D. The A major cluster may emerge from two reasons: the common use of the *Picardy* third, or the dominant of *dorian* mode pieces. It is worth noting that seventh chords and passing chords are merged into appropriate clusters, *e.g.*, $(0, 2, 4, 7)$ in the same state for $(0, 4, 7)$ and $(2, 5, 7, 11)$ in the same state for $(2, 7, 11)$. And also, the $(2, 5, 11)$ in the same state for $(2, 5, 9)$ would be understandable since it usually used in the first inversion.

Though employing the same Neural HSMM, the worst perplexity model (in the middle of Fig. 6) seems to be less appropriate than the best model. For example, a is mixed up with F and E in s_7, and C and passing chords around it are separated. Although we would be able to choose a model by the perplexity, we admit that difficulty for obtaining the global optimum is still existing even though we employ the efficient gradient based optimizer.

The best baseline model (the evaluation perplexity of which is 9.92) is still worse than the worst Neural HSMM (9.18). Not only the best baseline model scored a worse perplexity, but it has more miscellaneous clusters than Neural HSMMs. However, we can observe that C and G7, or G and D, in the same tonality would tend to be merged into the same category in the baseline HSMM.

5.3 State Transitions

Although we trained models without distinction of major and minor pieces, we can observe that obtained state transitions reflect the difference of tonalities. Note that, instead of examining the state transition probability directly, we counted the number of transitions after decoding the sequence of hidden states by the Viterbi algorithm since the state transition probability changes by contexts.

Now, we count the state transitions on major pieces, minor pieces and *dorian* pieces separately to see the model's ability to predict a next state which should fit contexts. We show the result of state transition properties by the best scored Neural HSMM in Fig. 7 (the emission probability of which is shown in the top of Fig. 6). In major pieces, the tendency that $\{s_4, s_2\}$(subdominant or secondary dominant) $\rightarrow s_7$(dominant) $\rightarrow s_3$(tonic) is noticeable. While in minor pieces, the tendency described above suggests an existence of the relative major keys, which is consistent with previous works [21,27]. In addition, a strong transition from s_5(dominant) $\rightarrow s_6$(tonic = I) and $s_5 \rightarrow s_4$(tonic = VI) are observed. Unlike in major pieces, the transition from s_2(subdominant) $\rightarrow s_5$(dominant) is increased in minor pieces. Finally, in *dorian* pieces, transitions from s_1(dominant) $\rightarrow$ s_2(tonic = I) and $s_1 \rightarrow s_4$(tonic = III) are observed.

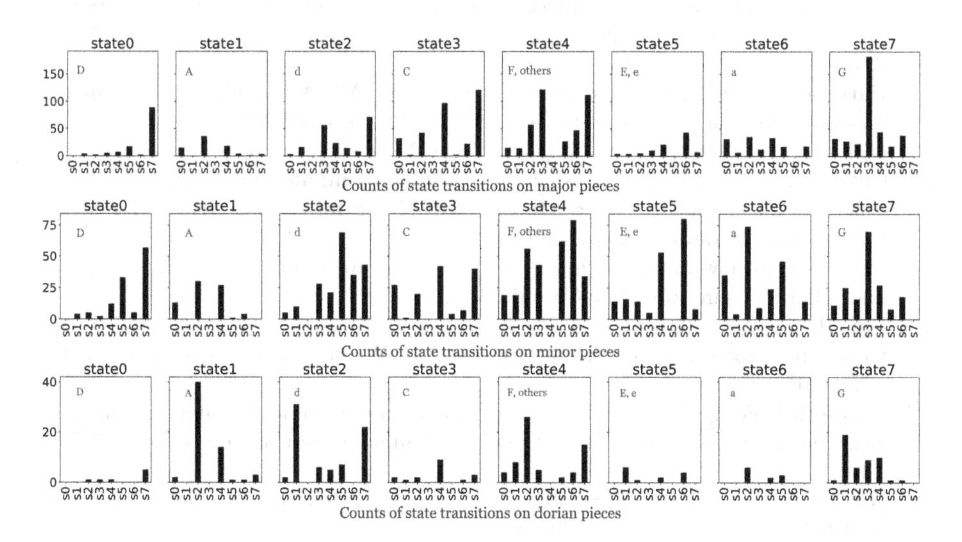

Fig. 7. Counts of state transitions on: (Top) major pieces, (Middle) minor pieces, and (Bottom) *dorian* pieces. Sequence of hidden states are calculated by the Viterbi algorithm.

6 Conclusion

In this paper, we proposed a new method to induce chord clusters with a Neural Hidden Semi-Markov model. Experimental results satisfactorily showed the efficacy of additional contexts in terms of perplexity by comparing the baseline and conducting ablation studies. Even though we trained the model with minimally pre-preprocessed dataset, we successfully obtained clear chord clusters with the best scored proposed model. We found that the acquired chord categories governed the progression of forthcoming chords to some extent, and thus they possessed the similar roles to the chord functions. Note that they are classified autonomously without textbook knowledge, and thus we may regard that they are the emergence of hidden chord functions.

Since the model is provided additional contexts, it has an ability to adjust the state transition probability by the contexts. We observed that the statistics of the state transition appropriately reflected the difference between major and minor pieces, despite that we trained the model on mixed pieces. Our future work includes the application of the method to different styles of music, *e.g.*, to those with metrically more complex pieces.

References

1. Anders, T., Miranda, E.R.: A computational model that generalises Schoenberg's guidelines for favourable chord progressions. In: 6th Sound and Music Computing Conference (2009)
2. Blunsom, P., Cohn, T.: A hierarchical Pitman-Yor process HMM for unsupervised part of speech induction. In: Proceedings of the 49th Annual Meeting of the Association for Computational Linguistics: Human Language Technologies, pp. 865–874 (2011)
3. Christodoulopoulos, C., Goldwater, S., Steedman, M.: A Bayesian mixture model for PoS induction using multiple features. In: Proceedings of the 2011 Conference on Empirical Methods in Natural Language Processing, pp. 638–647 (2011)
4. Cuthbert, M.S., Ariza, C.: music21: a toolkit for computer-aided musicology and symbolic music data. In: Proceedings of the 11th International Society for Music Information Retrieval Conference (2010)
5. Dahn, L.: So how many Bach four-part chorales are there? (2018). http://www.bach-chorales.com/HowManyChorales.htm
6. Dai, H., Dai, B., Zhang, Y.M., Li, S., Song, L.: Recurrent hidden semi-Markov model. In: Proceedings of the International Conference on Learning Representations (2017)
7. Granroth-Wilding, M., Steedman, M.: Statistical parsing for harmonic analysis of Jazz chord sequences. In: International Computer Music Conference, pp. 478–485 (2012)
8. Groves, R.: Automatic harmonization using a hidden semi-Markov model. In: AIIDE Workshop, pp. 48–54 (2013)
9. Harasim, D., O'Donnell, T.J., Rohrmeier, M.: Harmonic syntax in time rhythm improves grammatical models of harmony. In: Proceedings of the 20th International Conference on Music Information Retrieval, pp. 335–342 (2019)

10. Hermansky, H., Ellis, D.P.W., Sharma, S.: Tandem connectionist feature extraction for conventional HMM systems. In: 2000 IEEE International Conference on Acoustics, Speech, and Signal Processing. Proceedings (Cat. No. 00CH37100), vol. 3, pp. 1635–1638 (2000)
11. Jacoby, N., Tishby, N., Tymoczko, D.: An information theoretic approach to chord categorization and functional harmony. J. New Music Res. **44**(3), 219–244 (2015)
12. Kim, Y., Wiseman, S., Rush, A.M.: A tutorial on deep latent variable models of natural language. arXiv preprint arXiv:1812.06834 (2018)
13. Koops, H.V., de Haas, W.B., Burgoyne, J.A., Bransen, J., Kent-Muller, A., Volk, A.: Annotator subjectivity in harmony annotations of popular music. J. New Music Res. **48**(3), 232–252 (2019)
14. Levinson, S.E., Rabiner, L.R., Sondhi, M.M.: An introduction to the application of the theory of probabilistic functions of a Markov process to automatic speech recognition. Bell Syst. Tech. J. **62**(4), 1035–1074 (1983)
15. Liu, L., et al.: On the variance of the adaptive learning rate and beyond. In: International Conference on Learning Representations (2020)
16. Navarro, M., Caetano, M., Bernardes, G., Nunes de Castro, L., Corchado, J.M.: Automatic generation of chord progressions with an artificial immune system. In: International Conference on Evolutionary and Biologically Inspired Music and Art, pp. 175–186 (2015)
17. Rabiner, L.R.: A tutorial on hidden Markov models and selected applications in speech recognition. Proc. IEEE **77**(2), 257–286 (1989)
18. Rameau, J.P.: Treatise on Harmony. Dover Publications, New York (1971)
19. Riemann, H.: Harmony Simplified: Or the Theory of the Tonal Functions of Chords. Augener, London (1896)
20. Rohrmeier, M.: Towards a generative syntax of tonal harmony. J. Math. Music **5**(1), 35–53 (2011)
21. Rohrmeier, M., Cross, I.: Statistical properties of tonal harmony in Bach's chorales. In: 10th International Conference on Music Perception and Cognition, pp. 619–627 (2008)
22. Schoenberg, A.: Structural Functions of Harmony (revised edition). W. W. Norton & Company, New York (1969)
23. Temperley, D.: The tonal properties of pitch-class sets: tonal implication, tonal ambiguity, and tonalness. Comput. Musicol. **15**, 24–38 (2007)
24. Tran, K., Bisk, Y., Vaswani, A., Marcu, D., Knight, K.: Unsupervised neural hidden Markov models. In: Proceedings of the Workshop on Structured Prediction for NLP, pp. 63–71 (2016)
25. Tsushima, H., Nakamura, E., Itoyama, K., Yoshii, K.: Function- and rhythm-aware melody harmonization based on tree-structured parsing and split-merge sampling of chord sequences. In: Proceedings of 18th International Society for Music Information Retrieval Conference, pp. 502–508 (2017)
26. Tsushima, H., Nakamura, E., Itoyama, K., Yoshii, K.: Generative statistical models with self-emergent grammar of chord sequences. J. New Music Res. **47**(3), 226–248 (2018)
27. Uehara, Y., Nakamura, E., Tojo, S.: Chord function identification with modulation detection based on HMM. In: Proceedings of 14th International Symposium on Computer Music Multidisciplinary Research, pp. 59–70 (2019)
28. White, C.W., Quinn, I.: Chord context and harmonic function in tonal music. Music Theory Spectr. **40**(2), 314–335O (2018)

29. Yu, D., Deng, L., Dahl, G.E.: Roles of pre-training and fine-tuning in context-dependent DBN-HMMs for real-world speech recognition. In: NIPS 2010 workshop on Deep Learning and Unsupervised Feature Learning (2010)
30. Yu, S.Z.: Hidden semi-Markov models. Artif. Intell. **174**(2), 215–243 (2010)

Incremental Evolution of Stylized Images

Florian Uhde(✉)

Faculty of Computer Science, Otto-von-Guericke-University, Magdeburg, Germany
florian.uhde@posteo.de
https://www.inf.ovgu.de/en/

Abstract. This paper examines and showcases a framework to generate artworks using evolutionary algorithms. Based on the idea of an abstract artistic process stylized images are generated from different input images without human supervision. After explaining the underlying concept, the solution space of different styles is explored and its properties are discussed. Given this insights into the framework, current shortcomings are evaluated and improvements are discussed.

Keywords: Genetic algorithm · Artistic rendering · Computational art

1 Introduction

Whereas evolutionary systems are often used to support optimization-focused, domain specific design tasks, the act of creating artistically pleasing artworks remains a challenge. Based on the idea of art creation as a series of composeable, stacked actions towards an desirable result, this work conceptualizes a framework of artistic creation. It uses a genetic algorithm and the means of evolution to produce artistic artifacts without human supervision. Using simple building blocks and their composition the algorithm exhibits a wide variety of parameters which allow to configure the emerging human-like painting process. A variety of different styles and expressions can be achieved, while each of those can be applied to different inputs, producing consistent results. The following section positions this work in the wider context of creative evolutionary systems and highlights similarities and core differences to existing approaches. Afterwards, Sect. 3 explains the concept and Sect. 4 explores aspects of its implementation (Fig. 1). Finally Sect. 5 highlights some improvements and possible next steps.

1.1 An Incremental Artistic Process

The process mimicked by the algorithm in this paper defines the creation of art as an overlapping series of actions. An artist, striving to express an object o_{truth}, will do so by firstly perceiving it as o_{artist} and secondly expressing this representation of the object in a medium, creating o_{art}. The perception of the artist is shaped by a multitude of factors: Inner convictions, social surroundings,

J. Romero et al. (Eds.): EvoMUSART 2021, LNCS 12693, pp. 281–296, 2021.
https://doi.org/10.1007/978-3-030-72914-1_19

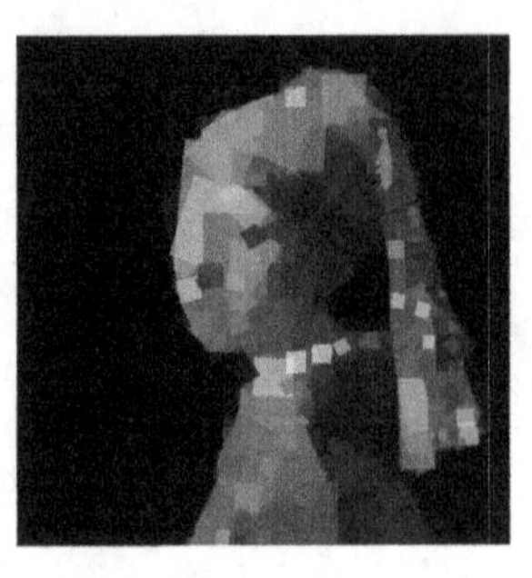

Fig. 1. Example results created within the scope of this work.

upbringing, education, ideology and so forth. This results in a transformation during the perception, turning o_{truth} into a personalized version of the object o_{artist}. When creating art based on o_{artist} the used medium and the skill of the artist influence the outcome o_{art}. The transformation of o throughout this process is what defines the style and signature of an artist. Both are embodied in the personalized way of viewing o_{truth}, as well as in the ability and limitations of the expression within the chosen medium.

While executing these transformations the artist takes a series of actions, each step being perceived as the current, most valuable one. As an example, when constructing a landscape painting, an artist might start with a rough composition, coloring large-scale features to give a backdrop and then further refine the outlines, adding details and fine grained shades further into the process. This abstract way of art creation is the foundation of the algorithm designed and explored in this work.

2 Related Work

At their core all evolutionary algorithms solve a search problem for a good candidate of a certain fitness function within a vast solution space. To achieve this, the algorithm utilizes two systems, one for creating and modifying such candidates, and another for rating them in terms of their fitness. After already being used for various tasks in high-knowledge domains like architecture and engineering, supporting the human knowledge workers in different applications, their usage as generative systems for art was pioneered by Dawkins [1] and later on popularized by Sims [2] and Todd [3]. In the context of this work the field of creating two dimensional artworks can be divided into *imaginative-* and *interpretative*-systems (for a much finer classification see [4]).

Imaginative systems try to evolve and create the very object that should be expressed as an artwork, while *Interpretative systems* strive to reinterpret an existing object artistically. Examples from the first category are often expression based systems, modeling the generated picture as a set of functions [2,5–7]. Approaches from the second category seek to replicate a given source image by *reinterpreting* it [8–11], shifting the focus from the generation of an interesting object, towards an interesting interpretation. The algorithm in this work follows a *interpretative* approach, implementing a simplified artistic process. A common problem, given the vast solution space and the subjectivity of artworks, is the rating of candidates. Evaluating the aesthetics of a generated image is hard using evolutionary algorithms, due to the complexity of fitness function that would incorporate the notion of aesthetics. The solution space also contains many undesirable results, either because of missing aesthetic features, or because they are unimpressive and just 'more-of-the-same'. One possible solution to this problem is to include human interaction in the design process [12]. Those *interactive evolutionary computing* systems are able to produce a variety of artifacts for images [6,7,12–14]. At the same time involving a human slows the generative process down and, due to the subjective nature, also comes at a cost for consistency and coverage [15]. While the mentioned problem of efficient solution space exploration is less prevalent in interpretative systems, as the content of the painting is defined by the input image, the generation of an interesting and artistically pleasing result remains a difficult task. Fully automated algorithms struggle to identify visually interesting, so called salient, elements of the source image, something that humans easily do [16]. Failing to identify those elements and creating a painting by some form of uniform optimization [10], "tends to produce a machine-generated signature in the resulting painterly renderings" [16]. More recent approaches therefore "[...] trend away from use of local low-level image processing operators towards the incorporation of mid-level computer vision techniques in stroke placement heuristics" [16]. Those techniques include color segmentation [17], analysis of interest by eye-tracking [18] and image salience mapping [19] to guide the algorithm in the generative process.

The approach of this work, outlined in the next section, draws inspiration from existing *interpretative* systems, especially the concept of composing the final image from a set of strokes [8]. While the problem of salience is not addressed as directly as in other works [16,19] it allows for some intrinsic benefits (see Subsect. 4.2). Other implementations exist, which explore a similar direction [20,21]. Contrary to a global generation and optimization of a final image, this works focus on a local, limited generation, combined with a global fitness function. This yields a composition of multiple optimized steps, which can only reach a certain fitness on their own, rather than a globally optimized result.

3 Evolutionary Artistic Rendering

As described in Subsect. 1.1 the idea of this work is to transform the art of painting into an incremental optimization problem. Instead of optimizing a number of fully evolved candidates globally, it limits the optimization process to a number of sub routines, each optimizing up to a certain fitness ceiling, before expanding the solution space. By transforming the problem of *"What is the best candidate"* into *"What are the best next n steps to take"* the system mimics an incremental process, constructing the painting piece by piece. By restricting the possible actions for the algorithm to choose from, a consistent and expressive style can be created and applied to input images.

3.1 Overview of the System

This section explains the high level workings of the algorithm and shows the different parts and how they interact with each other. This serves as a foundation for the following sections. The overall system used for the generative process is shown in Fig. 2 and the process can be separated into three logical stages:

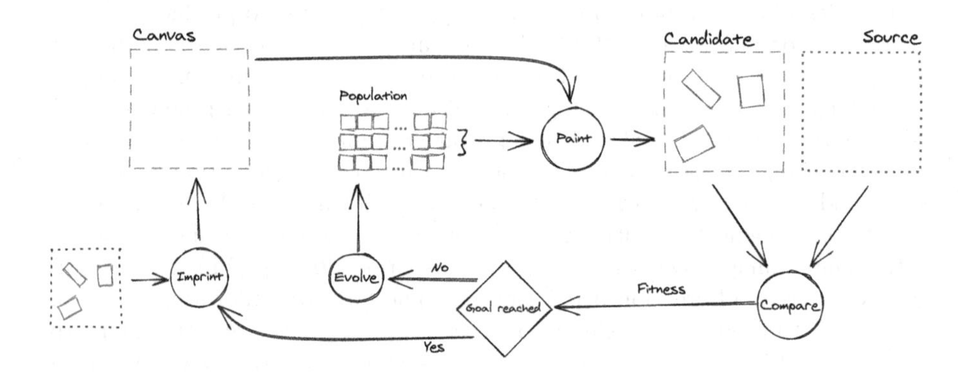

Fig. 2. High-level overview of the procedure. (Color figure online)

Setup. Given a source image (·····), as well as a configuration, an initial population (—) of individuals is created. Each individual encodes n brush strokes within its genome (see Sect. 3.3). The canvas (- - -) for the result is initialized as an empty texture with the same size as the source image.

Generation. For a single generation, each individual candidate is plotted on the current canvas and compared to the source image using a fitness function f based on the pixel. The fitness feeds back into the evolutionary process of mutation, recombination and selection. Based on each individual fitness the algorithm generates the next generation of the population and this steps repeat.

Iteration. New generations are evolved until the goal, or finishing criteria, is reached. Then the n brush strokes of the best individual are permanently imprinted into the canvas. This finishes the iteration, recreates a new population and starts again, painting onto the changed canvas. Given n strokes per iteration and i iterations the total number of strokes on the canvas after finishing the process is $i * n$.

3.2 Implementation

The software is implemented in C# and HLSL[1] using the Unity3d Engine[2] for rendering and GeneticSharp [22] for the evolutionary optimization. The source code is publicly available at github.com/floAr/EvolutionaryArtistUnity. The genome structure is a custom implementation to allow for very fast normalization and adaptive pruning of genes, that are set to constant values. One important concept is that a single gene, for the purpose of genetic algorithms, holds multiple bits of data, which map to a single property of a brush stroke. This allows operations like mutation and crossover to either operate on bit level (operate the numeric value of a property) or gene level (operate on the whole property). By default gene values are represented using 16bit unsigned integers which are mapped from [0, 65.535] to [0,1][3], resulting in a minimum step size of $1.52e^{-5}$ between possible values. If desired, this could be decreased to use 8bit unsigned integers, with a step size of $3.9e^{-2}$ and a smaller memory footprint, or increased up to 64bit, resulting in a step size of only $5.5e^{-20}$. The computation of heavy operations like painting a candidate, imprinting the canvas and comparing a candidate with the source image is implemented using shaders and executed on the GPU, to allow parallelization.

3.3 Evolutionary Algorithm

To generate a desirable image the process uses a genetic algorithm to optimize candidates. A decent familiarity with evolutionary-, especially genetic algorithms, is assumed. A more complete introduction into genetic algorithms is given in [4] or [23]. This section features key areas of interests of the artistic process. First the genotype and phenotype representation, defining how candidates are stored and rendered, are explained, then the manipulation and selection strategies and lastly the possible parameter space for the image generation, as well as some aspects of the implementation.

Genotype and Phenotype. The genome for a single candidate consists of n sub-sections, each describing a single brush stroke, where n equals the number of strokes per candidate defined globally for the generation process. The maximal configuration of a single brush stroke part is shown in Fig. 3. As mentioned in Subsect. 3.2, values that are removed from the evolutionary process, for example by fixing their value, are pruned and will not show up in the genome. Each gene provides the interpreter with a value between 0 and 1, which then in turn is translated to form the phenotype of a brush stroke. The phenotype of a single brush stroke, as shown in context in Fig. 4, translates the [0,1] values from

[1] https://docs.microsoft.com/en-us/windows/win32/direct3dhlsl/dx-graphics-hlsl, accessed 27.10.2020.

[2] https://unity.com/, accessed 27.10.2020.

[3] https://docs.microsoft.com/en-us/dotnet/csharp/language-reference/builtin-types/integral-numeric-types, accessed 22.10.2020.

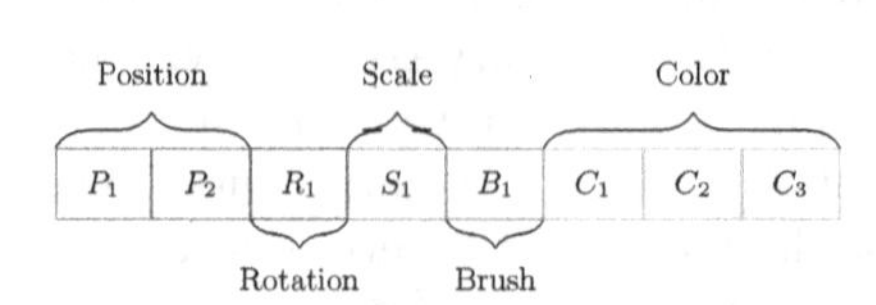

Fig. 3. Maximum genome configuration for a single brush stroke. Each box corresponds to one gene, which yields a value within [0,1].

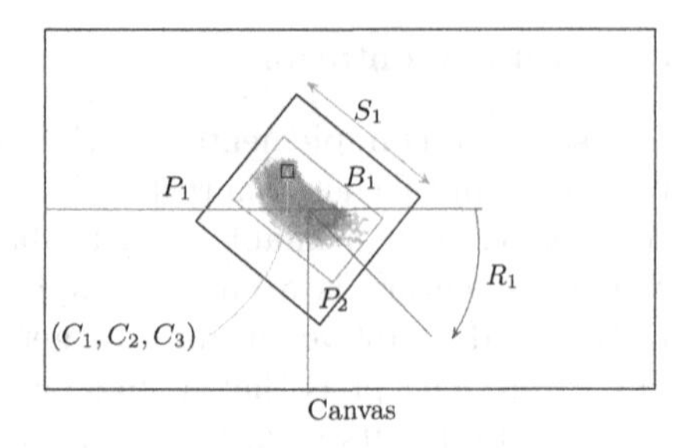

Fig. 4. Phenotype of a single brush stroke, with a certain color, placed and transformed on the canvas.

the genotype into the transformation and shape to be imprinted on the canvas. Technically a brush stroke is a transparent quad, with a certain texture, which is scaled, rotated and positioned on the canvas. The position is encoded as a vector within the canvas object space. This means $(P_1, P_2) = (0, 0)$ corresponds to the lower left corner, (1,1) to the upper-right respectively. The quad can be rotated clockwise, e.g. a value $R_1 = 0.25$ yielding a 90 degree rotation. The quads uniform scale is defined by a base size (see Subsect. 4.2), which is calculated by the overall algorithm and not evolved with the candidates. The scale gene can be used to deviate from this base size by [-5%, 5%]. This mechanic and its effect is explained further in Subsect. 4.2. The actual texture content is selected from a texture array of size t. The array holds multiple brush strokes, allowing the algorithm to evolve the stroke textures used by each candidate, as shown in Subsect. 4.1. To allow this, B_1 gets translated into an index of the array, indicating which texture to pick. If the array only contains a single texture (see Sect. 5: *Cubic* brush pack) B_1 has no effect and is pruned from the genome. Lastly the texture selected is tinted by a RGB color defined by (C_1, C_2, C_3) multiplied by the transparency value of the texture.

Manipulation. One important part of evolutionary algorithms are the means of manipulating candidates. Two general concepts for this are *mutation* and *recombination*. The proposed algorithm implements rather simple variants of both mechanics to change the genome of candidates. *Mutation* is controlled by an overall mutation rate [0,1] expressing the chance to mutate for each candidate. If a mutation occurs a two step process selects a gene within the candidate and then a single bit within this gene to flip. While this method is quite robust and can execute reasonably fast, it has the drawback of high interdependence between the number of brush strokes, as well as the precision (see Subsect. 3.2) of the genome representation. A higher number of strokes n gives a longer genome per candidate and as only a single mutation is carried out, the chance of each brush to be mutated is $\frac{1}{n}$. Furthermore, due to the binary representation of numbers, the position of each gene that is mutated has a strong influence on the resulting numerical change. A mutation on position c in our gene will change the value of the gene by $\frac{2^c}{maxValue}$, where $maxValue$ depends on the chosen data type and is used to normalize the gene. The current implementation balances the first shortcoming by having a generally high mutation rate of 90%, which declines until 20% over the first ~800 generations. Those values have been chosen after test runs and proven to be effective for this setup. *Recombination* executes a uniform crossover strategy between two candidates A and B [24]. For each gene index of A and B a swap is performed with a likelihood of 50%, exchanging the gene of A with the gene of B and vice versa. As this crossover is performed on each individual gene it is independent of the genome length. As a single gene fully encodes a phenotype's property as mentioned in Subsect. 3.2, crossover will never change the value of a gene, instead only swap properties between two candidates.

Selection and Fitness. The last component of the evolutionary algorithm is the selection of candidates to create the next generation for the population. This approach uses Tournament Selection [25]. The fitness function is used by the evolutionary algorithm to rate the performance of different individuals. In this work the fitness function compares the artifact generated by painting the candidates strokes onto the current canvas with the original image. One of the simplest fitness functions is the negated sum of the difference for each pixel. This work uses a slightly improved variation of this function, scaling the difference of each color channel to approximate the visual sensitivity of human perception [26]. As discussed in Sect. 5 various improvements can be made in this aspect, which might unlock more sophisticated image analogies.

4 Evaluation

In this chapter the properties and artistic capabilities of the system are explored and evaluated. This work focuses on the expressiveness of the system (see Subsect. 4.1), as it is fundamental to help its user to express an artistic idea. Figure 5 shows the input images used for the experiments. Each image is scaled to 512 by 512 pixels and used in the experiments without further modification. The first two images are paintings by Johannes Vermeer [27] and Vincent Van Gogh [28], the third and fourth images are photographs, with minor modifications (some houses on the horizon were cropped out) and the last image serves as a benchmark for stroke precision and color fastness.

In total four different brush texture packs were used: *Watercolor*, *Droplets*, *Cubic* and *Stroked*. The Appendix lists the brush textures used in each pack. Unless specified otherwise the experiment settings used for the genetic algorithm are 50 iterations, with 7 strokes per candidate. Mutation is handled by flipping a random bit, the mutation chance starts at 90% and decreases by 10% every 50 generations, down to a minimum of 20%. Crossover uses the *Uniform Crossover* strategy [24] with a fixed probability of 50%. Selection is done with *Tournament*

Selection [25], with a tournament size of two genomes. Termination criteria is a stagnation of the fitness value for 20 generations. For the experiment all features of a brush stroke are evolved as described in Sect. 3.3. Other default settings are base opacity value for each brush, set as $\alpha = 0.7$ (70% opacity) and a brush base size which interpolates between 0.8 (which corresponds to 80% of the canvas) and 0.025 over all iterations. The functions used to interpolate are shown in Fig. 9, the red graph being the default. The effective base size is calculated using the current iteration i, by sampling the interpolation function at $\frac{i}{\text{total iterations}}$ and using the value to interpolate between the maximum and minimum size value. The generative process starts from a white canvas for each image, except the *Benchmark* image, which starts with a black canvas.

4.1 Expressiveness

A core requirement of an artistic system is the ability to express the users desired artistic outcome. Therefore this section evaluates the expressiveness of the algorithm. For this, two different aspects are to be considered: *Style Variety*, which is the scope of different styles that can be generated, as well as *Style Consistency*, which means generating consistent results, when applying a selected style to different input images. Both features are important to model the signature and style described in Subsect. 1.1. To model a signature and style of an artist the system needs to be able to generate diverse, but also consistent styles.

Style Variety. The simplest way of creating different styles is adjusting the brushes used to paint the image. In Fig. 6 three different examples of possible styles, which vary only in the selected brush texture, are shown on four different images. Each row shows how a single image can be represented differently by changing the brush texture, creating variety in plasticity and style, while preserving the content. This is in line with [8] which found that brush characteristics are a mayor factor influencing the outcome of the generative process.

The first two columns show how relatively similar brushes, both are opaque and laminar, can invoke very different detail textures. The difference in shape (stretched vs circular) as well as in border structure (smooth vs rigged) translate nicely into the structure of the painting, without creating noticeable artifacts.

Fig. 5. Input images used. From left to right: Girl with a Pearl Earring [27], Wheat Field With Cypresses [28], Photos of a dog under tree and a church (2020) and benchmark image.

The third column shows how far a style can differ, given a more diverse brush set. The thin strokes create a hatched and sketchy look, overshooting the content target (especially noticeable in the benchmark image on the very bottom), as their inherit error for this is way smaller compared to more laminar brushes.

Other means of variation are shown in Fig. 7: Using the same brush texture different images can be generated by setting other constrains. The first column shows the image generated by the *Cubic* brush packs, which uses a single, white square as the only brush texture. The upper image uses the default configuration and is able to replicate the image quite well using rotation and scale to vary the single texture. The lower image shows a result with fixed rotation. In this case the R_1 gene (see Sect. 3.3) is removed for each brush and instead provided as a fixed value. This leads to axis aligned blocks, which can not approximate the geometry of the source as well as before, yielding a mosaic like style. The center column shows the difference between the default opacity value (70%) in the first row and 20% opacity below. The image with lower alpha, albeit using the same brush pack, appears smoother and more continuous. The last column shows the default result for the *Watercolor* brush and a variant where the color genes where removed below. Instead a single gene was used in the genome to evolve a one-of-n color selection. Just like brush selection this allows the algorithm to only evolve colors from a predefined set. The color space consists of seven colors sampled from the image (shown in upper image) as well as black and white. Furthermore the α for the brushes was fixed on 100% opacity for this experiment, to prevent color mixing due to blending. This limitation results in a posterized look of the image, bringing forth sharp contrasts and cutting smaller features due to missing means of approximation. This examples show the capabilities of the system to express multitudes of styles given the possible combinations of different restrictions.

Style Consistency. Given the ability to generate a variety of different styles, enabling a wide spectrum of styles to realize a custom reinterpretation, consistency is as important when considering an artistic tool. As described in Subsect. 1.1 the combination of both factors allow to mimic the transformative process of creating an artwork from an internal representation. Looking at Fig. 6 each column shows that a style produces consistent results over a variety of different images, creating an equable look and feel. Between different types of input images, style elements remain noticeable and create a recognizable set of interpretations. This effect increases as the difference between the used style restrictions does. The more regulated and therefore specific a certain style is, the easier it is to recognize those peculiarities in the generated artifacts.

4.2 Image Saliency

Within approaches that aim to transform an input image into stylized artefacts, a common problem is detection of salient regions. One way to counteract this problem is to employ more complex operators in the fitness function, like higher order computer vision mechanics [17,19] to preprocess the image, or to be used

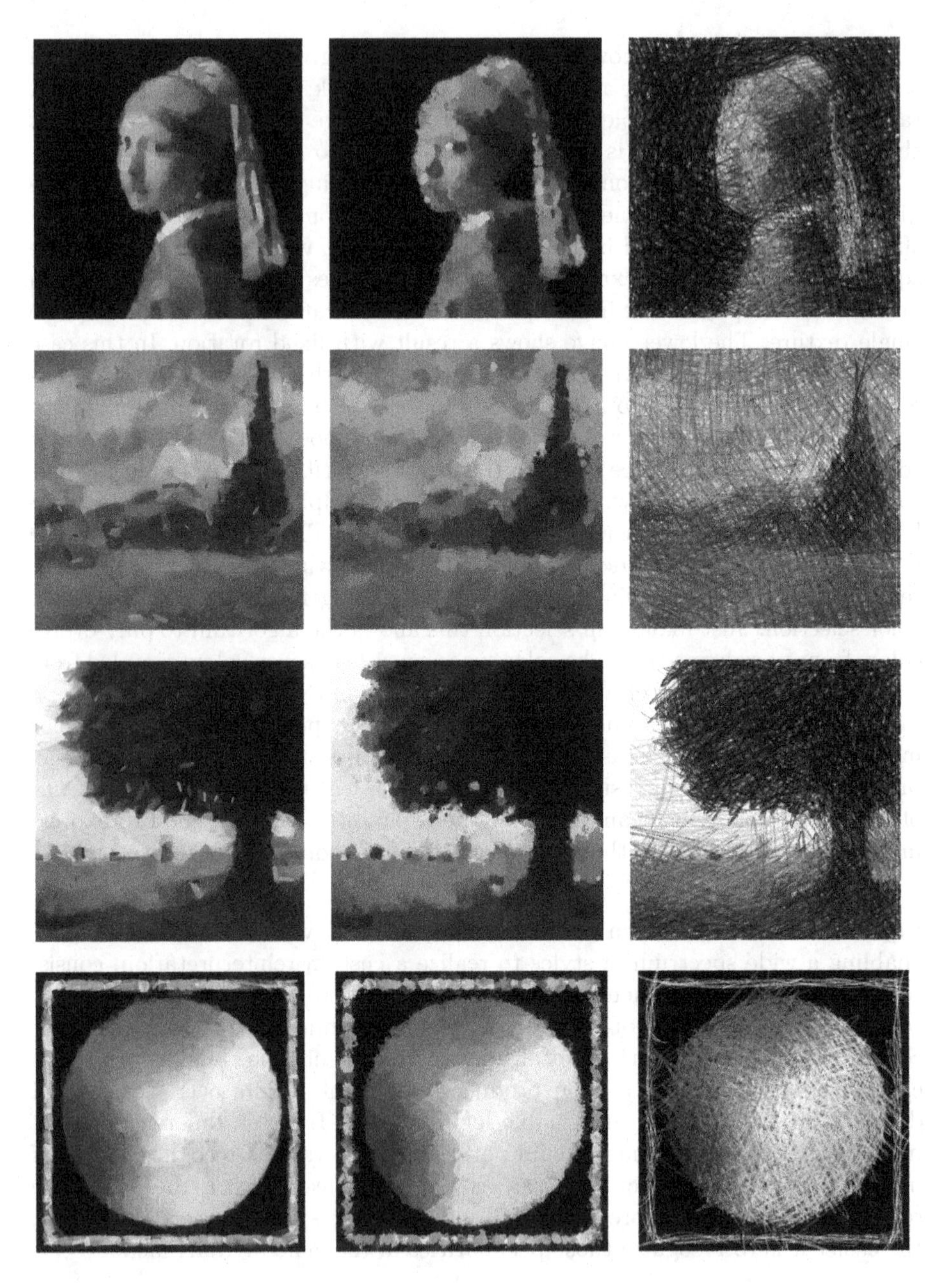

Fig. 6. Different brushes yield varying results. From left to right: The *Watercolor*, *Droplet* and *Strokes* packs were used (see Sect. 5)

in the fitness function. This approach instead makes use of emerging properties of the image generation itself: The way an image is constructed by the algorithm resembles the construction of a painting as done by humans (see Subsect. 1.1).

Fig. 7. Example of fine grained style variation. The upper image is the 'default', the lower one a more restricted result. The leftmost column has constrained rotation, the center one uses a lower transparency value and the right one is limited to a set of seven colors.

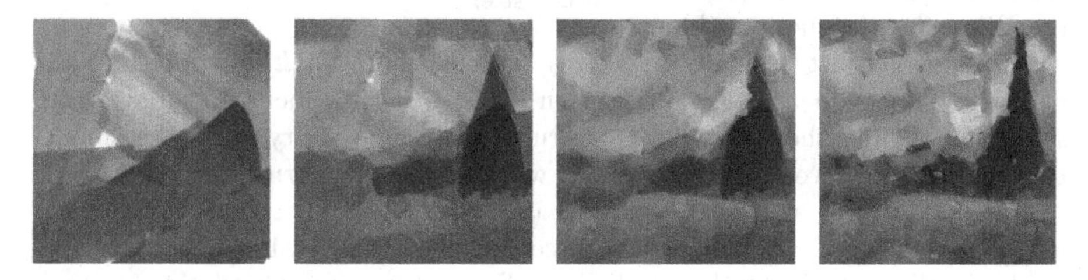

Fig. 8. Different stages of the generation of a single image. Left to right: 1, 5, 25 and 50 iterations.

This behaviour emerges because of two properties of the algorithm: By only being able to place a limited number of brush strokes the algorithm has a fitness ceiling for each iteration. Given seven strokes in the first iteration the target image can only be approximated to a certain amount. This leads to the construction of the artwork from coarse to fine, as filling the most erroneous large areas will yield the highest fitness gain. Furthermore the adjustment of the brush size over the course of the process strengthens the behaviour to start with larger features and move to more detailed adjustments later on, starting already from a more sophisticated representation. This progression from coarse to fine features forces the algorithm to only add smaller details, after the overall color composition has been already executed, which favors regions with smaller details in later iterations. This behaviour is visible in Fig. 8, which shows the canvas at different stages in the evolution process. The first iteration provides the overall shading, as the seven brush strokes are used to cover the white background. Four iterations later the outline of the image is roughly sketched, and the following iterations keep on refining the outline and adding finer shading. After 50 iterations the overall outline of the image is clearly defined, with the latest and smallest strokes adding highlights and details. While this does not directly map to salience in all cases, it puts focus onto adding high contrast details, which correlates with visually highly interesting regions [16].

A more detailed examination of different brush size progression and its influence on the generated result can be seen in Fig. 10. The target image contains many details, such as small shaded areas, color gradients and tiny features, which makes it hard to replicate truthfully. Between the three results the only difference is how the brush size was interpolated between 0.8 and 0.025. Figure 9 shows the three functions used: Image (A) uses the *Bias Small* function (blue), which converges fast to small brush sizes, (B) uses the *Default* function (red) and image (C) uses the *Bias Large* function (green), with a focus on larger brush sizes. The results show the effect of brush size on plasticity, with the smaller brushes creating a very 'rough' surface, whereas the large brushes tend to blend together, exhibiting smoother gradients and softer edges. By adjusting the brush stroke sizes available to the algorithm certain focus on detail can be triggered, as the smaller strokes tend to embed smaller features into the design, as their effect to the fitness function is greater on small, high contrast features, than on large areas. While this helps to alleviate some of the problems in regards to missing salience detection, it does not solve the problem in itself, as all these intrinsic optimizations operate on a uniform level. This can be seen in Fig. 6, when comparing the landscape painting with the portrait: While the landscape painting has a relatively even distribution of salience, the portrait has specific details, like the eyes and lips, which are not captured well with this approach. A hybrid is the photo of the dog under the tree where the algorithm adds detail to the landscape, but also fails to add enough detail to the dog itself, reducing its presence in the final outcome.

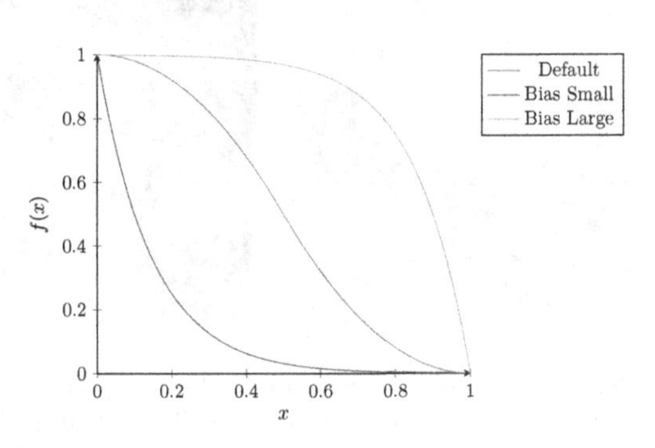

Fig. 9. Interpolation functions to calculate brush size.

Fig. 10. Influence of different size compositions. From left to right: (A) Bias towards small brush strokes, (B) default and (C) bias towards large brush strokes.

5 Conclusion and Future Work

This work has shown a generative system based on an abstract art generating process and a genetic algorithm. Different features have been explored and evaluated in regards of their possible solution space and shortcomings. As mentioned by Collomosse [16] detection of salient regions is a core feature of unsupervised generative systems. While this work employs techniques to improve detail placement for uniformly salient images, it remains a problem for source images with small salient regions. Other techniques exist and can be paired with this approach, like manually authored guidance maps for stroke placement [20] or higher order computer vision methods, like edge detection or color segmentation [17–19]. Recent advances in neural network research allow for high fidelity and automated extraction of salient regions [29]. Other neural network research areas like style transfer [30,31] might also provide benefits and allow for interesting results by generating more complex brush strokes and enriching the details of the final artwork. Conceptually this work is based on an abstract art generating process (see Subsect. 1.1) but focuses mainly on the second transformation, the expression of the internal representation into a medium. The first transformation, the personalized perception of objects is modelled by the calculation of the fitness function as it, defines how the system (artist) can perceive the ground truth (input image). By basing the fitness only on pixel errors, the algorithm always compares against o_{truth}, our input image itself. A more sophisticated method could also introduce additional artistic traits, that are found for example in abstract or expressionist art. Various error metrics, like Wasserstein distance [32] or style transfer loss metrics [30] allow the comparison of images within a higher order space and could yield artifacts that go beyond simple pixel similarity. Further improvements can be made in regards to the current implementation of the system. The Sect. 3.3 mentions the high interdependence between genome length and mutation chance, especially with the brush texture. Normalizing those values would allow to reuse settings between different styles more robustly and lessen the required user input when exploring different interesting styles. Another aspect is the continuity of the generation process. Iterations are decoupled from each others, the algorithm has no means to determine how many of them happened already, or how many are left. Currently each iteration starts with an imprinted canvas, onto which the strokes of the best candidates are added. In regards of the candidates itself, 'catastrophic forgetting' [33] occurs after each generation. Instead of imprinting only the best candidates, it would also be possible to build the stack of actions in memory, enabling parallel processing of the best n candidates of each generation. While this will exponentially increase memory consumption and time, it will allow the buildup of more complex patterns, that take several layered brush strokes, and therefore more iterations to evolve. Given these improvements, different paths ahead are possible, reinforcing the autonomous capabilities of the system presented in this work, or fusing them with user controlled input and turning it into a semi-supervised content creation tool. Other exotic use cases can be found in emerging properties of the generated artifact. One particular exotic use case would be compression of source images, given that a genome representation of an image is roughly 10 times smaller than its pixel data. Yet to enable those use cases more sophisticated methods of salience detection and non-uniform detail preservation are to be implemented. The approach explored in this work provides a solid framework for the creation of stylised artworks. Given the improvements above future work allows for different applications, especially with improved salience detection (supervised or unsupervised) and the infusion of higher order operations like methods from artistic style transfer.

Brush Packs. All brush textures, unless stated otherwise, where created for this work using Gimp 2.10.4[4] (Figs. 11, 12, 13 and 14).

Fig. 11. *Watercolor* from [20]

Fig. 12. *Droplets* from [34]

Fig. 13. *Strokes*, ball-point pen and scanned

Fig. 14. *Cubic*

References

1. Dawkins, R.: The Blind Watchmaker - Evidence of Evolution Reveals a Universe without Design. WW Norton & Company (1986)
2. Sims, K.: Artificial evolution for computer graphics. In: Proceedings of the 18th Annual Conference on Computer Graphics and Interactive Techniques. SIGGRAPH'91, Association for Computing Machinery, New York, NY, USA, pp. 319–328 (1991). https://doi.org/10.1145/122718.122752

[4] https://www.gimp.org/, accessed 28.10.2020.

3. Todd, S., Latham, W.: Evolutionary Art and Computers. Academic Press Inc., USA (1992). http://portal.acm.org/citation.cfm?id=561831
4. Lewis, M.: Evolutionary visual art and design. In: Romero, J., Machado, P. (eds.) The Art of Artificial Evolution. Natural Computing Series, vol. xx, pp. 3–37. Springer, Berlin, Heidelberg (2007). https://doi.org/10.1007/978-3-540-72877-1_1. https://link.springer.com/chapter/10.1007%2F978-3-540-72877-1_1
5. Unemi, T.: SBART 2.4: breeding 2D CG images and movies and creating a type of collage. In: International Conference on Knowledge-Based Intelligent Electronic Systems, Proceedings, KES, pp. 288–291 (1999)
6. Rooke, S.: Chapter 13 - Eons of genetically evolved algorithmic images. In: Bentley, P.J., Corne, D.W. (eds.) Creative Evolutionary Systems. The Morgan Kaufmann Series in Artificial Intelligence, pp. 339–365. Morgan Kaufmann, San Francisco (2002). http://www.sciencedirect.com/science/article/pii/B9781558606739500525
7. Hart, D.: Toward greater artistic control for interactive evolution of images and animation. In: Giacobini, M. (ed.) SIGGRAPH'06. LNCS, vol. 4448, pp. 527–536. Springer, Berlin, Heidelberg (2006). https://doi.org/10.1007/978-3-540-71805-5_58
8. Haeberli, P.: Paint by numbers: abstract image representations. Comput. Graph. (ACM) **24**(4), 207–214 (1990). https://doi.org/10.1145/97880.97902
9. Hertzmann, A., Jacobs, C.E., Oliver, N., Curless, B., Salesin, D.H.: Image analogies. In: Proceedings of the 28th Annual Conference on Computer Graphics and Interactive Techniques SIGGRAPH 01 Aug 2001, pp. 327–340 (2001)
10. Hertzmann, A.: Paint by relaxation. In: Proceedings of Computer Graphics International Conference, CGI, pp. 47–54 (2001)
11. Szirányi, T., Toth, Z.: Random paintbrush transformation. In: 15th ICPR, Barcelona, Spain (2000)
12. Takagi, H.: Interactive evolutionary computation: fusion of the capabilities of EC optimization and human evaluation. Proc. IEEE **89**(9), 1275–1296 (2001)
13. Machado, P., Cardoso, A.: All the truth about NEvAr. Appl. Intell. **16**(2), 101–118 (2002)
14. Secretan, J., et al.: Picbreeder: a case study in collaborative evolutionary exploration of design space. Evol. Comput. **19**(3), 373–403 (2010). https://doi.org/10.1162/EVCO_a_00030
15. Bentley, P.J., Corne, D.W.: An introduction to creative evolutionary systems. In: Bentley, P.J., Corne, D.W. (eds.) The Morgan Kaufmann Series in Artificial Intelligence, pp. 1–75. Morgan Kaufmann, San Francisco (2002). http://www.sciencedirect.com/science/article/pii/B9781558606739500355
16. Collomosse, J.P.: Evolutionary search for the artistic rendering of photographs. In: Romero, J., Machado, P. (eds.) The Art of Artificial Evolution. Natural Computing Series, vol. xx, pp. 39–62. Springer, Berlin, Heidelberg (2007). https://doi.org/10.1007/978-3-540-72877-1_2. https://link.springer.com/chapter/10.1007%2F978-3-540-72877-1_2
17. Gooch, B., Coombe, G., Shirley, P.: Artistic vision: painterly rendering using computer vision techniques. In: NPAR Symposium on Non-Photorealistic Animation and Rendering (2003)
18. Santella, A., Decarlo, D.: Abstracted painterly renderings using eye-tracking data. In: NPAR Symposium on Non-Photorealistic Animation and Rendering (2002)
19. Collomosse, J.P., Hall, P.M.: Genetic paint: a search for salient paintings. In: Rothlauf, F., et al. (eds.) Applications of Evolutionary Computing. Lecture Notes in Computer Science, vol. 3449, pp. 437–447. Springer, Berlin, Heidelberg (2005). https://doi.org/10.1007/978-3-540-32003-6_44

20. Opara, A.: Genetic Drawing Project (2020). https://github.com/anopara/genetic-drawing
21. Shahriar, S.: Procedural Paintings with Genetic Evolution Algorithm (2020). https://github.com/IRCSS/Procedural-painting
22. Giacomelli, D.: GeneticSharp (2013). https://github.com/giacomelli/GeneticSharp/
23. Johnson, C.G., Cardalda, J.J.R.: Genetic algorithms in visual art and music. Leonardo **35**(2), 175–184 (2002). https://doi.org/10.1162/00240940252940559
24. Syswerda, G.: Uniform crossover in genetic algorithms. In: Proceedings of the 3rd International Conference on Genetic Algorithms, pp. 2–9. Morgan Kaufmann Publishers Inc., San Francisco, CA, USA (1991)
25. Goldberg, D.E., Deb, K.: A comparative analysis of selection schemes used in genetic algorithms. In: Rawlins, G.J.E. (ed.) Foundations of Genetic Algorithms, vol. 1, pp. 69–93. Elsevier (1991). http://www.sciencedirect.com/science/article/pii/B9780080506845500082
26. CompuPhase: Colour Metric (2019). https://www.compuphase.com/cmetric.htm
27. Vermeer, J.: No Meisje met de parel (1665). https://www.mauritshuis.nl/en/explore/the-collection/artworks/girl-with-a-pearl-earring-670/
28. Van Gogh, V.: Wheat Field With Cypresses (1889). https://www.vincentvangogh.org/wheat-field-with-cypresses.jsp
29. Wei, J., Wang, S., Wu, Z., Su, C., Huang, Q., Tian, Q.: Label Decoupling Framework for Salient Object Detection (2020)
30. Uhde, F., Mostaghim, S.: Towards a general framework for artistic style transfer. In: Liapis, A., Romero Cardalda, J.J., Ekárt, A. (eds.) Computational Intelligence in Music, Sound, Art and Design. LNCS, vol. 10783, pp. 177–193. Springer, Cham (2018). https://doi.org/10.1007/978-3-319-77583-8_12
31. Lee, H.Y., Tseng, H.Y., Huang, J.B., Singh, M., Yang, M.H.: Diverse image-to-image translation via disentangled representations. In: Ferrari, V., Hebert, M., Sminchisescu, C., Weiss, Y. (eds.) Computer Vision – ECCV 2018. Lecture Notes in Computer Science (Lecture Notes in Artificial Intelligence and Lecture Notes in Bioinformatics), vol. 11205, pp. 36–52. Springer, Cham (2018). https://doi.org/10.1007/978-3-030-01246-5_3
32. Arjovsky, M., Chintala, S., Bottou, L.: Wasserstein generative adversarial networks. In: Precup, D., Teh, Y.W. (eds.) 34th International Conference on Machine Learning, ICML 2017. Proceedings of Machine Learning Research. PMLR, International Convention Centre, Sydney, Australia, vol. 1, pp. 298–321 (2017). http://proceedings.mlr.press/v70/arjovsky17a.html
33. Kirkpatrick, J., et al.: Overcoming catastrophic forgetting in neural networks. Proc. Nat. Acad. Sci. US Am. **114**(13), 3521–3526 (3 2017). http://www.pnas.org/content/114/13/3521.abstract
34. Starline: Watercolor Splatters. https://www.freepik.com/free-vector/black-ink-watercolor-splatters-drips_10555025.htm

Dissecting Neural Networks Filter Responses for Artistic Style Transfer

Florian Uhde(✉) and Sanaz Mostaghim

Faculty of Computer Science, Otto-von-Guericke-University, Magdeburg, Germany
florian.uhde@posteo.de
https://www.inf.ovgu.de/en/

Abstract. Current developments in the field of Artistic Style Transfer use the information encoded in pre-trained neural networks to extract properties from images in an unsupervised process. This neural style transfer works well with art and paintings but only produces limited results when dealing with highly structured data. Characteristics of the extracted information directly define the quality of the generated artifact and traditionally require the user to do manual fine-tuning. This paper uses current methods of deep learning to analyze the properties embedded in the network, group filter responses into semantic classes and extract an optimized layer set for artistic style transfer, to improve the artifact generation with a potentially unsupervised preprocessing step.

Keywords: Artistic style transfer · Filter response · Neural network

1 Introduction

Recent works of artistic generation, using neural networks [1–5], exhibit properties classically associated with creative work: Artistic style transfer [4] aims to extract content and style of different source pictures and synthesize a new artifact, combining the extracted properties. An image I is described by (1), as the combination of style (I^S) and content (I^C). Style Transfer takes two guidance images, I_{style} and $I_{content}$, extracts the respective element and combines them into a novel image, as defined in (2) [6].

$$I = combine(I^S, I^C) \tag{1}$$

$$I_{result} = ST(I_{style}, I_{content}) = combine(I^S_{style}, I^C_{content}) \tag{2}$$

Neural Style Transfer in particular, uses information encoded in neural networks trained for image processing to analyze images for style and content and extract the relevant elements to generate novel images; the filter responses within a layer of the neural network encode the image I [7]. The quality of a generated artifact is estimated by loss-functions operating on the filter response sets of the input image and the generated image. This allows optimization of the result to yield very similar filter responses [4,7].

J. Romero et al. (Eds.): EvoMUSART 2021, LNCS 12693, pp. 297–312, 2021.
https://doi.org/10.1007/978-3-030-72914-1_20

This paper proposes a method to increase the quality of generated artifacts by improving and automating the feature selection process through a data pre-processing step. Given a set of training images for the objects that should be style transferred, the approach described in this paper aims to select of relevant weights and network layers, to yield a desirable outcome of the style transfer without the requirement of human fine-tuning beforehand, especially when dealing with more structured and less artistic input images. Outcomes of this paper are also interesting for generative techniques, like work by Umetani [8], that rely on constructing a latent space from the training data, as the process prepares training data to span a compact and informative latent space.

2 Related Work

Artistic style transfer describes the process of analyzing two input images to create a new artifact containing the content of one input, rendered in the style of the other. Contrary to classical approaches, which rely on a patch-based synthesis of texture parts by sampling color data from source images [9–11], neural network driven techniques utilize the data encoded in a trained network. [1–5]. These approaches use the representation of information in trained weights along the hierarchy of the network [12], whereas deeper layers encapsulate broader features of the source image while being relatively invariant to precise structures [7]. Similar the *style* encodes within the correlation of filter responses of the neural network [7]. By retrieving this information over multiple layers of the network, texture and local distribution can be obtained, while leaving the global arrangement of this information out [7,13]. The algorithm generates an initial candidate, improving it to minimize the content- and style-loss functions [13]. Recent works have made improvements to the original algorithm, by incorporating statistical features of the input image [5,14], pre-training specific networks per style [2,15], or developed new methods using different processes for feature extraction and synthesis [1,16]. Especially the work of Li and Wand [17] is interesting for this paper, as it focuses on the problem of keeping larger structural features of images intact, which traditionally is problematic for artistic style transfer, yet important when dealing with highly structured data, like cars or building.

This paper also relies on work by Mordvintsev et al. and Zeiler and Fergus [18,19] to visualize different filter responses within neural networks.

3 Method

As mentioned above, artistic neural style transfer uses different layers in a pre-trained neural network to extract the filter responses of both input and target image and calculates a loss depending on the difference between both images [4]. As the granularity and level of abstraction of these encodings vary, depending on the current depth of the network [7], different approaches define custom sets of layers, which are relevant to extract style information and content information from the network. These often pair with a weight associated with each layer, forming the loss function to optimize the result. The choice of layers and therefore filters is a core influencing factor for the outcome and is often made by trial and error, or based on previous papers. To highlight the effect layer selection has on the final result Fig. 1 shows a style transfer conducted using the Visual Geometry Group VGG19 network [20]. The same input images are used to conduct style transfer, for the first result layers from the higher part of the network (block_3) calculate the loss metric, whereas for the second result layers of a deeper part (block_5) are used.

The approach of this paper aims to do an unsupervised preprocessing of the available training data, to generate a small, but at the same time descriptive feature set. In neural style transfer, this feature set represents filter activation within different layers of the neural network. Using a small set of layers with a high level of information benefits training stability and speed and improves solutions that rely on custom learned networks [2]. Instead of a user-designed selection of layers for both sets, the proposed mechanic ingests training data for each of the image classes and processes it with a trained neural network (Visual Geometry Group VGG19 network [20]), trained for to perform object recognition and localization [21]. The method aims to categorize filters into certain classes and sub-types to automatically select layers containing suitable filters for style transfer. To achieve this a set of images is provided as training data, divided into different object classes c. For every filter f in each convolutional layer l the activations are measured and averaged over all images I from different classes c: $A(f, c)$. Further calculation on those stored activations allows for the categorization of a filter into significant (Subsect. 3.1) and diverse filters (Sect. 3.1). In Subsect. 3.3 five types of filters are defined based on descriptive properties of a filter, which are used in Sect. 4 to automatically generate a layer set for style transfer.

Fig. 1. Different results of style transfer. From left to right: *Content input*, *style input*, *result using block 3* and *result using block 5*

3.1 Significant Filters

The first class, *significant filters*, contains filters that hold high information describing one class of the training data. Those filters have a relative constant activation for all images within one class. To select the relevant filters we calculate the coefficient of variation over all activation within class c with

$var(f,c) = \frac{\sigma A(f,c)}{\mu A(f,c)}$. All filters that do not exceed a threshold t_{sig} are considered to be significant for class c. Adjusting t_{sig} gives control over the number of selected significant filter responses to provide a certain spread over the images, but at the same time limits the results in a way that insignificant filter are discarded. In the experiments conducted in Sect. 4 the threshold is dynamically calculated to include n percent of the filters for each combination of classes. Generating those filters over multiple classes from the training set, instead of only a single class, allows further diversification into two sub-categories: Intersecting filters, that contain information shared between classes from the training set and disjoint filters, which are helpful in making distinctions between classes.

Intersecting Filters. These are all filters that are significant to more than one class in the training set. To find these filters the training data is preprocessed, as described above, class-by-class and out of those filters, we select all that are significant for at least two classes from the training data. Those filters contain information that is relevant for diversifying or grouping different classes, depending on if they exhibit varying or similar levels of activation between them. They are interesting for artistic style transfer, as they can be used to translate style information between classes, given the fact that it describes a feature that occurs in multiple classes in potentially different specificity.

Disjoint Filters. Contrary to the intersecting filters mentioned in Sect. 3.1 this set contains all filters that show similar levels of activation only for a single class of the training data. Those filters are thereby unique descriptors of properties present in this class. Similar to the approach above the filters are created by generating significant filters on a class basis and then selecting all filters that only occur for one class. Given that these filters are only significant for a single class in the training dataset they are well suited to describe this class in the style transfer process, highlighting unique features of a given class in comparison to other classes of the training data.

3.2 Diversifying Filters

Diverse filters describe a class of filters that have different numerical activation values between the different classes and are useful in making a differentiation between them. This concept is similar to the filters described in Sect. 3.1 but with a focus on a high variance between classes, instead of low variance within a class. They contain information about how to distinguish between different classes and are especially helpful for content loss. To calculate those filters the difference of activation strength $A(f,c)$ is calculated for each filter over all training data classes and used as a metric to sort them. A threshold t_{div} can be used to control the number of diverse filters extracted. As these filters potentially contain very diverse information, they are especially interesting as a dimension in a constructed latent space (see: [8]), to increase the variance of generated artifacts.

3.3 Semantic Classes

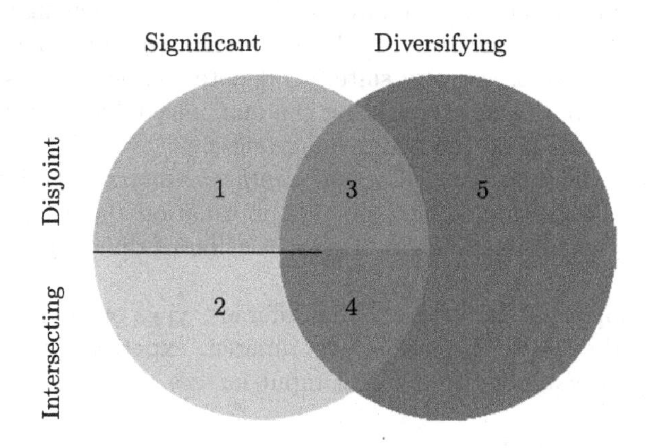

Fig. 2. Five different semantic types of filter responses

With those classes of the filter responses, five different semantic types for filter are derived as shown in Fig. 2. Significant filters are either intersecting or disjoint, and a filter can be significant or diversifying or both. Furthermore, as that semantics operates under the assumption that artistic style transfer deals with two input images of different classes, with the goal of transferring stylistic elements of one class onto the other, the sets of *Intersecting* and *Diversifying* filters are the same for both classes.

- *Type 1*: Significant for a single class but not for any other class from the training data set. The feature encoded in the filter describes a property of one class, which is not characteristic for any other class. Note that it means this feature is neither typical nor atypical for any other class.
- *Type 2*: Significant for both classes, which means activation levels are stable within both classes. Contrary to type 1 these filters describe the features that are either very typical or atypical for both classes.
- *Type 3*: Only significant for one of the classes, yet noticeable diverse activation levels between the classes. Similar to type 1, as the feature is only significant for a single class, but the difference in activation levels is strong enough to count as a diversifier. This property makes the filters valuable for content-loss.
- *Type 4*: Significant for both classes, with different activation levels per class. Filters that are both significant and diverse in their activation provide a high value for style transfer, especially for content loss. They support elements of the content image, while at the same time discouraging content elements of the style image.
- *Type 5*: Diverse activation levels, but low significance per class. This type contains filters producing a high spread between different classes, yet not being significantly constant per class. These are most likely outliners by very unusual elements in the training images.

Given these semantic types, desirable combinations for content- and style-loss emerge: When conducting style transfer, content-loss aims to create an artifact that yields the same filter activation over defined layers, preferring to pick layers deeper within the neural network [7]. Layers, which contain a high number of *type* 3 and *type 4* filters, are very suited to denote content-loss, as they act as diversifier between both classes, resulting in a maximized difference in loss values when creating objects from the wrong image class.

For style-loss the filters, which act as *significant-intersecting* filters (*class 2* and *class 3*) are interesting, as they provide information about features existing in both classes and by this allowing for translation of different peculiarities of those features.

To give examples of the usage of the different types of filter responses and semantic types the next chapter conducts different experiments, showing those filter response types applied to different input images.

4 Experiments

For this paper, the method was implemented and used with data from [22] and [23]. The purpose is to explore the effects of selecting specific filters to conduct style transfer. In accordance with Sect. 3 five different groups of filter activations can be generated per class. The classes used for the experiments are the Volkswagen Beetle from 2012 ('beetle'), the VW Golf from 2012 ('golf_new') and the VW Golf from 1991 ('golf_old') from the *Stanford Car Dataset* [22], as well as the leopard class ('leopard') from the *Open Images Dataset V4* [23]. The network used in all experiments is the VGG19 network, pre-trained on imagenet data [20]. The network consists of 5 stacked blocks of convolutional layers, denominated by a combination of block index and layer number. *Block2_conv1* means the first convolutional layer in the second block within the network [20]. The filters visualization technique is based on the deep dream algorithm [24] and the approach described in [25]. For filter visualization, the gradient-descent algorithm optimizes random noise to converge into an image, which yields a high filter activation. This section consists of two parts. The first part, Subsect. 4.1 investigates the information that can be extracted about the different classes in preparation of style transfer, discussing the different finding and their applicability to improve the style transfer process. The second part uses information highlighted before to compare a classical approach [4] with this approach infused with calculated data in Subsect. 4.2.

4.1 Analyzing Test Classes

When comparing similar object classes, in this case, the three different Volkswagen cars, one interesting insight is which significant class filters are disjoint to each other, as they contain information unique to the specific class and are therefore helpful in identifying this specific class. Calculating the top 10% of *significant* filters with a threshold 0.22 for the coefficient of variance yields ~500 filters for each class. Splitting those into disjoint and intersecting filter results

Table 1. Percentage of *significant-disjoints* filter responses per layer. Layers with <1% responses have been omitted.

Layer	'beetle'	'golf_old'	'golf_new'
block3_conv3	8%	6%	4%
block3_conv4	8%	7%	6%
block4_conv1	24%	20%	29%
block4_conv2	0%	0%	2%
block4_conv3	12%	8%	11%
block4_conv4	16%	24%	11%
block5_conv1	10%	19%	25%
block5_conv3	18%	11%	9%
block5_conv4	4%	2%	0%

in 50 *significant-disjoint* filters for the class 'beetle', 193 for 'golf_old' and 44 'golf_new'. This distribution already contains interesting information about the design language of Volkswagen: Images of class 'beetle' and 'golf_new' contain cars manufactured in 2012. They share the modern form factor used at this time, which means filters, trained to respond to elements from this design language, overlap instead of being disjoint to each other, as those elements occur in both car types. Contrary to the images showing the old golf from 1991 contain artifacts designed using an older form factor and therefore do not share as many significant filters with the other two classes.

To gain further information individual filters are mapped back to VGG19 layers, to rank them according to their information density for deciding between classes. Table 1 shows the percentage of *significant-disjoint* filters per layer, omitting the layers without any filter response. Interesting layers are *block4_conv4* where the old golf shows a significantly higher hit count, as well as *block4_conv3*, where both new cars dominate. As the table contains the percentage of *significant-disjoint* filters per layer, which generate a near constant response only for one certain class, it suggests that those layers contain filters usable to identify elements of the times design language. Furthermore it seems like *block5_conv1* and *block5_conv2* contain information about model-specific design, contrary to time-specific design, as there is a high difference between *'beetle'* and both *'golf'* classes.

To apply those findings within artistic style transfer in Subsect. 4.2, images from two different classes are preprocessed: *'beetle'* and *'leopard'*. The results of filter response generation from Sect. 3 and their mapping into different layers is shown in Table 2. Only two classes of training data are used, which means each of the sets of *significant-intersecting* and *diversifying* filters are the same for both classes. *Significant-disjoints* filters are shown as a tuple, with the value from class *'beetle'* in the first and for the class *'leopard'* in the second position.

This distribution already provides first insights about the usefulness of different layers for style transfer. The layers *block4_conv3* and *block5_conv2* contain

Table 2. Percentage of different filter response classes per layer. First entry of the disjoint tuple is for class *'beetle'*, second entry is *'leopard'*. Layers with <1% responses have been omitted.

Layer	Significant intersecting	Significant disjoint	Diversifying
block1_conv1	1%	(1%, 0%)	3%
block3_conv1	0%	(0%, 11%)	0%
block3_conv2	5%	(6%, 11%)	20%
block3_conv3	6%	(6%, 0%)	20%
block3_conv4	0%	(2%, 0%)	0%
block4_conv1	0%	(6%, 0%)	0%
block4_conv2	17%	(17%, 0%)	31%
block4_conv3	9%	(13%, 33%)	15%
block4_conv4	0%	(3%, 0%)	0%
block5_conv2	23%	(15%, 33%)	3%
block5_conv3	2%	(13%, 0%)	2%
block5_conv4	32%	(11%, 11%)	0%

a large number of the extracted filters and are prime candidates for usage in a style transfer. To gain further understanding of the usefulness of each layer the semantic types (see: Subsect. 3.3) are generated and evaluated.

In Fig. 3 the first four filter reconstructions of the *type 4* filter set are visualized. Those filters have a low coefficient of variation of their activation for each class while yielding different numeric activation values per class. All filters activate on organic patterns, either spots or stripes, which would result in a high overall activation for *'leopard'* and a low overall activation for images from the *'beetle'* class, making those filters a good metric to differentiate between both classes. The filters of each type are mapped back to their layer, then counting the total number of filters within each layer. After normalizing this value, it denotes a factor that can be used to select layers for style transfer, to generate a set of layers based on semantic types.

4.2 Artistic Style Transfer with Prominent Filters

The idea for artistic style transfer is to pick filters that have a high information density regarding certain classes, to increase the effect and stability of the style transfer. To enable unsupervised execution of this method a set of desirable semantic classes for content- and style-loss is defined below. The percentages of filter responses per layer for those semantic classes are calculated and normalized to have a mean of 0 and a standard deviation of 1. The algorithm then selects layers with a positive coefficient for style transfer.

For content-loss, the most interesting filters are *diversifying* and *significant* filters for the object class of our content, as a similar activation between con-

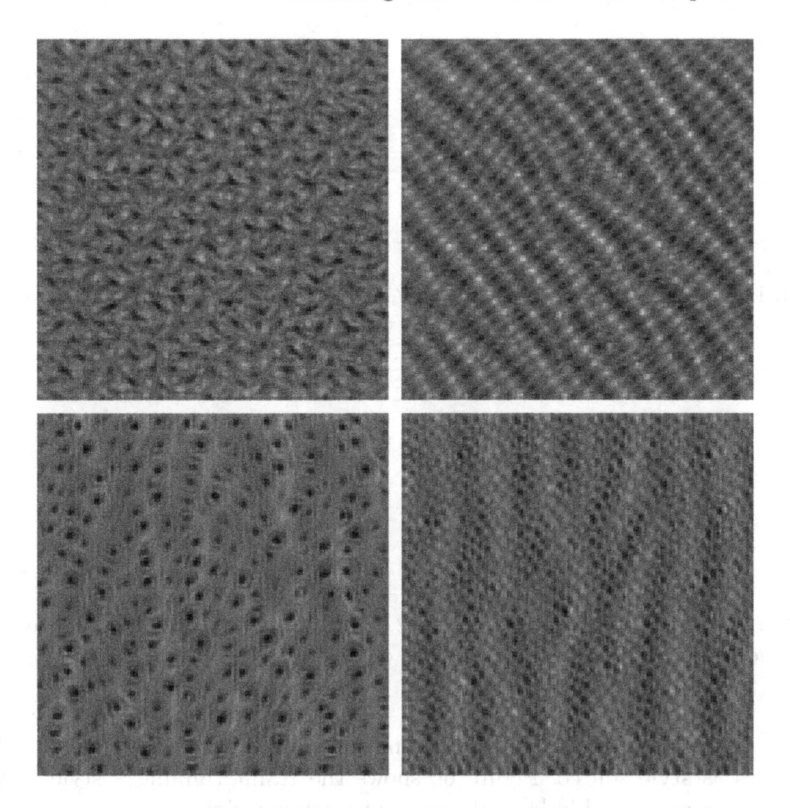

Fig. 3. Sample for filter responses classified as *type 4*

tent input and artifact is likely to correlate with the transference of the correct elements. Filters with these properties reside in semantic sets *type 3* and *type 4*.

Figure 4 shows the calculated contribution factor of each layer for *type 3* and *type 4* filters (see: Subsect. 3.3), as described above. The graph shows the highest contribution factor for layer *block4_conv2*, followed by *block3_conv3*, *block3_conv2* and *block4_conv3*.

For style-loss, the interesting layers are the *significant-intersecting*, as they contain information about both classes. Note that a significant-intersecting filter does not imply a similar activation for both classes, but rather a similar consistency of activation. This comprises filters from *type 2* and *type 4*, with *type 4* filter providing greater benefit as they additionally act as a diversifier. After calculating the contribution factor for each layer in regards to semantic *type 2* and *4*, analogous to content-loss, the layers *block5_conv4*, *block5_conv2*, *block4_conv2* and *block4_conv3* are selected for the style-loss (Fig. 5).

Comparing this to the work of Gatys et al. [4], which used the layer *block4_conv2* for content-loss and *block1_conv1*, *block2_conv1*, *block3_conv1*, *block4_conv1* and *block5_conv1* for style-loss [4] a style transfers is conducted. Figure 6 shows the result of the style transfer: The two input images are an

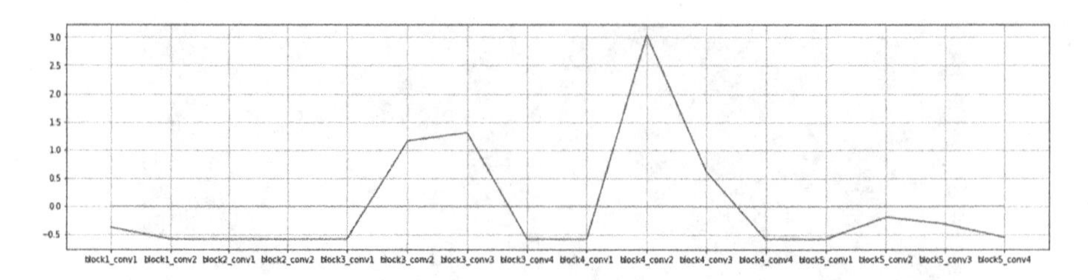

Fig. 4. Layer contribution factor for *type 3* and *4* filters. Each x-segment denotes one layer, each y-segment a difference in factor of 0.5.

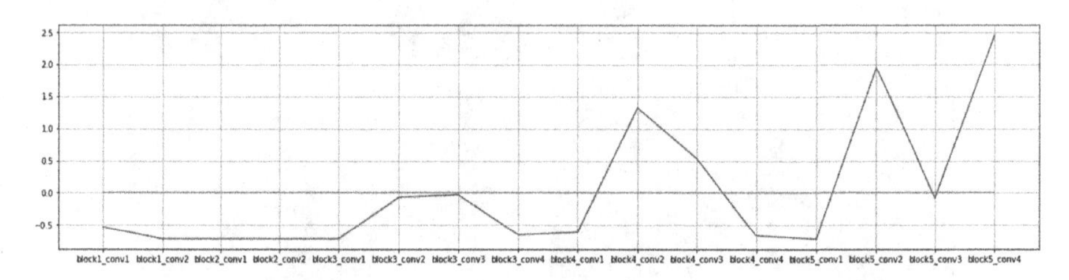

Fig. 5. Layer contribution factor for *type 2* and *4* filters. Each x-segment denotes one layer, each y-segment a difference in factor of 0.5.

image of the class *'beetle'* as the content source and an image from the *'leopard'* class as style source. Figure 6c shows the result running a style transfer using the layers defined in [4] and a content to style weight $\alpha_{contentweight} = 250$, $\beta_{styleweight} = 1$ [4].

Running the same process using the layers picked based on preprocessed filter responses yields Fig. 6d. The content weight α was adjusted to 65, to compensate for the higher number of layers influencing content-loss; all other parameters remain the same. Compared to the classical approach the infused result shows a better constraint of the style transfer onto the car. Figure 6c shows generated artifacts in the lower right corner, with the grass from the style source 'bleeding' into the car and the generation of dark patches from the style source in the background. The result using pre-selected layers also suffers from generative artifacts, but to a noticeable lesser extent. Both results can transfer certain novel *style elements*, like the grass in the foreground, whereas in Fig. 6d some elements from the content image, like the shadow of side-view mirror and structure of the rims, seem to preserve better. Conservation of features relates to the higher number of content layers, resulting in a more diverse spatial understanding of the content image features and their relation between each other. In combination with the adaptive nature of artistic style transfer, this works data preprocessing allows to identify relevant filters, to fine tune generated artifacts, which is required, when working with highly structured data, like car construction data, in which parts of the image need to be fully preserved, while still infusing it with custom style elements.

(a) Content input (b) Style input

(c) Classical result [4] (d) Our result

Fig. 6. Comparison Artistic Style Transfer

5 Discussion

This paper proposed a method of data preparation to improve the quality of Artistic Style Transfer. By pre-processing training data of images from the used classes, unsupervised parametrization of the style transfer process is possible. Different classes of filter responses were introduced in Sect. 3 and their attributes discussed. Furthermore semantic classes of those filter responses where described in Subsect. 3.3. Then Sect. 4 highlights some of these classes and their benefits with examples and presents a comparison between classical style transfer and the variation making use of prepared data. While showing promising first results, several questions remain to be investigated and different variations of this approach to be tested. One immediate limitation, or cost factor, of this method, is the requirement of training images. Classical artistic style transfer only requires two input images, as it solely relies on information encoded in the pre-trained neural network. While this does not pose a problem when dealing with style transfer of everyday objects, where enough training data exists, classes like Van-Gogh painting, where only a single instance of each artifact exists, need to be further examined.

The implementation used in this paper assumes some constant parameters: One example is the image size used in preprocessing (300px × 300px). Further research is required to investigate the effect of variable resolution, as a larger input image yields more granular filter responses, while smaller images improve generalization of classes and runtime. An extension of this technique would be an adaptive schema that used high-resolution images early to select essential filters and the gradually scales down to allow for better generalization.

Furthermore, the method in its current implementation does not differentiate between different levels of activation. A constant low activation means a filter is significant for an object class in this case. For generative models, which aim to describe what an object is, contrary to what an object is not, it should be investigated if restricting filter to high activation levels is beneficial for the outcome. Additionally, while this paper focuses on Artistic Style Transfer, recent works on generative methods [8,26] rely on building a latent space to train generative models. Those n-dimensional spaces encode the training data, used by variational autoencoders to extrapolate other results [27]. As shown in Sect. 4 the method of this paper allows identifying and selecting filters with a high information density. Using this approach to identify most significant filters, using them as axes for the latent space and positioning training samples based on their activation could produce a dense latent space, allowing for fast and stable training, while still encoding interesting novel artifacts.

Given the information of how many filters of a particular response type lie in a layer, procedural generation weights-per-layer can be used during style transfer. In this paper, the experiments all assume a constant weight for each layer within the style- and content-loss function, analogous to the loss function in [4] and its generalized form in [13] where only content- and style-loss are weighted against each other. Given the information of the percentage of significant and diversifying filters in each layer a more granular weighting, giving an individual factor to each layer would improve the users' control over the outcome.

In summary, this paper proposed a method of data preparation, to allow for an enhanced artistic style transfer. It yielded first successful results, but remain to be tested and evaluated, especially in combination with other recent improvements made to artistic style transfer [1,16,17]. The process used is running fully unsupervised and allows for an automatic adaption of artistic style transfer to image classes, given a user-designed reward function, to yield fine-tuned results.

6 Appendix

Below four images from each of the used classes in this work. Images of class *'beetle'*,*'golf_old'* and *'golf_new'* are from the *Stanford Car Dataset* [22], images from the *'leopard'* class are taken from the *Open Image Dataset* [23] (Figs. 7, 8, 9 and 10).

Fig. 7. Excerpt from *'golf_old'* (92 images total)

Fig. 8. Excerpt from *'golf_new'* (86)

Fig. 9. Excerpt from *'beetle'* (85 images total)

Fig. 10. Excerpt from *'leopard'* (361 images total)

References

1. Chen, T.Q., Schmidt, M.: Fast Patch-based Style Transfer of Arbitrary Style. arXiv preprint arXiv:1612.04337 (2016). (NIPS) (2016). http://arxiv.org/abs/1612.04337
2. Dumoulin, V., Shlens, J., Kudlur, M.: A learned representation for artistic style. In: Proceedings of ICLR (2017). http://arxiv.org/abs/1610.07629
3. Elad, M., Milanfar, P.: Style-transfer via texture-synthesis. IEEE Trans. Image Process. 26(5) (2017). http://arxiv.org/abs/1609.03057
4. Gatys, L.A., Ecker, A.S., Bethge, M.: A Neural Algorithm of Artistic Style, pp. 3–7 (2015)
5. Li, S., Xu, X., Nie, L., Chua, T.S.: Laplacian-Steered Neural Style Transfer (2017). ArXiv:1707.01253
6. Uhde, F.: Applicability of Convolutional Neural Network Artistic Style Transfer Algorithms (2019)
7. Gatys, L.A., Ecker, A.S., Bethge, M.: Image style transfer using convolutional neural networks. In: The IEEE Conference on Computer Vision and Pattern Recognition, pp. 2414–2423 (2016)
8. Umetani, N.: Exploring generative 3D shapes using autoencoder networks. In: SIGGRAPH Asia 2017 Technical Briefs on - SA'17, pp. 1–4 (2017). http://dl.acm.org/citation.cfm?doid=3145749.3145758
9. Efros, A., Freeman, W.: Image quilting for texture synthesis and transfer. In: Proceedings of the 28th Annual Conference on Computer Graphics and Interactive Techniques (August), pp. 1–6 (2001)
10. Hertzmann, A., Jacobs, C.E., Oliver, N., Curless, B., Salesin, D.H.: Image analogies. In: Proceedings of the 28th annual conference on Computer graphics and interactive techniques SIGGRAPH'01 2001(August), pp. 327–340 (2001)
11. Kuri, D., Root, E., Theisel, H.: Hexagonal Image Quilting for Texture Synthesis, pp. 67–76 (2014)
12. Gatys, L.A., Ecker, A.S., Bethge, M.: Texture Synthesis Using Convolutional Neural Networks. In: Neural Image Processing Systems, pp. 1–10 (2015)
13. Uhde, F., Mostaghim, S.: Towards a general framework for artistic style transfer. In: Liapis, A., Romero Cardalda, J.J., Ekárt, A. (eds.) Computational Intelligence in Music, Sound, Art and Design. LNCS, pp. 177–193. Springer, Cham (2018)
14. Li, C., Wand, M.: Precomputed real-time texture synthesis with Markovian generative adversarial networks. In: Leibe, B., Matas, J., Sebe, N., Welling, M. (eds.) Computer Vision - ECCV 2016. Lecture Notes in Computer Science(Lecture Notes in Artificial Intelligence and Lecture Notes in Bioinformatics), vol. 9907, pp. 702–716. Springer, Cham (2016). https://doi.org/10.1007/978-3-319-46487-9_43
15. Lim, I., Gehre, A., Kobbelt, L.: Identifying style of 3D shapes using deep metric learning. Comput. Graph. Forum **35**(5), 207–215 (2016)
16. Luan, F., Paris, S., Shechtman, E., Bala, K.: Deep Painterly Harmonization. Comput. Graph. Forum **37**(4), 95–106 (2018)
17. Li, C., Wand, M.: Combining Markov random fields and convolutional neural networks for image synthesis. CVPR **2016**, 9 (2016)
18. Mordvintsev, A., Christopher Olah, M.T.: Google AI Blog: Inceptionism: Going Deeper into Neural Networks (2015). https://ai.googleblog.com/2015/06/inceptionism-going-deeper-into-neural.html
19. Zeiler, M.D., Fergus, R.: Visualizing and Understanding Convolutional Networks. Lecture Notes in Computer Science (Lecture Notes in Artificial Intelligence and Lecture Notes in Bioinformatics). LNCS, PART 1, vol. 8689, pp. 818–833 . Springer, Cham (2014)

20. Simonyan, K., Zisserman, A.: Very deep convolutional networks for large-scale image recognition. ImageNet Challenge **96**(2), 1–10 (2014)
21. Russakovsky, O., et al.: ImageNet large scale visual recognition challenge. Int. J. Comput. Vis. **115**(3), 211–252 (2015)
22. Krause, J., Stark, M., Deng, J., Fei-Fei, L.: 3D object representations for fine-grained categorization. In: 4th International IEEE Workshop on 3D Representation and Recognition (3dRR-13), Sydney, Australia (2013)
23. Kuznetsova, A., et al.: The Open Images Dataset V4: unified image classification, object detection, and visual relationship detection at scale, pp. 1–20 (2018). http://arxiv.org/abs/1811.00982
24. Mordvintsev, A., Olah, C., Tyka, M.: Research blog: DeepDream - a code example for visualizing Neural Networks (2015). http://googleresearch.blogspot.ie/2015/07/deepdream-code-example-for-visualizing.html
25. Chollet, F.: Deep Learning with Python. Manning Publications Company (2017). https://books.google.de/books?id=Yo3CAQAACAAJ
26. Hoshen, Y., Malik, J.: Non-Adversarial Image Synthesis with Generative Latent Nearest Neighbors. Technical report, Facebook AI Research (2018). https://arxiv.org/abs/1812.08985
27. Schmidhuber, J.: Deep Learning in Neural Networks : An Overview, pp. 1–88 (2014)

A Fusion of Deep and Shallow Learning to Predict Genres Based on Instrument and Timbre Features

Igor Vatolkin(✉), Benedikt Adrian, and Jurij Kuzmic

Department of Computer Science, TU Dortmund, Dortmund, Germany
{igor.vatolkin,benedikt.adrian,jurij.kuzmic}@tu-dortmund.de

Abstract. Deep neural networks have recently received a lot of attention and have very successfully contributed to many music classification tasks. However, they have also drawbacks compared to the traditional methods: a very high number of parameters, a decreased performance for small training sets, lack of model interpretability, long training time, and hence a larger environmental impact with regard to computing resources. Therefore, it can still be a better choice to apply shallow classifiers for a particular application scenario with specific evaluation criteria, like the size of the training set or a required interpretability of models. In this work, we propose an approach based on both deep and shallow classifiers for music genre classification: The convolutional neural networks are trained once to predict instruments, and their outputs are used as features to predict music genres with a shallow classifier. The results show that the individual performance of such descriptors is comparable to other instrument-related features and they are even better for more than half of 19 genre categories.

Keywords: Instrument recognition · Genre recognition · Supervised music classification · Deep and shallow learning

1 Introduction

Music genre recognition belongs to the most often applied classification tasks in Music Information Retrieval. For instance, [29] has referred in 2012 to more than 400 related studies. High-level categories like genres help to organise personal music collections, select music for listening to in specific situations like working and learning, car driving, party, sports, etc., but also to recommend and discover new music. Early approaches like [30] focused on manual engineering of audio features which were later used to train supervised classification models with the help of machine learning algorithms. More recent approaches introduced end-to-end learning which simultaneously estimates both features and predictions. This

This work was partly funded by the German Research Foundation (DFG), project 336599081 "Evolutionary optimisation for interpretable music segmentation and music categorisation based on discretised semantic metafeatures".

J. Romero et al. (Eds.): EvoMUSART 2021, LNCS 12693, pp. 313–326, 2021.
https://doi.org/10.1007/978-3-030-72914-1_21

is often done with the help of artificial neural networks. These very successful methods received a lot of attention for many different tasks including speech and image recognition [18].

However, keeping in mind different real-world scenarios and various relevant evaluation criteria, in particular, Deep Neural Networks (DNNs) with a large number of hidden layers are still not always the best choice. Their application must be thoroughly justified; besides, No Free Lunch theorems claim that "any two algorithms are equivalent when their performance is averaged across all possible problems" [36].

To name a few issues, first, DNNs with many layers and neurons require a HUGE NUMBER OF PARAMETERS to learn. This means that the danger of overfitting towards the training data set is increased, and successful DNN models often benefit at most from very large data sets ("big data") or data sets increased with augmentation [19], having a DECREASED PERFORMANCE FOR SMALL TRAINING SETS. However, in applications like genre or personal music category recognition, a listener would sometimes prefer to define an own category based on a limited number of music tracks, to avoid fatigue annotating a large number of music pieces. Another problematic issue is the INTERPRETABILITY of classification models. If a listener or a music psychologist would like to understand *why* a music track is assigned to a particular genre or which meaningful musical properties are typical for a particular composer, composing time, country origin, etc., it is rather hard to find meaningful information in complex NN models. Finally, the training of DNNs often requires LARGE AMOUNT OF TIME, and even with the availability of new highly efficient GPUs one should consider large ENVIRONMENTAL COSTS required to train a deep neural network for each new personal category.

Our proposal is neither to stick to an "older" approach using shallow classifiers together with manually engineered features nor to focus only on the newer "disruptive" technology to predict everything with DNNs, but to combine both approaches with the target to increase user satisfaction, deriving classification models which still focus on the highest classification quality—but also on the evaluation issues from the previous paragraph. For music classification, this can be done when a set of "mid-level" musical properties will be predicted only once per music track with DNNs, but the last and often repeating step to create new high-level personal categories will be done with the help of a rather simple, fast, and robust classifier.

Various "mid-level" descriptors with backgrounds in music theory can be considered: instrumental, harmonic, melodic, dynamics, temporal, rhythmic, and other properties, as sketched in Fig. 1. However, to limit our initial study, we focus on descriptors of the instrumental texture, as instruments belong to the most relevant properties for the definition of genres and are also relevant for music transcription [16].

The results show that instrument predictions based on previously trained CNNs are better than another set of mid-level instrumental descriptors and a set of timbre-related signal features for more than half of 19 tested genre

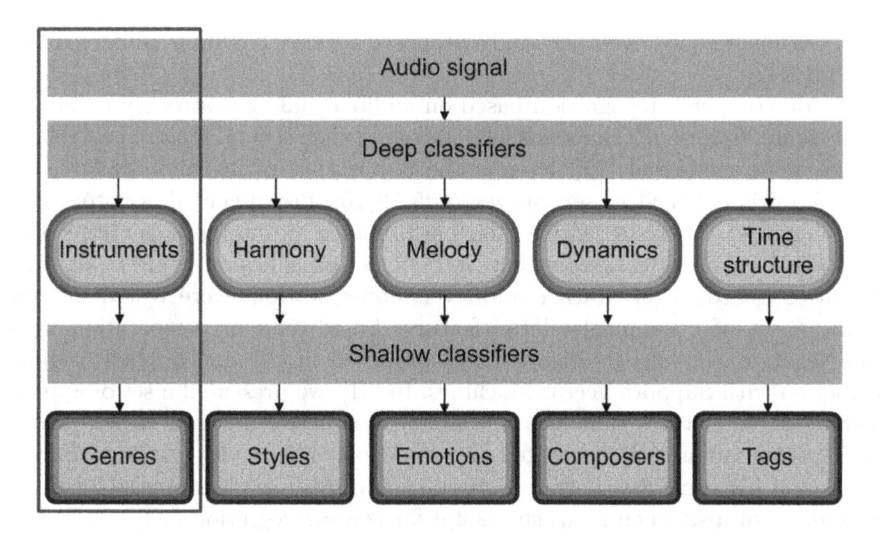

Fig. 1. A fusion of deep and shallow classifiers for prediction of high-level music categories based on mid-level semantic properties. The focus of the study in this paper is marked with a red frame.

categories. However, there still remains room for improvements, because of differences between selected instruments in the database to train a neural network and a "real-world" collection of music tracks with a significantly larger number of various instrument bodies, playing techniques, and applied digital effects during the mix production.

In Sect. 2, we provide an overview of related work on instrument and genre recognition. In Sect. 3, we introduce the implemented methods to extract features which describe instrumental and timbral texture (a deep convolutional neural network that directly predicts the instruments, a method based on evolutionary approximation of polyphonic recordings, and a set of manually engineered timbral descriptors). Section 4 deals with the setup of experiments and the discussion of results. Finally, we conclude with the most relevant findings of our study and ideas for future work.

2 Related Work

The first studies on instrument recognition focused on individual samples and manually engineered features [7,9], later approaches compared effective features for instrument recognition in polyphonic recordings [10,11,32]. In [20], music signals were decomposed into so-called sound atoms with instrument labels.

More recent studies applied deep neural networks. [16] proposed a CNN architecture for the classification task of recognising predominant instruments in a Mel-spectrogram feature space. The proposed architecture is also used in similar works using the IRMAS data set [28] and on a smaller data set containing jazz instruments [13]. Another study proposed a CNN based on Hilbert-Huang transform [21].

Automatic genre recognition based on audio signals was strongly influenced by the seminal work of Tzanetakis [30], and since then several hundreds of related studies were conducted [29]. Even if the importance of instruments for genre recognition is outlined in several works [5,12,15], instrument descriptors were only seldom explicitly tested for their performance in automatic genre recognition. The impact of features which characterise basslines on genre recognition was investigated in [3]. Rather mid-level representations were tested in some works. A set of instrument-related features based on source separation using Non-Negative Matrix Factorisation was proposed in [26] and applied to genre prediction with Support Vector Machines. In [31], we presented a set of approximative music features that measure similarities between onsets in polyphonic music pieces and artificially created mixtures from the samples of a large number of different instruments, showing that these features were less precise for direct recognition of instruments but successful for genre recognition.

3 Descriptors of Instrument Texture

3.1 Predictions of a Deep Neural Network

The applied Convolutional Neural Network (CNN) architecture is derived from [16], which implements a CNN consisting of four convolutional blocks with an increasing number of channels, each followed by a max-pooling layer and the whole CNN ending with a global-max-pooling layer, fully-connected layer, and a final layer with sigmoid activations. To fit smaller input dimensions, several sizes of pooling-kernels have been altered in the architecture. The sizes of the pooling-kernels in the first two convolutional blocks have been changed from (3,3) to (1,2). For the third and fourth block, an additional pooling-layer of size (2,2) has been inserted between the two convolutional layers. Furthermore, the size of the existing pooling-layers in blocks three and four has also been changed to a size of (2,2). Finally, the architecture has also been extended with batch-normalisation before each max-pooling layer, which resulted in overall better performance compared to previous work.

Let the layers be represented as follows:

- Convolutional(3,3): C
- Batch normalisation: B
- MaxPooling(1,2): p
- MaxPooling(2,2): P
- Dropout for 25% of inputs: d
- Dropout for 50% of inputs: D
- GlobalMaxPooling: G
- Fully connected: F

Then, the network architecture is described as:

CCBpd CCBpd CBPCBPd CBPCBGD FDF.

The Per-Channel Energy Normalisation (PCEN) [35] is applied to the input CNN features extracted by Librosa [23] for the first 2s of instrument samples and mixes (see below): magnitude and power Mel spectrograms, each represented with 128 dimensions, resulting in a two-channel input space.

The main challenge to predict instruments with DNNs for a further genre recognition is a required large training set with annotated instruments. This set must be large enough to prevent overfitting of a network with a high number of parameters, but also have a sufficiently large number of different instruments. Publicly available data sets with genre annotations like 1517-Artists [27] (see Sect. 4) do not have instrument annotations, multitrack data sets like MedleyDB [4] contain a rather small number of music pieces and also a limited number of different instruments.

Therefore, to train a CNN, we created a new data set based on mixes of 51 types of instrument samples used in previous work [31] and compiled from several instrument sample databases: the Ethno World 5 Professional and Voices samples [1], Komplete 11 Ultimate samples [2], McGill University Master Sample collection [8], RWC database [14], and University of Iowa instrument samples[1]: acoustic and electric guitar, balalaika, bandura, banjo framus, banjolin, bass and electric bass, bassoon, bawu, bouzuki, cello, ceylon guitar, clarinet, contrabassoon, cumbus, dallape accordeon, dilruba, domra, drums, dung dkar trumpet, egyptian fiddle, erhu, flute, fujara, horn, jinghu opera violin, kantele, melodica, morin khuur violin, oboe, oud, organ, panflute, piano and electric piano, pinkillo, pivana flute, saxophone, scale changer harmonium, shakuhachi, sitar, tampura, tanbur, trombone, trumpet, tuba, turkey saz, ukulele, viola, and violin. Many instruments were represented with several individual instrument bodies with an overall number of 165 individual instrument bodies. For all instruments except for acoustic and electric guitar, at least one body was taken from commercial high-quality samples of [1,2].

For the new data set, random pitches were selected to build the following subsets: 1000 sole samples, 1000 mixes of two samples, 1000 of three, 1000 of four, and 1000 of five samples. The network was trained not to predict the instrument occurrence but its dominance or relative share. For all 4000 mixes, we stored the share of root mean square energy of each individual instrument sample which contributed to the mix in relation to the energy of the complete mix, and set this value to 1 for instruments which contributed to 1000 individual samples.

After some preliminary experiments, we have created two CNN-based feature sets: the first one with a single CNN trained with 3750 randomly selected recordings and 1250 used for the validation of parameters (we refer to this set later in the table of results as CNN), and the second one based on an ensemble

[1] http://theremin.music.uiowa.edu. Accessed on 03.02.2021.

of six CNNs (CNN-E): (1) trained and validated only with 1000 individual samples, (2) with 1000 two-sample mixes, (3) with 1000 three-sample mixes, (4) with 1000 four-sample mixes, (5) with 1000 five-sample mixes, and (6) with all 5000 samples and mixes. For genre prediction, we applied the trained CNN models to each 0.5s of music tracks and estimated the mean and the standard deviation of the predicted instrument dominances for a classification window of 4s with 2s overlap. Thus, the CNN instrument feature set contained 102 final dimensions, and CNN-E feature set 612 dimensions.

3.2 Approximative Features

The approximative features were extracted after [31]. Here, polyphonic audio signals were approximated by samples of 51 aforementioned instruments, which were mixed with the help of an evolutionary algorithm, in assumption, that in some feature space (here the Mel spectrum), the distance between an unknown chord to approximate and a candidate mix would be smaller when the candidate mix will be generated with the same or similar instruments and pitches. For example, the mean absolute distance between all Mel spectrum values of the original recording of viola E4 and violin B5 and a candidate mix of violin E4 and violin B5 (possibly played with another instrument body and using another loudness level) should be smaller than the distance between the feature values of the original recording and the mix of violin E4 and piano B5 or viola E4 and violin A#5.

For the approximation of onsets (i.e., the beginnings of played notes) in an unknown music recording, a population of randomly generated mixes was evolved with the help of an evolutionary algorithm which minimised the distances to unknown chords and altered the mixes with musically meaningful mutation operators like the pitch shift or the exchange of one instrument with another one. During the optimisation process, all distances between candidate mixes of various instruments and short time segments around onset events from the complete music tracks were stored in an archive. A population of candidate mixes approximated simultaneously all onsets in a music piece to provide a more efficient optimisation, because individual music pieces typically contain many similar notes or chords. Further, it was distinguished between several subpopulations. “Specialist” mixes should approximate well only a small number of similar onsets from individual segments like verse or chorus, and “allrounder” mixes as many onsets as possible, approximating the general instrumental texture of the complete piece (for details we refer to [31]). After the optimisation, those mixes were saved, which could approximate the unknown time segments as perfect as possible. The output was the vector of similarities to 51 instrument types.

According to [31], several features were extracted: the minimum, mean, and maximum instrument similarities in a classification frame of 4s, as well as the normalised dominance instrument rank: The instruments were sorted according to the estimated similarities which were normalised to the interval $[0, 1]$. After the preliminary instruments, two feature groups remained: the dominance rank (referred later to as APPR-R) and the combination of all approximative features (APPR-A). As for CNN features, the mean and the standard deviation were calculated for classification windows, leading to 102 feature dimensions for the first group and 408 for the second one.

3.3 Timbre-Related Signal Characteristics

The timbre-related audio signal characteristics and the references to extraction tools are listed in Table 1. All features were extracted with AMUSE framework [34] from a 22050 Hz mono signal using extraction time frames of 512 samples.

Table 1. Timbre-related features.

Feature	Ref.	Dim.
Delta MFCCs	[17]	13
Low energy	[22]	1
MFCCs	[17]	13
Spectral bandwidth	[24]	1
Spectral centroid	[24]	1
Spectral crest factor	[24]	4
Spectral extent	[24]	1
Spectral flatness measure	[24]	4
Spectral flux	[24]	1
Spectral irregularity	[17]	1
Spectral kurtosis	[24]	1
Spectral skewness	[24]	1
Spectral slope	[24]	1
Sub-band energy ratio	[24]	4
Zero-crossing rate	[22]	1

Again, the mean and the standard deviation were estimated for 4s classification windows, leading to the set of 96 dimensions after processing.

4 Experiments

4.1 Data Set

For the prediction of music genres, we use 1517-Artists data set [27], because all audio tracks are available for academic research and reproducible study comparison, and the tracks are complete, in contrast to, e.g., widely used GTZAN data set [30] which contains only 30s excerpts of audio tracks which may be not enough to represent more complex genres. Furthermore, 1517-Artists has a large number of 19 different genres, and the tracks are assigned to them exclusively. Some of genres are related to each other (e.g., Alternative and punk vs. Rock and pop) or are compiled from very different tracks (Soundtracks and more, World), so that the identification of some genres could be particularly hard and close to our real-world scenario, in which a listener defines a personal genre category. All genres were recognised in a "one-vs.-all" binary approach which helps to identify genres which are particularly easy or hard to classify (in contrast to a multi-class approach, where a classification model assigns a label from more than two different classes).

From the original 3180 tracks, we selected 1674 after the application of an artist filter, so that only tracks of different artists remained (for the danger of learning rather artists or albums and not genres, we refer to [25,33]). For the 10-fold stratified cross-validation, we shuffled the track order and assigned the tracks to 10 non-overlapping parts with an equal genre distribution, leading to 160 tracks / part.

During 10 validation folds, each part served exactly once as a test set. The training set tracks for supervised learning were selected from the remaining 9 parts. To guarantee the balance and equal size of all training sets, the number of positive tracks (i.e., belonging to the genre to recognise) was always set to 36, because the overall number of Classical tracks was the smallest (40). Thus, each part contained 4 Classical tracks, so that 36 could be used for training and 4 for validation in each fold. The 36 negative training tracks (which belonged to other genres) were selected in the way that each remaining genre was represented by exactly two tracks (recall that the overall number of genres is 19).

Please note that the training of models with a relatively small number of 72 tracks is very close to our scenario when a listener wants to select only a limited number of music pieces to represent a desired category. It is obvious that the classification errors achieved in our experiments could not be as small as produced by models that use a significantly larger number of training examples, but much better represent the real situation.

4.2 Analysis of Results

Two common algorithms are used as shallow classifiers for genre recognition: Random Forest [6] with 200 trees and Support Vector Machine [37] with a linear kernel. Because the test sets were unbalanced, the balanced relative error is used as an evaluation measure which equally rates the relative error on all positive tracks belonging to genre (a sum of true positives TP and false negatives FN) and the relative error on all negative tracks belonging to other genres (a sum of true negatives TN and false positives FP):

$$e = \frac{1}{2}\left(\frac{FN}{TP+FN} + \frac{FP}{TN+FP}\right). \quad (1)$$

Table 2 contains all errors and standard deviations for 19 genres produced with Random Forest models, and Table 3 with Support Vector Machines. The

Table 2. Classification errors e and standard deviations σ with Random Forests. CNN: a single network; CNN-E: ensemble (see Sect. 3.1); Appr-R: Approximative dominance instrument rank; Appr-A: Combination of all approximative features (see Sect. 3.2); Timb: Signal-based descriptors (see Sect. 3.3). Combined feature groups: C: CNN-E predictions; A: Approximative features Appr-A; T: Timbre descriptors.

Genre		Individual feature groups					Combined feature groups			
		CNN	CNN-E	App-R	App-A	Timb	CT	CA	AT	CAT
Alternat	e	0.2094	**0.1980**	0.2776	0.2102	0.2340	**0.1928**	0.1989	0.2017	0.2054
	σ	0.0567	0.0546	0.0769	0.0674	0.0510	0.0535	0.0558	0.0492	0.0663
Blues	e	0.3227	**0.3170**	0.4490	0.4203	0.3427	**0.3180**	0.3390	0.3657	0.3297
	σ	0.0721	0.0634	0.0697	0.0661	0.0846	0.0651	0.0561	0.0575	0.0654
Childrens	e	0.3968	0.4309	**0.3937**	0.4154	0.4029	0.4137	0.3946	**0.3880**	0.3949
	σ	0.0913	0.0899	0.1136	0.0818	0.1072	0.0652	0.0491	0.0649	0.0581
Classical	e	0.1590	0.1516	0.1712	0.1683	**0.1054**	0.1240	0.1218	**0.0929**	0.1090
	σ	0.0759	0.0791	0.0660	0.0753	0.0584	0.0736	0.0657	0.0493	0.0614
Comedy	e	0.2866	0.2602	0.2945	0.2818	**0.2231**	0.2244	0.2494	**0.2214**	0.2402
	σ	0.0824	0.0676	0.1021	0.1183	0.0878	0.0749	0.0799	0.1152	0.0935
Country	e	0.2533	**0.2477**	0.3980	0.3873	0.2540	**0.2350**	0.2447	0.3023	0.2410
	σ	0.0534	0.0266	0.0615	0.0564	0.0441	0.0379	0.0466	0.0492	0.0347
Easy list	e	0.3282	**0.3091**	0.4253	0.3799	0.3260	**0.2904**	0.3479	0.3246	0.3245
	σ	0.0846	0.1092	0.0882	0.0825	0.0990	0.0891	0.1030	0.0916	0.0844
Electron	e	0.1512	**0.1490**	0.2510	0.2518	0.1613	0.1599	0.1539	0.2146	**0.1487**
	σ	0.0541	0.0449	0.0770	0.0852	0.0344	0.0410	0.0539	0.0643	0.0559
Folk	e	0.2795	**0.2686**	0.3558	0.3891	0.2950	**0.2682**	0.2861	0.3294	0.2706
	σ	0.0322	0.0390	0.0671	0.1041	0.0387	0.0551	0.0974	0.0963	0.0861
Hip-Hop	e	0.1299	0.1467	0.3480	0.2099	**0.1283**	0.1276	**0.1240**	0.1451	0.1319
	σ	0.0322	0.0547	0.0945	0.0564	0.0716	0.0540	0.0580	0.0588	0.0744
Jazz	e	0.3877	0.3527	0.4407	0.4393	**0.3123**	**0.3263**	0.3843	0.3543	0.3427
	σ	0.0960	0.0544	0.0677	0.0486	0.0576	0.0682	0.0691	0.0502	0.0619
Latin	e	0.3655	**0.3266**	0.4615	0.4444	0.3569	**0.3049**	0.3579	0.3865	0.3566
	σ	0.0722	0.0950	0.0693	0.0996	0.0625	0.0787	0.0677	0.0858	0.0711
New age	e	0.2796	0.2757	0.3576	0.3046	**0.2424**	**0.2349**	0.2724	0.2546	0.2701
	σ	0.0659	0.0902	0.0904	0.0716	0.0678	0.0870	0.0599	0.0667	0.0785
R‘n’B	e	0.2655	**0.2600**	0.4046	0.4061	0.3084	**0.2534**	0.2939	0.3369	0.2973
	σ	0.0548	0.0565	0.0529	0.0787	0.0696	0.0556	0.0553	0.0558	0.0499
Reggae	e	0.2332	0.2224	0.3780	0.3280	**0.1997**	**0.1941**	0.2168	0.2303	0.2125
	σ	0.0644	0.0800	0.0936	0.0513	0.0641	0.0664	0.0590	0.0788	0.0616
Religious	e	0.3795	**0.3759**	0.4866	0.4197	0.3870	**0.3833**	0.4104	0.4338	0.4042
	σ	0.0617	0.0880	0.0837	0.0812	0.0618	0.0964	0.0764	0.0810	0.0779
Rock	e	0.2728	**0.2458**	0.3251	0.2944	0.2549	**0.2346**	0.2622	0.2761	0.2445
	σ	0.0609	0.0528	0.0642	0.0563	0.0651	0.0661	0.0543	0.0754	0.0559
Soundtr	e	0.3044	0.2979	0.5121	0.4548	**0.2652**	**0.2871**	0.3213	0.3139	0.3077
	σ	0.0756	0.0674	0.1106	0.1169	0.1118	0.0710	0.0601	0.0797	0.0725
World	e	0.4365	**0.4327**	0.4726	0.4570	0.4633	0.4099	0.4426	**0.4059**	0.4059
	σ	0.0930	0.0756	0.1082	0.0956	0.0609	0.1059	0.0516	0.0798	0.0644

Table 3. Classification errors e and standard deviations σ with SVMs. CNN: a single network; CNN-E: ensemble (see Sect. 3.1); Appr-R: Approximative dominance instrument rank; Appr-A: Combination of all approximative features (see Sect. 3.2); Timb: Signal-based descriptors (see Sect. 3.3). Combined feature groups: C: CNN-E predictions; A: Approximative features Appr-A; T: Timbre descriptors.

Genre		Individual feature groups					Combined feature groups			
		CNN	CNN-E	App-R	App-A	Timb	CT	CA	AT	CAT
Alternat	e	0.3057	**0.1787**	0.3329	0.2974	0.2044	**0.1656**	0.2463	0.2498	0.2153
	σ	0.0430	0.0518	0.0729	0.068	0.0639	0.0443	0.0685	0.0617	0.0608
Blues	e	0.4097	**0.3180**	0.4857	0.4670	0.3620	**0.3030**	0.3820	0.4263	0.3190
	σ	0.0675	0.0525	0.0772	0.0831	0.0816	0.0405	0.0575	0.096	0.0697
Childrens	e	0.4872	0.4115	0.4471	0.4324	**0.3501**	**0.3366**	0.3911	0.3823	0.3677
	σ	0.0661	0.0628	0.0728	0.0938	0.1321	0.1012	0.1045	0.0981	0.1071
Classical	e	0.1817	0.1372	0.1359	0.1346	**0.1019**	**0.0885**	0.1016	0.1003	0.0987
	σ	0.1154	0.0736	0.0784	0.0674	0.084	0.0532	0.0603	0.0548	0.0612
Comedy	e	0.2968	**0.2396**	0.3443	0.3416	0.2456	**0.2360**	0.2823	0.2522	0.2469
	σ	0.0844	0.0722	0.0862	0.1037	0.086	0.0706	0.0872	0.1155	0.1198
Country	e	0.3907	**0.2247**	0.4207	0.4127	0.2977	**0.2260**	0.2993	0.3560	0.2773
	σ	0.064	0.0475	0.1013	0.0882	0.0872	0.0483	0.0985	0.1095	0.0893
Easy list	e	0.3681	0.3138	0.4380	0.4366	**0.3006**	**0.2980**	0.3603	0.3435	0.3192
	σ	0.0897	0.1002	0.0798	0.0787	0.1145	0.0788	0.0650	0.083	0.0937
Electron	e	0.1914	**0.1448**	0.3471	0.2926	0.1938	**0.1607**	0.1672	0.2336	0.1792
	σ	0.0571	0.0584	0.0681	0.0834	0.0655	0.0587	0.079	0.0861	0.0712
Folk	e	0.3357	**0.2621**	0.3641	0.3549	0.3205	0.2789	**0.2785**	0.3144	0.2977
	σ	0.0770	0.0569	0.0882	0.0768	0.0547	0.0480	0.0508	0.0520	0.0476
Hip-Hop	e	0.2336	0.1385	0.3812	0.2776	**0.1217**	**0.1201**	0.1388	0.1753	0.1513
	σ	0.0545	0.0761	0.0799	0.0927	0.0599	0.0957	0.0527	0.0597	0.0457
Jazz	e	0.4710	0.3430	0.4523	0.4340	**0.2753**	**0.2680**	0.3667	0.3343	0.3227
	σ	0.050	0.0428	0.0792	0.0734	0.0648	0.0557	0.0682	0.0781	0.0629
Latin	e	0.4753	**0.3303**	0.4648	0.4454	0.3931	**0.3168**	0.3704	0.4168	0.3539
	σ	0.0578	0.0783	0.0439	0.0442	0.0537	0.0886	0.0945	0.0915	0.0787
New age	e	0.3082	**0.2204**	0.3230	0.3043	0.2214	**0.2122**	0.2599	0.2635	0.2303
	σ	0.0816	0.0689	0.1025	0.1014	0.0527	0.0598	0.0896	0.1074	0.0811
R'n'B	e	0.3011	**0.2608**	0.4265	0.3982	0.3097	**0.2594**	0.3256	0.3549	0.3177
	σ	0.0616	0.0348	0.0838	0.1133	0.1153	0.0543	0.0960	0.1166	0.0780
Reggae	e	0.3164	**0.2102**	0.4010	0.3457	0.2355	**0.1872**	0.2592	0.2931	0.2437
	σ	0.0361	0.0566	0.0607	0.0562	0.0472	0.0740	0.0950	0.0863	0.1001
Religious	e	**0.4095**	0.4265	0.4869	0.4548	0.4559	**0.3751**	0.4236	0.4697	0.3914
	σ	0.0745	0.1004	0.0987	0.0946	0.0905	0.0868	0.0983	0.1458	0.1005
Rock	e	0.3536	0.2631	0.3373	0.3367	**0.2444**	**0.2390**	0.2957	0.3168	0.2958
	σ	0.0507	0.0566	0.0535	0.0712	0.0746	0.0564	0.0788	0.0851	0.0807
Soundtr	e	0.3188	0.3198	0.5299	0.4990	**0.3108**	**0.3308**	0.3789	0.4155	0.3689
	σ	0.0764	0.0779	0.0996	0.1150	0.0741	0.0876	0.0776	0.1028	0.0798
World	e	0.5466	0.4451	0.4669	0.4717	**0.4353**	0.3853	0.4225	0.4312	**0.3604**
	σ	0.0836	0.0818	0.0805	0.1079	0.0942	0.0978	0.0706	0.1068	0.0832

complete genre names for some abbreviations are: Alternat: Alternative and punk; Comedy: Comedy and spoken word; Easy list: Easy listening and vocals; Electron: Electronic and dance; R'n'b: R'n'b and soul; Rock: Rock and pop; Soundtr: Soundtracks and more. In the left half of each table, *e*-values averaged across 10 folds are reported for individual feature groups, and in the right half for their combinations. The smallest mean error across individual feature groups for each genre, as well as the smallest mean error across the combined feature groups, is marked with a bold font. The ever smallest error for each genre is underlined.

Concerning individual feature groups, CNN-E seems to perform at best, however, only for 11 categories, following with the timbre descriptors. We may observe that particularly for genres with a high share of speech (Comedy and spoken word with Random Forest, Hip-Hop with both classifiers), but also for genres with a typically rich number of possible different instruments (Classical, Jazz, Soundtracks and more with both classifiers) the timbre descriptors perform at best. This can be explained by different reasons. MFCC descriptors were developed for speech recognition and may help to recognise genres with a high share of speech. Even if we used a large number of instruments to train CNNs, it is still not enough for genres with a more complex instrument structure, so the hope is that with an increasing number of instruments and individual bodies to train the CNN models it would be possible to further reduce and overcome the performance gap between this method and signal-level descriptors.

Assuming that different individual feature groups may describe some complementary music characteristics relevant for genre recognition, we trained shallow classification models also with all combinations of three feature groups: CNN-E, APPR-A (both performed better than CNN and APPR-R), and timbre. Indeed, for 15 of 19 genres and both for Random Forest and Support Vector Machines, the best-combined feature sets performed better than the best individual feature sets. In particular, the combination of CNN ensemble features and timbre descriptors successfully contributed to feature sets with the smallest errors (9 of 15 using Random Forest and 14 of 15 using Support Vector Machines). A combination of all three feature groups was the best one only for a single genre. A possible reason is a significantly increased number of overall feature dimensions; an application of feature selection can theoretically help to achieve smaller errors using combined sets of all available features.

For 13 of 19 genres, the best models created with Support Vector Machines performed better than the best models created with Random Forest. However, the differences were sometimes marginal, and other settings of parameters may change the results. An exhaustive comparison of classifier hyperparameters was beyond the scope of our experiments.

5 Conclusions and Outlook

In this paper, we have compared three approaches to predict music genres based on instrument-related features: predictions based on a Convolutional Neural Network, similarities to instruments in approximations of polyphonic mixtures, and manually engineered signal-level descriptors. We have also proposed a general concept to combine deep and shallow learning for the prediction of music high-level categories, which has several advantages. The creation of complex and less interpretable models for the prediction of mid-level music properties can be done only once, and the classification of music into categories like genres, personal preferences, or emotions can be repeated with classifiers which are faster, have less parameters to set up, and can provide acceptable classification performance using significantly smaller training sets than required for deep neural networks.

The approach based on the combination of deep and shallow learning performed quite good for individual feature groups, producing smaller errors for 11 of 19 genre categories, and contributed to the best-combined feature sets compiled from all three feature sources for recognition of 13 of 19 genres using Random Forest, 18 of 19 genres using Support Vector Machines, and 17 of 19 genres when only the best model of both classifiers Random Forest and Support Vector Machines was taken into account.

However, we suppose that there is still a lot of place for further improvements, not only because timbre features performed quite good for many genres, but also because the selection of instruments, individual instrument bodies, and loudness levels in the data set to train instrument prediction models was very limited to represent all existing popular and classical music with roots in different Western and ethnic cultures. The application of digital effects and other augmentations to the instrument training data set can further improve the classification performance.

As our study was restricted to instruments and genres, another promising extension to our general approach is to increase the number of mid-level music characteristics predicted with deep classifiers (harmony, melody, tempo, structure, etc.), and also to validate the models created for the prediction of other high-level music categories: music styles and personal preferences, emotions and moods, music for specific situations and purposes (game and movie soundtracks, sports, background music during shopping, learning, etc.).

References

1. Best Service. Ethno World 5 Professional and Voices (2010). https://www.youtube.com/watch?v=-9F3q8kAb00. Accessed 03 Feb 2021
2. Native Instruments. Komplete 11 Ultimate (2016). https://www.youtube.com/watch?v=WEfxP0-YZgQ. Accessed 03 Feb 2021
3. Abeßer, J., Lukashevich, H., Bräuer, P.: Classification of music genres based on repetitive basslines. J. New Music Res. **41**(3), 239–257 (2012)
4. Bittner, R.M., Salamon, J., Tierney, M., Mauch, M., Cannam, C., Bello, J.P.: Medleydb: a multitrack dataset for annotation-intensive MIR research. In: Wang, H., Yang, Y., Lee, J.H. (eds.) Proceedings of the 15th International Society for Music Information Retrieval Conference, ISMIR, pp. 155–160 (2014)
5. Bosch, J.J., Janer, J., Fuhrmann, F., Herrera, P.: A comparison of sound segregation techniques for predominant instrument recognition in musical audio signals. In: Gouyon, F., Herrera, P., Martins, L.G., Müller, M. (eds.) Proceedings of the 13th International Society for Music Information Retrieval Conference, ISMIR, pp. 559–564 (2012)

6. Breiman, L.: Random forests. Mach. Learn. **45**(1), 5–32 (2001)
7. Brown, J.C., Houix, O., McAdams, S.: Feature dependence in the automatic identification of musical woodwind instruments. J. Acoust. Soc. Am. **109**(3), 1064–1072 (2001)
8. Eerola, T., Ferrer, R.: Instrument library (MUMS) revised. Music Percept. **25**(3), 253–255 (2008)
9. Eronen, A.: Musical instrument recognition using ICA-based transform of features and discriminatively trained HMMs. In: Proceedings of the 7th International Symposium on Signal Processing and Its Applications, ISSPA, pp. 133–136 (2003)
10. Essid, S., Richard, G., David, B.: Instrument recognition in polyphonic music based on automatic taxonomies. In: IEEE Transactions on Audio, Speech, and Language Processing, pp. 68–80 (2006)
11. Fuhrmann, F.: Automatic musical instrument recognition from polyphonic music audio signals. Ph.D. thesis, Universitat Pompeu Fabra, Department of Information and Communication Technologies (2012)
12. Fuhrmann, F., Herrera, P.: Polyphonic instrument recognition for exploring semantic similarities in music. In: Proceedings of the 13th International Conference on Digital Audio Effects, DAFx (2010)
13. Gómez, J.S., Abeßer, J., Cano, E.: Jazz Solo instrument classification with convolutional neural networks, source separation, and transfer learning. In: Proceedings of the 19th International Society for Music Information Retrieval Conference, ISMIR, pp. 577–584 (2018)
14. Goto, M., Hashiguchi, H., Nishimura, T., Oka, R.: RWC music database: music genre database and musical instrument sound database. In: Proceedings of the 4th International Conference on Music Information Retrieval, ISMIR, pp. 229–230 (2003)
15. Gururani, S., Sharma, M., Lerch, A.: An attention mechanism for musical instrument recognition. In: Flexer, A., Peeters, G., Urbano, J., Volk, A. (eds.) Proceedings of the 20th International Society for Music Information Retrieval Conference, ISMIR, pp. 83–90 (2019)
16. Han, Y., Kim, J., Lee, K.: Deep convolutional neural networks for predominant instrument recognition in polyphonic music. IEEE ACM Trans. Audio, Speech, Lang. Process. **25**(1), 208–221 (2017)
17. Lartillot, O., Toiviainen, P.: MIR in Matlab (II): a toolbox for musical feature extraction from audio. In: Dixon, S., Bainbridge, D., Typke, R. (eds.) Proceedings of the 8th International Conference on Music Information Retrieval, ISMIR, pp. 127–130. Austrian Computer Society (2007)
18. LeCun, Y., Bengio, Y., Hinton, G.E.: Deep learning. Nature **521**(7553), 436–444 (2015)
19. Lemley, J., Bazrafkan, S., Corcoran, P.: Smart augmentation learning an optimal data augmentation strategy. IEEE Access **5**, 5858–5869 (2017)
20. Leveau, P., Vincent, E., Richard, G., Daudet, L.: Instrument-specific harmonic atoms for mid-level music representation. IEEE Trans. Audio, Speech Lang. Process. **16**(1), 116–128 (2008)
21. Li, X., Wang, K., Soraghan, J.J., Ren, J.: Fusion of hilbert-huang transform and deep convolutional neural network for predominant musical instruments recognition. In: Romero, J., Ekárt, A., Martins, T., Correia, J. (eds.) EvoMUSART. LNCS, vol. 12103, pp. 80–89. Springer, Cham (2020). https://doi.org/10.1007/978-3-030-43859-3_6

22. McEnnis, D., McKay, C., Fujinaga, I.: jAudio: additions and improvements. In: Proceedings of the 7th International Conference on Music Information Retrieval, ISMIR, pp. 385–386 (2006)
23. McFee, B., et al..: Librosa: audio and music signal analysis in python. In: Proceedings of the Python Science Conference, pp. 18–24 (2015)
24. Mierswa, I., Wurst, M., Klinkenberg, R., Scholz, M., Euler, T.: YALE: rapid prototyping for complex data mining tasks. In: Eliassi-Rad, T., Ungar, L.H., Craven, M., Gunopulos, D. (eds.) Proceedings of the 12th ACM SIGKDD International Conference on Knowledge Discovery and Data Mining, KDD, pp. 935–940. ACM (2006)
25. Pampalk, E., Flexer, A., Widmer, G.: Improvements of audio-based music similarity and genre classificaton. In: Proceedings of the 6th International Conference on Music Information Retrieval, ISMIR, pp. 628–633 (2005)
26. Rosner, A., Kostek, B.: Automatic music genre classification based on musical instrument track separation. J. Intell. Inf. Syst. **50**(2), 363–384 (2018)
27. Seyerlehner, K., Widmer, G., Knees, P.: Frame level audio similarity - a codebook approach. In: Proceedings of the 11th International Conference on Digital Audio Effects, DAFx (2008)
28. Solanki, A., Pandey, S.: Music instrument recognition using deep convolutional neural networks. Int. J. Inf. Technol. (2019)
29. Sturm, B.L.: A survey of evaluation in music genre recognition. In: Proceedings of the 10th International Workshop on Adaptive Multimedia Retrieval: Semantics, Context, and Adaptation, AMR, pp. 29–66 (2012)
30. Tzanetakis, G., Cook, P.: Musical genre classification of audio signals. IEEE Trans. Speech Audio Process. **10**(5), 293–302 (2002)
31. Vatolkin, I.: Evolutionary approximation of instrumental texture in polyphonic audio recordings. In: Proceedings of the IEEE Congress on Evolutionary Computation (CEC), pp. 1–8 (2020)
32. Vatolkin, I., Rudolph, G.: Comparison of audio features for recognition of western and ethnic instruments in polyphonic mixtures. In: Gómez, E., Hu, X., Humphrey, E., Benetos, E. (eds.) Proceedings of the 19th International Society for Music Information Retrieval Conference, ISMIR, pp. 554–560 (2018)
33. Vatolkin, I., Rudolph, G., Weihs, C.: Evaluation of album effect for feature selection in music genre recognition. In: Müller, M., Wiering, F. (eds.) Proceedings of the 16th International Society for Music Information Retrieval Conference, ISMIR, pp. 169–175 (2015)
34. Vatolkin, I., Theimer, W., Botteck, M.: AMUSE (Advanced MUSic Explorer) - a multitool framework for music data analysis. In: Downie, J.S., Veltkamp, R.C. (eds.) Proceedings of the 11th International Society on Music Information Retrieval Conference (ISMIR), pp. 33–38 (2010)
35. Wang, Y., Getreuer, P., Hughes, T., Lyon, R.F., Saurous, R.A.: Trainable frontend for robust and far-field keyword spotting. In: Proceedings of the 2017 IEEE International Conference on Acoustics, Speech and Signal Processing, ICASSP, pp. 5670–5674 (2017)
36. Wolpert, D.H., Macready, W.G.: Coevolutionary free lunches. IEEE Trans. Evol. Comput. **9**(6), 721–735 (2005)
37. Yu, H., Kim, S.: SVM tutorial - classification, regression and ranking. In: Rozenberg, G., Bäck, T., Kok, J.N. (eds.) Handbook of Natural Computing, vol. 1, pp. 479–506. Springer, Berlin, Heidelberg (2012). https://doi.org/10.1007/978-3-540-92910-9_15

A Multi-objective Evolutionary Approach to Identify Relevant Audio Features for Music Segmentation

Igor Vatolkin[1(✉)], Marcel Koch[1], and Meinard Müller[2]

[1] Department of Computer Science, TU Dortmund, Germany
{igor.vatolkin,marcel3.koch}@tu-dortmund.de
[2] International Audio Laboratories Erlangen, Erlangen, Germany
meinard.mueller@audiolabs-erlangen.de

Abstract. The goal of automatic music segmentation is to calculate boundaries between musical parts or sections that are perceived as semantic entities. Such sections are often characterized by specific musical properties such as instrumentation, dynamics, tempo, or rhythm. Recent data-driven approaches often phrase music segmentation as a binary classification problem, where musical cues for identifying boundaries are learned implicitly. Complementary to such methods, we present in this paper an approach for identifying relevant audio features that explain the presence of musical boundaries. In particular, we describe a multi-objective evolutionary feature selection strategy, which simultaneously optimizes two objectives. In a first setting, we reduce the number of features while maximizing an F-measure. In a second setting, we jointly maximize precision and recall values. Furthermore, we present extensive experiments based on six different feature sets covering different musical aspects. We show that feature selection allows for reducing the overall dimensionality while increasing the segmentation quality compared to full feature sets, with timbre-related features performing best.

Keywords: Music segmentation · Evolutionary multi-objective feature selection

1 Introduction

The goal of music segmentation is to identify boundaries between parts of music pieces which are perceived as entities. Music segmentation can be defined as a

This work was funded by the German Research Foundation (DFG), project 336599081 "Evolutionary optimisation for interpretable music segmentation and music categorisation based on discretised semantic metafeatures". The experiments were carried out on the Linux HPC cluster at TU Dortmund (LiDO3), partially funded in the course of the Large-Scale Equipment Initiative by the German Research Foundation (DFG) as project 271512359. The International Audio Laboratories Erlangen are a joint institution of the Friedrich-Alexander-Universität Erlangen-Nürnberg (FAU) and the Fraunhofer-Institut für Integrierte Schaltungen IIS.

J. Romero et al. (Eds.): EvoMUSART 2021, LNCS 12693, pp. 327–343, 2021.
https://doi.org/10.1007/978-3-030-72914-1_22

binary supervised classification problem and can be applied on different levels–from the prediction of longer vocal segments on the verse or chorus level to short note sequences on the measure level. In contrast to music pieces given as digital score, the segmentation of audio tracks is a more challenging application because many relevant musical properties (instrumentation, tempo, harmonic structure) are only encoded implicitly in the signal. Even if the classification performance of music segmentation has continuously increased in recent years—in particular with regard to improvements done by deep neural networks—there still remain some issues: a higher interpretability of classification models to understand why a boundary was detected, a reduction of high computing efforts, and, in general, the optimization of algorithms with regard to several evaluation criteria beyond classification performance.

In this work, we propose an approach to identify relevant audio signal features resp. their individual dimensions for music segmentation based on multi-objective evolutionary feature selection. In a first setting, multiple trade-off feature sets are identified, which have a possibly low number of features and a possibly high F-measure. In a second setting, the feature sets are optimized with regard to precision and recall. We show that the application of feature selection does not only lead to a better classification performance compared to full feature sets, but also helps to measure the relevance of individual feature dimensions.

In Sect. 2, we provide a brief overview of related studies on feature selection for music classification and segmentation. Section 3 introduces feature groups involved in the study, which represent several musical aspects (timbre, harmony, temporal properties). Furthermore, we describe the applied novelty-based segmentation method, feature selection approach, and baseline music segmentation algorithms. Section 4 presents our experiments and their analysis. We conclude with the most relevant observations and ideas for future work in Sect. 5.

2 Related Work

Feature selection for classification tasks removes irrelevant and redundant features, helping to identify the most important ones. For an introduction to this topic we refer to [7]. The advantages of feature selection are an improvement of classification performance, a reduction of computing efforts to extract, process, and store the features, and the retaining of interpretability of the original feature dimensions compared to statistical feature reduction methods like principal component analysis.

Feature selection was applied for various music classification tasks like instrument [5], genre [1], and mood [20] recognition. However, there exist only a few studies which applied or discussed feature selection for music segmentation. [19] proposed a method to estimate feature weights for the computation of self-similarity matrices (SSMs). [22] introduced a method to learn feature relevance for SSM estimation by means of quadratic programming. [23] suggested to build "a knowledge-based segmentation system where the selection of the features, algorithms and their parameter configurations can be carried out automatically based on the source knowledge of the music signals," in particular with respect to limitations of music segmentation systems optimized only for Western pop music. In [9], no feature selection was applied directly, but three feature groups rhythm, timbre, and harmony were compared, leading to the outcome that timbre features individually performed at best for the boundary detection.

A multi-objective approach which simultaneously minimizes the dimensionality of selected features while optimizing one or more other criteria may help to identify the most important feature dimensions. Although this method was applied to genre [25] and instrument [24] recognition, to our knowledge, it was not used until now for music segmentation.

3 Algorithms

3.1 Feature Groups

For the estimation of individual relevance of feature dimensions[1], we considered features with relation to different musical properties: timbre, harmony, and temporal structure. The six selected feature groups are listed in Tables 1 and 2, together with individual features, references, sizes of extraction time frames, and numbers of dimensions. All features were extracted from mono signals downsampled 22050 Hz. Please note that the selected settings are based on previous research and a complete validation of all possible frame sizes and other extraction parameters is not possible in the scope of this study, so that other related descriptors may perform better for music segmentation. Jointly selecting features and optimizing their extraction parameters could be an interesting research direction for the future.

3.2 Segmentation Based on Novelty Function for Self-Similarity Matrix

Because the main scope of our work was to maintain the interpretability and to identify the most relevant feature dimensions for music segmentation, we do not focus on methods which learn features by themselves (e.g., deep neural networks) or transform them to new high-dimensional spaces (like support vector machines). Our segmentation algorithm is a novelty-based approach as described in [16, Chapter 4.4] based on the implementation [18]. Let the signal be represented with a feature matrix $X = (\boldsymbol{x}_1, \boldsymbol{x}_2, ..., \boldsymbol{x}_T)$, where $\boldsymbol{x}_i$ is the vector of all feature dimensions selected as described below in Sect. 3.3 and T the number

[1] The terminology within the scope of this paper is as follows: feature selection keeps individual *feature dimensions* (e.g., the 2nd MFCC) from *feature vectors* (e.g., a 13-dimensional MFCC vector) which exclusively belong to *feature groups* like timbre. A *feature set* selected for music segmentation is then constructed with various dimensions of various features which however belong to the same group in the current setup—the combination of features from different groups remains a promising future work.

Table 1. Timbre audio features used in this study.

Feature	Ref.	Frame	Dim.
Angles and distances in phase domain	[15]	512	2
Average distance between extremal spectral values and its variance	[15]	512	2
Average distance between zero-crossings of the time-domain signal and its variance	[15]	512	2
Bark scale magnitudes	[11]	512	23
Delta MFCCs	[11]	512	13
Discrepancy	[15]	512	1
ERB bands: root mean square, spectral centroid, zero-crossing rate	[11]	512	30
Linear prediction coefficients	[13]	512	10
Low energy	[13]	512	1
Mel frequency cepstral coefficients (MFCCs)	[10,11]	512	33
Normalized energy of harmonic components	[26]	512	1
Root mean square	[13]	512	1
Root mean square peak number above half of maximum peak in 3s	[26]	66150	1
Root mean square peak number in 3s	[26]	66150	1
Sensory roughness	[11]	1024	1
Spectral bandwidth, centroid, extent, flux, kurtosis, skewness, slope	[15]	512	7
Spectral brightness	[11]	512	1
Spectral crest factor	[15]	512	4
Spectral flatness measure	[15]	512	4
Spectral irregularity	[15]	512	4
Sub-band energy ratio	[15]	512	4
Tristimulus	[26]	512	2
Y-axis intercept	[15]	512	1
Zero-crossing rate	[13]	512	1

of time frames. Then, a self-similarity matrix [4] $S \in \mathbb{R}^{T \times T}$ is computed. Each cell $S(n,m)$ with $n,m \in [1:T] := \{1,2,\ldots,T\}$ is obtained by computing the inner product $S(n,m) = |\langle x_n, x_m \rangle|$ of each combination of two elements in X. In the following step, the SSM matrix is correlated along its main diagonal with a checkerboard kernel. The idea behind this procedure is to find two sequential, individually coherent block structures along the main diagonal which are different to each other. To prioritize local over global structures, thus being more insensitive to similarity values further from the center of the kernel, we use a Gaussian checkerboard kernel.

Table 2. Audio features from the groups beyond timbre used in this study.

Feature	Ref.	Frame	Dim.
Mel spectrum			
Mel spectrum	[14]	2048	128
Chroma			
Bass chroma	[12]	2048	12
Chroma	[11]	512, 4096	12
Chroma	[12]	2048	12
Chroma energy normalized statistics (CENS)	[17]	4410	12
Chroma DCT-reduced log pitch (CRP)	[17]	4410	12
Semitone spectrum			
Semitone spectrum	[12]	2048	84
Harmony			
Amplitude of maximum in the chromagram	[15]	512	1
Consonance	[12]	2048	1
Fundamental frequency	[8]	512	1
Harmonic change	[12]	2048	1
Harmonic change detection function	[11]	512, 4096	2
Inharmonicity	[11]	512	1
Key and its clarity	[11]	512, 4096	4
Local tuning	[12]	8192	1
Major/minor alignment	[11]	512, 4096	2
Strengths of major/minor keys	[11]	512, 4096	48
Tonal centroid vector	[11]	512, 4096	12
Tone with maximum strength in chroma	[15]	512	1
Rhythm and tempo			
Estimated onset number per minute	[11]	229376	1
Estimated beat and tatum number per minute	[10]	229376	2
First relative periodicity amplitude peak	[13]	512	1
First periodicity peak in bpm	[15]	512	1
Fluctuation patterns	[10]	32768	7
Peaks of fluctuation curves summed across all bands	[11]	229376	5
Rhythmic clarity	[11]	66150	1
Second periodicity peak in bpm	[15]	512	1
Sum of correlated components	[13]	512	1
Tempo based on onset times	[11]	66150	1

$$K_{\text{Gauss}}(k,l) = e^{-\varepsilon^2(k^2+l^2)} \cdot (\text{sig}(k) \cdot \text{sig}(l)) \tag{1}$$

with $\text{sig}(x)$ being the sign function, $k, l \in [-L : L]$, $\varepsilon = 0.5$ being the degree of tapering and $L = 30$ being the selected kernel size after preliminary tests.

The correlation between K and S yields the novelty function $\Delta : [1 : T] \rightarrow \mathbb{R}$ defined as:

$$\Delta(n) = \sum_{k,l \in [-L:L]} K_{\text{Gauss}}(k,l) \cdot S(n+k, n+l), \tag{2}$$

which corresponds to a measure of novelty for each frame where local maxima indicate probable segment boundaries. Figure 1 visualizes each step of this procedure.

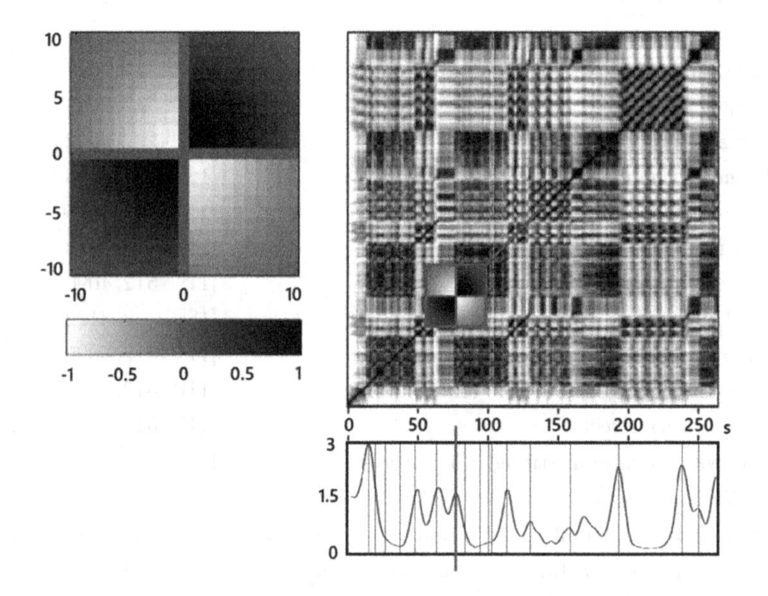

Fig. 1. Left: Gaussian checkerboard kernel with $L = 10$ [16, P. 208]. Upper right: self-similarity matrix with checkerboard kernel. Lower right: resulting novelty function with ground truth indicators for segment boundaries (gray vertical lines).

To further improve the quality of Δ by decreasing its fluctuations and thus reducing incorrect segment boundary predictions, the novelty curve is smoothed by applying a Gaussian filter with the standard deviation $\sigma = 3$ and subsequently analyzed for local maxima to retrieve segment boundary candidates. All values have been derived from preliminary tests to yield reliable results for a maximum variety of features, to identify the most relevant ones. However, for some fixed features, the optimal values may need a further adjustment.

3.3 Feature Selection

Let $\mathcal{F}$ be the full feature set, $\boldsymbol{q}$ the binary vector which indicates selected feature dimensions ($q_i = 1$ if and only if feature dimension x_i is selected, $i \in [1 : |\mathcal{F}|]$), and $X(\mathcal{F}, \boldsymbol{q})$ the feature matrix after feature selection from the complete feature set $\mathcal{F}$. Let $m_1, ..., m_C$ be optimization criteria like precision or recall. Then, the multi-objective feature selection can be defined as:

$$\boldsymbol{q}^* = \arg\min_{\boldsymbol{q}} \{m_i (\boldsymbol{y}, \hat{\boldsymbol{y}}, X(\mathcal{F}, \boldsymbol{q})) : i = 1, ..., C\}, \tag{3}$$

where $\boldsymbol{y}$ is the ground truth (segment boundaries), and $\hat{\boldsymbol{y}}$ Foote-segmentation result based on $X(\mathcal{F}, q)$.

In the first setting, we minimize the normalized number of selected feature dimensions $N \in [0 : 1]$ and maximize the F-measure. In the second setting, we maximize precision and recall. Both measures are calculated with 3s tolerance boundary. Because of criteria pairs, some feature sets are not comparable (consider a set with high precision and low recall compared to a set with low precision and high recall), while other feature sets dominate others, i.e. are "better" than others with respect to at least one criterion and "equal or better" with respect to all other criteria (a set with high precision and high recall compared to a set with low precision and low recall or a set with an equally high precision but lower recall). Therefore, the result of a multi-objective optimizer is typically a set of "solutions" (here feature sets) which are not dominated by any other solution estimated during the optimization process.

Because evolutionary algorithms simultaneously operate on several solutions (a population), and the non-deterministic mutation helps to overcome local optima, they are particularly suited for multi-objective optimization [2]. In this paper, evolutionary multi-objective feature selection is applied using an $\mathcal{S}$ metric selection evolutionary multi-objective algorithm (SMS-EMOA)[3] evolutionary algorithm which selects individuals based on their hypervolume contribution. While the general setup follows our previous work [24], the parameters were adjusted after preliminary experiments: population size was set to 30, initialization with an equally distributed random expected number of feature dimensions between 1 and $|\mathcal{F}|$, asymmetric mutation with a higher probability to remove feature dimensions, number of generations to 1500, and number of statistical repetitions to 10.

3.4 Baselines

Even if the main focus of our experiment is to identify the most relevant feature dimensions and not to achieve the best segmentation result, the performance is nevertheless important. The optimized feature sets are compared to four different baselines: (1) complete feature sets; (2) boundary detection based on a 13-dimensional MFCC vector; (3) feature reduction based on principal component analysis (PCA), and (4) a convolutional neural network (CNN). For the third baseline, the number of selected components was equal to the number of features in the sets with the highest F-measure after feature selection. For the fourth baseline, we applied a CNN from [6].

4 Experiments

4.1 Datasets

For our music segmentation task, we used the SALAMI dataset [21] which contains 1383 music tracks from various genres representing Classical, Jazz, Popular, World, and Live music. The first coarse annotation of boundaries provides the ground truth for evaluation. For a better robustness estimation, the track order was shuffled and they were assigned to five parts. In each experiment fold, exactly one part served as the optimization set for the evaluation of selected feature dimensions (the size of optimization sets was limited with respect to computing time). Another non-overlapping part served as the test set for the final independent evaluation of feature dimensions which have been identified as the best for the optimization set. The assignment of parts to five folds is sketched in Fig. 2.

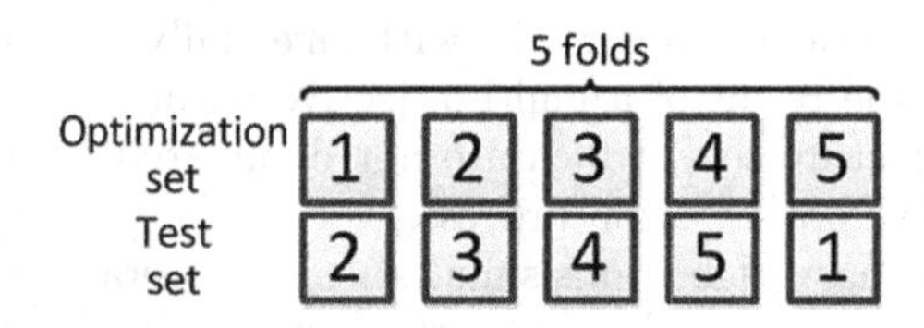

Fig. 2. Assignment of five SALAMI parts to the five folds.

4.2 Non-Dominated Fronts

Figure 3 visualizes the results after the feature selection with 10 statistical repetitions minimizing the normalized number of selected feature dimensions (N) and maximizing the F-measure (F). The top plot shows the measures for the optimization set, the bottom plot for the test set. Feature sets from the non-dominated fronts (ND-fronts) of the last populations are marked with large circles. Small points mark all feature sets analyzed during feature selection. Their "age" is marked with a brightness level: while early solutions are marked darker, later solutions become brighter. The shaded gray areas correspond to all points which are dominated by the non-dominated front.

Both plots provide important insights to the impact of feature selection on music segmentation. The complete feature sets ($N = 1$) do not only require a significantly larger number of feature dimensions, resulting in higher computing costs, but also have smaller F-values than many other feature sets with less dimensions. However, there is the tendency that more dimensions usually lead to

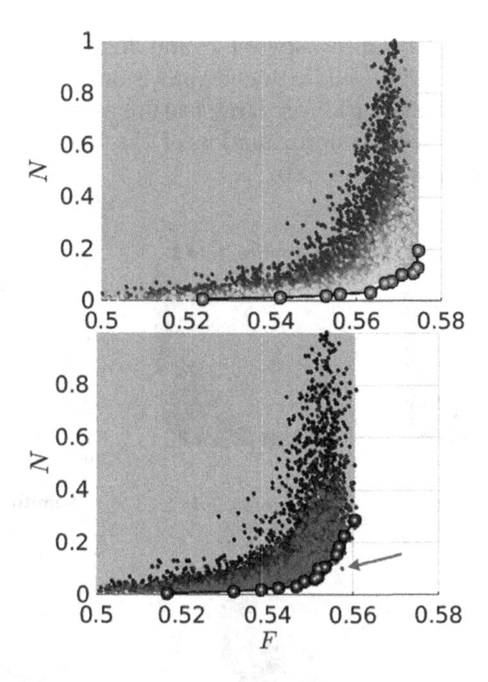

Fig. 3. Results using the 1st SALAMI part as the optimization set (top) and the 2nd part as the test set (bottom), segmentation with timbre features.

higher F-values. The performance on the independent test set is—as expected—lower but still better than for full feature sets in the bottom subfigure, and it can be seen that brighter solutions estimated after more optimization efforts are closer to the non-dominated front also for the test set. However, a couple of solutions exist which dominate feature sets from the non-dominated front of the test set (e.g., a feature set marked with an arrow). This feature set does not belong to the final feature sets after all 1500 evolutionary generations of feature selection.

Figure 4 shows different final ND-fronts evaluated on the test sets. In the upper row, we observe the feature sets across three different folds (2, 3, 4) for timbral features. Although the best F of the last population slightly varies between 0.5636 and 0.5757, the overall differences between the folds are not large. In the middle row, the results achieved with harmony, chroma, and semitone spectrum for the first fold are plotted. For the semitone spectrum, almost all feature dimensions seem to be important to achieve the highest F-values. With respect to the highest F-values, all feature groups in this row perform worse than timbre features from the upper row. The bottom row presents results using three feature groups for the first fold and the criteria P and R. The ND-fronts are smaller than those of the first optimization scenario, and the overall difference between the values of both measures across all analyzed feature sets is low. However, the feature sets from the non-dominated fronts of the last population occupy rather different areas of balance between P and R: A higher recall is achieved with the semitone spectrum, and a higher precision with timbre features. So, for the future, experiments with very large feature sets compiled from different groups may achieve a better approximation of the true Pareto front with the best possible compromise feature sets.

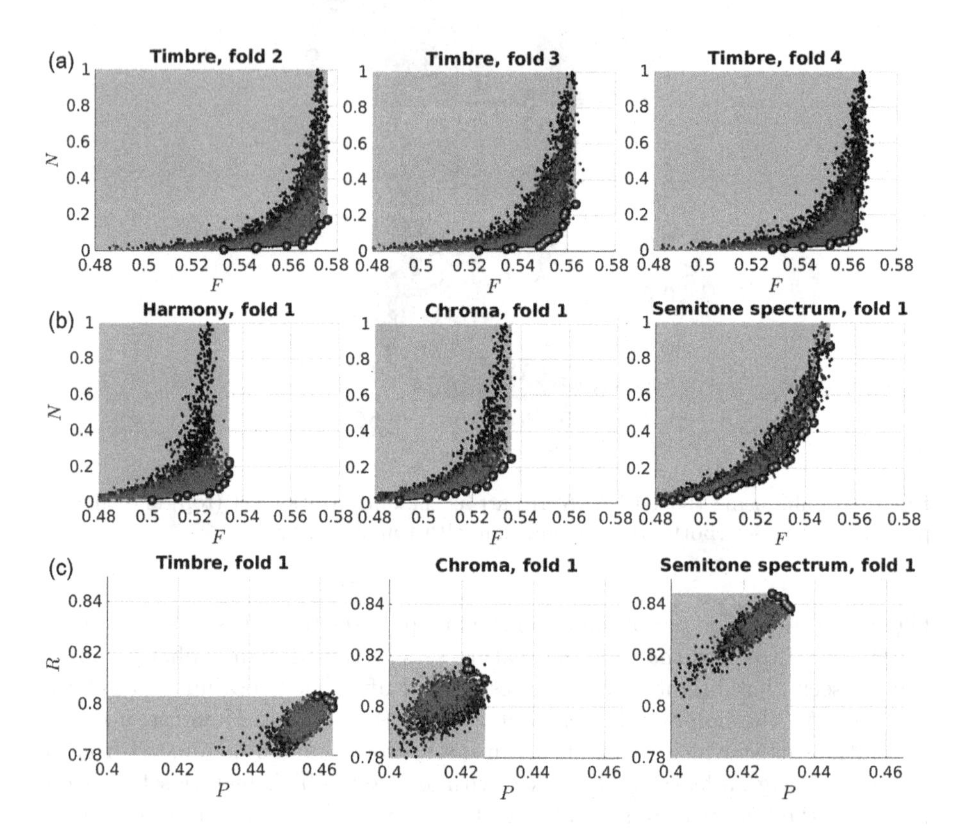

Fig. 4. Results for several folds, feature groups, and criteria. (a): Examples of different folds for the same feature set optimizing N, F; (b): Examples of different feature sets for the same fold using N, F; (c): Examples of different feature sets for the same fold using P, R.

4.3 Optimization Results

Tables 3 and 4 list measures for border solutions of ND-fronts after feature selection. Algorithms and feature sets are described in the first column. The first two lines report results with the baselines using the CNN-based approach of [6] and the novelty-based segmentation using only MFCCs. The following 18 lines correspond to the 6 tested feature groups. For each group, the results are shown for the full original feature set, for the "best" selected set, and the set based on feature reduction with PCA, where the number of selected components is equal to the number of feature dimensions in the "best" set. In Table 3, the highest $F_{\max}$ for each group is marked with a bold font. "Best" is dependent on the used feature selection scenario: results in columns 2 and 3 correspond to feature sets after minimizing of the number of feature dimensions and maximization of F. $F_{\max}$ is the best F-measure averaged for five folds (solutions in the upper right corner of the ND-fronts in Fig. 3), and $No(F_{\max})$ the corresponding number of feature dimensions. Standard deviations are provided after "±". Table 4 lists the results after the maximization of precision and recall: Columns 2 and 3 represent feature sets with the highest precision (cf. lower right corner in the fronts of Fig. 4c) and columns 4 and 5 with the highest recall (upper left corner in the fronts from the bottom subfigures in Fig. 4c).

Table 3. Music segmentation results (averaged over five folds) after the optimization of N and F. For details please see the text.

Features/Algorithm	$F_{\max}$	$No(F_{\max})$
CNN	0.5155 ± 0.0111	–
MFCCs	0.5471 ± 0.0069	13.0 ± 0.00
Timbre: full	0.5644 ± 0.0070	147.0 ± 0.00
Timbre: best	0.5673 ± 0.0060	44.6 ± 16.38
Timbre: PCA	**0.5760 ± 0.0065**	44.6 ± 16.38
Mel spectrum: full	0.5251 ± 0.0070	128.0 ± 0.00
Mel spectrum: best	**0.5254 ± 0.0063**	55.4 ± 36.99
Mel spectrum: PCA	0.5237 ± 0.0085	55.4 ± 36.99
Chroma: full	0.5331 ± 0.0088	72.0 ± 0.00
Chroma: best	**0.5373 ± 0.0087**	27.2 ± 9.44
Chroma: PCA	0.5373 ± 0.0097	27.2 ± 9.44
Semitones: full	0.5528 ± 0.0103	84.0 ± 0.00
Semitones: best	**0.5553 ± 0.0100**	71.6 ± 7.47
Semitones: PCA	0.5497 ± 0.0117	71.6 ± 7.47
Harmony: full	0.5240 ± 0.0126	75.0 ± 0.00
Harmony: best	**0.5368 ± 0.0071**	9.2 ± 4.38
Harmony: PCA	0.5228 ± 0.0111	9.2 ± 4.38
Rhythm/tempo: full	0.4611 ± 0.0028	21.0 ± 0.00
Rhythm/tempo: best	**0.5092 ± 0.0032**	3.2 ± 1.30
Rhythm/tempo: PCA	0.4665 ± 0.0079	3.2 ± 1.30

The results show that selected dimensions are always better than full feature sets with regard to both optimization settings, but the difference is not significant in every case. The best feature group for the first pair of optimization criteria is

Table 4. Music segmentation results (averaged over five folds) after the optimization of P and R. For details please see the text.

Features/Algorithm	P_{max}	$R(P_{max})$	$P(R_{max})$	R_{max}
CNN	0.522 ± 0.051	–	–	0.591 ± 0.061
MFCCs	0.446 ± 0.007	–	–	0.795 ± 0.005
Timbre: full	0.466 ± 0.007	–	–	0.803 ± 0.007
Timbre: best	$\mathbf{0.469 \pm 0.006}$	0.808 ± 0.006	0.466 ± 0.009	$\mathbf{0.813 \pm 0.006}$
Mel spectrum: full	0.421 ± 0.0063	–	–	0.771 ± 0.008
Mel spectrum: best	$\mathbf{0.422 \pm 0.005}$	0.769 ± 0.009	0.419 ± 0.006	$\mathbf{0.777 \pm 0.009}$
Chroma: full	0.426 ± 0.0076	–	–	0.803 ± 0.008
Chroma: best	$\mathbf{0.430 \pm 0.009}$	0.816 ± 0.008	0.424 ± 0.010	$\mathbf{0.822 \pm 0.007}$
Semitones: full	0.436 ± 0.0093	–	–	0.846 ± 0.011
Semitones: best	$\mathbf{0.438 \pm 0.009}$	0.848 ± 0.012	0.4356 ± 0.012	$\mathbf{0.851 \pm 0.010}$
Harmony: full	0.420 ± 0.0115	–	–	0.783 ± 0.012
Harmony: best	$\mathbf{0.431 \pm 0.009}$	0.795 ± 0.004	0.425 ± 0.007	$\mathbf{0.801 \pm 0.004}$
Rhythm/tempo: full	0.367 ± 0.0040	–	–	0.692 ± 0.009
Rhythm/tempo: best	$\mathbf{0.418 \pm 0.008}$	0.637 ± 0.098	0.396 ± 0.004	$\mathbf{0.772 \pm 0.003}$

timbre, for the second pair timbre (for P) and semitone spectrum (for R). PCA performs better than evolutionary feature selection only for timbre features and is able to achieve the best F-value of all feature groups in this study. However, it is important to mention that PCA requires all original feature dimensions and has disadvantages with regard to both computing resources and interpretability compared to the selection of individual features. The CNN baseline has a higher P, but lower R and F compared to the novelty-based segmentation. Please note that a fair comparison to this baseline was not possible in the current setup: for a dependable performance, deep neural networks typically require larger training sets. The MFCC baseline is the 2nd best with regard to P, 3rd best w.r.t. F, and 4th best w.r.t. R (compared to the "best" selected feature sets evaluated by novelty segmentation).

4.4 Analysis of the Best Selected Features

For the first optimization setting (the reduction of N and maximization of F), not only the analysis of border feature sets from ND-fronts with F_{max}, but also from the complete ND-fronts provides important insights into the meaning of individual features for music segmentation. These features also contribute to sets with as few dimensions as possible, but still maximize F. For each dimension, we measure its normalized occurrence Ω in all non-dominated feature sets in relation to the number of all feature sets in the non-dominated front (averaged over $M = 5$ folds):

Table 5. Importance of features in the ND-fronts after the optimization of N and F.

Feature: Dimension	Ω
Timbre	
Y-axis intercept: 1	0.6153
Angles in phase domain: 1	0.4816
Mel frequency cepstral coefficients: 2	0.3616
Mel frequency cepstral coefficients: 3	0.3443
Spectral centroid: 1	0.2569
Mel spectrum	
Mel spectrum: 98	0.4177
Mel spectrum: 107	0.3098
Mel spectrum: 100	0.2846
Mel spectrum: 79	0.2729
Mel spectrum: 104	0.2642
Chroma	
Chroma(512): 3	0.4990
Chroma(512): 12	0.4605
Chroma(512): 9	0.4385
Chroma(4096): 5	0.3929
Chroma(512): 6	0.3817
Semitone spectrum	
Semitone spectrum: 54	0.6172
Semitone spectrum: 49	0.5397
Semitone spectrum: 83	0.4903
Semitone spectrum: 78	0.4779
Semitone spectrum: 11	0.4569
Harmony	
Ampl. of chroma max.: 1	0.8732
Key and its clarity(512): 2	0.3603
Harmonic change detection function(512): 1	0.2673
Key and its clarity(4096): 2	0.2511
Harmonic change detection function(4096): 1	0.0911
Rhythm and tempo	
Fluctuation patterns: 4	0.9960
Fluctuation patterns: 5	0.0632
First relative periodicity amplitude peak: 1	0.0200
Fluctuation patterns: 7	0.0085
Fluctuation patterns: 3	0.0040

Table 6. Importance of features in the ND-fronts after the optimization of P and R.

Feature: Dimension	Ω
Timbre	
Linear prediction coefficients: 3	0.8933
Low energy: 1	0.8333
Mel frequency cepstral coefficients: 2	0.7933
Mel frequency cepstral coefficients: 3	0.7933
Spectral flatness: 1	0.7600
Mel spectrum	
Mel spectrum: 9	0.9500
Mel spectrum: 21	0.8333
Mel spectrum: 84	0.8000
Mel spectrum: 18	0.7833
Mel spectrum: 22	0.7833
Chroma	
Chroma(512): 3	0.7500
Chroma(4096): 5	0.7500
Chroma(512): 6	0.7167
Chroma(512): 5	0.7167
Bass chroma: 4	0.7000
Semitone spectrum	
Semitone spectrum: 54	1.0000
Semitone spectrum: 79	1.0000
Semitone spectrum: 49	0.9750
Semitone spectrum: 53	0.9750
Semitone spectrum: 81	0.9750
Harmony	
Key and its clarity(4096): 2	0.9500
Ampl. of chroma max.: 1	0.9333
Harmonic change detection function: 1	0.8700
Key and its clarity: 2	0.8433
Harmonic change detection function(4096): 1	0.4933
Rhythm and tempo	
Fluctuation patterns: 4	0.4135
Fluctuation patterns: 7	0.3768
Beats per minute: 1	0.3688
Fluctuation patterns: 5	0.3530
Fluctuation patterns: 1	0.2251

$$\Omega(k) = \frac{1}{M} \cdot \sum_{j=1}^{M} \left(\frac{1}{D_j} \cdot \sum_{i=1}^{D_j} a_k(i,j) \right), \tag{4}$$

where $a_k(i,j) = 1$ if and only if dimension k is selected in the i-th non-dominated feature set from j-th fold and D_j is the number of non-dominated feature sets in fold j.

The best value of 1 means that the dimension belongs to all non-dominated feature sets in all folds. Table 5 reports the top dimensions with the corresponding Ω-values. The 4th fluctuation pattern dimension has the highest Ω, but other dimensions of fluctuation patterns have significantly lower values. Such analysis means that for a more efficient and presumably even more precise segmentation it is not necessarily to store and process all individual dimensions of a single feature. The Ω-values for timbre features and the individual importance of the top-five reported features can be considered particularly high because the timbre set had the highest number of 147 original dimensions and left more room to deselect less relevant features—while proving to be a challenge for feature selection because of a significantly higher number of possible feature combinations (2^{147}-1). Considering the semitone spectrum, many different dimensions have a high share in non-dominated feature sets, cf. a rather high mean value of selected features $No(F_{\max}) = 71.6$ in Table 3.

Table 6 reports Ω-values for the optimization of precision and recall. The most often selected dimensions of the non-dominated sets are not the same as in Table 5. However, there are still some features that seem relevant for both scenarios: the 3rd chroma dimension, the 54th dimension of the semitone spectrum, and the 4th dimension of the fluctuation pattern vector are the most often selected features in the related groups.

5 Conclusions and Outlook

In this paper, we have proposed a multi-objective feature selection approach for music segmentation using six different groups of audio features. The results showed that with strongly reduced feature sets it was possible even to increase the classification quality. The timbre-related features performed the best with regard to F-measure. However, a more important contribution of this study is that it helps to identify the most relevant features and their individual dimensions, allowing for a better analysis of classification models.

Further work should consider segmentation with particularly interpretable features like instrument and chord properties and can be separately applied for different scenarios like segmentation in specific genres (also for non-Western music) or on different levels. Besides, other segmentation algorithm parameters like the checkerboard kernel size can be simultaneously optimized with the feature selection. Also, a more strict tolerance boundary for the evaluation of music segmentation (like 0.5 s) can be used for evaluation.

References

1. Burred, J.J., Lerch, A.: A hierarchical approach to automatic musical genre classification. In: Proceedings of the 6th International Conference on Digital Audio Effects (DAFx), pp. 8–11 (2003)
2. Coello, C.A.C., Lamont, G.B., Veldhuizen, D.A.V.: Evolutionary Algorithms for Solving Multi-Objective Problems. Springer, New York (2007). https://doi.org/10.1007/978-0-387-36797-2
3. Emmerich, M., Beume, N., Naujoks, B.: An EMO algorithm using the hypervolume measure as selection criterion. In: Coello Coello, C.A., Hernández Aguirre, A., Zitzler, E. (eds.) EMO 2005. LNCS, vol. 3410, pp. 62–76. Springer, Heidelberg (2005). https://doi.org/10.1007/978-3-540-31880-4_5
4. Foote, J.: Visualizing music and audio using self-similarity. In: Proceedings of the 7th ACM International Conference on Multimedia, pp. 77–80 (1999)
5. Fujinaga, I.: Machine recognition of timbre using steady-state tone of acoustic musical instruments. In: Proceedings of the International Computer Music Conference (ICMC), pp. 207–210 (1998)
6. Grill, T., Schlüter, J.: Music boundary detection using neural networks on combined features and two-level annotations. In: Müller, M., Wiering, F. (eds.) Proceedings of the 16th International Society for Music Information Retrieval Conference (ISMIR), pp. 531–537 (2015)
7. Guyon, I., Nikravesh, M., Gunn, S., Zadeh, L.A. (eds.): Feature Extraction. In: Foundations and Applications, Studies in Fuzziness and Soft Computing, vol. 207. Springer, Heidelberg (2006). https://doi.org/10.1007/978-3-540-35488-8
8. Jensen, K.: Timbre models of musical sounds - from the model of one sound to the model of one instrument. Ph.D. Thesis, University of Copenhagen, Denmark (1999)
9. Jensen, K.: Multiple scale music segmentation using rhythm, timbre, and harmony. EURASIP J. Adv. Sig. Process. (2007). https://doi.org/10.1155/2007/73205
10. Klapuri, A., Eronen, A.J., Astola, J.: Analysis of the meter of acoustic musical signals. IEEE Trans. Audio Speech Lang. Process. **14**(1), 342–355 (2006)
11. Lartillot, O., Toiviainen, P.: MIR in MATLAB (II): a toolbox for musical feature extraction from audio. In: Proceedings of the 8th International Conference on Music Information Retrieval (ISMIR), pp. 127–130. Austrian Computer Society (2007)
12. Mauch, M., Dixon, S.: Approximate note transcription for the improved identification of difficult chords. In: Downie, J.S., Veltkamp, R.C. (eds.) Proceedings of the 11th International Society for Music Information Retrieval Conference (ISMIR), pp. 135–140 (2010)
13. McEnnis, D., McKay, C., Fujinaga, I.: jAudio: additions and improvements. In: Proceedings of the 7th International Conference on Music Information Retrieval (ISMIR), pp. 385–386 (2006)
14. McFee, B., et al.: Librosa: audio and music signal analysis in python. In: Proceedings the Python Science Conference, pp. 18–25 (2015)
15. Mierswa, I., Wurst, M., Klinkenberg, R., Scholz, M., Euler, T.: YALE: rapid prototyping for complex data mining tasks. In: Eliassi-Rad, T., Ungar, L.H., Craven, M., Gunopulos, D. (eds.) Proceedings of the 12th ACM SIGKDD International Conference on Knowledge Discovery and Data Mining (KDD), pp. 935–940. ACM (2006)

16. Müller, M.: Fundamentals of Music Processing: Audio, Analysis, Algorithms, Applications. Springer, Heidelberg (2015). https://doi.org/10.1007/978-3-319-21945-5
17. Müller, M., Ewert, S.: Chroma Toolbox: MATLAB implementations for extracting variants of chroma-based audio features. In: Klapuri, A., Leider, C. (eds.) Proceedings of the 12th International Conference on Music Information Retrieval (ISMIR), pp. 215–220. University of Miami (2011)
18. Müller, M., Zalkow, F.: FMP notebooks: educational material for teaching and learning fundamentals of music processing. In: Proceedings of the 20th International Conference on Music Information Retrieval (ISMIR). Delft, The Netherlands, November 2019
19. Parry, R.M., Essa, I.A.: Feature weighting for segmentation. In: Proceedings of the 5th International Conference on Music Information Retrieval (ISMIR) (2004)
20. Saari, P., Eerola, T., Lartillot, O.: Generalizability and simplicity as criteria in feature selection: application to mood classification in music. IEEE Trans. Audio Speech Lang. Process. **19**(6), 1802–1812 (2011)
21. Smith, J.B.L., Burgoyne, J.A., Fujinaga, I., Roure, D.D., Downie, J.S.: Design and creation of a large-scale database of structural annotations. In: Klapuri, A., Leider, C. (eds.) Proceedings of the 12th International Society for Music Information Retrieval Conference (ISMIR), pp. 555–560. University of Miami (2011)
22. Smith, J.B.L., Chew, E.: Using quadratic programming to estimate feature relevance in structural analyses of music. In: Proceedings of ACM Multimedia Conference, pp. 113–122. ACM (2013)
23. Tian, M.: A cross-cultural analysis of music structure. Ph.D. Thesis, Queen Mary University of London, UK (2017)
24. Vatolkin, I.: Generalisation performance of western instrument recognition models in polyphonic mixtures with ethnic samples. In: Correia, J., Ciesielski, V., Liapis, A. (eds.) EvoMUSART 2017. LNCS, vol. 10198, pp. 304–320. Springer, Cham (2017). https://doi.org/10.1007/978-3-319-55750-2_21
25. Vatolkin, I., Preuß, M., Rudolph, G.: Multi-objective feature selection in music genre and style recognition tasks. In: Krasnogor, N., Lanzi, P.L. (eds.) Proceedings of the 13th Annual Genetic and Evolutionary Computation Conference (GECCO), pp. 411–418. ACM Press (2011)
26. Vatolkin, I., Theimer, W., Botteck, M.: AMUSE (Advanced MUSic Explorer) - a multitool framework for music data analysis. In: Downie, J.S., Veltkamp, R.C. (eds.) Proceedings of the 11th International Society on Music Information Retrieval Conference (ISMIR), pp. 33–38 (2010)

Exploring the Effect of Sampling Strategy on Movement Generation with Generative Neural Networks

Benedikte Wallace[1(✉)], Charles P. Martin[2], Jim Tørresen[1], and Kristian Nymoen[1(✉)]

[1] RITMO Centre for Interdisciplinary Studies in Rhythm, Time and Motion, Department of Informatics, University of Oslo, Oslo, Norway
{benediwa,krisny}@ifi.uio.no
[2] Australian National University, Canberra, Australia

Abstract. When using generative deep neural networks for creative applications it is common to explore multiple sampling approaches. This sampling stage is a crucial step, as choosing suitable sampling parameters can make or break the realism and perceived creative merit of the output. The process of selecting the correct sampling parameters is often task-specific and under-reported in many publications, which can make the reproducibility of the results challenging. We explore some of the most common sampling techniques in the context of generating human body movement, specifically dance movement, and attempt to shine a light on their advantages and limitations. This work presents a Mixture Density Recurrent Neural Network (MDRNN) trained on a dataset of improvised dance motion capture data from which it is possible to generate novel movement sequences. We outline several common sampling strategies for MDRNNs and examine these strategies systematically to further understand the effects of sampling parameters on motion generation. This analysis provides evidence that the choice of sampling strategy significantly affects the output of the model and supports the use of this model in creative applications. Building an understanding of the relationship between sampling parameters and creative machine-learning outputs could aid when deciding between different approaches in generation of dance motion and other creative applications.

Keywords: Mixture density networks · Movement generation · Generative networks · Creative prediction

1 Introduction

Expressive movement is an intrinsic part of human life. Simple movement patterns such as gait or arm movement can efficiently convey an emotional state and

This work was partially supported by the Research Council of Norway through its Centres of Excellence scheme, project number 262762.

J. Romero et al. (Eds.): EvoMUSART 2021, LNCS 12693, pp. 344–359, 2021.
https://doi.org/10.1007/978-3-030-72914-1_23

may allow us to detect characteristics such as personality or mood [15,18,19]. As such, motion analysis and generation have the potential to improve human-robot interaction, human activity recognition and artificial agent design [12]. In this paper, we present a deep recurrent neural network trained on dance movement and systematically examine a series of common sampling strategies in order to explore their effect on the movement generation process.

While the objective criteria of success for a predictive model will vary depending on the application, one would usually expect a model intended for the generation of creative data to be able to generate novel and varied outputs, while still ensuring a reliable level of realism. This gives rise to several challenges. For any predictive system, a loss function would typically include some metric indicating the distance from the model output to some ground truth. This approach produces good results when generating basic human motion such as gait [1]. However, when generating dance movements, or indeed any creative data, there is no single correct answer or ground truth to which the output can be compared as we may consider several predictions equally valid. Thus, rather than attempting to predict a single truth a good model for generating creative data should be able to generate a variety of likely outputs. Mixture density networks (MDNs) [2] have been used for various creative prediction tasks previously, as they are particularly suited for modelling these kinds of data. The MDN treats the outputs of a neural network as the parameters of a Gaussian mixture model (GMM), which can be sampled to generate real-valued predictions and allow us to explore several strategies for generating variations on the model's predictions. For the experiments outlined here, we will focus on three strategies: *priming*, *isolation of mixture components* and *temperature adjustments.*

Including a temperature parameter which is used to reweight the learned probability distribution is a common way to approach the challenge of generating variations on model output. This is often done by experimenting with sampling from the probability distribution with more or less "randomness", often referred to as stochastic sampling, until the output is found to be satisfactory [6–8,14]. If the model consistently outputs the most likely prediction, the output can become repetitive. Conversely, a flat probability distribution where all outcomes are equally likely would be equivalent to sampling at random, invariant to the learned distribution. Reweighting the distribution allows the model to generate less likely predictions some of the time. Allowing for this occasional unexpected predictions by implementing stochastic sampling, will in many cases improve the realism of the output and elicit more interesting predictions thereby improving the perceived creative capacity of the model. However, there are many possible temperature values and finding the "sweet spot" can be challenging.

As the MDN learns several distributions of the data simultaneously, it is further possible to explore each of the learned Gaussian distributions in isolation. This idea has been explored by Ellefsen et al. [6] in the context of world models [9], who found evidence to support the theory that each component will model different stochastic events. In this work, we examine the distributions learned by a model trained with 4 mixture components and discuss the potential of this approach as a means of simultaneously modelling several possible outcomes for a given input sequence.

Another common approach to vary the output of a trained model is to prime the model on different unseen examples. Priming consists of running an unseen sequence through the model before generating a prediction for the next time step. Graves [7] demonstrated interesting results using this strategy to generate handwriting examples in various styles by priming a model on examples by different subjects. With our priming experiments, we aim to examine to what extent the results obtained for handwriting are transferable to full-body motion.

While variation is a central aspect of generating creative data, there has been little focus on the sampling strategies and how they affect variation in the generated data. We aim to systematically explore the effect of applying the three sampling strategies outlined above: adjusting temperature, priming, and isolating mixture components. The code used to sample from the trained model has been made available in the interest of reproducibility. The following sections describe the training parameters and motion capture recordings used to train the MDRNN. Section 5 presents the results of sampling from the trained MDRNN using each of the aforementioned strategies. The results are further discussed in Sect. 6 where we present our recommendations for applying the different strategies in creative applications.

2 Motion Data

Working with expressive human movement presents a series of challenges. In the section below, we have outlined some considerations relevant to working with creative data and, more specifically, dance. In the following section, details of the data collection process and post-processing of the motion capture data are presented.

2.1 Quantifying Movement

Movement of a human body may be represented as real-valued multi-dimensional time-series data unfolding in space and time. High-precision recordings of movement may be recorded with marker-based optical motion capture technology, where each marker put on the body provides a 3D position vector over time. Dance is a particularly complex variant of human movement, with subtle features which can communicate important information. As such our network must be able to retain the nuances in our dance recordings as real-valued time-series. While other creative data sources such as music, text and images can be simplified into a discrete representation, it is clear that doing so would severely reduce the expressiveness of human movement.

In examining the output of a model which generates creative data the results are often evaluated according to whether or not they are typical examples of their type, be it a piece of music perhaps belonging to a specific genre [20], or a painting in a certain style [21], or in our case a realistic sequence of movement. Realism is naturally a quality that has some degree of importance when generating any type of data, but perhaps especially so for movement; any deviation from what is physically possible for a human to achieve is instantly recognisable. Another important aspect to consider is whether the model is able to produce novel output, different from the examples used in training.

In order to determine the novelty or realism of a generated motion sequence, it is useful to define a notion of similarity. However, the seemingly simple concept of movement similarity becomes quite complicated. In many cases, it may be sufficient to consider motions as similar if they only differ with respect to global rotation or spatial and temporal scaling [16], but for the purpose of comparing dance, such a definition may be unsatisfactory. Consider, for example, one dance sequence where the left arm is raised, and another the right arm. The semantic similarity between the two would not be reflected in metrics like the euclidean distance between global positions or between joint angles in consecutive frames. Further exploration of movement similarity is outside the scope of this article. Instead of comparing generated motion sequences to the motion capture recordings by use of distance metrics, we will describe the generated motion sequences in reference to the conceptual class of artefacts that this data belongs to, namely dance, and evaluate them by visual inspection.

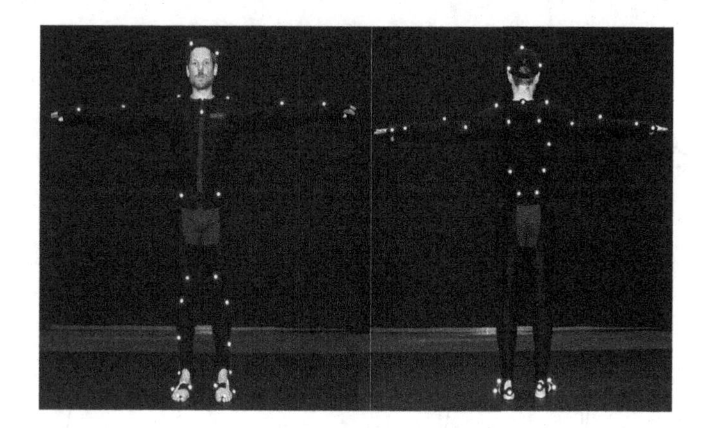

Fig. 1. Marker setup worn by the dancers during motion capture data collection

2.2 Dataset

Our dataset contains 54 one minute motion capture recordings of improvised dance performed by three female dancers. Their average age is 23 and each has more than 10 years experience in modern, jazz and ballet. Each dancer performs three one minute improvisations to six different musical stimuli which vary in terms of tempo, tonality and beat clarity. The dataset was recorded using a Qualisys optical motion capture system with 12 Oqus 300/400 series cameras which capture 43 reflective markers worn by the dancers. Figure 1 shows the front and back of a dancer wearing the motion capture suit with markers. Using the MoCap Toolbox 1.5 [3] these 43 markers are translated to 22 joint positions, see Fig. 2. Small gaps in the data were spline-filled using Qualisys Track Manager 2019.3 and a 2nd degree Butterworth filter with a 7.2Hz cutoff was applied to remove any marker jitter. Recordings in our dataset have been normalized so that the root marker (a weighted average of markers 41, 42, 6 and 7 in Fig. 2) is centred at the origin where x, y, and z position is 0. Body segment lengths are averaged across the three dancers ensuring that the data is invariant to global position and individual body dimensions. The data was captured 240 Hz and downsampled 30 Hz before model training to reduce the size of each example. The resulting 54 data tensors consist of 1800 frames (60 s 30 Hz) with 3-dimensional positions for each of the 22 points.

Two full motion capture recordings are withheld for testing while the remaining 52 examples are split into two sets, 80% are used for training and 10% for validation. Each example is sliced into overlapping sequences of 256 frames and the spatial dimensions of each of the 22 points are scaled using min-max normalisation. The input to our model thereby consists of 80288 overlapping sequences of 256 frames. The target value of each training example is the 3D position of each point in the following frame.

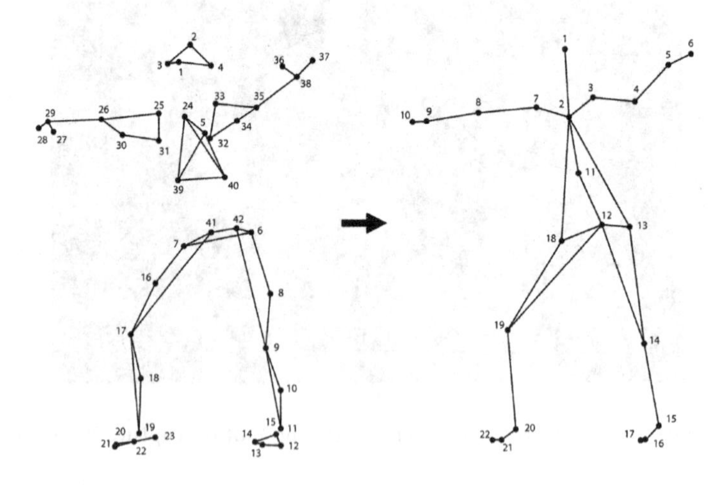

Fig. 2. The 43 reflective markers are translated to 22 points

3 Movement Prediction with MDRNNs

Mixture density networks (MDNs) [2] treat the outputs of a neural network as the parameters of a Gaussian mixture model (GMM), which can be sampled to generate real-valued predictions. MDRNNs are becoming well-established tools in the generation of creative data. They have previously been applied to musical sketches in two dimensions as part of a smartphone app [14], to sketches [8], and handwriting [7]. MDRNNs have also previously been applied to motion capture data [5,17,22] leading to promising results.

Figure 3 shows a simplified mixture distribution with 4 components. A GMM can be derived using the mean, weight and standard deviation of each component. The number of components needed to accurately represent the data is not known and is treated as a hyperparameter for our model. For the study outlined here, we have used 4 components. We can interpret these components as each representing different possible future movement. By combining a recurrent neural network (RNN) with an MDN to form an MDRNN we can make real-valued predictions based on a sequence of inputs. Figure 4 shows the model architecture of the MDRNN used in this work. The RNN consists of three layers of LSTM cells [10], known to be effective in modelling temporal sequences such as music, text and speech. The three LSTM layers contain 1024, 512 and 256 hidden units respectively. The outputs of the third LSTM layer are in turn connected to an MDN. The LSTM layers learn to estimate the mean, standard deviation and weight of the 4 Gaussian distributions.

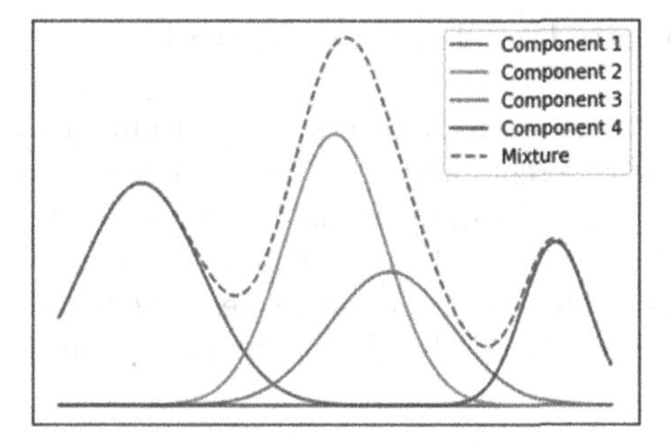

Fig. 3. Simplified mixture distribution example. The MDRNN presented in this work learns the mean (μ), standard deviation (σ) and weight (π) of 4 components, each having 66 dimensions, one for each of the 22 3D points.

To optimize an MDN, we minimize the negative log-likelihood of sampling true values from the predicted GMM for each example. A probability density function (PDF) is used to obtain this likelihood value. In our case, the GMM consists of $K = 4$ n-variate Gaussian distributions. For simplicity in the PDF, these distributions are restricted to having a diagonal covariance matrix, and thus the PDF has the form:

$$p(\theta; x) = \sum_{k=1}^{K} \pi_k \mathcal{N}(\mu_k, \Sigma_k; x) \tag{1}$$

where π are the mixing coefficients, μ, the Gaussian distribution centres, Σ the covariance matrices and n is the 66 position values (22 points * 3 dimensions) contained in each frame. This configuration corresponds to 8,540,692 parameters. The loss function in our system is calculated by the `keras-mdn-layer` [13] Python package which makes use of Tensorflow's probability distributions package to construct the PDF. The model is trained using the Adam optimizer [11] until the loss on the validation set failed to improve for 10 consecutive epochs.

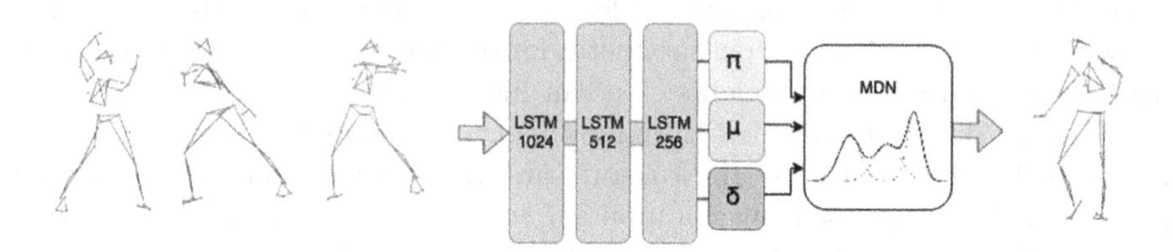

Fig. 4. Sampling from the MDRNN. A sequence of preceding frames is sent through the model which outputs the parameters of a mixture distribution. By sampling from this distribution the next frame is generated.

4 Sampling from the Trained Model

In the following sections, we explore how the output of the trained MDRNN can be affected using different sampling strategies. There are, of course, countless parameter settings and combinations to examine. Here, we have focused on outlining the extremes as well as the threshold points wherein the effect of each strategy changes. Three different strategies for examining the creative representational capacity of the MDRNN and their effect on the model output are explored:

Temperature Adjustment: When sampling from our trained model we can choose to alter the value of two temperature parameters, the π-temperature and the σ-temperature. The σ-temperature is used to scale the covariance matrix of each mixture component's multivariate normal distribution. Adjusting the σ-temperature affects the width of each mixture component. A high σ-temperature allows for samples further from the learned mean of each mixture component to be selected. For our experiments, we explored a range of σ values between 0 and 1 to locate at which temperatures the effect on the output changes.

The π-temperature adjusts the probability of sampling from individual mixture components. High π-temperatures reweight the probability of sampling from each component in such a way that each component is an equally likely choice, while sufficiently low temperatures will ensure that only a single component is selected. Here, we examine π-temperatures between 0 and 10. By selecting a π-temperature close to or higher than 1.0 we create a higher likelihood of sampling from a different mixture component at every time step. Sampling with

π-temperature close to 0, on the other hand, ensures that only the component which has the highest learned weight will be sampled from. This is equivalent to isolating this component and sampling from it at every time step.

Isolating Mixture Components: One of the advantages of the MDRNN architecture is its inherent ability to model a range of possible outcomes simultaneously. The MDN used in this work builds a GMM from 4 components, in the following experiments we isolate each of them by enforcing that sampling can happen from only a single component. This allows us to further examine what kind of movement is learned by each component. While the most realistic movement sequences would most probably be gained from sampling from the component with the highest π value, the other components contain variations on the predicted movement which might also produce interesting results.

Primed Sampling: Being able to generate motion in a certain style is useful both in the context of animation and as a creative tool for choreography. By feeding the model an unseen example we can evaluate to what extent the model is able to accurately predict a continuation of the sequence. Carlson et al. [4] have recently shown how the characteristics of a participant's individual style of movement are sufficiently unique to allow for classification using machine learning and as such, it is intriguing to investigate whether priming a model on an example may allow the model to generate movements in the style of a specific participant or performance. Previous work using MDRNNs to generate handwriting [7] show that it is possible to produce examples which display the characteristics of a particular writer when the model is primed on an example. In the experiments presented below, two motion sequences are used to prime the model during sampling. The first contains slow and flowing movements while the second was performed to an uptempo song and contains faster and more movement overall. By comparing the generated movements we examine the perceived likeness between the primer sequence and the model predictions.

5 Results

The code used to sample from the trained model can be found at our working repository together with video examples.[1]

5.1 Adjusting Temperature

Sampling with σ-temperature close to or higher than 1 results in movement sequences that change rapidly between frames. The movements are "shaky" and jump in unnatural ways between time steps. We observe that this shaking effect becomes noticeable with $\sigma > 0.05$. Conversely, a σ-temperature closer to 0 allows for smoother motions. An example of the extremes can be seen in Figures 5a and

[1] https://github.com/benediktewallace/Motion-MDRNN.

5b. These figures show the trajectories of hand and toe markers, with horizontal displacement indicating time evolution. Figure 5c shows a sequence wherein the temperature rises over time. σ-temperatures between 10^{-7} and 10^{-4} cause small variations in the motion sequence while remaining realistic. Higher values result in sequences which contain a fair amount of noise which can also distort the form of the body, while lower values contain less overall movement.

Figure 6b shows a sequence generated with a π-temperature close to 0. Figure 6c shows how the output changes as the π-temperature is increased over time. At temperatures around 1.0 the component with the highest weight is still sampled from most frequently but, we also intermittently sample from the other 3 components according to their respective weights. Figure 6a shows the result of sampling with a high π-temperature. Here, we sample from a different component at almost every time step. Shifting between mixture components when generating motion sequences causes abrupt changes in the position of the body, alluding to the notion that each component may learn a slightly different movement sequence. We examine this more closely in the upcoming section.

5.2 Isolating Mixture Components

For these experiments, we disregard the π-temperature and instead manually select which of the 4 mixture components to sample from. This ensures that each new frame is sampled from a single component. We observed in the previous section that the entire position of the body changed as we sampled with a higher π-temperature, indicating that individual components emphasise different features.

In order to examine this more closely the σ-temperature is kept at a low value to make certain that we sample close to the mean of each component and each sequence is given the same starting position. Figure 7 shows the 100th frame from sequences sampled from each of the 4 components using the same starting frames and temperature parameters. These figures show how each component has predicted a different outcome, with component 1 being the most dissimilar to the other components. Sequences sampled from component 1 contain almost no movement at all, while component 2 and 3 produces examples which show the highest amount of realism. The third component consistently produces the most realistic output and is the component with the highest learned π-value. Thereby making it the component which is most likely sampled from when no adjustments are made using the π-temperature parameter. Sampling from component 1 and 4 shows little overall movement and a fair amount of distortion in the body, as joints are placed at unnatural angles. From these experiments, it seems that each component learns a slightly different movement pattern. Thereby, selecting to sample from a single component when generating a movement sequence ensures that certain learned patterns appear in the predicted movement.

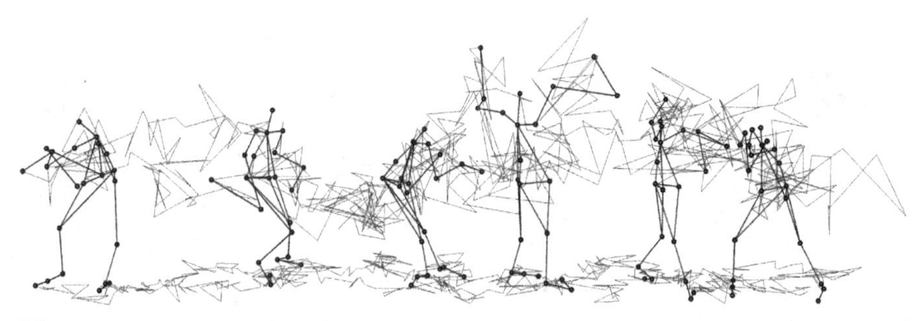

(a) Movement generated with σ=1.0. Movements are noisy, positions of points jerk between frames.

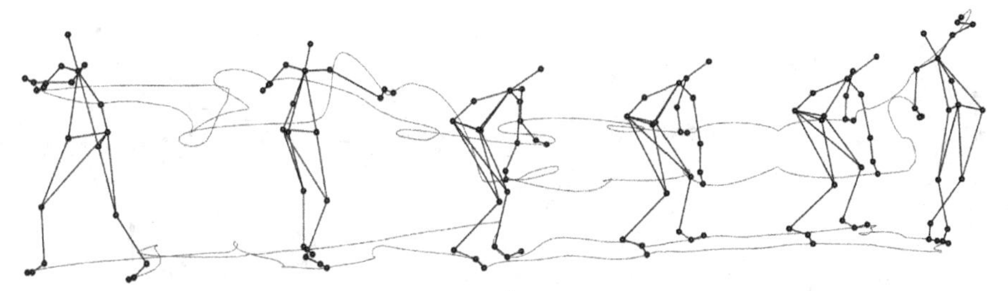

(b) Movement generated with $\sigma = 10^{-9}$ Movements are smooth and realistic.

(c) Movement generated with σ-temperature rising over time.

Fig. 5. Sequences generated using different σ-temperatures.

5.3 Primed Sampling

When generating motion with priming, a movement sequence which has not been used in training is given as input to the model. The next frame is then generated and the process is repeated. The model always predicts the next frame for a previously unseen real sequence, as opposed to non-primed sampling, wherein

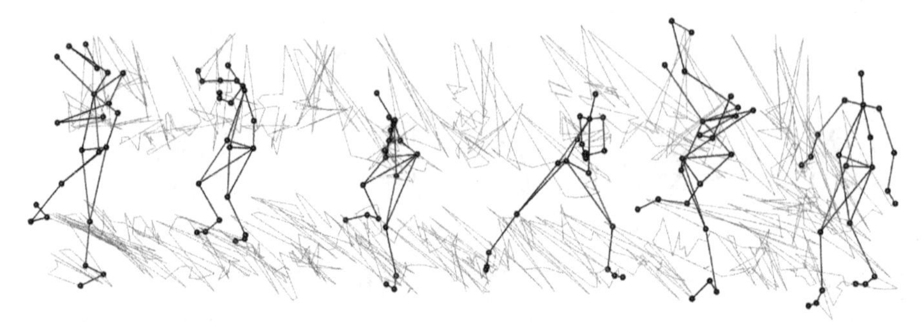

(a) Movement generated with π=10. Positions of the body jerk between frames as the we alternate between mixture components.

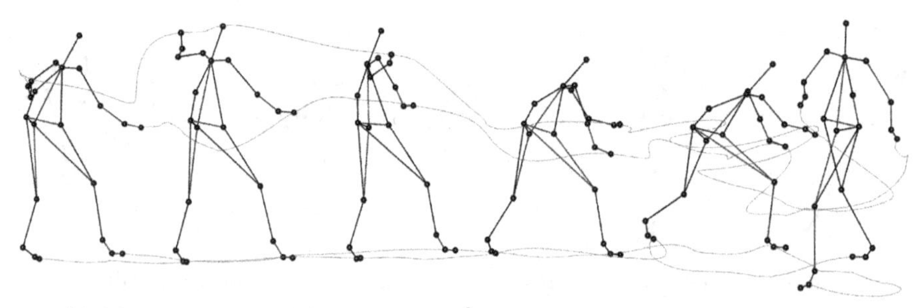

(b) Movement generated with $\pi = 10^{-9}$ Movements are smooth and realistic.

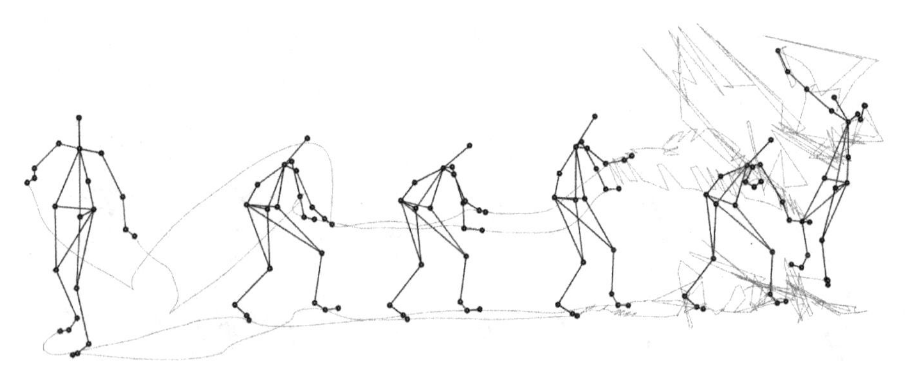

(c) Movement generated with π-temperature rising over time.

Fig. 6. Sequences sampled with different π-temperatures.

the models' previous predictions become part of the sequence used to generate the following frame. We explore the effect of priming on two performances by different individuals using the examples that were withheld during training. The first example hereafter referred to as *primer A*, was performed to rhythmical musical stimuli with a strong beat presence. The second example, *primer B*, was performed to slow, non-rhythmic musical stimuli. As such, the two prim-

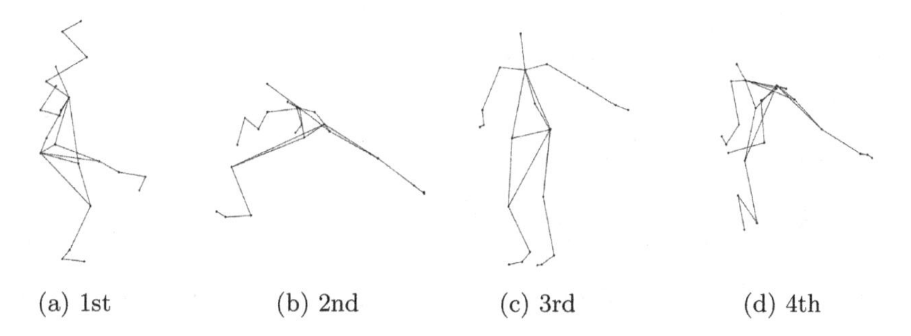

Fig. 7. 100th frame from sequences generated by 4 different mixture components. Keeping all other parameters the same, each component generates a different movement sequence.

ing examples have differing characteristics. Primer A contains more movement overall and a higher speed of movement, while primer B consists of slower movements. For these experiments, π and σ-temperatures are both set to values close to zero ensuring that we sample from close to the mean of the component with the highest weight, component 3.

Figure 8a shows a series of frames from primer A. Figure 8b shows the corresponding movement sequence generated using primer A. As can be seen in the figure, the generated movement to a large degree follows the movement of the priming sequence. Movement aspects such as the speed of motion and the overall amount of movement are similar to the priming sequence as well. Similar results were also found for primer B.

6 Discussion

We have implemented and systematically explored common sampling strategies which have previously been used in various generative tasks. However, the importance of realism is perhaps more important in the generation of movement than in generating other creative data. Even slight deviations from what is feasible for the human form breaks the perceived realism of the generated movement.

When sampling from only the component with the highest weight, either by sufficiently lowering the π-temperature or by isolating the component with the highest learned π value, we ensure that the model outputs smooth and realistic motion sequences. While stochastic sampling of mixture components is possible, our results indicate that this approach, at least for movement, does not necessarily improve the output as the movements modelled by each component are sufficiently dissimilar. Thereby, intermittent changes in the choice of mixture component cause abrupt changes and more jagged movement.

Sampling from a single mixture component for a given sequence and altering the amount of deviation we allow away from the mean using the σ-temperature instead can give varied and realistic results. While larger adjustments to σ-temperature result in noisy and irregular predictions, slight adjustments allow

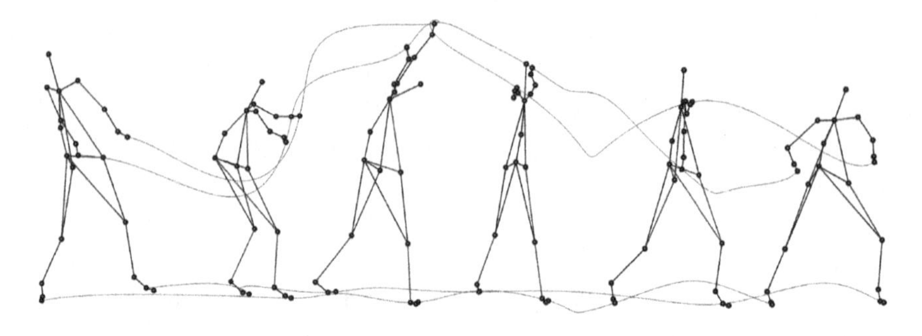

(a) Frames from priming sequence A.

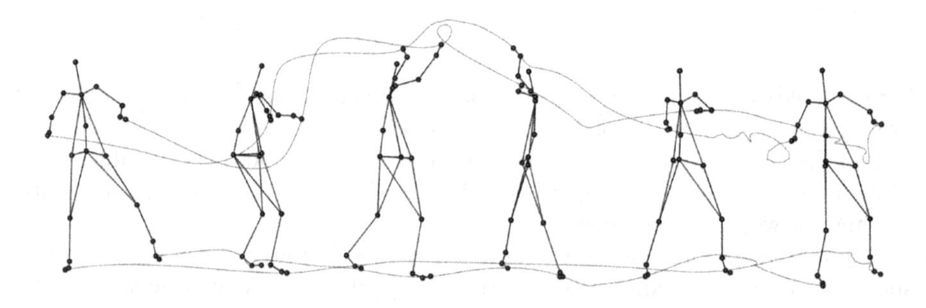

(b) Motion sequence generated when priming on priming sequence A.

Fig. 8. Priming sequence and corresponding output. The sequence generated by the model shows similar motion unfolding over time.

for some interesting variations to occur. This could have useful applications as a user-controlled parameter, both in a creative tool and for animation.

When examining the output generated by each mixture component in isolation it is apparent that the components which achieve the highest learned weight also produce the most realistic movement sequences, as would be expected. By examining each component individually, however, we found that two of the four components were able to produce interesting sequences which also achieved relatively high realism. It is possible that training on a larger dataset would cause the other components, which for our experiments show distorted body poses with little movement, to learn more realistic and useful features. This would then be an interesting parameter in a creative tool in order to create variations on a theme. Examining the output from each component also shows us the capabilities of this particular model to produce several possible futures with different degrees of variation and realism.

From our experiments with priming, we found that the output of the model was able to continue the motion of unseen examples with some variation. This indicates that the style of movement is indeed continued when sampling with priming. Priming the model on unseen data could be useful when using the model as a creative tool to assist choreographers and animators, as it allows for priming the model with their own data and achieve predictions in a desirable style closer to their own. The goal of this research is to critically examine sampling of a motion generation MDRNN with a view towards the production of novel and realistic data for creative applications. From our findings, we suggest the following recommendations for applying the different types of sampling we have explored in creative motion generation:

Temperature Adjustment. Adjusting π and σ during sampling allow variations to be produced within a narrow band of useful values. These need to be determined by trial and error.

Isolating Mixture Components. Generating motion from individual mixture components is a particularly useful tool. As we found, some mixture components with lower learned weights can produce useful outputs. Isolating these components can ensure that they are not overwhelmed by more likely parts of the mixture model.

Primed Sampling. Using new movement sequences to prime the model can ensure that the generated sequence follows the over-all movement shape and style of the priming example. By further combining priming and σ-temperature adjustments, one could produce several variations on a theme.

7 Conclusion

While MDRNNs have previously been applied to creative processes including motion generation, the process of *sampling*, or generating creative outputs, has so far not been critically examined. In this research, we have presented results from exploring three common sampling techniques (temperature adjustment, isolating mixture components, and primed sampling) that can be used when generating output from MDRNNs. We have analysed how these techniques affect generated dance movement sequences.

Our findings show that the priming technique allows the output to be shaped by an unseen example, indicating that the model could be used to generate movement sequences in a certain style. Further, the results show that changing the learned probability distribution by including temperature parameters has the potential to greatly affect the output. However, the range of viable parameter values is small, as any unnatural movement is easily detected by an observer. As such, temperature adjustment may be less suited for human motion than for other creative data. Alternating between mixture components when generating a motion sequence results in unnatural shifts in body positions between consecutive frames. Still, the components with a lower learned weight may also contain interesting movement sequences. Therefore, we believe this approach warrants further exploration and analysis.

Evaluations of the results of the different sampling strategies have been qualitative. Arguably, one might prefer a more quantitative approach to the evaluation of machine learning results, however, this is an inherently challenging aspect of creative computing in general, and particularly so for data representing human behaviour. In future work, we will conduct a perceptual study in order to quantify the qualitative evaluations conducted in this paper.

We feel that the findings of the present research, in confirming and exploring the capacity of generative MDRNN models of human motion to create a variety of outputs will assist with the production of more novel and more realistic motion sequences for use in artistic and scientific work.

References

1. Alemi, O., Pasquier, P.: WalkNet: a neural-network-based interactive walking controller. IVA 2017. LNCS (LNAI), vol. 10498, pp. 15–24. Springer, Cham (2017). https://doi.org/10.1007/978-3-319-67401-8_2
2. Bishop, C.M.: Mixture density networks. Technical report, Aston University, Birmingham, UK (1994). http://publications.aston.ac.uk/373/
3. Burger, B., Toiviainen, P.: MoCap toolbox-a Matlab toolbox for computational analysis of movement data. In: 10th Sound and Music Computing Conference, SMC 2013, Stockholm, Sweden. Logos Verlag Berlin (2013)
4. Carlson, E., Saari, P., Burger, B., Toiviainen, P.: Dance to your own drum: identification of musical genre and individual dancer from motion capture using machine learning. J. New Music Res. **49**, 1–16 (2020)
5. Crnkovic-Friis, L., Crnkovic-Friis, L.: Generative choreography using deep learning. In: Proceedings of the Seventh International Conference on Computational Creativity (2016)
6. Ellefsen, K.O., Martin, C.P., Torresen, J.: How do mixture density RNNs predict the future? arXiv preprint arXiv:1901.07859 (2019)
7. Graves, A.: Generating sequences with recurrent neural networks. arXiv preprint arXiv:1308.0850 (2013)
8. Ha, D., Eck, D.: A neural representation of sketch drawings. arXiv preprint arXiv:1704.03477 (2017)
9. Ha, D., Schmidhuber, J.: Recurrent world models facilitate policy evolution. In: Advances in Neural Information Processing Systems. pp. 2450–2462 (2018)
10. Hochreiter, S., Schmidhuber, J.: Long short-term memory. Neural Comput. **9**, 1735–80 (1997). https://doi.org/10.1162/neco.1997.9.8.1735
11. Kingma, D.P., Ba, J.: Adam: a method for stochastic optimization. arXiv preprint arXiv:1412.6980 (2014)
12. Lee, J., Marsella, S.: Nonverbal behavior generator for embodied conversational agents. In: Gratch, J., Young, M., Aylett, R., Ballin, D., Olivier, P. (eds.) IVA 2006. LNCS (LNAI), vol. 4133, pp. 243–255. Springer, Heidelberg (2006). https://doi.org/10.1007/11821830_20
13. Martin, C.P.: Keras-MDN-layer: Python Package, November 2018. https://doi.org/10.5281/zenodo.1482348
14. Martin, C.P., Torresen, J.: RoboJam: a musical mixture density network for collaborative touchscreen interaction. In: Liapis, A., Romero Cardalda, J.J., Ekárt, A. (eds.) EvoMUSART 2018. LNCS, vol. 10783, pp. 161–176. Springer, Cham (2018). https://doi.org/10.1007/978-3-319-77583-8_11

15. Michalak, J., Troje, N.F., Fischer, J., Vollmar, P., Heidenreich, T., Schulte, D.: Embodiment of sadness and depression-gait patterns associated with dysphoric mood. Psychosom. Med. **71**(5), 580–587 (2009)
16. Müller, M.: Information Retrieval for Music and Motion, vol. 2. Springer, Heidelberg (2007). https://doi.org/10.1007/978-3-540-74048-3
17. Pettee, M., Shimmin, C., Duhaime, D., Vidrin, I.: Beyond imitation: generative and variational choreography via machine learning. In: 10th International Conference on Computational Creativity (2019)
18. Pollick, F.E., Paterson, H.M., Bruderlin, A., Sanford, A.J.: Perceiving affect from arm movement. Cognition **82**(2), B51–B61 (2001)
19. Satchell, L., Morris, P., Mills, C., O'Reilly, L., Marshman, P., Akehurst, L.: Evidence of big five and aggressive personalities in gait biomechanics. J. Nonverbal Behav. **41**(1), 35–44 (2017)
20. Sturm, B., Santos, J.F., Korshunova, I.: Folk music style modelling by recurrent neural networks with long short term memory units. In: 16th International Society for Music Information Retrieval Conference (2015)
21. Tan, W.R., Chan, C.S., Aguirre, H.E., Tanaka, K.: ArtGAN: artwork synthesis with conditional categorical GANs. In: 2017 IEEE International Conference on Image Processing (ICIP), pp. 3760–3764. IEEE (2017)
22. Wallace, B., Martin, C.P., Torresen, J., Nymoen, K.: Towards movement generation with audio features. In: Proceedings of the 11th International Conference on Computational Creativity (2020)

"A Good Algorithm Does Not Steal – It Imitates": The Originality Report as a Means of Measuring When a Music Generation Algorithm Copies Too Much

Zongyu Yin[1(✉)], Federico Reuben[2], Susan Stepney[1], and Tom Collins[2]

[1] Department of Computer Science, University of York, York, UK
{zy728,susan.stepney}@york.ac.uk
[2] Music, Science and Technology Research Cluster, Department of Music, University of York, York, UK
{federico.reuben,tom.collins}@york.ac.uk
https://mstrcyork.org

Abstract. Research on automatic music generation lacks consideration of the originality of musical outputs, creating risks of plagiarism and/or copyright infringement. We present the originality report – a set of analyses for measuring the extent to which an algorithm copies from the input music on which it is trained. First, a baseline is constructed, determining the extent to which human composers borrow from themselves and each other in some existing music corpus. Second, we apply a similar analysis to musical outputs of runs of MAIA Markov and Music Transformer generation algorithms, and compare the results to the baseline. Third, we investigate how originality varies as a function of Transformer's training epoch. Results from the second analysis indicate that the originality of Transformer's output is below the 95%-confidence interval of the baseline. Musicological interpretation of the analyses shows that the Transformer model obtained via the conventional stopping criteria produces single-note repetition patterns, resulting in outputs of low quality and originality, while in later training epochs, the model tends to overfit, producing copies of excerpts of input pieces. We recommend the originality report as a new means of evaluating algorithm training processes and outputs in future, and question the reported success of language-based deep learning models for music generation. Supporting materials (code, dataset) will be made available via https://osf.io/96emr/.

Keywords: Music generation · Machine learning · Originality evaluation

1 Introduction

A quotation from Igor Stravinsky reads: "A good composer does not imitate, he steals" [39]. The quotation, while made in relation to a serial work, reflects

J. Romero et al. (Eds.): EvoMUSART 2021, LNCS 12693, pp. 360–375, 2021.
https://doi.org/10.1007/978-3-030-72914-1_24

Stravinsky's general interest in incorporating melodies, harmonic language, and forms from previous periods into new works such as his *Pulcinella Suite* (1922). Stravinsky uses the term "imitate" with a negative connotation: he would rather steal, say, a melody wholesale and rework it in a contemporary piece, than he would make mere allusions to (imitate) the work of past or contemporary composers. With respect to the current paper's context – the rise of AI music generation algorithms – we instead use the term "imitate" with a positive connotation and the term "steal" with a negative connotation. As we show, some deep learning algorithms for music generation [17] are copying chunks of original input material in their output, and we would count it as a success if an algorithm – from the deep learning literature or otherwise – could generate output that *sounded like* (imitated) – but did not *copy from* – pieces in a specific style.

Research on artificial intelligence (AI) has achieved various feats of simulating human perception (e.g., [15]) and production (e.g., [28]). A number of music generation models have been developed in recent decades, many predating or outside of deep learning [7,34] and some espousing a belief in the superiority of deep learning [12,29]. We have observed, with increasing alarm, that deep learning papers on music generation tend to rely solely or primarily on loss and accuracy as a means of evaluation [17,29]. If there are listening studies, they employ listeners with inadequate expertise, and there is little or no musicological analysis of outputs, and no analysis of whether generated material plagiarises (steals from) the training data. As an increasing number of musicians are now incorporating AI into their creative workflows, checking an AI's output for plagiarism is now a paramount challenge in this area. To this end, this paper considers the topic of **automatic stylistic composition** – a branch of automatic music generation where there is a stated stylistic aim with regards to the algorithm output, and a corpus of existing pieces in the target style.

In this context, we aim to establish a framework for checking the originality of auto-generated music with a specified style. We introduce and exemplify the originality report as a means of measuring when a music generation algorithm copies too much. We discuss how to calculate a distribution for the extent to which human composers borrow from themselves or each other in some corpus of pieces in a specific style; then we discuss how to use this as a baseline while moving a sliding window across a generated passage and measuring originality as a function of time in the generated material, complementing this with a musicological analysis of outputs from prominent deep [17] and non-deep [7] learning models. Finally, for the deep learning model, we interrogate how originality varies with the training epoch.

2 Related Work

2.1 Music Plagiarism

Music plagiarism is said to have occurred when there is demonstrable and perceivable similarity between two songs or pieces of music (hereafter, pieces), and when there is circumstantial evidence to indicate that the composer(s) of the latest piece would have been familiar with the existing piece. Stav [32] describes how the musical dimensions of melody, harmony, and rhythm contribute to music plagiarism, and gives an example-based explanation of how these dimensions have been used in handling music copyright disputes. Based on the features of melodies involved in selected plagiarism cases, Müllensiefen and Pendzich [22] derive an algorithm for predicting the associated court decision, and it identifies the correct outcome with 90% success rate. Recent failed or overturned cases also indicate that while music similarity and circumstantial evidence are necessary for delivering a verdict in favour of plagiarism having occurred, they are not sufficient, in that the distinctiveness of the music with respect to some larger corpus plays an important role too [5,9,24]: melodies that share contours and begin and end on the same scale steps may well point to potential cases of plagiarism, but it is likely that other melodies will have these same characteristics too [24]; drum beats, where the initial space of possibilities is smaller compared to pitched material, have been less successful as bases for music plagiarism convictions [26].

Recently, discussions on ethical issues surrounding AI have attracted widespread attention. Anon [8] uses a note-counting approach to show that twenty bars of computer-generated musical output from an algorithm by Cope [11] have 63% coincidence in pitch-rhythm combinations with a piece by Frédéric Chopin. In [33], a music generation algorithm's output and its tendency to copy original input pieces motivates the posing of open questions with respect to AI and music copyright law. As such generative models learn from existing music data, the copyright status of the output is unclear. Additionally, the evaluation of these models' outputs tends to be narrow – i.e. it does not involve any kind of **originality analysis** with respect to the human-composed pieces used for training – which creates copyright or plagiarism infringement risks for musicians who are using these algorithms as part of their creative workflows.

2.2 Cognitive-Computational Approaches to Music Similarity

Largely outside of the role played by similarity in determining cases of music plagiarism, the systematic study of music similarity has a relatively long lineage [31] and continues to be of interest to scholars [37]. One challenging aspect of studying the phenomenon is that two excerpts of music can be similar to one another in myriad ways (genre, instrumentation, timbre, tempo, dynamics, texture, form, lyrics, and mentioned above, melody, harmony, rhythm). This challenge interacts with variability in use cases too. Take a single paradigm such as query-based search in the form of music identification, which relies on some implementation of music similarity. Even for this one paradigm, there are various use cases: Shazam addresses the need for exact matching [38], a variant of SoundHound addresses query-by-humming (the user sings or hums at the interface and expects "successful" results),[1] and Folk Tune Finder allows lyrics or notes to be input and, as with SoundHound's query-by-humming variant, the user's expectation of Folk Tune Finder is that the sought-after song will be

[1] https://www.midomi.com/.

found, or at least something relevant or interesting will be returned.[2] Of these use cases, only the one addressed by Shazam is clear cut – the other two are made more challenging by variation in cognitive and music-production capabilities of users, and there not necessarily being one "right answer".

Here, we are concerned with a more reductive view of music similarity – the type of note-counting approaches mentioned above. This is the characterisation of music similarity that a teacher might employ if a student's composition appears to draw too heavily on or copy directly from a known piece. For instance, "Why do 90% of the pitch-rhythm combinations in bars 1–20 of your piece occur also in this string quartet movement by Haydn?!" The representations and calculations required to reason this way, especially in algorithmic fashion, began in [19] and have been implemented in various forms since [1,5,35]. In the note-counting vein, in Sect. 3.1 we define a similarity measure based on the P3 algorithm [35]. We finish this section of the review with some remarks about choices of music representation and comparison methods.

Generally, researchers take **sequential** (e.g., [9,10]) or **geometric** (e.g., [5,21]) approaches to the representation and comparison of music. There are pros and cons to each approach. With the sequential approach, if one chooses to focus on MIDI note number alone and two melodies have the same MIDI notes (up to transposition) but different rhythms, a sequential representation (specifically, difference calculations between consecutive notes) will recognise these melodies as similar, whereas a geometric representation may not. However, with a sequential representation, it is less obvious how to handle polyphony (multiple notes beginning and ending at possibly different times), whereas a geometric representation can encode a polyphonic piece as easily as it encodes a monophonic piece. For instance, in the sequential representation shown in Fig. 1(d) (which is Music Transformer's [17] chosen input representation, see next section), the tokens encoding the occurrence of the F♯4 and second F♯5 are ten indices apart, even though the notes sound together. So any parameter that allows these events to be recognised as related has to be large enough to span this gap in indices. Moreover, an embellished (or, on the other hand, reduced) variation of some melody may not be recognised by the sequential representation as similar because the relationships between adjacent notes will be altered by the added or removed notes, even though the "melodic scaffold" remains intact. A geometric representation may be more robust to this kind of variation.

2.3 Music Generation Models

Recently, a large number of deep learning models have been proposed for symbolic music generation [13,17,29]. Several of them regard music as a sequence of tokens, where generation involves predicting the next token based on previous tokens [17,29]. Oore et al. [25] introduce a way to serialise polyphonic music and apply recurrent neural networks (RNNs) to generate output with expressive timing and velocity (loudness) levels. Huang et al. [17] use this same

[2] https://www.folktunefinder.com/.

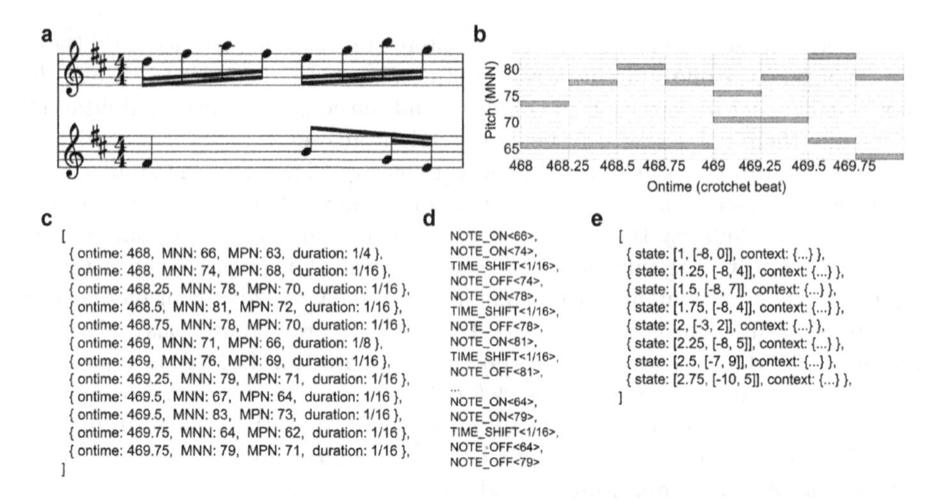

Fig. 1. Examples of symbolic music representations, starting from the same excerpt. (a) half a bar of music; (b) the so-called piano-roll representation indicating some of the music's numeric properties; (c) a 4-dimensional representation of the music as a set of points; (d) one sequential representation that handles polyphony; (e) another sequential representation that handles polyphony.

serialisation to adapt a transformer model [36] to generating music. Benefiting from the self-attention mechanism, it achieved lower validation loss compared to the RNN of [25] and also longer-term stylistic consistency than previous RNNs-based approaches. In other work, based on the assumption that each musical output can be sampled from a normal distribution, [29] use variational autoencoders (VAEs) combined with long short-term memory networks (LSTMs). The application of generative adversarial networks (GANs) and convolutional neural networks (CNNs) to music generation has been explored also [13], using the piano-roll representation as in Fig. 1(b) and treating music as images that can be generated in a hierarchical manner.

An issue with all the above deep learning approaches to music generation is that there has been inadequate consideration of music plagiarism in the algorithms' outputs. One user of the Music Transformer algorithm, Ruiz, writes:

> The thing is that I ran the code on my machine and it overfits. It needs a way to check that it isn't stealing from the dataset say no more than 6 or 8 continuous notes. If it can't do that it's useless. I mean your piano dataset is huge but after running the program for 20 times I found it composes note by note music of well known classical melodies. That's not OK. That should be avoided [30].

Simon, a member of the Google Magenta team, replies:

> In the checkpoints we've released, we tried hard to reduce the ability of the model to perform pieces from the training set. And in the samples

we released, we tried hard to remove any samples that are too similar to an existing piece of music. But it's difficult to get to 100% on these for a number of reasons, including the lack of a clear definition for "too similar" [30].

Members of the general public can make use of Music Transformer for laudable reasons – Google Magenta have open-sourced the code – but their attempt to guard against music plagiarism appears problematic, and whatever constitutes "trying hard" in the above quotation has not been open-sourced, leaving general musicians who use Magenta algorithms in their creative workflows at risk of copyright infringement.

A non-deep learning approach to music generation that uses Markov models, pattern discovery, and pattern inheritance to ensure that generated material evidences long-term, hierarchical repetitive structure, also constitutes the first use of an **originality or creativity analysis** to assess the extent to which the model plagiarises human-composed works by J.S. Bach and Chopin on which it is based [7]. This algorithm, called MAIA Markov, uses the representation given in Fig. 1(e), where each state consists of a beat of the bar and the MIDI note numbers relative to tonal centre occurring on that beat.

The remainder of this paper studies two of the most promising models for music generation, Music Transformer [17] and MAIA Markov [7], and focuses on the concept of originality, and methodologies for measuring it, which are then implemented and discussed.

3 Method

This section introduces the method we use to analyse the originality of one set of symbolically encoded music excerpts relative to another. We begin by defining the two sets of excerpts: the queries (the excerpts we are testing for originality), $\mathcal{Q}$, and the targets (the excerpts we are testing against), $\mathcal{R}$. Depending on the use case, $\mathcal{Q}$ may contain one or more excerpts from one or more pieces of music, and $\mathcal{R}$ usually contains overlapping excerpts from multiple pieces of music. The use cases for wanting to produce an **originality report**, which we explore further and exemplify, are as follows:

1. The user wants to determine the "baseline" level of originality within a corpus $\mathcal{C}$. In this instance, the queries $\mathcal{Q}$ are a (pseudo-)random sample from $\mathcal{C}$, drawn from pieces $\mathcal{Q}^*$, and the targets $\mathcal{R}$ are the set complement, $\mathcal{R} = \mathcal{C} \setminus \mathcal{Q}^*$.[3] The outcome is a sample of N originality scores, from which estimates of the underlying distribution can be made, such as mean originality and confidence intervals about this mean.

[3] We distinguish between $\mathcal{Q}$ and $\mathcal{Q}^*$ because if some excerpt $q \in \mathcal{Q}$ repeats or substantially recurs elsewhere in the piece q^* from which it is drawn, and we leave this repetition in the set of targets $\mathcal{R}$, then q would be considered trivially unoriginal. Therefore, it is sensible to hold out entire pieces from which queries are selected.

2. The user wants to determine the level of originality of an algorithm's output relative to a corpus whose contents the algorithm is designed to imitate. In this instance, the queries $\mathcal{Q}$ would be overlapping excerpts of the algorithm's output, and we plot the originality of elements of $\mathcal{Q}$ as a function of time in the output, relative to elements of the target corpus $\mathcal{R}$. As well as plotting, we could also compare the mean or minimum originality found for the algorithm output to the distribution mentioned in the previous point, in which case the distribution acts as a "baseline" and can be used to address the question, "Is this algorithm's output sufficiently original?".
3. The user wants to incorporate originality reports into the modelling process itself, in order to analyse or steer/halt that process. The details are similar to the previous point, but the deployment of the method is during training or generation rather than after the fact.

3.1 Originality, Similarity, and Set of Points

To implement the originality reports that are associated with each of the use cases, it is necessary to employ at least one similarity measure – that is, some function $c : \mathcal{Q} \times \mathcal{R} \to [0, 1]$, which takes two symbolically encoded music excerpts q and r and returns a value in the range $[0, 1]$, indicating q and r are relatively similar (value near one) or dissimilar (value near zero). The measure ought to be commutative, $c(q, r) = c(r, q)$, and have an symmetry-like property that $c(q, q) = 1$. The choice of similarity metric influences subsequent decisions with respect to addressing questions such as "Is this algorithm's output sufficiently original?", but the contributions of this paper are the delineation of the use cases and the originality reports themselves, rather than the definition or use of any one similarity metric in particular.

Each originality report centres on calculating an **originality score**, OS, for some query q in relation to the set of targets $\mathcal{R}$. In particular, we find the element $r \in \mathcal{R}$ that maximises the similarity score $c(q, r)$, and subtract it from 1:

$$OS(q, \mathcal{R}) = 1 - \max\{c(q, r) \mid r \in \mathcal{R}\} \tag{1}$$

So the originality score $OS : \mathcal{Q} \times \mathbb{P}\mathcal{R} \to [0, 1]$, where $\mathbb{P}\mathcal{R}$ is the power set of $\mathcal{R}$, is also a measure in the range $[0, 1]$ indicating q is original relative to $\mathcal{R}$ (value near one) or unoriginal (value near zero). If q is a copy of something that occurs in $\mathcal{R}$, then the originality score will give $OS(q, \mathcal{R}) = 0$.

In this paper, we use a similarity measure called the cardinality score, cs [4,6, 35]. To calculate it, we represent each music excerpt as a set of points containing the start time (in crotchet beats) and numeric pitch representation (morphetic pitch number [20]) of each note.[4] So an element of the query set $\mathcal{Q} : \mathbb{P}(N \times N)$ is represented as $q = \{(x_1, y_1), (x_2, y_2), \ldots, (x_n, y_n)\}$, and an element of the target set $\mathcal{R} : \mathbb{P}(N \times N)$ is represented as $r = \{(x'_1, y'_1), (x'_2, y'_2), \ldots, (x'_{n'}, y'_{n'})\}$. An example of this representation is provided in Fig. 2(c) and (d). The bottom-left

[4] We use morphetic pitch in preference to MIDI note number here because the former is robust to major/minor alterations.

point in (c) has the value ($x_1 = 468, y_1 = 53$), representing a start time at the beginning of the excerpt ($x_1 = 468$) and the morphetic pitch for C3 ($y_1 = 53$). The viola and cello have coincident notes at this moment, which project to a single point in our representation.

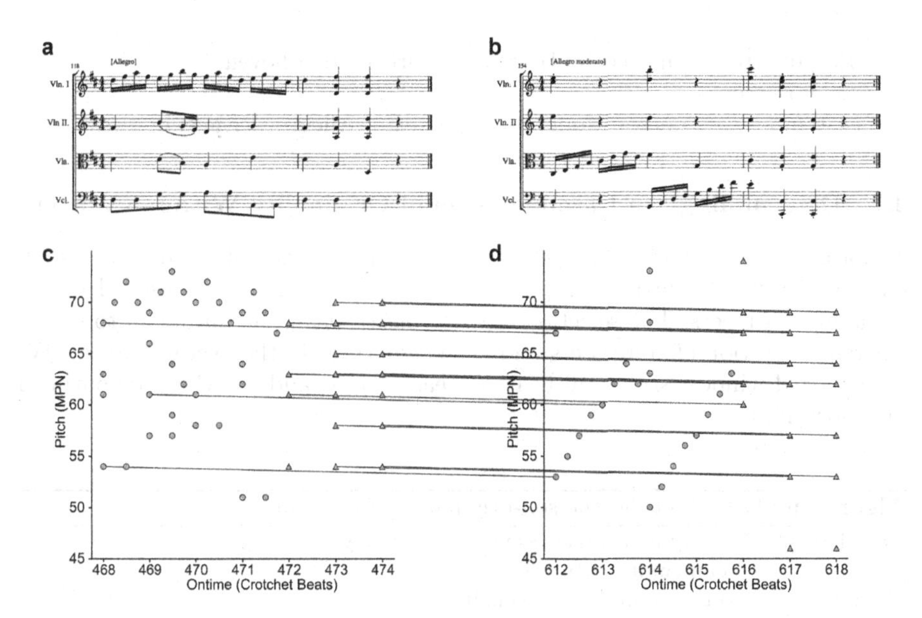

Fig. 2. Visualisation of the cardinality score between two excerpts. (a) a 2-bar excerpt from Mozart; (b) a 2-bar excerpt from Haydn; (c) mapping notes in the excerpt (a) to a set of points; (d) mapping notes in the excerpt (b) to a set of points. For clarity, notes in the first/second bars are shown as circles/triangles.

Letting $\boldsymbol{t}$ be the translation vector that gives rise to the maximum cardinality of the intersection $(q + \boldsymbol{t}) \cap r$, we define the **cardinality score** as

$$cs(q, r) = |(q + \boldsymbol{t}) \cap r| / \max\{|q|, |r|\} \tag{2}$$

where $|q|$ is the size of the set of points q. We demonstrate calculations of the cardinality score with reference to the examples in Fig. 2. In the top half of this figure, there are two excerpts of string quartets: (a) is by Mozart and (b) is by Haydn. Considering the set of points corresponding to bars 119 of the Mozart and 155 of the Haydn (second bars in both Figs. 2(a) and (b), with corresponding points shown as triangles), the vector $\boldsymbol{t} = (-144, -2)$ translates 15 points from the Mozart excerpt to points in the Haydn, and the more numerous of the two sets is the Haydn excerpt, with 19 points, so the cardinality score is $cs(q, r) = 15/19 \approx 0.7895$. As a second example, considering larger point sets corresponding to bars 118–119 of the Mozart and 154–155 of the Haydn, the vector $\boldsymbol{t} = (-144, -2)$ translates 18 points from the Mozart excerpt to points in the Haydn, and the more numerous of the two sets is the Mozart excerpt, with 52 points, so the cardinality score is $cs(Q, R) = 18/52 \approx 0.3462$.

4 Originality Reports

The dataset we use in this paper contains 71 Classical string quartets in MIDI format from KernScores.[5] The dataset was formed according to the following filters and constraints:

- string quartet composed by Haydn, Mozart, or Beethoven;
- first movement;
- fast tempo, e.g., one of Moderato, Allegretto, Allegro, Vivace, or Presto.

4.1 Determining the Baseline Level of Originality Within a Corpus

To form the query and target sets, we divide the 71 excerpts into two sets: 50 queries $\mathcal{Q}$ were drawn from 7 pieces $\mathcal{Q}^*$, and the targets $\mathcal{R}$ consisted of the remaining 64 pieces. The selection of the 7 pieces was pseudo-random to reflect the representation of composers and time signatures in the overall dataset. We used a fixed window size of 16 beats for each query, and ran the code outlined in Algorithm 1.

Algorithm 1. Estimating the self-originality of a corpus

Require: $\mathcal{Q}$, $\mathcal{R}$, query and target corpus, respectively.
1: Initialize O to an empty output list.
2: Initialize N to the number of originality scores.
3: **for** $(i := 0, i < N, i++)$ **do**
4: $q := \text{sample}(\mathcal{Q})$
5: Initialise C as an empty list to store cardinality scores.
6: **for each** $q \in \mathcal{Q}$ **do**
7: $C.append(cs(q, \mathcal{R}))$
8: **end for**
9: $O.append([1 - \max(C)])$
10: **end for**

For our sample of $N := 50$ excerpts from Haydn, Mozart, and Beethoven string quartets, the mean originality was $\text{mean}(OS) = 0.699$, with bootstrap 95%-confidence interval 0.672 and 0.725. We interpret this to mean that for the current corpus and sample (space of fast, first-movement Classical string quartets), composers wrote music that was 69.9% original, at least according to the note-counting music similarity measure employed here.

4.2 Is This Algorithm's Output Sufficiently Original?

We can use the mean and confidence interval calculated above to help address the question of whether an algorithm's output is sufficiently original. Let us

[5] See https://osf.io/96emr/ for the dataset, algorithms, and analyses.

suppose we have a passage generated by an algorithm, and we traverse that output, collecting n-beat excerpts with 50% overlap, say, into a query set $\mathcal{Q}$. In this paper, we use $n := 8, 16$ beats, which corresponds to 2- and 4-bar excerpts in 4-4 time, respectively. It is advisable to use at least two different window sizes, to probe the assumption that originality should increase with window size. In other words, different window sizes can be used to determine whether a worrisome-looking instance of low originality at the 2-bar level increases – and so becomes less worrisome – at a longer 4-bar window size.

We ran the MAIA Markov [7] and Music Transformer algorithms [17] to explore this question of sufficient originality, based on the training data of 64 string quartet movements described above. Markov model was built on the representation shown in Fig. 1(e). The Music Transformer model's training data was augmented by transposing the original pieces in the range $[-5, 6]$ MIDI notes, and then we sliced these into subsequences of fixed size 2,048 for batch training, giving a training set of 4,128 and a validation dataset of 564 subsequences. The model, with six layers, eight heads, and hidden size of 512, was trained with smoothed cross entropy loss [23] and the Adam optimiser [18] with custom learning rate schedule [2]. In keeping with the standard approach, the training process was stopped at epoch (checkpoint) 3, where the validation loss reached a minimum value of 1.183. Afterwards, 30 excerpts were generated by MAIA Markov and Music Transformer to form the query set $\mathcal{Q}$, based on which the mean value of originality scores were obtained by following the same method in Sect. 3.1, now for each time window as the excerpt is traversed.

For both algorithms, we see in Figs. 3(a) and (b) that the originality at the 2-bar level is low relative to the mean and 95%-confidence interval for the baseline, but this is to be expected because the baseline was calculated at the 4-bar level. What we expect to see is the solid line – indicating algorithm originality at the 4-bar level – lie entirely inside that confidence interval. This is the case for MAIA Markov [7], but Music Transformer's [17] mean originality level is entirely below this confidence interval, indicating it has issues with borrowing too heavily from the input on which it is trained.

Three typical worst case examples of copying are shown in Figs. 3(c), (d), and (e), with generated outputs on the left and original excerpts on the right. Figure 3(c) shows one from MAIA Markov having 57.1% originality associated with Beethoven's String quartet no.6 in B-flat major, op.18, mvt.1, bars 61–64. And then, Fig. 3(d) shows one generated by the "best" checkpoint (checkpoint 3 with the minimum validation loss) of Music Transformer having 51.2% originality associated with Mozart's String quartet no.13 in D minor, K.173, mvt.1, bars 125–128. We found most of the generated outputs in this stage with less than 50% originality are due to repeating the same note, which is also frequently found in Classical string quartets, and the model tends to start by reproducing this simple kind of pattern. Finally, Fig. 3(e) shows one generated by checkpoint 15 of Music Transformer having 90.1% originality associated with Beethoven's String quartet no. 1 in F major, op.18, mvt.1, bars 233–234. Generally, the model appears to be over-fit at this checkpoint. We infer from these original-

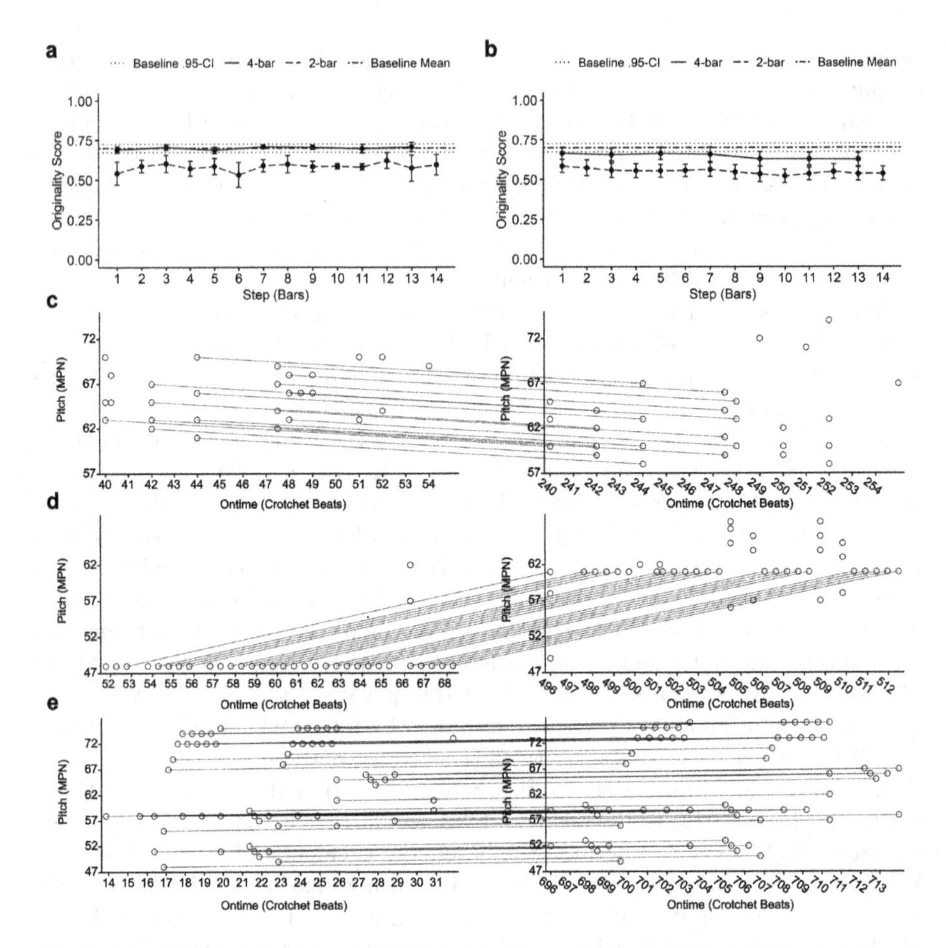

Fig. 3. Originality report for the MAIA Markov and Music Transformer algorithms. (a) and (b) show the change in originality scores over the course of the excerpts obtained for MAIA Markov and Music Transformer, respectively, at 2- and 4-bar levels compared to the baseline mean and 95%-confidence interval; (c), (d) and (e) show worst-case examples of copying by MAIA Markov and Music Transformer at checkpoints 3 and 15, respectively, where the generated outputs are on the left and the human-composed excerpts are on the right.

ity reports and basic musicological interpretations that the results generated by Music Transformer gradually morph during training from reproduction of simple patterns (e.g., repeated notes) to verbatim use of more distinctive note sequences.

4.3 Incorporating Originality Reports into an Algorithmic Process: Originality Decreases as Epoch Increases

Here, we demonstrate the use of an originality report in the modelling process itself, as a means of analysing changes in originality as a function of model training or validation epoch. Music Transformer was used as an example of a deep learning model, with the train/validation split as in Sect. 4.1. To monitor the originality change along with the training process, 10 checkpoints including the initial point were saved. Again, we used each of them to generate 30 excerpts, to which the aforementioned originality report was applied. Afterwards, we calculated the mean value of those 30 originality scores for each checkpoint.

Figure 4(a) and (b) show the change of loss and accuracy respectively over training. As mentioned previously, the standard training process would stop at epoch (checkpoint) 3, where the validation loss reaches a minimum, but we extended the training process further to more fully investigate the effect of training on originality. Figure 4(c) and (d) contain a dashed line indicating the baseline originality level of 0.699 for the string quartet dataset. In Fig. 4(c), mean originality score decreases as a function of model training epoch, but remains largely in the 95%-confidence interval of the baseline originality level of the corpus. Figure 4(d) is more concerning, indicating that minimum originality score decreases to well below the 95%-confidence interval of the baseline originality level of the corpus. Originality decreases until epoch 3, and then it stays relatively flat afterwards. However, as with the discussion of Figs. 3(c) and (d) in the previous subsection, we found that the model's borrowing still becomes more verbatim (or distinctive) after epoch 3, thus originality in a more general sense is still decreasing, a fact that is not immediately evident from Fig. 3 because the cardinality score does not consider distinctiveness, discussed further below.

5 Discussion

This paper puts forward the notion that AI for music generation should result in outputs that imitate instead of merely copying original pieces, and highlights that checks of whether this is the case – what we refer to as the originality report – are often omitted. We introduce the methodology of the originality report for baselining and evaluating the extent to which a generative model copies from training data. By substituting in different similarity metrics, it would be possible to adapt the methodology to have emphases on different musical dimensions, but here we take a relatively straightforward note-counting approach based on the cardinality score [6,35]. We analyse outputs from two example models, one deep learning algorithm called Music Transformer [17] and one non-deep learning model called MAIA Markov [7], to illustrate the use of this methodology and the existence of music plagiarism in recent research.

We recognise Google Magenta for making their source code (e.g., for Music Transformer) publicly available, because it enables a level of scrutiny that has not always been possible for previous work in this field [27]. That said, the results indicate a phenomenon wherein this type of deep learning language model

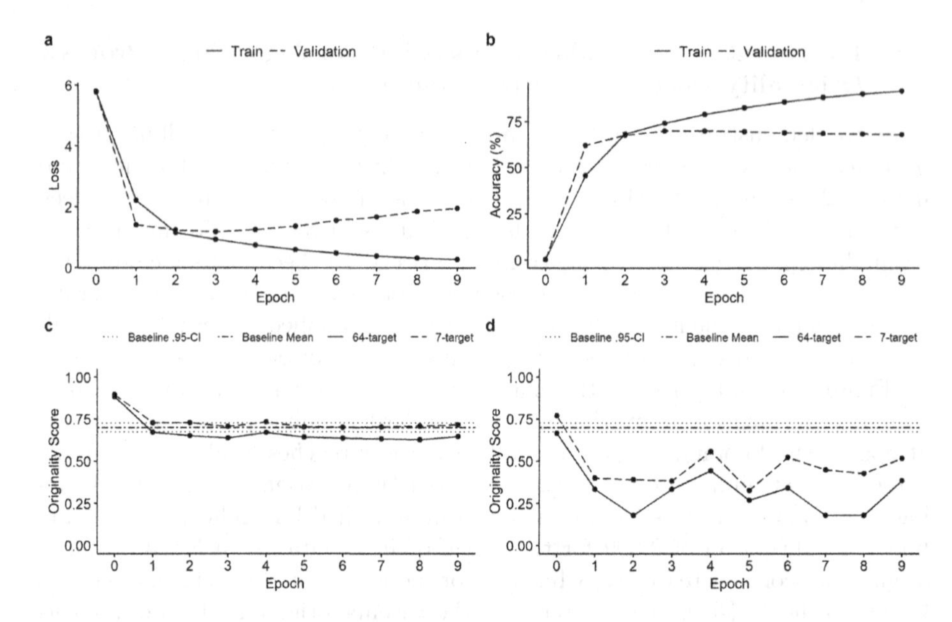

Fig. 4. (a) the loss curve of train and validation; (b) the accuracy curve of train and validation; (c) the mean originality curve for 64-target and 7-target sets; (d) the minimum originality score curve for 64-target and 7-target sets.

gradually copies increasingly distinctive chunks from pieces in the training set, calling into question whether it really *learns* to generate. More recent research found the information in training data can be retrieved from large language models, which highlights various issues of memorization [3]. Furthermore, using the conventional stopping criteria for the training process, the "best" model not only has a low level of originality, but also the quality of generated excerpts is low in the sense that the same note is repeated most of time (see Fig. 3(d)). Going forward, the field of deep learning needs to reconsider in what situations the conventional stopping criteria are appropriate: perhaps loss and accuracy should no longer be the only criteria when evaluating the model, because we need to prevent these models copying training data, especially when they are used increasingly in a "black-box" manner by practising musicians.

5.1 Limitations

The size of the dataset used above is smaller than that used for the original work on Music Transformer [17], and it is also quantised to a smaller set of time values. A potential solution is to pretrain the model with a large dataset to gain "general musical knowledge" and then finetune with a smaller style-specific dataset [10,12]. However, our dataset still represents a substantial amount of Classical music, and certainly enough to give a human music student an idea of the intended style, so if deep learning algorithms cannot operate on datasets of this size, then it should be considered a weakness of the deep learning approach rather than a limitation of our methodology.

The simplicity of the cardinality score has some appeal, but as noted in the review of existing work, it can mean that subtle variations along some musical dimension destroy any translational equivalence, giving a low cardinality score that is at odds with high perceived similarity. For instance, the expressive timing in the MAESTRO dataset [16] constitutes such subtle variations along the dimension of ontime, and make it ill-advised to use the cardinality score to assess the originality of an algorithm trained on these data. In addition, the cardinality score shares some general advantages of the geometric approach. But in its current use, it is also not able to take into account the distinctiveness of excerpts being compared [5,9]. For instance, Fig. 2 indicated an instance of similarity between Mozart and Haydn, but when we take into account how many Classical pieces end in this way, it is not a particularly distinctive or interesting example.

5.2 Future Work

We would like to see the originality report method that we have developed be embedded into the training processes of various music generation algorithms, to play a role as an advanced stopping criterion. Meanwhile, we will need to ensure that this criterion can still maintain the generalisability asserted by standard stopping criteria. Additionally, we will investigate the compatibility of the originality report method with model selection, which is often conducted as an outer loop of model training. Loss function engineering is a topic addressed in recent novel generating strategies (e.g., [14]), so it should be possible to merge high/low originality scores as rewards/penalties in training loss, to further investigate the problem that we have identified of language-based deep learning models appearing to be little more than powerful memorisers.

We will also explore the weighting of shift errors [6], a fingerprinting approach [1], and distinctiveness [9] to address the limitations mentioned above, arising from the simplicity of the cardinality score. This should mean originality reports can be generated for an algorithm trained on *any* music data, including those with expressive timing information, and taking into account distinctiveness with respect to an underlying corpus.

References

1. Arzt, A., Böck, S., Widmer, G.: Fast identification of piece and score position via symbolic fingerprinting. In: ISMIR, pp. 433–438 (2012)
2. Bengio, Y.: Practical recommendations for gradient-based training of deep architectures. In: Montavon, G., Orr, G.B., Müller, K.-R. (eds.) Neural Networks: Tricks of the Trade. LNCS, vol. 7700, pp. 437–478. Springer, Heidelberg (2012). https://doi.org/10.1007/978-3-642-35289-8_26
3. Carlini, N., et al.: Extracting training data from large language models. arXiv preprint arXiv:2012.07805 (2020)

4. Collins, T.: Discovery of repeated themes and sections. Accessed 4 May 2013 (2013). https://www.music-ir.org/mirex/wiki/2013:Discovery_of_Repeated_Themes_&_Sections
5. Collins, T., Arzt, A., Frostel, H., Widmer, G.: Using geometric symbolic fingerprinting to discover distinctive patterns in polyphonic music corpora. Computational Music Analysis, pp. 445–474. Springer, Cham (2016). https://doi.org/10.1007/978-3-319-25931-4_17
6. Collins, T., Böck, S., Krebs, F., Widmer, G.: Bridging the audio-symbolic gap: the discovery of repeated note content directly from polyphonic music audio. In: 53rd International Conference: Semantic Audio. Audio Engineering Society (2014)
7. Collins, T., Laney, R.: Computer-generated stylistic compositions with long-term repetitive and phrasal structure. J. Creat. Music Syst. **1**(2) (2017). https://doi.org/10.5920/JCMS.2017.02
8. Collins, T., Laney, R., Willis, A., Garthwaite, P.H.: Developing and evaluating computational models of musical style. AI EDAM **30**(1), 16–43 (2016)
9. Conklin, D.: Discovery of distinctive patterns in music. Intell. Data Anal. **14**(5), 547–554 (2010)
10. Conklin, D., Witten, I.H.: Multiple viewpoint systems for music prediction. J. New Music Res. **24**(1), 51–73 (1995)
11. Cope, D.: Computer Models of Musical Creativity. MIT Press, Cambridge (2005)
12. Donahue, C., Mao, H.H., Li, Y.E., Cottrell, G.W., McAuley, J.: LakhNES: improving multi-instrumental music generation with cross-domain pre-training. In: ISMIR, pp. 685–692 (2019)
13. Dong, H.W., Hsiao, W.Y., Yang, L.C., Yang, Y.H.: MuseGAN: multi-track sequential generative adversarial networks for symbolic music generation and accompaniment. In: Thirty-Second AAAI Conference on Artificial Intelligence (2018)
14. Elgammal, A., Liu, B., Elhoseiny, M., Mazzone, M.: CAN: creative adversarial networks generating "art" by learning about styles and deviating from style norms. In: 8th International Conference on Computational Creativity, ICCC 2017. Georgia Institute of Technology (2017)
15. Graves, A., Mohamed, A., Hinton, G.: Speech recognition with deep recurrent neural networks. In: 2013 IEEE International Conference on Acoustics, Speech and Signal Processing, pp. 6645–6649. IEEE (2013)
16. Hawthorne, C., et al.: Enabling factorized piano music modeling and generation with the MAESTRO dataset. In: International Conference on Learning Representations (2019)
17. Huang, C.Z.A., et al.: Music transformer: generating music with long-term structure. In: ICLR (2018)
18. Kingma, D.P., Ba, J.: Adam: a method for stochastic optimization. arXiv preprint arXiv:1412.6980 (2014)
19. Lewin, D.: Generalized musical intervals and transformations. Oxford University Press, New York (2007). Originally Published by Yale University Press, New Haven (1987)
20. Meredith, D.: The ps13 pitch spelling algorithm. J. New Music Res. **35**(2), 121–159 (2006)
21. Meredith, D., Lemström, K., Wiggins, G.A.: Algorithms for discovering repeated patterns in multidimensional representations of polyphonic music. J. New Music Res. **31**(4), 321–345 (2002)
22. Müllensiefen, D., Pendzich, M.: Court decisions on music plagiarism and the predictive value of similarity algorithms. Musicae Scientiae **13**(1_suppl), 257–295 (2009)

23. Müller, R., Kornblith, S., Hinton, G.E.: When does label smoothing help? In: Advances in Neural Information Processing Systems, pp. 4694–4703 (2019)
24. Neely, A.: Why the Katy Perry/Flame lawsuit makes no sense. https://www.youtube.com/watch?v=0ytoUuO-qvg. Accessed 30 Oct 2020
25. Oore, S., Simon, I., Dieleman, S., Eck, D., Simonyan, K.: This time with feeling: Learning expressive musical performance. Neural Comput. Appl. **32**, 1–13 (2018)
26. Otzen, E.: Six seconds that shaped 1,500 songs. https://www.bbc.co.uk/news/magazine-32087287. Accessed 30 Oct 2020
27. Pachet, F., Papadopoulos, A., Roy, P.: Sampling variations of sequences for structured music generation. In: ISMIR, pp. 167–173 (2017)
28. Radford, A., Metz, L., Chintala, S.: Unsupervised representation learning with deep convolutional generative adversarial networks. arXiv preprint arXiv:1511.06434 (2015)
29. Roberts, A., Engel, J., Raffel, C., Hawthorne, C., Eck, D.: A hierarchical latent vector model for learning long-term structure in music. In: International Conference on Machine Learning, pp. 4364–4373. PMLR (2018)
30. Ruiz, A., Simon, I.: My only problem with Magenta's Transformer. Magenta Discuss Google Group. https://groups.google.com/a/tensorflow.org/g/magenta-discuss/c/Oxiq-Gdaavk/m/uHIsQZKtBwAJ. Accessed 30 Oct 2020
31. Selfridge-Field, E.: Conceptual and representational issues in melodic comparison. In: Computing in Musicology (1999)
32. Stav, I.: Musical plagiarism: a true challenge for the copyright law. DePaul J. Art Tech. Intellect. Prop. Law **25**, 1 (2014)
33. Sturm, B.L., Iglesias, M., Ben-Tal, O., Miron, M., Gómez, E.: Artificial intelligence and music: open questions of copyright law and engineering praxis. In: Arts, vol. 8, p. 115. Multidisciplinary Digital Publishing Institute (2019)
34. Todd, P.M.: A connectionist approach to algorithmic composition. Comput. Music. J. **13**(4), 27–43 (1989)
35. Ukkonen, E., Lemström, K., Mäkinen, V.: Geometric algorithms for transposition invariant content-based music retrieval. In: Proceedings of the International Symposium on Music Information Retrieval, pp. 193–199 (2003)
36. Vaswani, A., et al.: Attention is all you need. In: Advances in Neural Information Processing Systems, pp. 5998–6008 (2017)
37. Volk, A., Chew, E., Margulis, E.H., Anagnostopoulou, C.: Music similarity: concepts, cognition and computation. J. New Music Res. **45**(3), 207–209 (2016)
38. Wang, A.L.C., Smith III, J.O.: System and methods for recognizing sound and music signals in high noise and distortion (2012). Patent US 8,190,435 B2. Continuation of provisional application from 2000
39. Yates, P.: Twentieth Century Music: Its Evolution from the End of the Harmonic Era into the Present Era of Sound. Allen & Unwin, Crows Nest (1968)

Short Talks

From Music to Image a Computational Creativity Approach

Luís Aleixo[1](✉), H. Sofia Pinto[1], and Nuno Correia[2]

[1] INESC-ID, Instituto Superior Técnico, Lisbon, Portugal
luis.aleixo@tecnico.ulisboa.pt, sofia@inesc-id.pt
[2] NOVA LINCS, Costa da Caparica, Portugal
nmc@fct.unl.pt

Abstract. In this paper we propose a possible approach for a cross-domain association between the musical and visual domains. We present a system that generates abstract images having as inspiration music files as the basis for the creative process. The system extracts available features from a MIDI music file given as input, associating them to visual characteristics, thus generating three different outputs. First, the Random and Associated Images - that result from the application of our approach considering different shape's distribution - and second, the Genetic Image, that is the result of the application of one Genetic Algorithm that considers music and color theory while searching for better results. The results of our evaluation conducted through online surveys demonstrate that our system is capable of generating abstract images from music, since a majority of users consider the images to be abstract, and that they have a relation with the music that served as the basis for the association process. Moreover, the majority of the participants ranked highest the Genetic Image.

Keywords: Computational creativity · Music analysis · Image generation · Cross-domain associations · Genetic algorithm

1 Introduction and Motivation

Computational Creativity (CC) emerged as a sub-field of Artificial Intelligence (AI), exploring the machine's ability to generate human-level creative artifacts, hence it focuses on the development of software that exhibits behavior that can be considered creative by humans.

Having as motivation processing music and image, our main contribution is the development of a system that exhibits creative behavior by generating abstract images inspired in musical artifacts. We define abstract images as visual artifacts that do not reflect or convey anything "concrete" or "real". Throughout the development of our system, we went through several phases, from research on music theory, to the study of color harmonies, shape assembling, and image generation techniques. In the end, to improve the aesthetic value of the generated images, as well as to search for better results, we went through research on Genetic Algorithms, implementing one from scratch.

J. Romero et al. (Eds.): EvoMUSART 2021, LNCS 12693, pp. 379–395, 2021.
https://doi.org/10.1007/978-3-030-72914-1_25

Regarding the final results, the system outputs three different images in response to each musical input. The first, called the **Random** Image, was generated assigning a random shape and texture to each instrument found on the music file. The second, called the **Associated** Image, was generated assigning a predefined association between musical instruments and respective shapes. Regarding the third version, called the **Genetic** Image, it results from the Genetic Algorithm that receives the two previous versions to generate the initial population.

We decided to assess four pairs of music and respective images through online surveys. The evaluation had good results since the majority of the participants consider the images to be abstract. Besides, regardless of the version, the participants believe that there is a relationship between the images and the music. Finally, it is also worth mentioning that all generated images were generally liked. Having these interesting results, we believe that our system achieved its goals.

This paper is organized as follows. In Sect. 2, we provide background knowledge related to the state of the art. In Sect. 3, we present our basic background on the musical domain, and in Sect. 4, we explore the visual domain by presenting an analysis of different elements of art. In Sect. 5.1, we explain our approach and in Sect. 5.2 we describe how we implemented our system. In Sect. 6, we describe how we evaluated our system, presenting its results. Finally, in Sect. 7, we present the conclusions of our work, as well as possible future work.

2 Related Work

Diverse research work has been pursued in the Computational Creativity field. NEvAR is an Evolutionary Art Tool proposed by Machado and Cardoso [8] that generates artworks through Evolutionary Computation. In this system, individuals are represented by trees, in which the genotype is a symbolic expression constructed from a lexicon of functions and terminals. The function set is mainly composed of simple functions such as arithmetic, trigonometric and logic operations. Terminals are variables and constants which can be scalar values or 3d-vectors. The interpretation of the genotype results on a phenotype, an image in this case. Through one Evolutionary process, the system is guided by a user to produce images compatible with the aesthetic and/or artistic principles of the user [8].

However, since our goal is to generate abstract images inspired by music, and not symbolic expressions based on mathematical functions, we focus our analysis on inspirational systems that establish possible analogies between the musical and visual domains, in which the characteristics of one domain are associated to characteristics of another domain.

The Visual Information Vases (VIV) is an AI-based generative art system proposed by Horn et al. [7] that focuses on the evocation of inspiration from a source domain to create an artifact in a different domain through cross-domain analogy mappings. The system uses as model of inspiration to produce 3D-printable vases 2D images uploaded by a user. It attempts to create a vase with similar aesthetic measures to those of the inspiring image through evolution. Results are diverse and functional creations.

Teixeira and Pinto [12] described a system that proposes an association between the visual and musical domains by generating music from images. This inspirational system generates musical artifacts given an image as input, by identifying a set of features to be extracted from an image. Therefore, by considering visual features, the system interprets and uses them as a starting point to translate into several components of the musical domain. The system outputs three different versions, using one possible approach between the two domains, and Genetic Algorithms to generate music with improved aesthetic value.

Moura [11] started in 2001 the first version of a system composed of little car robots that have the ability of line drawing with colors (red, green, blue) in a white canvas, leaving an ink trail as they go. For sixteen years, the system has been improved to perceive movement and nearby activity through sensors that influence the robot's path and activity. Negative feedback determines the finalization of the robots' activity - robots stop reacting because a specific density of color is achieved [11]. As with any artist, these robots are stimulated by everything that surrounds them, either the environment, sound, or interactivity with people.

There are other systems that generate art having as a source of inspiration artifacts from other domains [2, 4]. Since the most used techniques are Learning Systems and Genetic Algorithms, we decided to implement our own Genetic Algorithm, as further explained in the following Sections.

3 Musical Domain

Music is considered a form of art that usually combines sounds, following arrangements over time - it has a duration and well-defined starting and ending points. In Western Cultures, musical compositions are usually divided into three main parts: **harmony**, **melody** and **rhythm**. For our work, we considered Classical music pieces as our starting point. Regarding the music's structure, we consider that symphonies are divided in movements, that are subdivided in sections, and subdivided in measures. Measures are subdivided in individual **notes** - the minimal unit of music - and **chords** - three or more notes played together. It is important to note that this division is relative and that there are elements that rule each of these parts.

Parts are generally represented by instruments (or voices) and usually refer to a single strand, or melody or harmony of music, within a larger ensemble musical composition. When analyzing sheet music, each line represents one part of the music, that is, one instrument (or group of similar instruments or voices).

In Western Music, each staff[1] is usually divided by vertical lines. This division is called a **measure** (or bar) and, when analyzing a musical sheet, it provides a way of quantitatively divide the sounds of a musical composition into groups, based on its beats.

The ***tempo*** indicates the speed or pace of a given music (or music subsection). In classical music, it is usually represented with an instruction at the start of

[1] Set of five horizontal lines found on music sheets where musical notes are placed.

a piece (often using conventional Italian or German terms). However, it can be represented by a numerical value measured in Beats per Minute (BPM).

Throughout the music, each measure has a **time signature** - that is, how many beats each measure contains. It is typically at the beginning of the staff, represented by two values written as a fraction. The top number is the number of beats a measure has. The bottom number is the note value, that is, the type of note to count.

Notes are the minimal unit in musical compositions, and represent sounds that are formed by one mode of vibration of the air. Each note has a **duration** and **pitch**: the former describes how long, temporally, the note lasts, while the latter is related to how high or low one sound is in comparison with another. Besides, the perceptual attribute that enables humans to distinguish among sounds that are playing the same tones, equally loud, is the **timbre**. The difference between two notes is the **interval**, measured in tones. Each note may also be represented by its pitch letter followed by a number, the **octave** in which the note is. It represents how high or low a note should be played. Notes can be played with different intensities, indicated by the **volume**. It represents the variance between a weak and a strong sound, according to its loudness.

In Western Music there are twelve ordered pitch-classes[2], designated by the letters of the alphabet, from A to G. These pitch-classes can be represented in a circular diagram, the **Chromatic Circle**. A **scale** is any set of musical notes that can be ascending or descending from the twelve pitch-class, according to its steps. A scale has a tonal center, a key, which is usually the first degree of a scale. In turn, the **Circle of Fifths** is a circular diagram that represents the relationship among the twelve pitch-classes of the Chromatic Scale, their corresponding key signatures, and the associated major and minor keys.

4 Visual Domain

For this work, the visual domain considered is made up of images that do not reflect or convey anything "concrete" or "real" that pertains to elements of the real world.

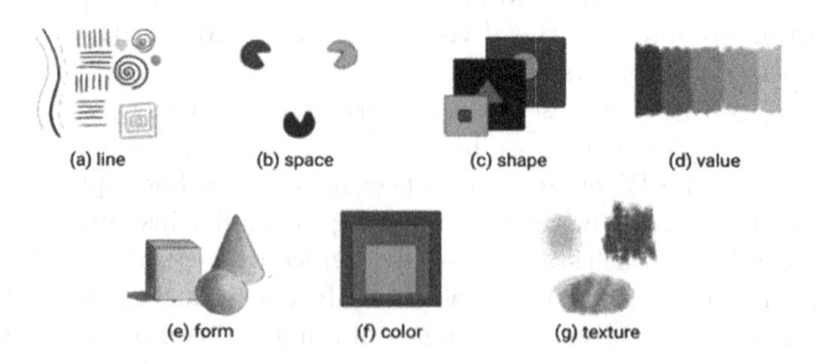

Fig. 1. 7 elements of art [6] (Color figure online)

[2] In musical notation, a pitch-class is the set of all pitches that are a whole number of octaves apart.

Art is a kind of collaboration between artist and viewer, and in each case the conditions of successful collaboration may vary [1]. A piece of artwork appears as a single entity that can be broken down into several constituents of visual representations. According to Esaak [6], there are Seven Elements of Art that serve as "building blocks" to create the visual representations: line, space, shape, value, form, color and texture, as depicted in Fig. 1.

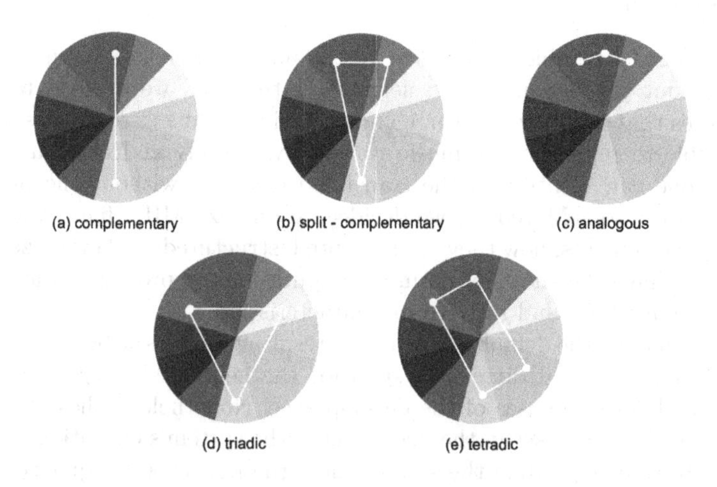

Fig. 2. Color schemes (Color figure online)

Colors are computationally defined by color models, such as the HSV model. In HSV, colors are represented by three measurable attributes: hue, value, and intensity. Hue stands for the color itself, value for the brightness of the color, and intensity for the quality that distinguishes a strong color from a weak one. Colors, as well as music, can be combined to create harmonic results. This harmony is expressed by color schemes, which can be monochromatic, complementary, split complementary, analogous, triadic, and tetradic, as depicted in Fig. 2, all of which are based on the Color Wheel [10].

When creating an image, a painting or a drawing, usually a two-dimensional space is generated by a shape in two dimensions: height and width. In the most basic approach, a shape is created when a line is enclosed - while the line forms the boundary, the shape is the form circumscribed by that boundary. Thus, a **line** is defined by a point moving in space between two other points. **Space** refers to the perspective (distance between and around) and proportion (size) among different elements. **Value** stands for the degree of perceivable color's lightness within an image. **Form** is a three-dimensional element of art that encloses volume - includes height, width and depth. Finally, the **texture** describes the surface quality of the artifact, related to the type of lines used [6].

5 Cross-Domain Associations

We aim to create a system that is capable of generating images inspired by music. To create an analogy from one domain to the other requires defining our starting language.

5.1 Approach

Since we want to be able to perceive all the musical elements in the generated image without the perception of elements' saturation, we decided to analyze music pieces that last between three and five minutes at most. Besides, this way it is possible to represent the music's harmony, melody and rhythm in such a way that one can identify all the translated elements without the perception of visual saturation. Moreover, we decided to analyze MIDI files to access the music's basic elements, how they are presented, structured, and organized. These files already have the sheet music in a computationally processed way where it is easy to manipulate and retrieve its characteristics.

We also defined the default value for the resulting images as being 1600×2700 pixels, although several dimension sizes for the generated images were tested throughout different phases of the development. Nonetheless, these dimensions can be defined by the user in the beginning of the system's execution, and there are no limitations regarding the screen where the images are displayed.

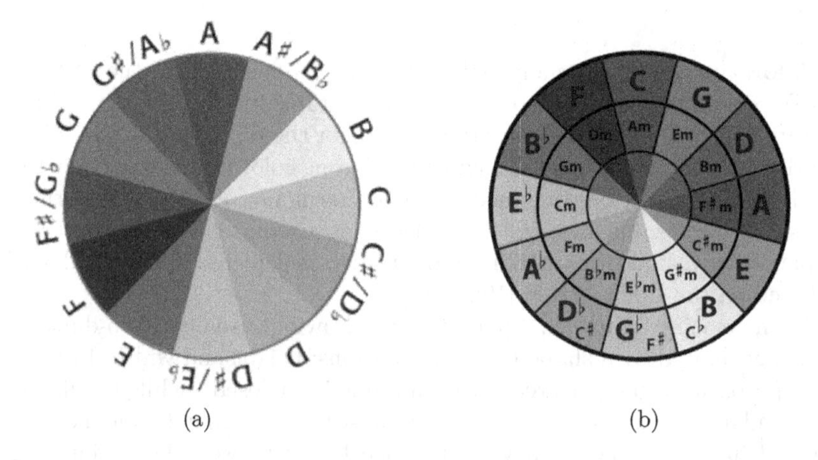

Fig. 3. (a) Overlapping the Chromatic Circle and the RGB Color Wheel (b) Overlapping the Circle of Fifths and the RGB Color Wheel (Color figure online)

Our approach's foundations lie in the association of the music's **melody** with the image's foreground, and the music's **harmony** with the image's background. In turn, the music's **rhythm** is associated with the size of the image's elements, as well as with the overall luminosity of the image's background. Having our high-level association for the system, we then need to obtain each of the image's elements from the music's structure. We started with the foreground.

One of the most basic analogies that can be created between the musical and visual domains is that a note can be translated into a color. Many different associations could be established, even random ones. We decided to overlap the Chromatic Scale and the Color Wheel - Fig. 3a - where each pitch from the Chromatic Scale is associated to one, and only one hue from the Color Wheel. Both the Chromatic Scale and the Color Wheel have twelve fractions organized sequentially, that seemed to be a fair enough decision for the mapping we were trying to achieve. The A pitch-class is considered as a reference standard, with a frequency 440 Hz, used to calibrate musical instruments: we defined it as the first pitch-class of the Chromatic Scale. Red can be considered the first hue of the Color Wheel: it is the first color in the visible spectrum, and, in the HSL model, it corresponds to $hue = 0$. To improve the color's quality definition, two other visual properties were added to the hue - saturation and lightness - leading us to have the HSL color model in our mind. From the note's octave, that represents how high or low a note is, we decided that this could give us how light or dark one color is. A lower octave corresponds to a darker color, with low value, whilst a higher octave corresponds to a lighter color, with a higher value. Finally, the saturation of the color is determined by the volume of the note. Just like the volume represents the strength of a note, the intensity represents the strength of a color.

Table 1. One possible shape association for each instrument family

Family Name	**Shape**
Piano	Circle
Chromatic Percussion	Smaller Circle
Organ	Rectangle
Guitar	Rectangular Spot
Bass	Bigger Rectangular Spot
Strings	Circular Spot
Ensemble	+10 Polygon
Brass	Irregular Square
Reed	Square
Pipe	Rhombus
Other	Triangle

At this point, we needed an association to define the **shape** of each visual element. We decided to search for an association that better approximates each instruments' family to a shape, one visual representation based on the sound to be heard, and what it mentally reminds us. Several different associations were tested. We developed an online platform to study and customize all the possible associations between instruments and shapes to help in this process - it is available on https://creativity-2020.nw.r.appspot.com. One possible association between each instrument's family and shapes is represented in Table 1, where each shape was defined using two elements of the Seven Elements of Art [6] - shape and texture - thus increasing the abstraction level. It is important to note that we made this association based only on our perception, and that a mapping between instruments and shapes that has a perfect rational is hard to achieve. Moreover, since this is a subjective task, there would always be, in our perception, instruments that do not fit in these rigid shapes. In the end, we decided to consider **two** different approaches for the output of our images: the first with a random shape and texture applied to each instrument found in the sheet music, and the second, where the associations presented in Table 1 were assigned.

Time signatures, plus information about *tempo*, are associated with the note's duration to define the element's **size**. These three concepts significantly influence the way we perceive a music piece, and so we decided they should affect the way our image's elements are represented. They were combined to achieve the best responsive size for all the elements that could fit in different screen sizes without the perception of visual elements saturation.

Finally, the **position** of each element in the foreground is directly related to the music's offset - the element's relative location in the music. This way, to build a dedicated output in which the elements are sequentially organized, the generated image is vertically divided into as many sections as the number of musical measures. Then, each vertical section is horizontally divided into as many sections as the number of events in the respective measure.

According to what is explained in Sect. 3, one can analyze the music piece as a whole to obtain the most probable tonality or scale used, thus obtaining a sense of its harmony [3]. This way, we performed the analysis of each measure to obtain its most likely tonality key, thus performing an in-depth analysis measure by measure. Having this in our mind, for each musical measure, we obtained the most probable tonality key to associate with characteristics of the background.

As previously stated, the most basic analogy that can be created between the musical and visual domains is that a note can be translated into a color. This time, having a set of tonality keys that belong to the music's harmony, we decided to overlap the Circle of Fifths and the Color Wheel, where each key from the circle is associated to one, and only one hue from the Color Wheel. Similarly to what happens with the Chromatic Circle, both the Circle of Fifths and the Color Wheel have twelve fractions organized in a sequential way, which is a fair enough decision for the mapping we are trying to achieve. Furthermore, we decided the A pitch-class with a major key as the starting point for the overlap with the first color of the visible spectrum - red. The overlapped circles and our chosen translation are represented in Fig. 3b. Likewise to the previous case, we decided to consider the HSL color model, thus two more visual properties can be added - saturation and lightness - to the hue to improve the color's quality definition. Since each measure's harmony is composed of chords, we decided to get both the volume and each tonic note [3], to retrieve its octave. The volume is associated with the color's saturation, while the octave defines the color's lightness, thus establishing a possible analogy between chords and colors.

Our approach for the background considers each measure as a **vertical stripe** that is placed sequentially from left to right in the generated image, following the sequential way of writing a musical sheet in Western Countries. Therefore, having the measure's respective color, our background consists of equally sized vertical stripes, whose position is directly mapped from the measure's offset. Each measure has its *tempo*, that directly influences how overlapped two stripes are, that is, the vertical stripe's **irregularity**: the higher the BPM, the less overlapped two stripes are, as well as the other way around. This decision was based on our visual perception of music - slower music (lower BPM value) is less "sharp" than faster music. Finally, to better differentiate the background from the foreground, we applied an **overall luminosity** filter, related to the *tempo* of the music: the lower the BPM, the darker the image, as well as the opposite, since slower music are more melancholic, therefore related to darker colors. In contrast, faster music have more energy, thus are related to lighter colors.

Having the image's foreground and background from the music's melody, harmony and the respective rhythm, we finally obtained the first two versions - **Random** and **Associated** Images. Besides these, we decided to implement a Genetic Algorithm to provide a way to search for better results while maintaining certain randomness that could lead to interesting results. Its implementation is described in the Sect. 5.2.4.

5.2 Implementation

Our system was developed in Python 3.8[3]. To perform the extraction of musical elements, the Music21 [3] library was used. For the generation of the pixels in the image, the Pycairo[4] module was used.

The architecture of our system is divided into several different modules to solve various problems individually, each containing different methods for a specific goal.

5.2.1 Interpreter Module

This module is responsible for extracting and processing the needed information from the music given as input. After receiving the MIDI file, we use the Music21 library to process it. With this, we perform a top-down analysis for each file, from its parts to its minimal unit, its notes: our system is organized in such a way that we first analyze the Parts, then Measures and finally, measure's Events, in a hierarchical way. When analyzing a **Part**, we get the correspondent

[3] https://www.python.org.

[4] https://www.cairographics.org/pycairo/.

Instrument name, as well as its program change[5]. When analyzing a **Measure**, we get its offset. When analyzing a measure's event, if we find an instance of **MetronomeMark**, it means that we are dealing with music's *tempo* for that measure: we get its value (BPM). If we find an instance of **TimeSignature**, it means that we found the music's time signature for that measure: we get its value. When we find an instance of **Note**, **Chord**, or **Rest**, first we get and store its offset. Then, if it is a Note, we get its correspondent Pitch, as well as its Volume and Octave. If it is a Chord, we get its duration, performing an analysis Note by Note, retrieving the Chord's characteristics.

5.2.2 Cross-Domain Association Module

Starting with the part analysis, we get the shape definition. For the **Random Image**, for each instrument found in the music, we decided to randomly assign a shape among 17 available: small circle, circle, circle spot, triangle, rectangle, rectangular spot, big rectangular spot, square, irregular square, rhombus, pentagon, hexagon, heptagon, octagon, eneagon, decagon and 10+ sides polygon. For the **Associated Image**, the associations represented in Table 1 are applied. Regarding the element's position on the canvas, we normalized the measure's offset to the x value, and the event's offset to the y value, according to the dimensions of the image. Each pitch is assigned to the hue definition, among 12 available - Fig. 3a. The note's volume, normalized between 0 and 1, is assigned to the color's saturation. The octave is assigned to the lightness and alpha value: after some research and experiments, we consider seven different octaves, from the zeroth to the sixth, and lightness values ranged between 0.1 and 0.7. While the former is a direct translation, the latter applies an inverse operation: the higher the octave, the less solid (more transparent) the element will be: $alpha = 1 - lightness$. Regarding the note's duration, time signature and *tempo*, we assigned them to the element's size, by applying Eq. 1.

$$size = \left(Duration \times \frac{1}{Tempo}\right) + \left(\frac{1}{Tempo} \times \frac{1}{TimeSignature}\right) \tag{1}$$

Finally, the *tempo* is assigned to the background's overall luminosity and stripe's irregularity by overlap. The music's *tempo* ranges from 60 (slow) to 150 (fast), while the luminosity from 0 (black) to 1 (white), and irregularity from 5 (regular) to 0 (irregular).

Regarding the background, we associate each measure's tonality key to the hue definition - Fig. 3b. For every chord found on each measure, we extract its root note. Then, having all the notes, we calculate an average for the volume to later associate with the color's saturation. In the end, for each chord on each measure, we extract its root note. Similarly, we calculate an average for the predominant octave to later associate with color's lightness and alpha value.

[5] MIDI message that identifies the instrumental sound the device uses when it plays a Note.

5.2.3 Generator Module

Using Pycairo, the system starts by creating a surface with the desired dimensions of the image. Then, it continues by defining the background as a black rectangle with the overall luminosity filter applied. Finally, using a linear gradient, it draws the vertical stripes considering the respective irregularity by overlap, color and position features. To fill out the foreground, since each element has already all the necessary features attached, the system sequentially traverses the array in which the data is stored and, considering each element, draws it using Pycairo drawing methods. In the end, the system outputs their first **two** final versions, the Random and Associated Images, that are saved as separate Portable Network Graphics (PNG) files.

5.2.4 Genetic Algorithm Module

To improve the quality of the generated images, we implemented one Genetic Algorithm (GA) mainly applied to the visual domain, although we consider some concepts of music harmony in the generated elements. The algorithm considers an individual an image whose characteristics are the position, color, shape, size of the elements of the foreground, as well as the musical notes that gave rise to them. The initial population of the algorithm consists of 24 individuals. The first two are the Random and Associated Images, and the remaining 22 are obtained from these previous, where the element's horizontal and vertical position in the canvas is shuffled.

For each generation, the fitness of each individual is calculated based on color harmonies between nearest elements, as well as visual perceptions of color, location of elements in the canvas, and shape distributions. For each element we calculate their five nearest neighbors based on the k-nearest neighbors algorithm (k-NN)[6]. Then we score each nearest neighbor according to the harmony between their colors, considering the color schemes presented in Sect. 4. In turn, we remove score if the color's perception to the human eye is visually close, according to the Delta E distance metric[7], that compares the difference between two colors. This way, we maximize the color diversity of generated elements considering its harmony and visual perception. Then, for each two nearest visual elements, we consider if the musical notes in which they were inspired on are next to each other in the Chromatic Scale. If so, we give a score according to their Euclidean distance, where closest elements are scored higher. Finally, we raise the score if the element's position lies within 10 pixels from its origin position, and if the association between mapped instrument and the element's shape follows the distributions in Table 1. In the end, we remove score if the element's shape is a regular polygon with more than eight sides, since we consider that it reassembles to a circle.

Subsequently, after having a fitness value for each individual, we implemented an **elitism factor**, in which for each new generation, exactly four individuals are passed on to the next generation so that the fitness value never decreases.

[6] https://scikit-learn.org/stable/modules/neighbors.html.

[7] https://python-colormath.readthedocs.io/en/latest/delta_e.html.

Then, the selection of each pair of individuals takes into account the **Roulette Wheel Selection**.

In our implementation, on each pair of selected individuals, there is a 90% probability of occurring **Single-Point Crossover** [9], where only the elements' characteristics of one individual are crossed with the elements' characteristics of another individual. These characteristics may be their color, position, shape or size. Then, the **Uniform Mutation** [9] involves changing one element characteristic, and similarly to crossover, several types may occur: Position Mutation, with a 10% chance to occur; Offset Mutation, with a 5% chance to occur; Color Mutation, with a 10% chance to occur; and Shape Mutation, with a 5% chance to occur.

Our algorithm finishes after 300 generations. The final solution is the best population with the fittest individual, that is returned to the Generator Module (Sect. 5.2.3), thus generating the **Genetic Image**.

6 Evaluation

Our dataset consists of fourteen music pieces from different eras (from the Baroque to the Modernism) and composers from the Classical tradition. However, it is important to note that our approach is independent of the musical style, and any could have been considered. Although we have executed the system with each of the MIDI files that compose our dataset, for the evaluation

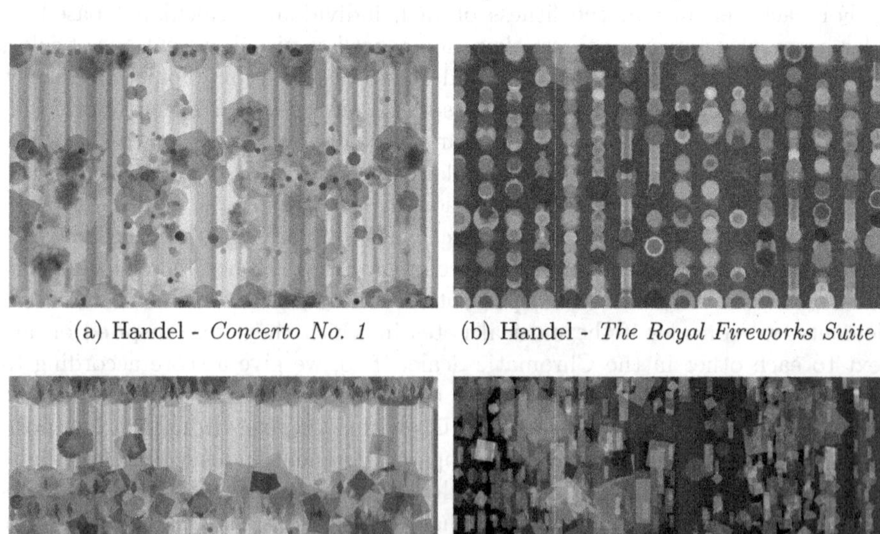

(a) Handel - *Concerto No. 1* (b) Handel - *The Royal Fireworks Suite*

(c) Mozart - *Symphony No. 40* (d) Stravinsky - *The Firebird*

Fig. 4. Some system's output images used in the evaluation process

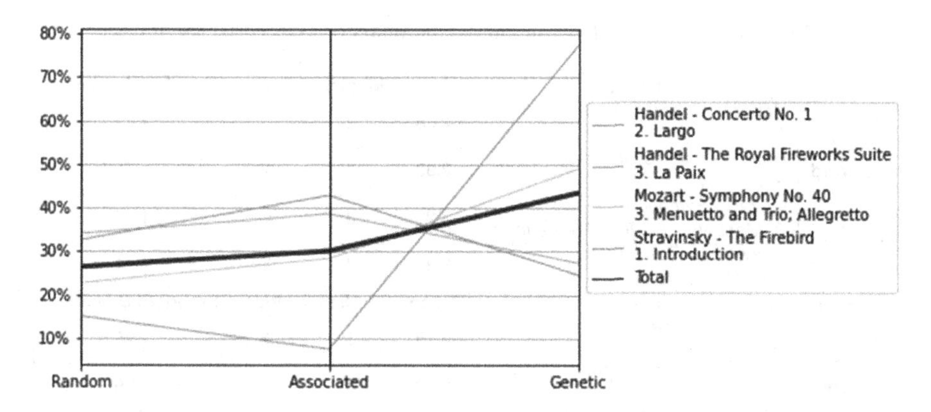

Fig. 5. Which image version do you prefer?

process we considered four of them, spread over four online surveys. Our dataset and respective system outputs can be seen on https://luisaleixo.github.io.

Each survey starts by asking the participants' age, gender, and how frequently they usually go to exhibitions (online, or before the COVID-19 pandemic). Afterward, the survey asks the participant to describe each generated image, using three sentences of their own. Besides, it demands the participants to rank from the most to the least preferred images, and whether participants consider them to be abstract. After that, the survey requests a description of each image version according to predefined adjectives: Exciting; Smooth; Happy; Enjoyable; Surprising; Contemptible; Sad; Aggressive; Disgusting; Boring; Angerful; Fearful. Then, the survey continues to the musical domain, repeating the process that was done with images. After that, for the preferred and least preferred images, the survey asks whether participants generally like them, and if they perceive any relation with the music artifact that served as inspiration for the creation. Finally, it cites the following sentence - "All the images were created having the previous music as inspiration." - and asks if the participants agree with it. These questions were evaluated with a Likert scale from 1 to 5.

Figure 4 displays the Genetic Images used in the evaluation process. In total, 93 people - 62 female and 31 male - were questioned through all the surveys. Both participants were aged between 18 and 29 years. Regarding how often participants go to exhibitions, the majority answered: "Once a year".

The relative frequency for each music and respective preferred image version can be seen in Fig. 5. From this, by observing the "Total" line, we conclude that the most preferred Image version is the Genetic one. Besides, we observe that participants prefer the Associated Image version over the Random one.

The mean, mode, and standard deviation for the results of the most and least preferred images can be seen in Table 2. At this point, we related these statistics with the adjectival description participants had on the artifacts.

Regarding the first case, Handel - *Concerto No. 1* and respective preferred version, we can see that, for both image and music, the opinions are much divided: the image was mainly perceived as happy and enjoyable, while the

Table 2. Statistics for "Do you think this image is related to the music?"

Version	Music	Mean	Median	Mode	Standard Deviation
Most Preferred	Handel - *Concerto No.1* *2. Largo*	2.38	2	2	1.27
	Handel - *The Royal Fireworks Suite* *3. La Paix*	2.93	3	4	1.21
	Mozart - *Symphony No. 40* *3. Menuetto and Trio; Allegretto*	3.23	3	4	1.09
	Stravinsky - *The Firebird* *1. Introduction*	3.33	3	2	1.41
Version	**Music**	**Mean**	**Median**	**Mode**	**Standard Deviation**
Least Preferred	Handel - *Concerto No.1* *2. Largo*	2.98	3	4	1.25
	Handel - *The Royal Fireworks Suite* *3. La Paix*	2.89	3	3	0.99
	Mozart - *Symphony No. 40* *3. Menuetto and Trio; Allegretto*	2.89	3	3	1.05
	Stravinsky - *The Firebird* *1. Introduction*	3.31	4	4	1.26

music was perceived as sad, smooth and fearful. These results are contradictory since they transmit opposite sensations, which can be confirmed by the results in Table 2. The mean and mode value for the relation between the music and image is 2, and the median 2.38. Since the results are under 3, we consider that participants do not perceive any relation between the music and image, which is a reasonable explanation for the contradictory opinions they have. For the least preferred image, the results are similar and can be confirmed in Table 2.

The second case is Handel - *The Royal Fireworks Suite*. Here, by looking at the results, participants mainly perceive the preferred image as happy and smooth, while the music as enjoyable and smooth. Regarding the relation with the music, since the mean value is 2.93, median value 3, and mode value 4, we consider that a better relationship could be perceived with it, since the used descriptors belong to the same type of emotions [5]. For the least preferred image, there is no slight difference, since most chosen descriptors were the same as with the preferred image. Once we are assessing the least preferred version, this similarity on the descriptors was not expected. However, by looking at Table 2, we observe that participants perceive a weaker relation between the music and the least preferred image. Anyway, these results are positive, since the chosen descriptors for both artifacts mostly match.

The third case is Mozart - *Symphony No. 40*. Regarding the preferred version, the used descriptors and statistics are similar to the preferred version of the second case. For the least preferred image, despite the peak on the exciting descriptor, participants also consider the image to be happy, surprising, and aggressive. By observing the statistics in Table 2, we consider that participants perceive the relation between the artifacts, specially between the most preferred image and respective music.

The fourth and last case is Stravinsky - *The Firebird*. As in the first case, we see that the opinions are contradictory for the music-image pair: participants perceived the music mainly as sad and boring, while the preferred image as enjoyable and happy. Considering the statistics regarding the relationship between both artifacts - Table 2 - we observe that the mode value is 2, the lowest of the four cases. This explains the opposite perceptions participants had. The least preferred image was perceived equally exciting, happy and enjoyable. However, its statistic values for the image rating are lower than the preferred version, which was expected. Nonetheless, by observing Table 2, we conclude that, for both versions, participants perceive a relation between the image and music.

We should not fail to point out that, despite the individual analysis, all the images were generally described with adjectives that belong to a positive type of emotions, such as happiness, excitement, or joy. The majority considered that images are colorful, geometric, somehow following one pattern. In some cases, people said that had a sensation of movement through the images, and that somehow they were a combination of each other. Having finished the adjectival description of the most and least preferred versions, we can move on to the question - "Do you consider these to be abstract images?" - that was applied to all the generated set of images. The results were positive since the majority of the participants - around 84.71% - consider the images to be abstract. Besides, "abstract" was one of the most used concepts to describe the images through all the versions.

Regarding the last question - "All the images were generated having the previous music as inspiration" - Do you agree with this statement? - the results are still positive and in line with our expectations and goals: by observing the statistics, we can see that the majority of the participants believe that there is a relation between the music and image.

Since our main objectives were to generate abstract images that have as inspiration musical files, we consider that the results are desirable given our goals. First, most of the participants consider that the generated images are abstract. Second, regarding the image-music association, we believe that the results were also positive, since the average for the relation between the artifacts is above 3. Finally, regarding the preferred images, the ratings are good - generally, participants seem to like the artifacts, as already seen. Regarding the least preferred images, the results are more neutral, but still good. From these, we can state that participants mostly liked all the images, even the least preferred ones.

7 Conclusions and Future Work

Inspired in the algorithmic way of processing music and image, the main objective of this work was to develop a system that exhibits creative behavior through inter-domain associations, by generating abstract images inspired from music. Our approach was based on the relation between the music's harmony, melody and rhythm, and visual elements that refer to both image's background and foreground. Our high-level approach is based on the association of the music's melody to the foreground, music's harmony to the background, and music's rhythm to the size of the generated elements. It is important to note that one correct association between two domains does not exist, since the developer's choice directly influences the analogies made.

Three images were generated from each music file given as input. The first, called the **Random** Image, was generated assigning a random shape and texture to each instrument found in the musical composition. The second, called the **Associated** image, was generated assigning a predefined association between instruments and respective shapes. This was entirely based on our personal perception - it is one possible approach for the association process, among others that could have been taken. It is also important to note that a study on these associations was made and deeply tested through an online platform developed for this goal. With this, users can test their different ideas by creating their inter-domain associations. The third, called the **Genetic** Image, results from the execution of a Genetic Algorithm that receives the two previous versions to generate the initial population.

Regarding evaluation, the majority of the participants considered that the generated images are abstract, and believe that they have a relation with the music that served as the basis for the inspiration process. Besides, the majority liked the presented images, ranking the genetic as the most preferred one. We consider that these results are favorable since they go in line with our goals.

The system we developed has some limitations, especially regarding the dimensions of each element in the canvas, and the color levels fluctuation among different screens. Therefore, we tried to overcome these by applying two image filters, as well as to give the user's the possibility to parameterize some variables related to the dimensions of the image.

As future work we consider that changes can be made to our approach regarding the associations made between instruments and shapes, so that a better reflection of the essence of each timbre could be perceived. Moreover, by using MP3 instead of MIDI files, it would be possible to analyze more characteristics of the timbre besides the instruments.

This work was supported by FCT project UIDB/50021/2020.

References

1. Bossart, W.H.: Form and meaning in the visual arts. Br. J. Aesthet. **6**(3), 259–271 (1966)
2. Colton, S.: The painting fool: stories from building an automated painter. In: McCormack, J., d'Inverno, M. (eds.) Computers and Creativity, pp. 3–38. Springer, Heidelberg (2012). https://doi.org/10.1007/978-3-642-31727-9_1
3. Cuthbert, M.S., Ariza, C.: music21: a toolkit for computer-aided musicology and symbolic music data (2010)
4. DiPaola, S., Gabora, L.: Incorporating characteristics of human creativity into an evolutionary art algorithm. Genet. Program. Evolvable Mach. **10**(2), 97–110 (2008)
5. Ekman, P.: Universal emotions. https://www.paulekman.com/universal-emotions/. Accessed 02 Jan 2020
6. Esaak, S.: The 7 elements of art and why knowing them is important. https://www.thoughtco.com/what-are-the-elements-of-art-182704. Accessed 16 July 2020

7. Horn, B., Smith, G., Masri, R., Stone, J.: Visual information vases: towards a framework for transmedia creative inspiration. In: ICCC, pp. 182–188 (2015)
8. Machado, P., Cardoso, A.: All the truth about NEvAR. Appl. Intell. **16**(2), 101–118 (2002)
9. McCall, J.: Genetic algorithms for modelling and optimisation. J. Comput. Appl. Math. **184**(1), 205–222 (2005)
10. Morton, J.: Basic color theory. https://www.colormatters.com/color-and-design/basic-color-theory. Accessed 01 Aug 2020
11. Moura, L.: Robot art: an interview with Leonel Moura. In: Arts, vol. 7, p. 28 (2018)
12. Teixeira, J., Pinto, H.S.: Cross-domain analogy: from image to music. In: Proceedings of the 5th International Workshop on Musical Metacreation (2017)

"What Is Human?" A Turing Test for Artistic Creativity

Antonio Daniele(✉), Caroline Di Bernardi Luft, and Nick Bryan-Kinns

Queen Mary University of London, London, UK
a.daniele@qmul.ac.uk

Abstract. This paper presents a study conducted in naturalistic setting with data collected from an interactive art installation. The audience is challenged in a Turing Test for artistic creativity involving recognising human-made versus AI-generated drawing strokes. In most cases, people were able to differentiate human-made strokes above chance. An analysis conducted on the images at the pixel level shows a significant difference between the symmetry of the AI-generated strokes and the human-made ones. However we argue that this feature alone was not key for the differentiation. Further behavioural analysis indicates that people judging more quickly were able to differentiate human-made strokes significantly better than the slower ones. We point to theories of embodiment as a possible explanation of our results.

Keywords: Drawing · Embodiment · Turing test · Bias · Artificial Intelligence · Computational Creativity · Interactive Art · Computational Art

1 Introduction

This paper presents the design of the interactive art installation "Grammar#1" by Antonio Daniele[1] which was exhibited in May 2019 at the Albumarte gallery in Rome, as one of the 10 winners of the open call *Re:Humanism*. The art installation is presented to the audience as a Turing Test (TT) for artistic creativity [2] where the goal is to recognise human-made artefacts among AI-generated ones. The art work is described to the audience as informing both our scientific research as well as artistic enquiry using their behavioural data collected during the interaction. By questioning what is human and what is artificial, the artist invites the audience to reflect on their own nature as human beings in relation to their understanding of the concept of "artificial". The artistic enquiry of this work addresses the relation between humans and technology in the era of super-human AI and deep fakes by exploiting one of the most ancient and simplest form of human communication: the drawing of lines. In fact, glyphs and symbols are considered one of the earliest forms of human communication

[1] Throughout the paper, the first author will also be referred to as the "artist".

J. Romero et al. (Eds.): EvoMUSART 2021, LNCS 12693, pp. 396–411, 2021.
https://doi.org/10.1007/978-3-030-72914-1_26

[10]. In addition, from more contemporary studies in experimental psychology, we know that abstract lines are successfully used to express moods [22] and spontaneous drawings can be used to interpret complex non-conscious states in clinical settings [5,6]. While the art work invites reflection about the relation between *human* and *technological* in the arts, the study explores the cognitive implications of interfacing with artificially generated content. This problem was previously investigated and discussed from a theoretical perspective [11,12]. In this paper, the approach is quantitative and focuses on the medium of the drawing, asking whether we perceive substantial differences between human-made and computer-generated strokes and how.

The first part of the paper describes the artist's creative and technical process for the making of the art installation. Initially, a collection of 300 abstract drawings were produced by using automatic drawing techniques [4]. Then, the individual strokes of these drawings were manipulated into a larger dataset and used to train SketchRNN [21] a Variational Autoencoder (VAE) capable of generating simple sketches. Finally, the interactive installation is designed as a TT for artistic creativity [2] where the audience is challenged to recognise human-made strokes amongst the AI-generated ones. The second part of this paper presents the results of the study consisting of the TT for artistic creativity and the analysis of the audience's behavioural patterns, more specifically, the audience response time in relation to their performance. Next, the strokes selected by the audience from both datasets (human-made and AI-generated) are analysed at a pixel level by comparing their respective average entropy and symmetry. Although we find a significant difference in symmetry between the two datasets, a deeper analysis of the audience behaviour shows that the visual feature alone is not crucial for the differentiation. We discuss our findings pointing at theories of embodiment [17,18,35].

2 Background

2.1 Turing Test

In 1950, Alan Turing [45] proposed the question "Can machines think?". He described this problem in terms of a game where an interrogator, located in an isolated room, has the objective to recognise the sex of a man and a woman by simply asking them questions. Then, Turing asks if the interrogator would guess wrongly more often if a machine took the part of the man. This game, originally named "the imitation game", became commonly known as the Turing Test, a way to assess a machine's ability to emulate human skills. In 2010, Margaret Boden [2, p. 409] argued that

> for an artistic program to pass the TT would be for it to produce artwork which was:
> 1. indistinguishable from one produced by a human being; and/or
> 2. was seen as having as much aesthetic value as one produced by a human being.

Pease and Colton [34] argue that this test might not be accurate enough to evaluate creativity in its true complexity. However, the first point proposed by Boden represents a necessary and sufficient step for avoiding the bias against machine generated content during the evaluation. Any evaluation of an AI-generated artefact indistinguishable from a human-made one would be the result of a choice unbiased against technology. In the next section, we present some of the most relevant studies investigating this problem.

2.2 Bias Against Machine Generated Content

Artists and scientists have investigated how people react to computer-generated content since the early age of computer arts. The "Mondrian Experiment" conducted in the Bell Labs by the computer art's pioneer Michael Noll [33], is one of the first attempts to investigate how people compare art made with machines to art made by humans. In this case, Noll generated an image with a program instructed to replicate the patterns from the painting "Composition with Lines" by Piet Mondrian (1917). The results show that only 28% of people were able to recognise the image created in the Bell Labs and that the 59% preferred the computer picture. As expected by Noll, the majority of subjects with a technical background were able to recognise the computer image. In contrast, the non-technical people were fooled by assuming that a computer would have built a more "ordered" image while humans would have expressed their creativity with more random patterns.

In Moffat and Kelly [30], the test was conducted on musicians and non-musicians, using music in the style of Bach or Jazz. The participants had to differentiate computer generated from human made music and give their preference to each track. Surprisingly, their results show that non-musicians were significantly better than musicians in recognising computer generated music. Furthermore, both groups preferred human made music, independently from knowing who or what created it.

A more recent experiment was conducted by Elgammal et al. [15] to test the efficiency of their generative model for images. The Creative Adversarial Network (CAN) is a generative model proposed by Elgammal et al. as a variation of the Generative Adversarial Network (GAN) [20]. The visual output generated with CAN was compared with the one generated by existing models (DCGAN) and with abstract art from famous painters or contemporary established artists from the Basel Art fair. Their results show that the participants rated CAN generated works more likeable than the ones produced with DCGAN as well as the works made by the emerging artists. Furthermore, the majority of the participants were tricked into believing that the works made by CAN were actually made by human artists.

In opposition to Elgammal results, Ragot and Martin [36] found a negative bias towards machine generated visual content. The experiment, conducted on a large group of 565 participants, showed images of landscapes and portraits generated by algorithms mixed with paintings of the same subjects made by human artists. According to their results, the 66% of participants were able to recognise human-made paintings and a significant majority preferred human paintings despite knowing who/what made them.

The experiments in this field usually focus on music or more complex visual representations like paintings, whereas the medium of the drawing is almost completely unexplored. To the best of our knowledge, the only study including artificially generated drawings is Chamberlain et al. [8]. The general findings of their larger study show a negative bias towards computer generated works. In the specific experiment with the drawings, the authors explore possible reasons for the negative bias by using the robotic system by the artist Patrick Tresset [43]. The results show that the audience evaluation was conditioned by how much "anthropomorphic" the robotic arm appeared to them, suggesting that the judgment was involving concepts such as "social engagement" or the "embodiment" of human gestures.

The following sections explain why drawings are important for our species and why they are a suitable medium to investigate the mechanisms involved in people's perception of artificially generated content versus human-made one.

2.3 Why Drawings?

From an evolutionary perspective, drawings have had a special role in shaping human cognition since prehistory. Drawing gestures allowed a physical embodiment of the inhabited environment, specifically, simple lines were used to illustrate the shapes of rivers and trees [10]. Drawings possess qualities that are processed by the human brain in a unique way, common to children, cavemen and monkeys [40]. They can be used to represent shapes, volumes and shadows with high accuracy and they are interpreted by our brain as realistic [7]. Furthermore, the action of drawing is essential during the developmental age [14,22]. For instance, representational flexibility in children facilitates the association of symbols with new meanings [14]. The same cognitive mechanisms allow children as young as four years of age to express emotions and moods by using abstract lines [22]. Furthermore, drawings are often used in therapeutic settings to help patients express non-conscious states [5,6].

2.4 Automatism and Automation

Automatic Drawings. In the history of art, spontaneous activities capable of revealing non-conscious thoughts and feelings are regarded as *automatisms* and are traditionally associated with the Surrealist movement [4,27]. More precisely, automatic drawings consist of a "pure and simple abandon to graphic impulse" [4, p. 274], an artistic approach that influenced art movements such as the American Abstract Expressionism as well as artists like Salvador Dali and Jean Miro, Jackson Pollock among others.

Let It Brain. The abstract drawings used in this art installation and for training SketchRNN are created with an automatic technique named by the artist as “Let it Brain” (LIB). LIB drawing technique has been developed by the artist over the past 20 years in a spontaneous way, independent from traditional automatic techniques from the Surrealist movement. Although LIB shares similarities with Surrealist automatic drawing, it can be better described as a sort of “human-generative” drawing or “enactive drawing”. In this technique the approach consists in enacting the drawing by reducing the time between thought and action, similarly to the “enactivism” described in Manning and Massumi [28]. In this method, the role of real-time and the access to intuitive reasoning is crucial for the expressive gesture.

Computational Drawings. While the concept of *automatism* in drawing is generally associated with an expressive activity conducted by humans, drawings made by or with machines might rather be associated to the concept of *automation*. However these two concepts are not necessarily in conflict and in some cases they can intersect. In fact, computers have been used to produce art since the early 60’s by computer scientists [32] as well as by artists that treated algorithms as a new artistic medium like Vera Molnar [31] or Harold Cohen [29]. Being able to produce geometric shapes and lines using programming languages was the gateway to produce more complex agents like Cohen’s AARON [29], the Painting Fool [9] or, more recently, Paul the drawing robot by Patrick Tresset [43] or D.O.U.G the collaborative system created by Sougwen Chung. These machines or software are considered by their creators either as tools, extensions, collaborators or as independent entities able to produce art of their own. As with automatic drawings, where non-conscious thinking is involved and the action results in almost random patterns, in computational arts, it is also possible to find approaches like generative arts or evolutionary art, involving a reduced control of the human artist, leaving the machine to operate in (almost) complete autonomy [12].

Drawings and Machine Learning. Research in computational drawing has boomed with the ever increasing availability of touchscreen devices and the recent advances in Machine Learning (ML). Nevertheless, the absence of colour and texture, the abstract nature of sketches and the variability in styles are still representing some of the greatest technical challenges [46]. Some attempts have been made in the direction of drawing abstraction at the stroke level [1,39], however, the target of these architectures is usually a figurative sketch or a photograph. Early attempts to generate drawings using computers mostly exploited perceptual properties of vision by modifying brightness and contrast of existing images, usually relying on features such as edge detection [19]. More recent generative models [1,13,24] have demonstrated great results by analysing drawings at a stroke level in their spatio-temporal distribution and classifying by purpose (e.g., shading or contour) [1] or by semantic grouping using deep learning techniques [24].

A milestone in sketch synthesis was reached with the release of SketchRNN [21] a Variational Auto-Encoder (VAE) that uses the “Quick,Draw!” (QD) dataset, the largest dataset of hand-drawn sketches to date. SketchRNN is made of a bidirectional RNN encoder and an autoregressive RNN decoder. At present, this architecture is the state-of-the-art in sketch synthesis. Whereas models based on raster data might be very effective for sketch analysis (e.g., CNN) or generation (e.g., GAN) by using the spatial domain, SketchRNN learns from the very process of drawing in its temporal domain at a fine-grained level.

3 Method

The motivation of this work is twofold. On the one hand, the art installation asks the question "What is human?", inviting the audience to reflect on how the line between the concepts of *human* and *artificial* is often blurred. The artistic exploration uses drawing strokes abstracted from any specific conceptual meaning. On the other hand, the aim of the study is to investigate people's ability to distinguish human-made and artificially generated content as well as exploring some of the factors involved in the differentiation. The following sections describe the process involved in the making of the art installation and the methods used in the study.

3.1 The LIB Dataset

For this study, the artist initially created a set of 300 automatic drawings using the LIB technique (see Sect. 2.4). The drawings consist only of abstract lines (no figurative subjects) and were created over a month-long period. In this text, we name this group of drawings as "LIB dataset" (Fig. 1). All drawings were realised on paper (size A5: 148 × 210 mm) using a digital tablet capable of translating the artists' gestures from pencil/pen to Scalable Vector Graphics (SVG) format. This guaranteed a more accurate and natural feeling in the making of the drawing.

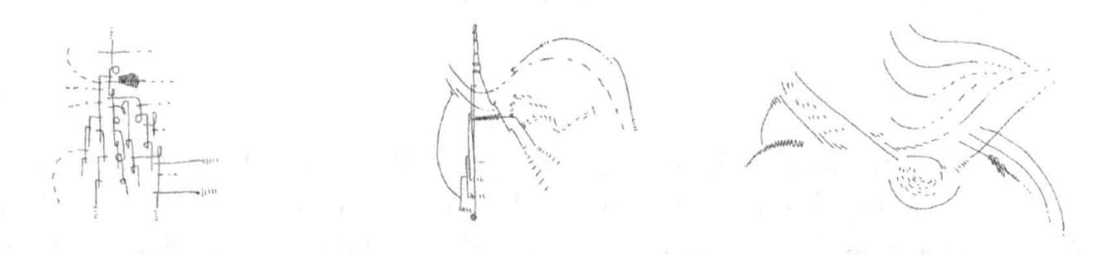

Fig. 1. Three examples of LIB drawings from the LIB dataset

3.2 Preparing the Drawings' Dataset

The first part of the work consisted in preparing a custom dataset for training SketchRNN. We converted each drawing from SVG format to a Numpy array where the coordinates of the drawing are stored in a specific format: ΔX, ΔY, p [21]. The ΔX and ΔY are the offset's distance of the 2D spatial coordinates in each point of the drawing on the canvas, whereas the p represents the binary state of the pen: 0 if the pen is on the paper, 1 if the pen is lifted.

Modelling the Dataset. The "Quick,Draw!" (QD) dataset [21] is made of 50 million drawings across 345 categories whereas our dataset (300 drawings) is not just smaller than any class of QD, it is also very different in terms of stroke distribution. While the maximum number of Δ offsets per drawing (ΔX, ΔY) that SketchRNN supports is 250, each drawing of our dataset counts a number of offsets much larger than the limit supported. Whereas both offsets Mean value and Standard Deviation per drawing in the LIB dataset are not in the same range when compared to the QD categories, the high sampling frequency (140 Hz) of the tablet produced a total number of offsets much more consistent with QD.

For this reason and because of the abstract nature of the drawings, we considered the whole set of coordinates as a long, whole drawing which we split into smaller units or strokes. This strategy addresses and solves 2 existing problems: 1) the new dataset so obtained will contain a number of strokes much higher than the initial 300 drawings which allows the use of deep learning techniques; 2) The strokes obtained from the split operation will be compatible with SketchRNN standards because their length can be tailored to a maximum of 250 offset points.

Training, Generating, Evaluating. Nine different splitting strategies were tried and nine respective datasets produced and then used to train SketchRNN. From each of the nine trained models we generated 60 strokes among which the artist selected the most similar to his drawing style and evaluated them from *"not acceptable; acceptable; good; best"*. The model that produced the highest number of *good* and *best* strokes was then used to generate a new dataset of artificially generated strokes.

The results of the process so far are two datasets of strokes: 1) *Human Strokes*, a collection of strokes obtained from the original drawings (LIB dataset) and 2) *Artificial Strokes*, a collection of generative strokes learnt from the original drawings. Both datasets contain the same number of strokes (n=20981).

3.3 The Installation

The installation was located in a room (D=1,65 × W=2,23 × H=3,27 mt) and all 300 drawings of the LIB dataset were displayed on three walls. In the middle of the room, the interactive experience is accessed via touch screen and a web-based interface (Fig. 2). The colour palette used for the installation was limited to black and white. One reason for this choice is to recall the style of LIB drawings. At the same time, the choice of binary colours is a play on the dichotomy between *human* and *artificial*, as commonly perceived in modern society [12]. The experience is designed so that neither the black or white colours can be linked to a specific category (*human*/*artificial*).

3.4 User Experience

The interactive experience consists of 6 stages accessed through a simple touch interface. The initial stage displays the question "What is Human?", which plays

Fig. 2. Pictures of the interactive art installation

as a philosophical question as well as an instruction. The following three stages consist of simple tasks, specifically three types of TTs for artistic creativity. Once the third task is completed, the results are presented as a data visualisation. Finally, the last touch directs the user to the end of the experience which restates the initial question: "What is human?". In this paper, we analyse and discuss the second task.

The Task 2 is designed as a matrix of 4 × 4 tiles where the artificial and human-made strokes are randomly displayed. The goal of this task is to recognise the human-made strokes among the artificially generated ones by clicking on the preferred tile. The grid always shows images from both sets but no duplicates are displayed to the user in the same section. The images are randomly taken from both *Human Strokes* and *Artificial Strokes* sets in random proportion. Therefore, each completed task has a variable probability of success determined by the maximum number of human-made strokes displayed in the matrix. The users are not aware of the number of human-made strokes available per session and each time a tile is selected, it disappears from the screen leaving a gap in the matrix. The task is completed when the user has given a number of choices equal to the maximum number of human-made strokes displayed, despite their choices being correct or wrong. The audience has no feedback on whether their selection is correct.

3.5 Materials

The interactive interface was created using P5js, an open source Javascript library based on Processing [38]. The interface is accessed via a touch screen from an 11,6″ Acer Chromebook R11. The datasets used for this installation are described in a previous section as *Human Strokes* and *Artificial Strokes.* Both datasets contain n=20981 strokes. Each of the 300 drawings from the LIB dataset was printed on an 11″ photographic paper. The whole dataset is divided into three 10 × 10 tiles composition and displayed on 3 adjacent walls. The drawings are placed in no particular order (Fig. 2).

3.6 Data Collection and Quality Control

This study was conducted in a naturalistic setting, in our case an art gallery. In this context, the participants are the visitors of the exhibition interacting with the installation without any structured introduction or controlled test procedures as it might happen in a lab setting. The behavioural data collected during the interaction consist of the images selected and the time response per selection. No personal data was recorded and the sessions were not monitored, so it can be assumed that one user could have interacted with the installation multiple times or that multiple users interacted within the same session.

To minimise the limitations described above, we conducted additional data quality control steps. In particular, among the two hundred twenty-five (n=225) completed user experiences collected in total, we decided to exclude the first response per user, where the person might be getting familiar with the interface and the task. Then, by controlling the distribution of the data we excluded the outliers completing the task above 30 s and with a time response Standard Deviation for each category (*AI* and *Human*) lower than 6 s.

Performance Score As explained in Sect. 3.4, in each task there is a different probability of success determined by the number of human-made strokes displayed in the matrix. Therefore, all 225 completed tasks represent a binomial distribution where each task is a series of independent Bernoulli distribution. Each independent Bernoulli distribution has two outcomes: 1) human-made or 2) artificially generated. For each completed task, we calculate how the user results compares to the expected probability of success. If we adopt the null hypothesis that the users are randomly guessing, we can reject that hypothesis in all those cases where the probability is below or equal to 0.05. This value tells us to what extent the user was randomly guessing between *human* and *artificial* options. For the behavioural analysis, we analyse only the tasks where the user's probability of randomly guessing is below or equal to 0.05, in which case we assumed that the participant was actively engaging with the task. This parameter is also used as a factor to calculate a performance score on a scale between 0 and 1, where a lower probability of randomly guessing means a better performance.

3.7 Behavioural Analysis

The data collected also included the response time for each selected tile. We analysed the differences between the average response time for the human-made strokes and the AI-generated ones. Then, we controlled how the performance related to the response time for each category. To do so, we split the mean response time by percentile (25%, 50%, 75%) and obtained three groups: *fast* below 528 ms (N=59; M=355 ms; SD=115 ms), *medium* between 528 and 1686 ms (N=119; M=952 ms; SD=338 ms), *slow* above 1686 ms (N=59; M=3256 ms; SD=2836 ms).

3.8 Strokes Analysis

We undertook further analysis at the pixel level of the strokes displayed to the audience during the exhibition. Out of 3600 strokes (225 trials × 16 tiles), the audience saw in total 3418 unique strokes (1700 human-made; 1718 AI-generated), meaning that 182 strokes were shown more than once in the matrix. The Table 1 shows the distribution of the strokes according to the audience's choices (Fig. 3).

Table 1. Distribution of strokes according to people's choices.

Stroke groups	Number of strokes
Human-made recognised	927
AI-generated mistaken for human	816
Human-made not recognised	773
AI-generated not selected	902

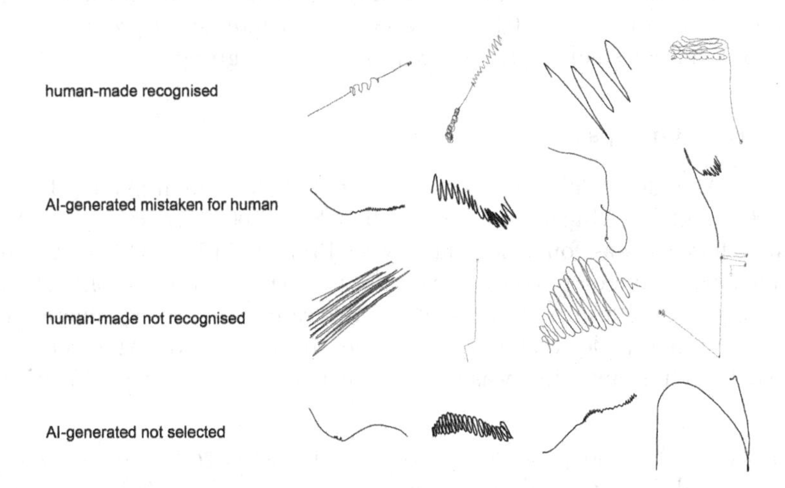

Fig. 3. Some samples from the 4 groups of strokes

For each image in the 4 stroke groups, we analyse and compare two visual features that could be involved in the audience differentiation strategy: the entropy and the symmetry. The entropy [44] is a statistical measure of randomness of the visual stimulus, calculated at the pixel level. Whereas, symmetry is a visual property particularly involved in aesthetic judgment [37] and that the human brain is highly specialised in detecting in nature [25]. We obtained the symmetry from the method proposed in Loy and Ekludun [26]. The feature extracted for each image is a polar coordinate that we converted into a Cartesian coordinate. Then, we calculated the distance from the origin (0, 0) and considered this as a scalable value of symmetry.

4 Results

4.1 Turing Test of Artistic Creativity

Considering the 225 completed tasks, in 63.7% of the cases, the audience was able to select the human-made strokes above the chance.

4.2 Behavioural Results

For each of the two classes of strokes (human-made and AI-generated) a one-way ANOVA was conducted to compare the effect of time response as Independent Variable (IV) on performance as Dependent Variable (DV) in the three groups *fast, medium, slow.* In the human-made class, there was a significant effect of time response $F(2,120)=3.158$, $p=0.04$. Post-hoc comparisons using the Games-Howell test indicated that the mean performance score for the *fast* group ($M=0.994$, $SD=0.007$) was significantly different than the one in the *slow* group ($M=0.987$, $SD=0.144$), $p=0.036$. However, the *medium speed* group ($M=0.990$, $SD=0.138$) did not significantly differ from the other groups.

4.3 Stroke Groups

Entropy. An independent t-test was conducted on the human-made ($M=15.997$; $SD=0.003$) and the AI-generated ($M=15.997$; $SD=0.002$) stroke groups. No significant difference was found for entropy $t(3158)=-0.512$, $p=0.609$. A one-way ANOVA was conducted to compare the effect of (DV) entropy on the (IV) audience choice in the four groups: a) human-made recognised; b) AI-generated mistaken for human-made; c) human-made not recognised; d) AI-generated not selected. No significant effect was found for entropy $F(3,3414)=0.267$, $p=0.849$.

Symmetry. An independent t-test is conducted on the human-made ($M=132.51$; $SD=60.26$) and the AI-generated ($M=151.84$; $SD=57.32$) stroke groups. A highly significant difference was found for symmetry $t(1253)=-5.225$, $p<0.001$. A one-way ANOVA was conducted to compare the effect of (DV) symmetry on the (IV) audience choice in the four groups: a) human-made recognised; b) AI-generated mistaken for human-made; c) human-made not recognised; d)AI-generated not selected. We find a highly significant effect for symmetry $F(3,1251)=9.368$, $p < 0.001$. Post-hoc comparisons using the Tukey (HSD) test indicated a highly significant difference in symmetry between the human-made strokes ($M=134.03$; $SD=57.79$) and the AI-generated mistaken for human-made ($M=150.36$; $SD=58.01$), $p=0.007$; a highly significant difference in symmetry between the human-made strokes ($M=134.03$; $SD=57.79$) and the AI-generated not selected (M=153.12; SD=56.74), p=0.001. Furthermore, a highly significant difference in symmetry between the AI-generated mistaken for human-made (M=150.36; SD=58.01) and the human-made strokes not recognised as such (M=130.45; SD=63.58), p=0.002; a highly significant difference in symmetry between the human-made strokes not recognised as such (M=130.45; SD=63.58) and the AI-generated not selected (M=153.12; SD=56.74), p<0.001 (Table 2).

Table 2. Comparison of significant differences in symmetry among the stroke groups

Human-made not recognised	AI-generated not selected	p<0.001
Human-made recognised	AI-generated not selected	p=0.001
AI-generated mistaken for Human	Human-made not recognised	P=0.002
AI-generated mistaken for Human	Human-made recognised	p=0.007

5 Discussion

The result of our TT of artistic creativity shows that, in a significant majority of cases, people were able to recognise human-made strokes among the AI-generated ones. We can exclude the idea that the entropy of the images displayed during the exhibition is involved in the audience judgement because the difference between the two groups of strokes was not significant. In contrast, the analysis shows a significant difference in symmetry between the strokes made by a human hand and the ones artificially generated with SketchRNN. Therefore, we might suggest that symmetry is one of the features used by the audience to recognise the human-made strokes. Nevertheless, a deeper analysis of the individual groups of strokes selected by the audience indicates that this is not always valid.

For instance, there is a highly significant difference in symmetry between the strokes generated with SketchRNN but never selected by the audience, and the human-made strokes recognised by the audience (p=0.001) as well as the AI-generated ones that fooled the audience and the human-made ones which were not recognised as such (p=0.007). This still points to symmetry as a key factor in the differentiation strategy. However, the most significant difference is found between the human-made strokes not recognised by the audience and the AI-generated strokes not selected (p<0.001). Although we cannot assert that the AI-generated strokes not selected by the audience were actually recognised as artificially generated, we can still ask why, if there is such an evident difference of symmetry between these two groups, the human-made strokes were not recognised? Furthermore, the difference in symmetry between the human-made strokes recognised and not recognised is statistically not significant (p=0.944), meaning that this feature might not be determinant for the final choice. Similarly, when we look at the images artificially generated and mistaken for humans', we can see that their symmetry is significantly different from the group of human-made strokes not recognised as such (p=0.002). However, the symmetry of the AI-generated images that fooled the audience are not significantly different from the ones that did not fool them (p=0.891).

Further analysis of the behavioural data might offer a deeper insight into this matter. Our results show that people responding faster performed significantly better than the slower ones. The average response time of the slow group was around 3.2 s, whereas the faster people average time response was just below 500 ms, a threshold usually associated with non-conscious processes [16] and compatible with an intuitive thinking [3,23]. Therefore, the results indicate that the differentiation might have happened before conscious thought.

One factor to take into account in our analysis is that the majority of the people interacting with the installation are art-goers. Therefore, it is likely they have a certain familiarity with drawings and this might contribute to the results of our TT of artistic creativity. However, in discussing empirical studies on drawing and perception, Pignocchi [35] explains how even people without formal training are able to recognise drawings made with dexterity or spontaneity, before activating any sort of propositional knowledge. This shows how complex visual features can be processed by the human brain even before activating conscious cognitive processes. Pignocchi creates a case for what he calls the "Motor Perception Hypothesis" (MPH) according to which the simple sight of a drawing can unconsciously inform the viewers about the artist's movements. In the same direction, more recent research in neuroaesthetics [41,42,47] demonstrated a strict correlation between drawings and the sensorimotor system, such that one can think of drawings as physical gestures that leave a trace behind [47]. Some of these studies show how simple exposure to static drawing strokes activates in the viewers' brain regions associated with the Mirror Neuron System [41,42], a group of neurons that are argued to be responsible for mechanisms of social cognition, empathy and Embodied Simulation (ES) [17,18]. Similarly to the MPH [35], the Embodied Simulation theory proposes that the viewers can non-consciously embody the artists' gestures by simply looking at their drawing strokes.

Considering our results, in particular how intuitive thinking affected the audience's ability to distinguish human-made strokes, we suggest that the properties of the lines involved in the stroke differentiation were processed by the viewers before the conscious cognition. In the ES framework, the brain of the viewer activates as if it was themselves executing the artist's gesture. Therefore, we can speculate that, at least in our case, the model was not able to learn and synthesise such properties in the generative strokes. If human-made strokes are embodied by viewers as the actions of another human, we should ask whether our generative strokes were processed on a non-conscious level as non-human, perhaps because they were not compatible with human gestures. In that case, the differences between human-made and artificially generated strokes might reside not in their mere visual features, rather in the very process of their making.

6 Conclusion

This paper presented the interactive art installation "Grammar#1" exhibited at the Albumarte gallery in Rome and a study conducted with data collected during the audience interactive experience. The art piece questions the audience about the concept of *human* and *artificial*, tapping into their primordial cognition by using abstract drawing strokes. The creative process involved in the making of the art work included the creation of 300 automatic drawings (LIB Dataset) which were used to obtain a larger dataset named in this text *Human Strokes* and to train SketchRNN to generate a second dataset called *Artificial Strokes*. The installation was designed as a TT of artistic creativity [2] where the audience were asked to recognise human-made strokes in different tasks. The data gathered from one of these tasks was used for a study conducted in naturalistic setting where we looked at the behavioural patterns of the audience. Finally, we analysed at pixel level two visual features of the strokes selected by the audience, specifically the entropy and symmetry. The results show that people were able to recognise human-made strokes above the chance in the 63.7% of the cases and we argued that the images' entropy was not involved in their judgment. Although there is a significant difference in symmetry between the human-made and AI-generated images, the discrepancies among the subgroups of selected images led us to suggest that this feature alone was not key for the distinction. Results suggests that the audience achieved better results when using intuitive judgment. We point to the theories of the "Motor Perception Hypothesis" [35] and "Embodied Simulation" [17,18] to explain the possible dynamics involved in the evaluation. In conclusion, considering our results, the model obtained from training SketchRNN with our dataset was only partially able to learn from the artist's drawings. Further research exploring the process of human-made drawings and people's perception of drawing at a stroke level might eventually produce generative results more difficult to differentiate for the human eye. At the same time, a model able to generate strokes as if they were drawn by human hands, might contribute to the knowledge of human drawing.

Acknowledgements. EPSRC and AHRC Centre for Doctoral Training in Media and Arts Technology (EP/L01632X/1).

References

1. Berger, I., Shamir, A., Mahler, M., Carter, E., Hodgins, J.: Style and abstraction in portrait sketching. ACM Trans. Graph. **32**(4), 1–12 (2013)
2. Bishop, M., Boden, M.A.: The Turing test and artistic creativity. Kybernetes **39**(3), 409–413 (2010)
3. Bowers, K.S., Regehr, G., Balthazard, C., Parker, K.: Intuition in the context of discovery. Cogn. Psychol. **22**(1), 72–110 (1990)
4. Breton, A.: Manifestoes of Surrealism, vol. 182. University of Michigan Press, Ann Arbor (1969)

5. Buk, A.: The mirror neuron system and embodied simulation: clinical implications for art therapists working with trauma survivors. Arts Psychother. **36**(2), 61–74 (2009)
6. Burns, R.C., Kaufman, S.: Kinetic Family Drawings (KFD), An introduction to understanding children through kinetic drawings. Brunner/Mazel (1970)
7. Casati, R., Pignocchi, A.: Communication advantages of line drawings. Pragmalingüística, pp. 000–000 (2007)
8. Chamberlain, R., Mullin, C., Scheerlinck, B., Wagemans, J.: Putting the art in artificial: aesthetic responses to computer-generated art. Psychol. Aesth. Creat. Arts **12**(2), 177 (2018)
9. Colton, S.: The painting fool: Stories from building an automated painter. In: Computers and Creativity, pp. 3–38. Springer (2012)
10. Crowther, P.: What Drawing and Painting Really Mean: The Phenomenology of Image and Gesture. Taylor and Francis, London (2017)
11. Daniele, A., Song, Y.: Artistic assemblage. In: xCoAx 2019: Proceedings of the Seventh Conference on Computation, Communication, Aesthetics and X. pp. 240–255. xCoAx 2019 (2019)
12. Daniele, A., Song, Y.Z.: Ai+Art=Human. In: Proceedings of the 2019 AAAI/ACM Conference on AI, Ethics, and Society. pp. 155–161 (2019)
13. Das, A., Yang, Y., Hospedales, T., Xiang, T., Song, Y.Z.: Béziersketch: a generative model for scalable vector sketches. In: Vedaldi, A., Bischof, H., Brox, T., Frahm, J.M. (eds.) Computer Vision - ECCV 2020, pp. 632–647. Springer International Publishing, Cham (2020)
14. DeLoache, J.S.: Becoming symbol-minded. Trends Cogn. Sci. **8**(2), 66–70 (2004)
15. Elgammal, A., Liu, B., Elhoseiny, M., Mazzone, M.: Can: Creative Adversarial Networks generating "art" by learning about styles and deviating from style norms. In: 8th International Conference on Computational Creativity, ICCC 2017. Georgia Institute of Technology (2017)
16. Fifel, K.: Readiness potential and neuronal determinism: new insights on Libet experiment. J. Neurosci. **38**(4), 784–786 (2018)
17. Freedberg, D., Gallese, V.: Motion, emotion and empathy in esthetic experience. Trends Cogn. Sci. **11**(5), 197–203 (2007)
18. Gallese, V.: Embodied simulation: from neurons to phenomenal experience. Phenomenol. Cognit. Sci. **4**(1), 23–48 (2005)
19. Gooch, B., Reinhard, E., Gooch, A.: Human facial illustrations: creation and psychophysical evaluation. ACM Trans. Graph. **23**(1), 27–44 (2004)
20. Goodfellow, I., et al.: Generative Adversarial Nets. In: Advances in Neural Information Processing Systems, pp. 2672–2680 (2014)
21. Ha, D., Eck, D.: A neural representation of sketch drawings. In: International Conference on Learning Representations (2018)
22. Ives, S.W.: The development of expressivity in drawing. Br. J. Educ. Psychol. **54**(2), 152–159 (1984)
23. Kahneman, D.: Thinking, Fast and Slow. Macmillan (2011)
24. Li, Y., Song, Y.Z., Hospedales, T.M., Gong, S.: Free-hand sketch synthesis with deformable stroke models. Int. J. Comput. Vis. **122**(1), 169–190 (2017)
25. Little, A.C., Jones, B.C.: Attraction independent of detection suggests special mechanisms for symmetry preferences in human face perception. Proc. R. Soc. B: Biol. Sci. **273**(1605), 3093–3099 (2006)
26. Loy, G., Eklundh, J.O.: Detecting symmetry and symmetric constellations of features. In: European Conference on Computer Vision, pp. 508–521. Springer (2006)

27. Maclagan, D.: Line Let Loose: Scribbling, Doodling and Automatic Drawing. Reaktion Books (2013)
28. Manning, E., Massumi, B.: Thought in the Act: Passages in the Ecology of Experience. U of Minnesota Press (2014)
29. McCorduck, P.: Aaron's Code: Meta-art, Artificial Intelligence, and the Work of Harold Cohen. Macmillan (1991)
30. Moffat, D., Kelly, M.: An investigation into people's bias against computational creativity in music composition. In: The Third Joint Workshop on Computational Creativity. ECAI 2006, Universita di Trento (2006)
31. Molnar, V.: Toward aesthetic guidelines for paintings with the aid of a computer. Leonardo **8**(3), 185–189 (1975)
32. Nake, F.: Computer art: a personal recollection. In: Proceedings of the 5th conference on Creativity and Cognition, pp. 54–62 (2005)
33. Noll, A.M.: Human or machine: a subjective comparison of Piet Mondrian's "composition with lines" (1917) and a computer-generated picture. Psychol. Rec. **16**(1), 1–10 (1966)
34. Pease, A., Colton, S.: On impact and evaluation in computational creativity: a discussion of the Turing test and an alternative proposal. In: Proceedings of the AISB Symposium on AI and Philosophy, vol. 39 (2011)
35. Pignocchi, A.: How the intentions of the draftsman shape perception of a drawing. Conscious. Cogn. **19**(4), 887–898 (2010)
36. Ragot, M., Martin, N., Cojean, S.: AI-generated vs. human artworks. a perception bias towards artificial intelligence? In: Extended Abstracts of the 2020 CHI Conference on Human Factors in Computing Systems, pp. 1–10 (2020)
37. Ramachandran, V.S., Hirstein, W.: The science of art: a neurological theory of aesthetic experience. J. Conscious. Stud. **6**(6–7), 15–51 (1999)
38. Reas, C., Fry, B.: Processing: A Programming Handbook for Visual Designers and Artists. MIT Press, London (2007)
39. Riaz Muhammad, U., Yang, Y., Song, Y.Z., Xiang, T., Hospedales, T.M.: Learning deep sketch abstraction. In: Proceedings of the IEEE Conference on Computer Vision and Pattern Recognition, pp. 8014–8023 (2018)
40. Sayim, B., Cavanagh, P.: What line drawings reveal about the visual brain. Front. Hum. Neurosci. **5**, 118 (2011)
41. Sbriscia-Fioretti, B., Berchio, C., Freedberg, D., Gallese, V., Umiltà, M.A.: ERP modulation during observation of abstract paintings by franz kline. PLoS ONE **8**(10), (2013)
42. Thakral, P.P., Moo, L.R., Slotnick, S.D.: A neural mechanism for aesthetic experience. NeuroReport **23**(5), 310–313 (2012)
43. Tresset, P., Leymarie, F.F.: Portrait drawing by Paul the robot. Comput. Graph. **37**(5), 348–363 (2013)
44. Tsai, D.Y., Lee, Y., Matsuyama, E.: Information entropy measure for evaluation of image quality. J. Digit. Imaging **21**(3), 338–347 (2008)
45. Turing, A.M.: Computing machinery and intelligence. In: Parsing the Turing Test, pp. 23–65. Springer (2009)
46. Yu, Q., Yang, Y., Liu, F., Song, Y.Z., Xiang, T., Hospedales, T.M.: Sketch-a-net: a deep neural network that beats humans. Int. J. Comput. Vis. **122**(3), 411–425 (2017)
47. Yuan, Y., Brown, S.: The neural basis of mark making: a functional MRI study of drawing. PLoS ONE **9**(10), (2014)

Mixed-Initiative Level Design with RL Brush

Omar Delarosa[1(✉)], Hang Dong[1], Mindy Ruan[1], Ahmed Khalifa[1,2], and Julian Togelius[1,2]

[1] New York University, Brooklyn, NY 11201, USA
{omar.delarosa,hd1191,mr4739,ahmed.khalifa}@nyu.edu
[2] modl.ai, Copenhagen, Denmark
julian@togelius.com

Abstract. This paper introduces *RL Brush*, a level-editing tool for tile-based games designed for mixed-initiative co-creation. The tool uses reinforcement-learning-based models to augment manual human level-design through the addition of AI-generated suggestions. Here, we apply *RL Brush* to designing levels for the classic puzzle game *Sokoban*. We put the tool online and tested it in 39 different sessions. The results show that users using the AI suggestions stay around longer and their created levels on average are more playable and more complex than without.

Keywords: Mixed initiative tools · Reinforcement learning · Procedural content generation

1 Introduction

Modern games often rely on procedural content generation (PCG) to create large amounts of content autonomously or with limited human input. PCG methods can achieve many different design goals as well as enable particular aesthetics. Incorporation of PCG methods can streamline time-intensive tasks such as designing thousands of unique tree assets for a forest environment. By off-loading these tasks to AI, the time constraints put on game developers and content creators can be relaxed freeing them up to work on other tasks for which AI may be less-well suited. Through such blending of AI and the human touch a system of human and AI co-creation yields not only unique game content the human designer may not have even considered alone but also enables new creative directions [14].

In Procedural Content Generation via Reinforcement Learning, or *PCGRL* [9], levels are first randomly generated and then incrementally improved. The generated levels are initially good enough that they could–though unlikely good enough that they *would*–be used by a human designer. That reluctance to use a level could arise from a level's misalignment with the human designer's needs and they would likely have to keep generating new levels until they find one that is satisfactory. Generally speaking, the human designer exerts minimal control

J. Romero et al. (Eds.): EvoMUSART 2021, LNCS 12693, pp. 412–426, 2021.
https://doi.org/10.1007/978-3-030-72914-1_27

over the resulting level's features and may end up generating many just to find one that suits their needs.

In order to make this level generation method more compatible with a human designer's workflow, we leverage the incremental nature of PCGRL in building a mixed-initiative level-editing tool. This paper presents *RL Brush*, a human-AI collaborative tool that balances user-intent and AI model suggestions. *RL Brush* allows a human designer to create levels as they please while continuously suggesting incremental modifications to improve the level using an ensemble of AI models. The human designer may choose to accept suggestions as they see fit or reject them. The tool thereby aims to assist and empower human designers to create levels that are good, unique, and suitable to the user's objectives.

2 Related Work

Procedurally generated content has been used in games since the early 1980s. Early PCG-based games like *Rogue* (Michael Toy 1980) used PCG to expand the overall depth of the game by generating dungeons as well as coping with the hardware limitations of the day [14,21]. This section will lay out more contemporary applications and methods of generating game content procedurally, with a focus on the use of reinforcement-learning-based approaches.

2.1 PCG via Reinforcement Learning

Reinforcement Learning (RL) is a Machine Learning technique where, typically, an agent takes action in an environment at each time-step and desirable actions are reinforced, interpreted as state and reward, from the environment [18]. Most of the RL work in games focuses on playing. We suspect it may be due to the direct and easy way of representing the game playing problems as Markov decision processes. On the other hand, representing content generation as a Markov decision process poses challenges and hence could explain the disproportionally smaller number of works using RL in game content generation problems.

Of the existing works describing RL approaches to game content generation, a few approaches stand out and demonstrate the breadth of possibilities. Chen et al. [5] demonstrate using Q-learning to create a card deck for collectable cards games. Guzdail et al. [7] have shown how active learning can be used to adjust an AI agent to adapt to user choices while creating levels for Super Mario Bros. PCGRL [9] introduces reinforcement learning into level generation by seeing the design process as a sequential task. Different types of games provide information on the design task as functions: an evaluation function that assesses the quality of the design and a function that determines whether the goal is reached. RL agents that *play* out the content generation task defines the state space, action space, and transition function. For typical 2D grid based games, the state can be represented as a 2D array or 2D tensor. Agents of varying representation observe and edit the map using different patterns. The work demonstrates how three types of agents, namely *narrow*, *turtle* and *wide*, can respectively edit tiles in a sequential manner, move on the map in a turtle-graphics-like way and modify the passed tiles, or have control to select and edit any tile in the entire map.

2.2 PCGRL Agents

The three RL-based level-design agents introduced in *PCGRL* [3,9] as *narrow*, *turtle* and *wide* have origins in search-based approaches to level-generation, however the primary focus of the subsequent sections will be on their RL-based implementations. This section describes these three canonical agent types.

Narrow. The *narrow* agent observes the state of the game and a *location* (x, y) on the 2D-array grid representation of the game level. Its action space consists of a *tile-change* action: whether to make a change or not at location (x, y) and what that change would be.

Turtle. Inspired by turtle graphics languages such as *Logo* [6,9], *turtle* agent also observes the state of the grid as a 2D array and a *location* (x, y) on that grid. Like *narrow* agent, one part of its action-space is defined as a *tile-change* action. Unlike *narrow*, its action space also includes a *movement-action* in which the agent changes the agent's current position on the grid to (x', y') by applying a 4-directional translation on its *location* moving it either *up*, *down*, *left* or *right*.

Wide. The *wide* agent also observes the state of the grid as a 2D array. However, it does not take a *location* parameter. Instead, its action space selects a *location* on the grid (x, y) as the *affected location* and a *tile-change* action.

2.3 PCG via Other Machine Learning Methods

In addition to RL, other machine learning (ML) approaches have also been applied to procedural content generation and mostly based on supervised or unsupervised learning; the generic term for this is Procedural Content Generation via Machine Learning (PCGML) [16]. *Mystical Tutor* [17], an iteration on the Twitter bot `@RoboRosewater`, generates never-before-seen *Magic: The Gathering* cards using an Long short-term memory (LSTM) neural network architecture. While Torrado et al. [19] demonstrate that *Legend of Zelda* (Nintendo 1986) levels can be generated using generative adversarial networks (GAN). However, one distinction that arises when comparing PCGRL with other types of PCGML: PCGRL does not strictly require ahead-of-time training data. RL-based models utilize a system of reward functions instead, which can be either manually designed or themselves learned. PCGRL's system of incremental approach to level-generation also distinguishes it from more holistic ML approaches such as many GAN-based PCGML approaches. At each time step, the agent takes an action such as moving to or selecting a certain position (for example, in 2D grid space) or changing the tile at the current position. This characteristic of PCGRL makes it well-suited for mixed-initiative design.

2.4 PCG via Mixed-Initiative Level Design

In mixed-initiative design, the human and an AI system work together to produce the final content [20,22]. Multiple mixed-initiative tools for game content creation have been developed in recent years. *Tanagra* [15] is a prototype mixed-initiative tool for platformer level design in which AI can either generate the entire level or fill in the gaps left by human designers. *Sentient Sketchbook* [10] is a tool for designing a *Starcraft*-like (Blizzard 1998) strategy game. Users can sketch in low-resolution and create an abstraction of the map in terms of player bases, resources, passable and impassable tiles. It uses feasible-infeasible two population GA (FI-2pop GA) for novelty search and generates several map suggestions as users are sketching. An example of a mixed-initiative PCG tool that generates levels for a specific game is *Ropossum*, which creates levels for the physics-based puzzle game *Cut the Rope*, based on a combination of grammatical genetic programming and logic-constrained tree search [12,13]. Another such example is the mixed-initiative design tool for the game *Refraction*, which teaches fractions; that tool is built around a constraint-solver which can create puzzles of specific difficulty [4].

More recently, Alvarez et al. [2] introduced Interactive Constrained MAP-Elites for dungeon design, which offers similar suggestion-based interaction supported by MAP-Elites algorithm and FI-2pop evolution. Guzdial et al. [8] proposed a framework for co-creative level design with PCGML agents. This framework uses a level editor for *Super Mario Bros* (Nintendo 1985), which allows the user to draw with a palette of level components or sprites. After finishing one turn of drawing, the user clicks the button to allow the previously trained agent to make additions sprite-by-sprite. This tool is also useful for collecting training data and for evaluating PCGML models. In a similar vein, Machado et al. used a recommender system trained on databases of existing games to recommend game elements including sprites and rules across games [11].

3 Methods

This section introduces *RL Brush*, a mixed-initiative level-editing tool for tile-based games that uses an ensemble of trained level-design agents to offer level-editing suggestions to a human user. Figure 1 shows a screenshot of the tool[1]. The present version of *RL Brush* is tailored for building levels for the classic puzzle game *Sokoban* (Thinking Rabbit 1982) and generating suggestions interactively.

3.1 Sokoban

Sokoban, or "warehouse keeper" in Japanese, is a classic 2-D puzzle game in which the player's goal is to push boxes to their designated locations within an enclosed space (called goals). The player can only push boxes horizontally or

[1] https://rlbrush.app/.

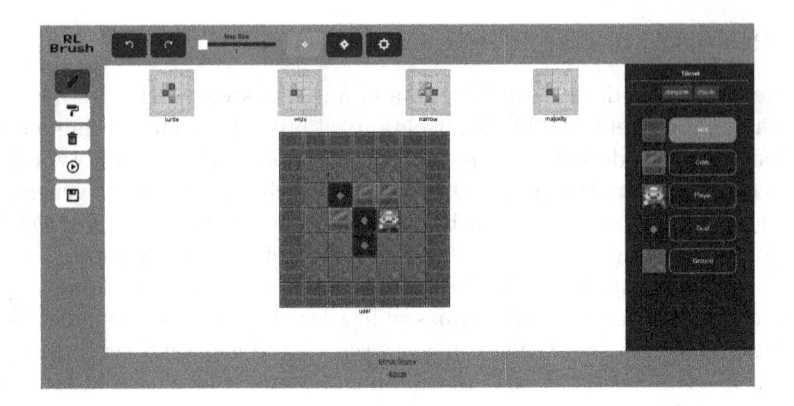

Fig. 1. RL brush screenshot of the Sokoban level editor.

vertically. The number of boxes is equal to the number of designated locations. The player wins when all boxes are in the correct locations.

3.2 RL Brush

In the spirit of human-AI co-creation of tools like *Evolutionary Dungeon Designer* [1] and *Sentient Sketchbook* [10], *RL Brush* interactively presents suggested edits in to a human level creator, 4 suggestions at a time. Instead of using search-based approaches to generate the suggestions *RL Brush* utilizes the reinforcement-learning-based level-design agents presented by [9]. *RL Brush* builds on the work introduced by *PCGRL* [9] by combining user-interactions with the level-designing *narrow*-, *turtle*- and *wide*-agents and an additional *majority*, meta-agent into a human-in-the-loop[2], interactive co-creation system.

3.3 Architecture Overview

Figure 2 shows the system architecture for our tool RL Brush. The system consists of 4 main components:

- **GridView** is responsible for rendering and modifying the current level state.
- **TileEditorView** allows the user to select tools to edit the current level viewed in the GridView.
- **SuggestionView** shows the different AI suggestions from the current level in the GridView.
- **ModelManager** updates all the suggestions viewed in SuggestionView if the current level changed in the GridView.

[2] These are a subclass of AI-based systems that are designed around human interaction being one of their components.

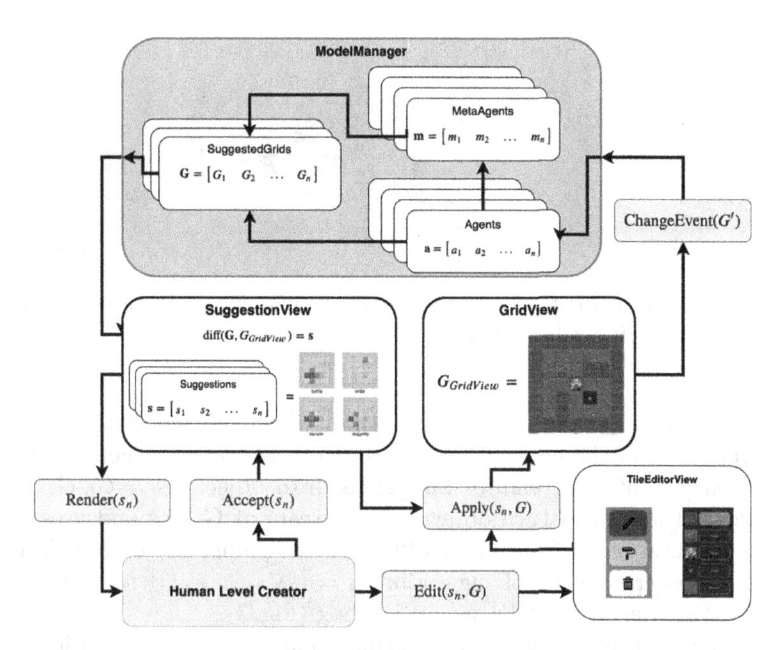

Fig. 2. RL brush system architecture.

The user can edit the current level (G) either by selecting a suggestion from the *SuggestionView* or by using a tool from the *TileEditorView* and modifying directly the map. This change emits a *ChangeEvent* signal to the *ModelManager* component with the new grid (G'). The ModelManager runs all the AI models and collects their results and send the results back to the *SuggestionView*. The *ModelManager* will be described in more details in subsequent section.

3.4 Human-Driven, AI-Augmented Design

Both the `TileEditorView` and the `SuggestionView` respond only to user-interactions in order to ultimately provide the human in the loop the final say on whether to accept the AI suggestions or override them through manual edits. The goal is to provide a *best-of-both-worlds* approach to human and AI co-creation in which the controls of a conventional level-editor can be augmented by AI suggestions without replacing the functionality a user would have expected from a manual tile editor. Instead, the human drives the entire level design process while taking on a more collaborative role with the ensemble of AI level-design agents.

3.5 `ModelManager` Data Flow

The `ModelManager` in Fig. 2 handles the interactions with the PCGRL agents $a = \left[a_0\, a_1\, ...\, a_x\right]$ (where x is the number of used PCGRL agents) and meta-

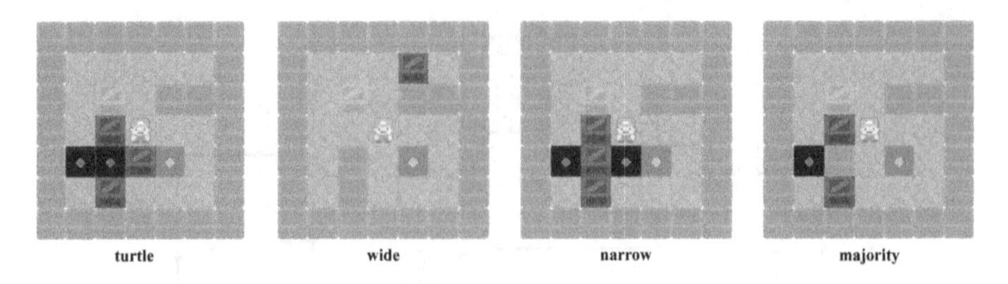

Fig. 3. Each suggestion in the UI is generated by a different agent whose name appears below its diff rendering. Clicking on the suggestion applies it to the grid G.

agents $m = [m_0\ m_1\ ...\ m_y]$ (where y is the number of used meta-agents). The `ModelManager` gets the current level state and sent to these agents where they edit it then it emits a stream of `SuggestedGrid` objects $G = G_0\ G_1\ ...\ G_{x+y}$. The `SuggestionView` in turn observes the stream of **G** lists and uses them to generates suggestions **s** from **G** by diffing them against the current level state $G_{\texttt{GridView}}$ to generate a list of suggestions $s = [s_0\ s_1\ ...\ s_{x+y}]$ for rendering and presenting the user in the UI's suggestion box (Fig. 3).

Meta-agents in **m** consist of agents that combine or aggregate the results of **a** in some way to generate their results. In *RL Brush*, the *majority* agent is an example of a meta-agent that aggregates one or more of the agents suggestions ($[G_{a_i}\ G_{a_{i+1}}\ ...\ G_{a_j}]$) to a new suggestion ($G_{m_i}$). The *majority* meta-agent is powered by a pure, rule-based model that only makes a suggestion of a tile mutation if the majority of the agents have the same tile mutation in their suggestions. In our case, we are using 3 different PCGRL agents (narrow, turtle, and wide) which means at least 2 agents have to agree on the same tile mutation.

3.6 `ModelManager`'s Hyper-Parameters

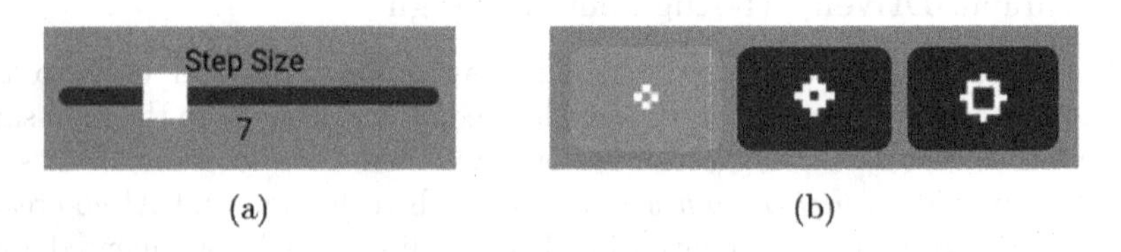

Fig. 4. These two UI elements *a* and *b* control the *step* and *tile radius* parameters respectively.

Two primary hyper-parameters exist in *RL Brush* for tuning the performance of `ModelManager`. One is the number of *steps* and the other is the *tool radius*. These are each controlled from the UI using the components in Fig. 4.

The *step* parameter controls how many times the `ModelManager` will call itself recursively (Fig. 5). For each *step* the `ModelManager` will call itself recursively n

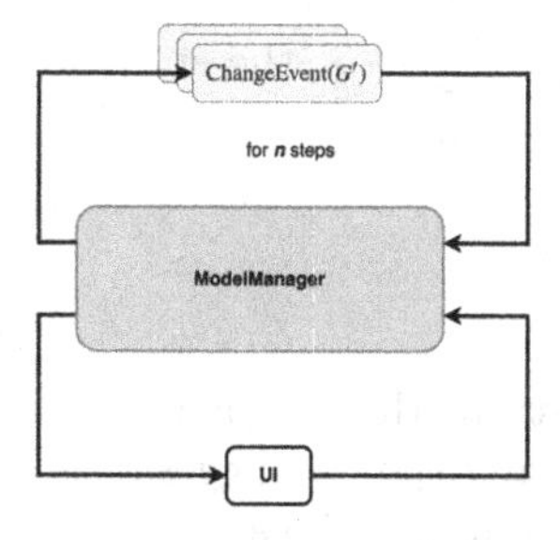

Fig. 5. The changes in the *step* parameter control the number of iterations n in the loop of recursive `ChangeEvent` objects that feed back into the `ModelManager`

times on a self-generated stream of `ChangeEvent` (G') objects. Having a higher step value allows agents to make more than one modification to the map. This is an important hyper-parameter because most of these agents are trained to not be greedy and try to do modification that requires long term edits. Limiting these agents to only see one step ahead will suffocate them and their suggestions might not be very interesting for the users.

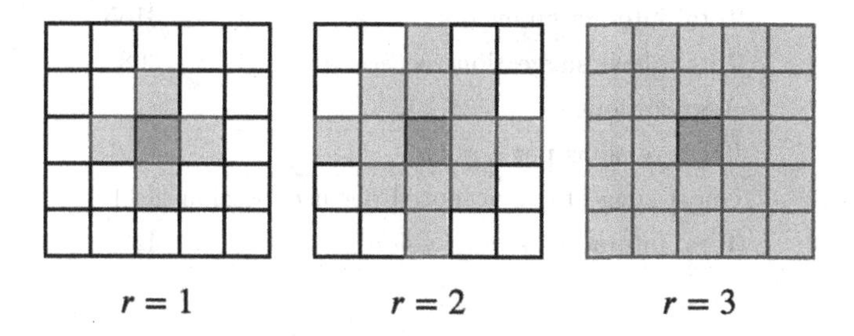

Fig. 6. The changes in the *tool radius* parameter control the size of the slice of grid G' that is visible to the agents as input.

The *tool radius* parameter controls how big the window of tiles are visible to the agent as input. Agents can't provide suggestions outside of this window. It focuses the suggestion to be around the area the user is modifying at the current step. In Fig. 6 the white tiles are padded as empty or as walls, depending on the agent. The red tiles represent the integer values of each tile on the grid G. The green tile represents the *pivot tile* or position on the grid G that the user last clicked on if a tile was added manually. In cases where no tile was clicked[3], the center of the grid G is used as the *pivot tile*. The radius r refers to the Von-Neuman neighborhood's radius with respect to the *pivot tile*. However, note that

[3] Such as the case in which the user accepted an AI suggestion.

for all grids G where $r \geq \left\lceil \frac{gridRadius}{2} \right\rceil$, the entire grid is used such as in cases of $r = 3$ on *microbans* of size 5×5.

4 Experiments

In this section we demonstrate through a user study conducted to study the interactions between users and the AI suggestions. We are primarily interested in answering the following four questions:

- Q1: Do users prefer to use the AI suggestions or not?
- Q2: Does the AI guide users to designing more playable levels?
- Q3: Which AI suggestions yield higher engagement from users?
- Q4: What is the effect of the AI suggestions on the playable levels?

Table 1. Interaction event summary

Total event counts	
Total user sessions	75
Total interaction events	3165
Total ghost suggestions accepted	308
Aggregations	
Level versions per session	10.6
Ghost suggestions accepted per user session	4.11
Total interactions per session	42.2

For the experiment, we published the *RL Brush* app[4] to the web, and shared the link with university students and faculty on a shared Slack channel as well as on social media platforms Twitter and Facebook. We then recorded anonymized user-interaction events to the application web server. During the course of about 2 weeks, 75 unique user sessions were logged in total. Figure 7 shows the final states of a few levels created using a combination of human and edits and AI suggestions in *RL Brush*. Table 1 shows the counts of key metrics that we used to measure the interactions of users and the *RL Brush* UI. For instance, each session resulted in an average of 10.6 level versions, defined as unique levels, throughout each user's total average of 42.2 interactions with the UI (i.e. button presses or clicks) during the course of the session. For example, a user may have generated 2 level versions and 100 interactions in a session by making a single edit and just clicking "Undo" and "Redo" over and over again. From these 10.6 level versions 4.11 were generated using the AI suggested edits or *ghost suggestions.*

[4] https://rlbrush.app/.

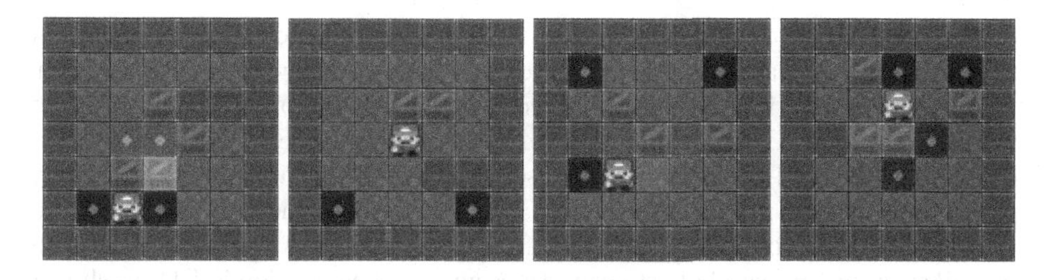

Fig. 7. User-generated levels

Table 2. Statistics on the 39 full session

	Used AI	Didn't use AI	Total
Playable	9	2	11
UnPlayable	8	20	28
Total	17	22	39

5 Results

From these 75 user sessions, 39 sessions were fully logged interaction events during the session from start to finish. We analyzed these sessions on an event-by-event basis and found a few trends. Table 2 shows the statistics about all these 39 fully-logged, sessions. The amount of people that didn't use the AI (22) is slightly higher than the ones used the AI (17). There might be a lot of different reasons that users never engaged with the system, but we suspect the absence of a formal tutorial could have impacted the results here. On the other hand, users that interacted with at least one AI suggestion yielded at more playable levels (9 out of 17) than users did not interact with AI suggestions at all (2 out of 22). There is multiple different factors that could reflect this higher percentage. We think that the main factor is that the AI suggestions nudge/inspire users toward building playable levels.

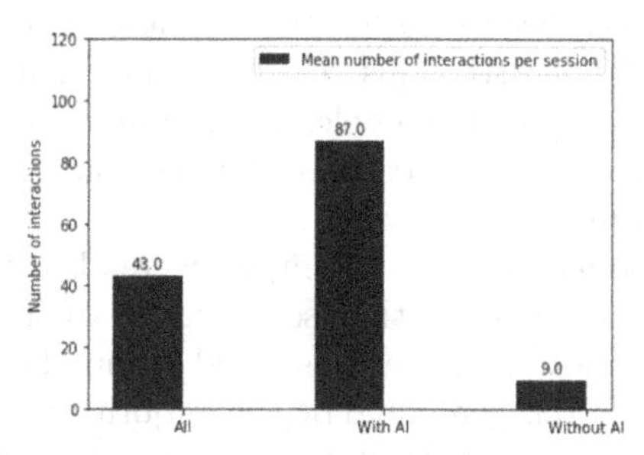

(a) Mean interactions by user cohort.

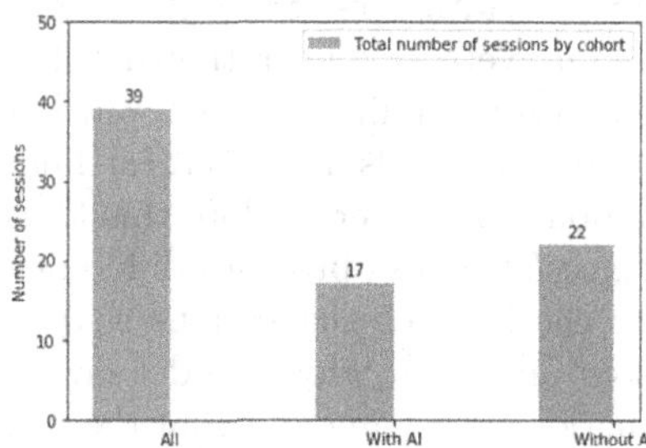

(b) Total sessions by user cohort.

Fig. 8. The number of interactions on average is greater in the cohort of users that accepted at least one AI suggestion during their session.

One such trend, as described in Fig. 8a, users working alongside AI (i.e. users that accepted at least one AI suggestion in their workflow) generate a higher average number of interactions when compared with the other cohort of users working without AI and the combined average of all users. The higher rate of interaction could be attributed to user-engagement, if interpreted positively, or perhaps, if interpreted negatively, that AI increases the complexity of the workflow and requires more clicks to fix any unwanted AI actions. Since the overall average number of ghost suggestions accepted per session is 4.11, as shown on Table 2, we interpret the increase in the interactions to positive engagement than an overwrought system, which we assume would have a higher number of accepted suggestions overall compared to average level versions of 10.6 as frustrated users might fight with the system and produce many more interactions per level version. Of course, other external factors could explain the trend. The small number of users that interacted with the AI at all, as seen in Fig. 8b, could point to a majority of users new to level editing and without having a formal tutorial may have quit before discovering the AI features at all.

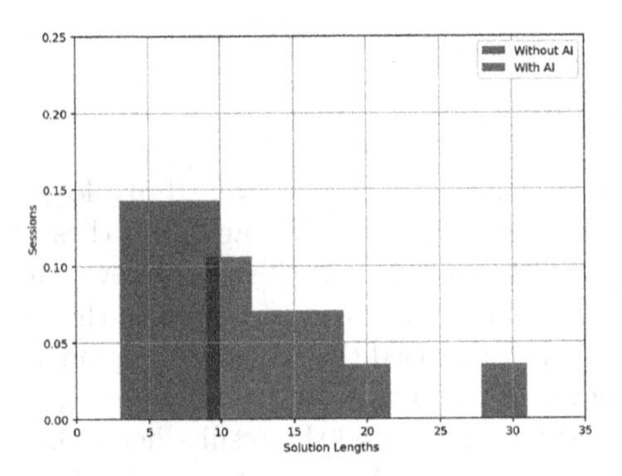

Fig. 9. Users that used AI suggestions seemed to create levels that required more steps in their solutions.

Another trend seems to be a relationship between level complexity and using AI suggestions, as seen in Fig. 9. There the solution length is calculated using a BFS (Breadth-First Search) solver. Each level created with the assistance of AI is, on average, longer than levels without AI. This could indicate that AI suggestions yield direct users toward creating more complex levels and therefore higher quality levels. However, further studies on a larger set of users would need to be done to further explore this trend more definitively.

Since *RL Brush* provides different models to choose from, we were also curious to check which suggestions were most useful for the users. Figure 10 shows a histogram about which model saw the most usage, as measured by number of interactions. We found out that the suggestions generated by the majority-voting meta-agent received the most interactions. One interpretation for its unexpected popularity could be that agents, in isolation, may have divergent lower-quality suggestions but collectively tend toward higher suggestion quality. In the *Discussion* section, we discuss plans for further investigations aggregated suggestions.

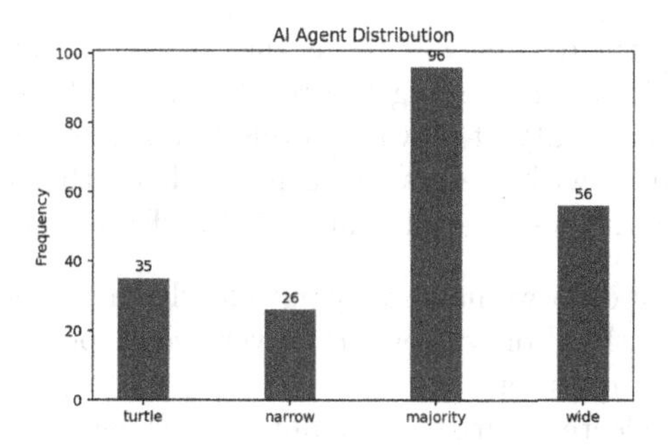

Fig. 10. The *majority* agent seems to be the most popular across sessions and received the most total interactions or clicks.

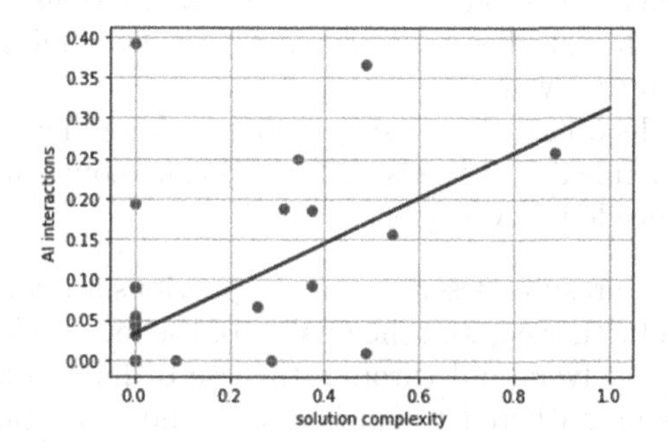

Fig. 11. Correlation between number of AI suggestions accepted and overall solution complexity.

Finally, we compute the correlation between the number of AI-accepted suggestions in a session and the solution length of the created level. Figure 11 shows a correlation (with coefficient equal to 0.279) between the number of AI suggestions used during level creation and the maximum level difficulty achieved during that session, in terms of solution length. This correlation could further make the case that AI suggestions nudge users towards creating more complex levels with longer solutions.

6 Discussion

The system described here can be seen as a proof of concept for the idea of building mixed-initiative interaction on PCG methods based on sequential decisions. Most search-based PCG methods, as well as most PCGML methods, outputs the whole level (or other type of content) as a unit. PCGRL, where the generator has learned to design one step at a time, might afford a mode of interaction more suited to how a human designs a level. It would be interesting to investigate whether tree search approaches to level generation could be harnessed similarly [3].

Looking back at the results, we can say that *RL Brush* introduces a few areas for further exploration surrounding the relationship between user engagement and perhaps the complexity of playable levels. We also noticed that more users seem to interact more with meta-agent compared to other models. Comparing these results with our questions introduced in the Experiments section:

- Q1: Based on the data we have, we can't clearly say if the users preferred to use the system with AI or without. However, we suspect that a further study could answer this question.
- Q2: From the collected statistics the amount of playable levels generated by users that incorporated AI into their workflow exceeds the number of those that did not interact with the AI suggestions at all.
- Q3: The majority agent received the most interactions when compared to all the rest. Further studies could explore new ideas for additional types of meta-agents in future work.
- Q4: The results lean towards AI suggestions yielding higher quality levels, as defined using complexity of levels and with longer solution lengths, but more data would be needed to verify that.

In addition to the results described in the previous section, a broader test of human users could further explore the quality of the levels generated beyond the scope of automated solvers and through the use of human play-testing. Additional metrics can be gathered to support this and more targeted, supervised user research can be done here. A more supervised user research will help us understand the different factor affecting our results. We would know if external factor such as game literacy and familiarity with games and level editor affects the user their engagement with AI. We believe that users with higher game literacy may find the tool less intimidating than ones with lower game literacy. Another important study could be to understand the influence of the AI suggestion on the final created levels: were the AI suggestions pivotal for the final created levels or merely inspiration for heavily hand-made levels?

Once the broader user studies have been conducted, additional client-side models can be added to *RL Brush* that learn the weights of meta-agents and continuously optimize them through online-model training. In this way, we could better leverage the `ModelManager`'s ensemble architecture's capabilities. Furthermore, the existing PCGRL models could be extended to continuously train online using reward functions incorporating parameters based on user actions. Similarly, novel client-side models specifically tailored to improve the UX (user experience) could be incorporated into future versions that better leverage the capabilities of TensorFlow.js, which *RL Brush* utilizes in its code already.

Subsequent versions would also add support for additional games, level-design agent types and $N \times M$ grids in order to increase the overall utility of *RL Brush* as a functional design tool.

7 Conclusion

In the previous sections we have introduced how *RL Brush* provides a way to seamlessly integrate human level editing with AI suggestions with an opt-in paradigm. The results of the user study suggest that using the AI suggestions in the context of level editing could impact the quality of the resulting levels. In general, using AI suggestions seemed to result in more highly playable levels per session and higher overall level quality, as measured by solution length.

There is clearly more work to do in this general discussion. We don't know yet to what types of levels and other content this method can be applied, and there are certainly other types of interaction possible with an RL-trained incremental PCG algorithm. *RL Brush* will hopefully serve as a nexus of discovery in the space of using PCGRL in game-level design.

References

1. Alvarez, A., Dahlskog, S., Font, J., Holmberg, J., Johansson, S.: Assessing aesthetic criteria in the evolutionary dungeon designer. In: Proceedings of the 13th International Conference on the Foundations of Digital Games. pp. 1–4 (2018)
2. Alvarez, A., Dahlskog, S., Font, J., Togelius, J.: Empowering quality diversity in dungeon design with interactive constrained map-elites. In: 2019 IEEE Conference on Games (CoG). pp. 1–8. IEEE (2019)
3. Bhaumik, D., Khalifa, A., Green, M.C., Togelius, J.: Tree search vs optimization approaches for map generation. arXiv preprint arXiv:1903.11678 (2019)
4. Butler, E., Smith, A.M., Liu, Y.E., Popovic, Z.: A mixed-initiative tool for designing level progressions in games. In: Proceedings of the 26th annual ACM symposium on User interface software and technology. pp. 377–386 (2013)
5. Chen, Z., Amato, C., Nguyen, T.H.D., Cooper, S., Sun, Y., El-Nasr, M.S.: Q-deckrec: A fast deck recommendation system for collectible card games. In: Computational Intelligence and Games. IEEE (2018)
6. Goldman, R., Schaefer, S., Ju, T.: Turtle geometry in computer graphics and computer-aided design. Computer-Aided Design **36**(14), 1471–1482 (2004)
7. Guzdial, M., et al.: Friend, collaborator, student, manager: How design of an ai-driven game level editor affects creators. In: Proceedings of the 2019 CHI Conference on Human Factors in Computing Systems. pp. 1–13 (2019)
8. Guzdial, M., Liao, N., Riedl, M.: Co-creative level design via machine learning. arXiv preprint arXiv:1809.09420 (2018)
9. Khalifa, A., Bontrager, P., Earle, S., Togelius, J.: Pcgrl: Procedural content generation via reinforcement learning. arXiv preprint arXiv:2001.09212 (2020)
10. Liapis, A., Yannakakis, G.N., Togelius, J.: Sentient sketchbook: computer-assisted game level authoring (2013)
11. Machado, T., Gopstein, D., Nealen, A., Togelius, J.: Pitako-recommending game design elements in cicero. In: 2019 IEEE Conference on Games (CoG). pp. 1–8. IEEE (2019)
12. Shaker, N., Shaker, M., Togelius, J.: Evolving playable content for cut the rope through a simulation-based approach. In: Ninth Artificial Intelligence and Interactive Digital Entertainment Conference (2013)

13. Shaker, N., Shaker, M., Togelius, J.: Ropossum: An authoring tool for designing, optimizing and solving cut the rope levels. In: Ninth Artificial Intelligence and Interactive Digital Entertainment Conference (2013)
14. Shaker, N., Togelius, J., Nelson, M.J.: Procedural content generation in games.Springer (2016)
15. Smith, G., Whitehead, J., Mateas, M.: Tanagra: A mixed-initiative level design tool pp. 209–216 (2010)
16. Summerville, A., et al.: Procedural content generation via machine learning (pcgml). IEEE Transactions on Games **10**(3), 257–270 (2018)
17. Summerville, A.J., Mateas, M.: Mystical tutor: A magic: the gathering design assistant via denoising sequence-to-sequence learning. In: Twelfth artificial intelligence and interactive digital entertainment conference (2016)
18. Sutton, R.S., Barto, A.G., et al.: Introduction to reinforcement learning, vol. 135. MIT press Cambridge (1998)
19. Torrado, R.R., Khalifa, A., Green, M.C., Justesen, N., Risi, S., Togelius, J.: Bootstrapping conditional gans for video game level generation (2019)
20. Yannakakis, G.N., Liapis, A., Alexopoulos, C.: Mixed-initiative co-creativity (2014)
21. Yannakakis, G.N., Togelius, J.: Artificial intelligence and games, vol. 2.Springer (2018)
22. Zhu, J., Liapis, A., Risi, S., Bidarra, R., Youngblood, G.M.: Explainable ai for designers: A human-centered perspective on mixed-initiative co-creation. In: 2018 IEEE Conference on Computational Intelligence and Games (CIG). pp. 1–8. IEEE (2018)

Creating a Digital Mirror of Creative Practice

Colin G. Johnson(✉)

School of Computer Science, University of Nottingham, Nottingham, UK
Colin.Johnson@nottingham.ac.uk

Abstract. This paper describes an ongoing project to create a "digital mirror" to my practice as a composer of contemporary classical music; that is, a system that takes descriptions (in code) of aspects of that practice, and reflects them back as computer-generated realisations. The paper describes the design process of this system, explains how it is implemented, and gives some examples of the material that it generates. The paper further discusses some broader issues about the technological approach to building creative systems, in particular the advantages and disadvantages of building bespoke algorithms for generating creative content vs. the use of optimisation or learning from examples.

Keywords: Music composition · Computational creativity · Optimisation · Creative practice · Computer composition · Reflective practice · Autoethnography · Research by design

1 Introduction

The aim of this paper is to discuss work-in-progress in creating a computer system that imitates aspects of my artistic practice as a composer of contemporary notated music for (primarily) acoustic instruments. Call this system a *digital mirror*, as it attempts to "mirror" an understanding of artistic practice back to me. A related concept is the *digital twin* [10] in engineering—a computer system that replicates a physical system in detail. Working on this system has encouraged me to reflect on different means of creating artistic generative systems, and the various advantages of these and how they can be combined together.

What are the purposes of this digital mirror? The first is to act as a concrete way of reflecting on practice and techniques. Articulating and discussing aspects of practice in words is a powerful way of understanding and developing that practice. Trying to build an implementation of that technique requires yet more rigour, because code is an unforgiving medium, requiring each assumption to be spelled out clearly and unambiguously. When writing or speaking about an aspect of a practice, it is possible to rely on the reader/listener to fill in gaps and assumptions from their knowledge—or to engage in dialogue to resolve those ambiguities and gaps. By contrast, building a description of aspects of a practice

J. Romero et al. (Eds.): EvoMUSART 2021, LNCS 12693, pp. 427–442, 2021.
https://doi.org/10.1007/978-3-030-72914-1_28

in terms of code cannot rely on any interpretive skill or contextual knowledge from the computer.

The second is to act as a source of inspiration. By having a machine mirror back an interpretation of a creative practice, the creator can understand how clear that articulation is. These realisations, particularly when they produce unexpected results, can help to refine the creator's understanding of their practice. If a machine attempts to realise an articulated aspect of a practice, and produces unexpected content, then that can expose aspects that were not clearly articulated. Furthermore, these outputs—particularly ones that are somewhat unexpected—can act as sources of inspiration for the creator, and push their practice in new directions that might not have arisen from traditional reflection.

This paper begins (Sect. 2) with a discussion of the process of articulating a practice, together with a specific set of examples. It then moves on in Sect. 3 to discuss methods for creating a digital mirror, including a discussion of the advantages and disadvantages of four approaches: writing algorithms to create content, deriving content from statistical distributions, using optimisation to generate content, and generating content by imitating examples. Section 4 then describes the design and implementation of such a system and gives a number of examples of its outputs. Finally, Sect. 5 concludes the paper with a summary, and directions for ongoing work.

2 Articulating Aspects of a Practice

A common way in which to develop a creative practice is to describe the ideas that underlie that practice. This can involve articulating the the techniques and processes used, sources of inspiration and material, and the meaning or emotion that the practice is designed to invoke in an audience. In this paper, I am focusing on the first of these, the articulation of process and technique.

A positive way to think about such processes and techniques is that they are the way in which a particular creative individual realises their distinctive voice as practitioner. It is common to say that a particular practitioner (or a "school") has a particular voice or style. One way in which this cashes out is in the use of specific techniques to shape, process and arrange material. This can be at a number of scales. For musical composition, such techniques can range from methods of the large scale organisation of musical material in specific forms, down to the detailed organisation of musical notes. This is not the only way in which ideas of voice and style cash out. For example, the *subject* of an artform, what it is *about*, can be an important aspect. In more abstract artforms, such as the "music alone" [13] of purely instrumental music, the importance of techniques and processes become more prominent.

There are also negative aspects to such patterns. Overuse of a technique, or use of a technique with little reflection on the wider aesthetics of the piece, can become clichéd. Of course, reflection on and articulation of practice can be used to avoid, develop, or move on from techniques that have become stale.

This articulation is also part of a wider turn in creative practice towards seeing the creative process as a *research* process. that is, not just that creative practice "might on occasion *depend* on research" [8], but that it is a research process in its own right [21]. Whether creative practice can be characterised in this way is controversial. Croft [8] argues that creative practice in musical composition lacks answerable research questions, clear research hypotheses, and a sense of generalisation from specific examples, and is therefore different to a research practice. Reeves [22] argues that insisting on a formulation of research that is focused on such a hypothetico-deductive method is too narrow a view of research, and that creative practice as research affords an opportunity to expand our view of research beyond a science-based paradigm. Similarly, Frayling [9] has emphasised that research in art and design areas can include research "through" creative practice, as well as "into" and "for" practice. An important part of any research process is making the objects of that research clearer through careful definitions and descriptions. This paper focuses on articulating these in the form of computer code.

2.1 Articulating My Practice

In my own practice as a composer of contemporary classical music for (primarily) acoustic instruments, I have developed a number of techniques. Often, such techniques have arisen out of abstractions from improvisatory practice or *bricolage*, particularly in the case of the practices that are concerned with detailed sequences of notes. These are then abstracted into techniques that are used and developed consciously. Larger scale structural elements arise less from such improvisatory processes, and more from reflection on the structure of composition, including what has been successful in my past work, and exemplars from elsewhere.

These techniques exist at various levels of granularity, from the details of note choice and the creation of textures through to large scale structural features. Here are a few prominent examples, with notated excerpts given in Figures 1 and 2.

Slapdash serialism of intervals. Musical lines created by successive notes, where over the course of the line, a large number of different intervals are found between successive notes (example in Fig. 1a).

Layered and mutated melodies. Melodic lines, i.e. sequences of notes with small intervals (usually tones/semitones) between successive notes, often with a regular rhythmic pattern. These are layered on a number of instruments/voices, with the melodic pattern repeating but with mutations—changes of pitch, notes missed out, short gestures repeated. The layerings are uneven, with entries coming in at a variety of times relative to the other instruments (example in Fig. 1b). Often, the melodies are locally tonal, but overall don't have a strong key centre.

Repeated chordal material. Repeated chordal material, sometimes antiphonal (bouncing back-and-forth) between two voices, to provide a static background against other material that is happening at the same time, or to provide a period of stasis in between two more active sections of music (example in Fig. 1c).

Contrasted held material against shorter/moving material. Some instruments/ voices will hold a chord, single note, or a melodic or textural material. Typically, this will be static material, but sometimes a complex texture is used instead—if the texture is sufficiently complex, it becomes a single, unified sound. At the same time, there are short gestures—individual notes, chords, flurries of notes, or spoken text. An example is given in Fig. 1d. Often there is a large difference in dynamics between very soft static material and very loud gestural material.

Trills and tremolo. Passage of trills (rapidly repeating pairs of notes), tremolo (rapid repetition of a single note), approximate trills, rapid repeated short scale passages, sometimes not accurately repeated (example in Fig. 2a, which also illustrates passing use of slapdash serialism of intervals and repeated chordal material).

Large-scale approximate repetition. Large blocks of material return later in a composition; sometimes these are exact repetitions, sometimes they are distorted by changes of instrumentation, dynamics, or harmony; sometimes these repetitions are short extracts from earlier material, typically from the middle of a block of material from earlier in the piece.

Silent blocks and cutouts. One large-scale structural element is the use of long silences in the middle of otherwise coherent musical material, rather than to mark divisions between different sections of contrasting material (example in Fig. 2b). This takes inspiration from ideas from Elliott Carter [3], particularly as realised in his First String Quartet. Relatedly, the repetition of material, but with blocks of notes cut out, perhaps just in some instrumental parts.

It should be noted that these articulations of practice are not necessarily accurate ones. In the midst of a complex process such as composition, it is easy to diverge from a specific practice that you have committed to; it is also notable that these are attempts to abstract from a complicated practice where material is not only influenced (consciously or not) by these practices, but also by reflective, listening processes and the qualia of listening. Furthermore, a single sonic event—say, a note—can be playing multiple roles at the same time: as part of a contrapuntal passage where different lines of music interact, as part of a harmonic structure, and as part of a larger scale structural block. Each of these different roles will impose constraints on groups of notes, meaning that they cannot fulfil their roles in these various techniques in a perfect way (not that that is always desired anyway; the techniques are a means to a musical end, not the end in itself). Indeed, this is one of the reasons for building a digital mirror—to see how the computer resolves these different constraints, and perhaps suggests new ways to combine them.

3 Methods of Creating a Digital Mirror

In creating a digital mirror of a creative practice, what techniques could be used? One approach is to train an AI system in an end-to-end way from a corpus of examples—i.e. a corpus of whole pieces, not examples of the specific practices.

(a) Bars 10–11 of *"Her Fingernails Struck Carillons"* (2001) for small ensemble, cello part.

(b) Bars 1–3 of *Juxtapositions* (2020) for string quartet, violin II and viola parts.

(c) Bars 29–32 of *In Medias Res* (2014) for solo pianist and ensemble, piano part.

(d) Bars 51–53 of *Ballyhoo* (2017) for brass ensemble and organ, brass parts.

Fig. 1. Examples of compositional practice (1)

(a) Bars 76–77 of *"Her Fingernails Struck Carillons"* (2001) for ensemble, percussion, violin and cello parts.

(b) Bars 5–9 of *Five Glimpses into Alternative Universes* (2020) for trombone and piano.

Fig. 2. Examples of compositional practice (2)

In this approach, the various aspects mentioned above would not be represented explicitly in the code of the mirror, but would (perhaps!) be learned by the system from the corpus of examples. This is the approach taken by researchers such as Cope [7] in building systems that pastiche historical music.

This is not the approach taken here. Instead, the approach will be to create a number of computational systems that generate material that is aligned with the description of the kind of material that is generated in the practice. This distillation can be seen as akin to an expert system, but an important difference is that an expert system usually distills the knowledge of a number of experts, whereas in this system there is a single "expert" on the compositional practice. It is a mirror not to the practice as seen externally in the material provided to performance and/or to audience, but a mirror to the practice as articulated from within. A related piece of work is the painting robot AARON [5,16], where the artist Harold Cohen worked over many years on a piece of painting software, modifying the software by reflecting on the successes, failures and serendipities of each version of the software.

There is a resonance here with the use of blackboard systems for poetry generation by Misztal-Radecka and Indurkhya [18]. In that system, various agents that encode expertise in different aspects of poetry (rhyme, metre, metaphor, etc.) take fragments of poetry from a common blackboard, modify that fragment, then return it to the blackboard. In that work, the agents have a rather generic idea of the kind of work that they are trying to create; by contrast, in this paper the focus is on experts that articulate specific aspects of my own practice.

3.1 Algorithmic Methods for Generating Material

An important consideration in building systems of this kind is to consider the advantages and disadvantages of different kinds of processes. For the purposes of this paper, we will group generative processes into four categories:

Algorithmic Generation. In this kind of generative process, code is written that directly generates the material. This could be a deterministic process where the outcomes of the process are easy to predict, a process involving random choices at the micro-level, or an emergent process (whether deterministic or stochastic) where the results are harder to predict (i.e. generative art in the sense of Galanter [11]).

Distributional Generation. In this kind of generative process, material is generated from a statistical distribution. The origins of this are in the observation by Xenakis [29] that for a formal generative system of sufficient complexity, a similar psychoacoustic effect can be generated by sampling from a statistical distribution. As a naive example, consider serialist pitch organisation. As Xenakis [28] has noted, complex compositional frameworks are self-destroying: "what one hears in reality is nothing but a mass of notes in various registers" [28] (translation in [29]). This can be distinguished from the use of randomness in the Algorithm Generation process in that here the probability distributions are used at a macro level to describe large sections of material, whereas in the former randomness is used at a micro-scale to generate individual notes or gestures.

Optimisation. In this approach, the desired outcome is specified in a declarative way—the creator specifies what outcome is wanted, but not how to generate it. This could, for example, be by specifying a fitness function. The actual material is then generated by optimisation guided by that fitness function. These approaches can use statistical distributions, but via a generate-and-optimise process. For example, rather than sampling from a distribution, a measure is constructed of how accurately a particular example matches a distribution, and that measure is used as a fitness function in an optimisation algorithm such as a GA.

Corpus Imitation. In this approach, a number of examples of the use of the technique are collected, and this corpus is then used to train a machine learning algorithm to imitate and generalise from it. In Ritchie's [23] terminology, this is an "inspiring set". However, this inspiring set would not be of whole artworks, but instead of particular exemplars of the use of a specific technique. The idea would be to use machine learning to build models that act as experts on that specific aspect of practice. This is distinguished from the Distributional Generation in that in this approach, the distribution (or other model) is learned from a corpus of examples; whereas in the former, the distribution is crafted directly.

3.2 Example: Slapdash Serialism of Intervals

Consider the example of *slapdash serialism of intervals*, which was one of the techniques that I articulated above. In short, this means generating musical material where, over a decent period of time, the intervallic leaps between successive notes in an instrumental part are, roughly, equally distributed; there are the same number of minor thirds as there are perfect fifths as there are major sixths, etc. How would this material be generated in each of the four algorithmic methods described above? The focus here is on generating a single line of music.

Algorithmic Generation. Notes would be generated one-by-one in sequence, initially at random within a wide range. After each note is generated, the interval between it and the previous note would be stored, and a list of the intervals used built up as the musical line is developed. As the musical line develops, the probability of generating each note would vary depending on its interval from the previous note: if an interval had been heavily used up to that point in the line, the probability of choosing that note would be small; if the interval had been hardly used, the probability would be very high. As a result of this process, over a long musical line, the expected distribution of the intervals would be equal.

Distributional Generation. A uniform distribution of intervals would be created, the length of the musical line to be created decided, and a sample of that length generated at random from the distribution. This would then be used to generate a sequence of notes by starting from a random note, and moving through that list of intervals and using it to generate each note in turn. Note the similarities between this and the previous approach. The contrast between the two approaches is more pronounced when the stochastic, distributional approach is used to generate similar material to a deterministic algorithmic generative approach, as in the Xenakis examples, where sampling from a distribution of pitches, dynamics, note-lengths etc. is used to produce a similar musical effect as deterministic total serialism. There is another important difference. With the distributional approach, it is possible to combine two distributions in a fairly straightforward way (e.g. by forming a joint or conditional distribution). By contrast, in the algorithmic approach, it is not always clear how to combine two algorithms.

Optimisation. A population of random musical lines is generated by sampling uniformly from a chromatic scale of notes in a wide interval. This is the starting population in a genetic algorithm. The fitness function takes the sequence of notes, calculates the successive intervals between notes, counts each time a note has been used, and compares that to a uniform distribution of intervals using a root-mean-square error measure. This is then iterated until a musical line is discovered that has a low value for that error measure.

Corpus Imitation. A number of examples would be provided of musical lines that use this kind of interval distribution, which could then be used as the transition matrix in a generative Markov chain. Alternatively, a machine learning system such as an autoencoder [27] would be used, which takes the corpus of examples and learns to imitate their features.

3.3 Advantages and Disadvantages

All four of the above approaches can, broadly speaking, be used to generate similar material. This section discusses the advantages and disadvantages of each. Is not necessary to choose a single method for the development of a system. Particularly if the system architecture is designed as a collection of agents that generate, evaluate, and transform the material, each aspect could be generated using an appropriate approach.

In particular, this is one of the advantages of optimisation over algorithmic generation. In algorithmic generation, code is written to generate material that uses a particular technique or process. If we subsequently wanted to generate material that uses two processes at once, we would have to rewrite the code to generate material that aligned with both of those processes.

For example, consider trying to combine the *layered and mutated melodies* discussed above with the *trills and tremolos.* To create this in an algorithmic generative way, we would need to write a new piece of code that generated material that had that melodic aspect, whilst simultaneously including appropriate amounts of the trills, tremolos and repeated scale passages. This is not impractical, but it requires a lot of additional work and thought. By contrast, in the optimisation approach, what would be needed would be some way of bringing together a measure of "melodicity" and a measure of "trilliness" using multi-criterion optimisation. Then, an optimisation algorithm could search the space of possible note-sequences that included both of those aspects, without having to have written much new code other than a specification of how that multi-criterion optimisation would be done. Relatedly, a distributional generative approach can often combine different desirable properties by combining the underlying probability distributions.

A related advantage of the optimisation approach over the other approaches is that it is possible to use multi-criterion optimisation to create material that satisfies desirable criteria at different timescales. For example, one problem with the commonly used Markov models for algorithmic generation of musical or textual material is that, whilst the material can have desirable local structure, it lacks global coherence. Trying to achieve these two aims in a generative way is difficult, but again a multi-criterion optimisation approach has the potential to combine desirable properties on different time scales.

One disadvantage of an optimisation approach is that it is is difficult to discover material that requires a very precise placement of items. E.g., it would be time-consuming for an optimisation process to converge precisely onto a long trill between two notes, or to create a period of silence in the middle an otherwise busy texture. To create an approximate repetition of a whole passage of music later in a piece by optimisation of sequences of notes is even more challenging. These problems are multiplied if the optimisation algorithm is also optimising on other criteria. Generating such material is, by contrast, very easy using the algorithmic generation approach. A design pattern that has a lot of promise is to use algorithmic generation to generate this kind of material in the first place, and then to use a fitness function to ensure that aspects of it are retained whilst other processes are acting on the material.

Another advantage of the optimisation approach is that different aspects of composition can be specified by different kinds of functions. As long as a function can return a fitness value, it can be incorporated into the system. By contrast, in the algorithmic generation approach, we have to choose up-front a "theory" of how our system is going to work. E.g., our theory might be that notes are generated successively, with each note value depending on a moving window of previous notes. This makes generation of certain kinds of material fairly straightforward (e.g. the *slapdash serialism* is easy to generate using this theory), but the large scale repetition of material isn't. The distributional generation is limited too—in that case we have already chosen what "theory" is going to be used, i.e. that of sampling from a distribution. If the desired process or technique can be realised as sampling from a probability distribution, then it is straightforward; otherwise, near impossible.

The choice of the word "theory" in the above discussion is designed to resonate with its use in the philosophy of concepts. Early theories of concepts rely on a single "theory" for all concepts [19]. For example, the prototype theory of concepts [24] is built on the theory that each concept is centred around a small number of prototypical examples of the concept, and that we assess whether an object is an example of that concept by closeness of that example to the prototype(s). By contrast, the more recent theory-theory of concepts [26] argues that each concept is associated with its own "theory". This fits amenably with the idea of fitness functions—each concept that we want to include in the generation has its own theory, realised in code. Multi-criterion optimisation does not require the same kind of theory to define each concept, only that it cashes out in the form of a scalar fitness measure. It would be interesting to try to drop even that requirement, and to see the process of multi-criterion optimisation to take the form of a "negotiation" between different desirable outcomes (e.g. based on computational argumentation [4]).

The final approach is learning from a corpus. One advantage that this has is that it can pick up on tacit features of the examples given. The other approaches are limited to those features that the system creator has articulated explicitly in the code, e.g. the generative code or the fitness function. By using machine learning from examples, aspects of those examples that are not obvious to the creator might be extracted; if the machine learning method used is an interpretable one, it may be clear what these features are.

4 Design, Implementation and Examples

Aspects of the above system have been implemented in a Python program, with the *Mido* package [17] being used for the generation of MIDI signals for playback via *Ableton Live* [1], and *MusicXML* [20] export being used to create notation which is realised in the *Sibelius* [25] notation system. A copy of the code can be downloaded from http://www.colinjohnson.me.uk/researchSoftware.php.

4.1 Structure of the System

The two main components of the system are *generators* and *drivers*. Generators are functions that create sequences of notes of a given length, and they (or expressions built from them) are used to create the initial population. A description of the generators available in the system are given in Table 1. Details of the implementations are given in the code mentioned above.

Table 1. Descriptions of generators.

Generator	Description
Create sequence	Creates a sequence of random notes in a given pitch-range (by default, midi notes 60–76), of random durations from 1–4 beats (quaver to minim).
Create even-rhythm sequence	Creates a sequence of random notes in a given pitch range (by default, midi notes 60–76), all quavers
Create trill	Chooses a random note from a given pitch range (by default, midi notes 60–76 and a second note a semitone or tone higher, and then creates a sequence of quavers Alternating between those two notes
Insert gaps	Chooses a number (a user-specified parameter; 2 in the example below) of points in the sequence and adds a silence of between 20–25 quavers length

Drivers are functions that take a sequence and return a number in the range $[0.0, 1.0]$, which measure some aspect of that sequence. These are designed so that fitness functions can be created either by using a single driver, or by combining drivers. A description of the drivers available in the system are given in Table 2.

Table 2. Descriptions of drivers.

Driver	Description
Melodicity	A measure of how closely the distribution of intervals between successive notes in the sequence is concentrated around smaller intervals, in particular whole tones (a similar measure is explored in [2])
Slapdash serialism	A measure of how closely the distribution of intervals between successive notes in the sequence is to a uniform distribution
Trillicity	A measure of how closely the sequence matches with a trill passage
Gappiness	A measure of how much the sequence contains large silent gaps

The main evolutionary system works as follows. The user chooses a generator, and a fitness function which is either one of the drivers or an expression made up from the drivers. Each member of the population consists of a sequence of notes, with the initial population being generated by the specified generator. A genetic algorithm is run, using tournament selection, elitism and a mutation operator only (no crossover). The mutation operator takes each note in the sequence and:

- If it is a note, with probability 0.1 changes the note value to a new value from a Gaussian distribution centred on the current note value and with s.d. 4 semitones, capped above and below by the MIDI notes 76 and 60.
- If it is a note and has not been changed by the above, with probability 0.1 changes it to a rest (if the "mutation to silence" parameter is true).
- If it is a rest, replaces it with a note with MIDI value drawn randomly in the range [60, 76].

The parameters are summarised in Table 3. The parameter values vary between examples, as determined by empirical experimentation. These can be seen as "curated" examples, in that they are designed to show specific features that are of help in understanding the outputs of the system. Systems with a high "curation coefficient" [6] such as these, where the user has to explore a large number of runs or parameter settings, have been identified with systems that have low levels of autonomous computational creativity. For an application such as this, where the point of the system is to facilitate autoethnographic reflective practice, such concerns are less. The point of such a system is not to autonomously generate imitations and pastiches of the human creator's work as such; instead, this exploration of parameter space is part of that reflective process.

Table 3. Genetic algorithm parameters.

Example	Sequence length	Population size	Max generations	Tournament size	Elitism	Mutation to silence?
Example 1	50	1000	100	7	no	no
Example 2	50	1000	100	7	no	no
Example 3	50	50	300	7	yes	yes
Example 4	50	50	1000	7	yes	yes

The system can be used to start from random sequences of notes, and to evolve towards a desired combination of features by specifying a fitness function that combines appropriate drivers. Alternatively, it can be used to create a transition between two kinds of material; as discussed by Magnus [15], one interesting way to use genetic algorithms (or other AI techniques) is to play out the evolutionary process as a performance; what Johnson [12] has called a "seeds and targets" approach.

4.2 Examples

This section of the paper gives a number of examples from the digital mirror. These are a mixture of examples to demonstrate particular points, and a final example that shows how multiple fitness functions can be brought together.

Example 1. This example tests the idea discussed earlier of whether it is difficult to use evolution to evolve an exact trill. The generator used in this experiment is *evenSequence*, which generates random notes in a certain interval, and the driver is *analyseTrillicity*, which measures how close the sequence is to a trill. For simplicity, we turn off the mutation that makes a note into a rest for this example. Samples from the 20th and 100th generations are shown in Fig. 3. There is some aspect of trill, but it has not evolved a perfect trill.

Example 2. This uses the generator *generateTrill* so that the initial population consists of trills on a randomly chosen note. The driver is *analyseMelodicity* + *analyseTrillicity*, that is, we are trying to keep the core trill behaviour, whilst adding melodic aspects to it. As we can see in Fig. 4a, even after 10 generations we can see a good combination of short melodic phrases whilst retaining the trill behaviour; after 100 generations a different example, but with largely similar characteristics (trill passages then some melodic variation) has emerged.

Example 3. The generator used here is *evenSequence* combined with *insertGaps* which inserts silences. The fitness function moves this random note distribution towards a more "melodic" distribution, whilst also retaining those gaps. After 100 generations (Fig. 5), the larger gaps have been retained, but the note patterns have a more melodic feel than the initial random note choices.

Example 4. This shows a more complex situation, where the generator used in is again *evenSequence* combined with *insertGaps*. However, the fitness function combines retaining gaps, creating trills, and creating a "melodic" distribution—all of which are retained/created in the example (Fig. 6).

5 Conclusions and Ongoing Work

This paper has presented ongoing work on an attempt to make a "digital mirror" to reflect my articulations of compositional practice. Creating this has forced me to clearly articulate some of my processes, because code is an unforgiving medium where it is necessary to formally state assumptions. The aim of this is not focused on generating whole pieces of music, but to facilitate reflection.

This is an ongoing project, and there are a number of future developments. One would be to expand the optimisation techniques used, in particular using crossover and multi-criterion optimisation. In terms of musical development, there is much to be developed in terms of dynamics and rhythm, which are very limited in the current version. Another limitation is that the current version is that there is a single line; it would be interesting to explore both multi-line notated music, and multiple interacting musical agents (as explored by Lewis [14]).

(a) Trying to evolve a trill: after 20 generations.

(b) Trying to evolve a trill: after 100 generations.

Fig. 3. Trying to evolve a trill (trill fragments bracketed).

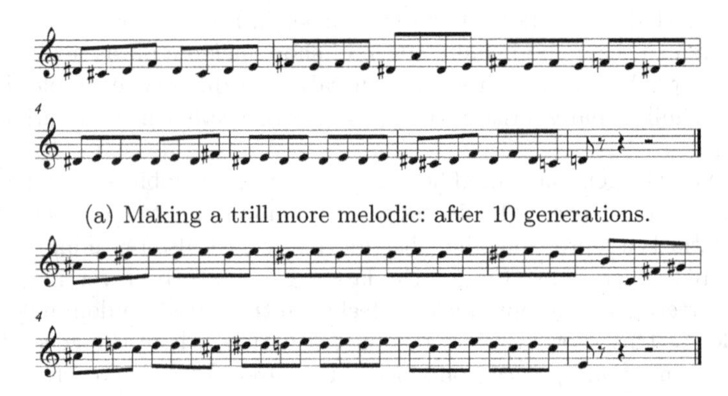

(a) Making a trill more melodic: after 10 generations.

(b) Making a trill more melodic: after 100 generations.

Fig. 4. Making a trill more melodic.

Fig. 5. Evolving from a random sequence of notes to a more "melodic" sequence, whilst retaining gaps: after 100 generations (gaps bracketed).

One key place where the digital mirror falls down is that it—obviously—doesn't replicate the conscious qualia of listening. Musical techniques almost always admit some *choice*, whether of note-level details, or high-level parameter

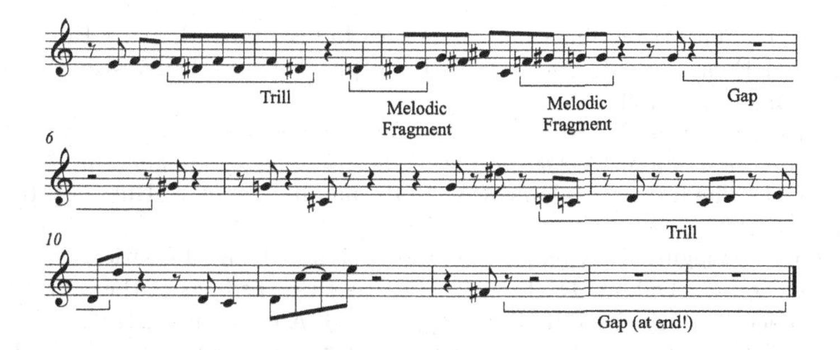

Fig. 6. Evolving with a more complex fitness function: after 800 generations (notable features bracketed).

choice. In a digital mirror, these choices are typically made randomly, with some filtering out by the fitness function. But, this doesn't replicate the reflective process of *listening* to a line of music, and getting a musical understanding. As Croft [8] puts it, there are plenty of ways of "solving problems" about creative practice, but in the end what is being sought are "striking or idiosyncratic *musical* solutions to problems of *musical* material that arise only during the process of composition". A merely formal solution to the question is not sufficient. Perhaps some kind of interactive evolution could afford some progress.

As a personal reflective process, I have overall found this useful. Formalising processes I thought I was using caused me to revise my understanding; e.g. I realised that what I was remembering as primarily trills and tremolos were more commonly rather rougher repetitions of short scale-fragments. In terms of the material that the mirror "reflected back", one example that surprised me was how short the material could be in between silent blocks. A couple of times the machine generated a passage consisting of just a beat or two of sound between long silences, and I found this musically effective. This is not something that I would have thought of without this. Perhaps there is future work to be done in this broad idea of using code as part of reflective practice and autoethnography.

Aside from the design and development of system, the paper has also considered the kinds of methods used; this is likely to be of interest to a wider constituency of people working on creative systems. In particular, the contrast between four methods of generating material in such a system has been discussed.

References

1. Ableton Live. November 2020. https://www.ableton.com/en/live/
2. Calderon Alvarado, F.H., Lee, W.H., Huang, Y.H., Chen, Y.S.: Melody similarity and tempo diversity as evolutionary factors for music variations by genetic algorithms. In: Cardoso, F.A., Machado, P., Veale, T., Cunha, J.M. (eds.) Proceedings of the 11th International Conference on Computational Creativity, pp. 251–254. Association for Computational Creativity (2020)

3. Carter, E.: The time dimension in music. In: Bernard, J. (ed.) Elliott Carter: Collected Essays and Lectures 1937–1995, pp. 224–228. University of Rochester Press, New York (1997)
4. Cocarascu, O., Toni, F.: Argumentation for machine learning: a survey. In: Baroni, P., et al. (eds.) COMMA 2016, pp. 219–230. IOS Press, Amsterdam (2016)
5. Cohen, H.: The further exploits of AARON, painter. Stanford Humanit. Rev. **4**(2), 141–158 (1995)
6. Colton, S., Wiggins, G.: Computational creativity: the final frontier? In: de Raedt, L., Bessiere, C., Dubois, D., Doherty, P. (eds.) Proceedings of the 20th European Conference on Artificial Intelligence, pp. 21–26. Amsterdam (2012)
7. Cope, D.: Computers and Musical Style. Oxford University Press, Oxford (1991)
8. Croft, J.: Composition is not research. Tempo **69**(272), 6–11 (2015). https://doi.org/10.1017/S0040298214000989
9. Frayling, C.: Research in art and design. Royal College of Art Research Papers, **1**(1) (1993–1994)
10. Fuller, A., Fan, Z., Day, C., Barlow, C.: Digital twin: enabling technologies, challenges and open research. IEEE Access **8**, 108952–108971 (2020). https://doi.org/10.1109/ACCESS.2020.2998358
11. Galanter, P.: What is generative art? Complexity theory as a context for art theory. In: Proceedings of the Sixth International Conference on Generative Art (2003). https://www.generativeart.com/on/cic/papersGA2003/a22.pdf
12. Johnson, C.G.: Fitness in evolutionary art and music: a taxonomy and future prospects. Int. J. Arts Technol. **9**(1), 4–25 (2016)
13. Kivy, P.: Music Alone Philosophical Reflections on the Purely Musical Experience. Cornell University Press, Ithaca (1991)
14. Lewis, G.: Too many notes: computers, complexity and culture in Voyager. Leonardo Music J. **10**, 33–39 (2000)
15. Magnus, C.: Evolutionary musique concrète. In: Rothlauf, F., et al. (eds.) EvoWorkshops 2006. LNCS, vol. 3907, pp. 688–695. Springer, Heidelberg (2006). https://doi.org/10.1007/11732242_65
16. McCorduck, P.: AARON's Code: Meta-art, Artificial Intelligence, and the work of Harold Cohen. W.H. Freeman, New York (1991)
17. Mido: MIDI objects for Python. https://mido.readthedocs.io. Accessed November 2020
18. Misztal-Radecka, J., Indurkhya, B.: A blackboard system for generating poetry. Comput. Sci. **17**(2), 265–294 (2016)
19. Murphy, G.L.: The Big Book of Concepts. MIT Press, Cambridge (2004)
20. MusicXML. https://www.musicxml.com. November 2020
21. Pace, I.: Composition and performance can be, and often have been, research. Tempo **70**(275), 60–70 (2016)
22. Reeves, C.: Composition, research and pseudo-science: a response to John Croft. Tempo **70**(275), 50–59 (2016)
23. Ritchie, G.: Assessing creativity. In: Proceedings of the AISB Symposium on Artificial Intelligence and Creativity in Arts and Science, pp. 3–11. AISB Press (2001)
24. Rosch, E.H.: Natural categories. Cogn. Psychol. **4**, 328–350 (1973)
25. Sibelius. https://www.avid.com/sibelius. November 2020
26. Weiskopf, D.A.: The plurality of concepts. Synthese **169**, 145–173 (2009)
27. Welling, M., Kingma, D.P.: An introduction to variational autoencoders. Found. Trends Mach. Learn. **12**(4), 307–392 (2019)
28. Xenakis, I.: La crise de la musique sérielle. Gravesaner Blätter **1**, 2–4 (1955)
29. Xenakis, I.: Formalized Music: Thought and Mathematics in Composition. Indiana University Press, Bloomington (1971)

An Application for Evolutionary Music Composition Using Autoencoders

Robert Neil McArthur(✉) and Charles Patrick Martin

Australian National University, Canberra, Australia
{u6956078,charles.martin}@anu.edu.au

Abstract. This paper presents a new interactive application that can generate music according to a user's preferences inspired by the process of biological evolution. The application composes sets of songs that the user can choose from as a basis for the algorithm to evolve new music. By selecting preferred songs over successive generations, the application allows the user to explore an evolutionary musical space. The system combines autoencoder neural networks and evolution with human feedback to produce music. The autoencoder component is used to capture the essence of musical structure from a known sample of songs in a lower-dimensional space. Evolution is then applied over this representation to create new pieces based upon previously generated songs the user enjoys. In this research, we introduce the application design and explore and analyse the autoencoder model. The songs produced by the application are also analysed to confirm that the underlying model has the ability to create a diverse range of music. The application can be used by composers working with dynamically generated music, such as for video games and interactive media.

Keywords: Algorithmic music composition · Interactive evolutionary computation · Autoencoder neural networks

1 Introduction

Evolutionary art (where art is generated through computation, usually with human feedback) has had a long history. There have been many successes in applying evolutionary algorithms over the years to generate, more commonly visual, but also aural forms of art [18]. This paper presents an interactive application that allows a user to create new songs inspired by the process of biological evolution, as shown in Fig. 1. The application repetitively generates sets of songs which the user can choose from so that the underlying algorithm can create new, similar pieces (each song is represented by a box in the upper left area of the application). Once the user reaches a tune they enjoy, they can save the music.

The application is based upon a novel method for algorithmic music composition, combining autoencoder neural networks and evolution with explicit human feedback to produce music. This method is related to latent variable evolution

J. Romero et al. (Eds.): EvoMUSART 2021, LNCS 12693, pp. 443–458, 2021.
https://doi.org/10.1007/978-3-030-72914-1_29

which was introduced in a paper presenting how the latent input variables of a trained generative adversarial network (GAN) could be evolved to create fingerprints [2]. Despite the successes autoencoders and evolution have had in making music, little work has been done in combining them for interactive generative composition.

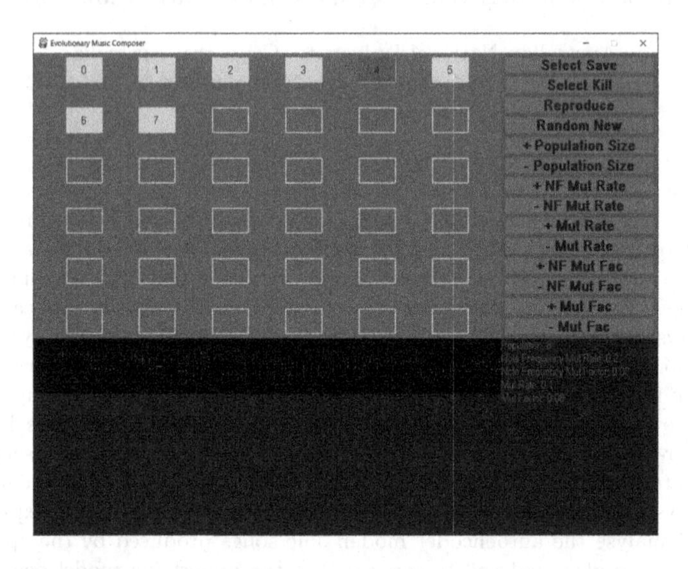

Fig. 1. The evolutionary music composition application. Each numbered square represents an individual song.

For the purposes of algorithmic music composition with human feedback, the algorithm should know some sense of musical structure. Otherwise, it can quickly become possible that the algorithm would become stuck forever in a random mess of notes. This leads to the question - what actually is music? The answer is much deeper than saying music is some ordering of groups of notes. After all, the vast majority of people would not classify something as structured as playing two major scales on a piano simultaneously (with the second scale starting a major seventh above the first - a notoriously dissonant combination) as music. Despite this, there exist some fantastic atonal masterpieces in the world (where the notes need not conform to any scale). The true definition of music is complex, to say the least.

Though often not perfect, autoencoders work exceptionally well at finding hidden structures within data. If we have a dataset of music, we can then use an autoencoder to learn some hidden structure within the songs represented at a much lower dimension. Following this, each of the features can be analysed to find the latent space of the encoding (the range of input which produces meaningful outputs - the features need not be 0-centred for example). Principal component analysis can then be applied to find a meaningful mapping from a standard normal distribution (the principal components with dimension equal to the feature size) to this latent space. For our purposes, the use of an autoencoder solves the problem of defining music. It provides a lower-dimensional representation of music, forming a genotype to which evolution can then be applied.

This application could be used by composers working with dynamically generated music, such as for video games and interactive media. Further, modifications to the underlying algorithm could allow for music to automatically change while it is being played. This could be applied in games so that the accompanying music transforms dependent on metrics such as mood.

In the next section, we discuss related works in the fields of evolutionary art and algorithmic music composition. We then describe our dataset, the autoencoder model, and how evolution has been applied before putting it together in the application. We perform experiments to confirm that the application has the capacity to create a diverse range of musical outputs.

2 Related Works

One of the most well-known works in the field of evolutionary art is the online collaborative service called Picbreeder (an application allowing its users to create a wide range of pictures) [20]. The mechanism behind Picbreeder is the usage of NEAT (NeuroEvolution of Augmenting Topologies) to successively add layers of complexity to each generation of generated images. This is achieved by the use of composition pattern producing networks (CPPNs), mapping the coordinate of a pixel to an intensity value (using a variety of functions to add regularities found in nature). The combination of NEAT and CPPNs allows users to select from a variety of lower complexity images and progressively add more detail through evolution. One of the primary outcomes of this work was the ability to overcome user fatigue (creating good images takes time). Their method countered this by allowing users to start generating new images from others which were made by the online community.

A similar work has shown promise for images by combining GANs with interactive evolution [3]. The paper shows how the latent input variables (to a trained GAN for some class of image) could be evolved to produce new images. The work intended to make it easy for users to evolve images to some conceived target as opposed to Picbreeder, which is more exploratory. Users were able to do as such more successfully for when the GAN was trained on shoes as opposed to faces. The paper gives merit to the idea of combining autoencoders with interactive evolution for generating music, though audio presents its own unique challenges compared to images.

In the field of algorithmic music composition, one recent paper [17] has started looking into this combining autoencoders with evolutionary algorithms. This work achieved success by using a variational autoencoder and evolutionary algorithm to generate new music (with genotype equivalent to the latent space of the variational autoencoder). The variational autoencoder used was Magenta's MusicVAE [16], an autoencoder which on its own can make a wide range of music. Instead of relying on human feedback, however, the fitness function for evolution is done by finding the distance from a generated song to some predetermined target state. The combination successfully composed music, albeit with some memory loss regarding tonality for longer compositions.

In the field of evolutionary music-making, there have been many successes in generating music. Two works [1,12] go into detail in explaining the mechanisms behind evolutionary music-making and potential solutions to problems which may arise. This includes dealing with issues surrounding musical structure (how elements including key, melody patterns and motifs could be embedded within a model). A more specific example is a paper which presents an interactive melody generation tool [5]. The paper breaks its technique down into three phases. In the first phase, a spectrum of melodies is generated in ABC notation using a genetic algorithm based on the similarity to a database of compositions. These results are passed on to a human who rates the songs from 0 to 100 for the second phase. These ratings are fed into a bidirectional LSTM which is used to mimic the human's scoring. The same genetic algorithm, as described as the first phase, runs again to maximise the trained bidirectional LSTM instead. The paper reported the results to be pleasurable and consistent, though, through personal experience from prototypes of our work, it is hard for a human to rate songs in such a manner.

Another important work for rhythmic evolutionary music-making was the creation of evoDrummer [13]. In the paper, the authors introduce the notion of rhythmic divergence (calculated as the mean relative distance between two feature vectors of drum music, for example, syncopation). A user selects a base rhythm and a desired divergence factor. Rhythms are initialised either randomly, copying the base rhythm, or anywhere in between (taking parts from each). The rhythms and intensities of the percussive elements are then evolved over time to reach the desired divergence. Another well-known work within evolutionary music-making is MaestroGenesis, an application for composing music built on a technique called functional scaffolding for musical composition. In a work extending the technique to produce multi-part accompaniments [9], two CPPNs are used to generate the accompaniment based on the rhythm and pitch information of the melody. One CPPN controls the pitch information, while the other controls rhythmic information for each part. The CPPN pairs are randomly initialised, and they combine and mutate through user selections. Further, the paper presents a layering technique whereby previously generated parts can serve as inputs to new CPNNs to relate separate layers more closely. Other proposed and successful techniques in evolutionary music-making worth mentioning include using particle swarm optimisation to evolve musical features to user ratings [14]; performing random walks on common chord progressions while evolving the pitches of a melody with lightly randomised rhythms [19], and evolving measures of music at a time with random fitness functions [21].

There have also been many methods regarding the use of autoencoder neural networks for composing new music. Choi et al. created an autoencoder with a transformer model to capture musical style from a sample and embed it in a base melody [4]. Tikhonov and Yamshchikov applied a variational recurrent autoencoder supported by history which is trained to compress a dataset of music into features which can then be used to generate original music [23]. Magenta's MusicVAE [16] is based on a multi-layered bidirectional LSTM to encode an input sequence of music. In this system, the decoder used a novel hierarchical recurrent neural network in order to separately produce a drum, bass and melody output from the compressed representation. The hierarchical method showed definite improvements over the flat counterparts for the reconstruction accuracy metric for each part that was produced. The hierarchical method was shown to outperform the flat model in a human-centred listening study consistently. These previous works show that autoencoder neural networks are an effective way of encoding and generating music.

3 Methodology

3.1 Dataset

Fig. 2. An example of the training data expressed in piano roll notation. Note that all notes have the same very short duration, a limitation of our model.

The dataset consisted of symbolic music in MIDI format. Initially, MIDI files were downloaded from the MAESTRO dataset [8] which contains over 200 h of piano performances. Unfortunately, our model did not perform well with this dataset, potentially due to the expressive timing that occurs in this dataset. Our second training attempts applied a dataset of video game music, which was more repetitive and included simpler rhythmic information. A dataset of more than 30000 MIDI files was scraped from VGMusic [15]. The dataset was further simplified by changing all notes to have the same very short duration; equivalent to instantly pressing and releasing a note on a piano.

The files need to be processed so that they could be fed into an autoencoder. The files were converted to piano roll notation, as demonstrated by Fig. 2. The horizontal axis represents time, and the vertical axis represents pitch - black indicating a note being played and white being the absence of a note. The range of the pitch was limited to be 88 notes, matching the range of a piano. Further, to restrict the horizontal axis, we decided that the autoencoder would produce 16 bars of music at a time, with a bar broken down into 48 ticks (which was enough to represent faster moving passages). This limited the horizontal axis to 768 across.

The 16-bar length training samples are represented by a 768×88 matrix of binary values. The VGMusic dataset yielded 75136 non-overlapping 16-bar segments for training.

3.2 Autoencoder Model

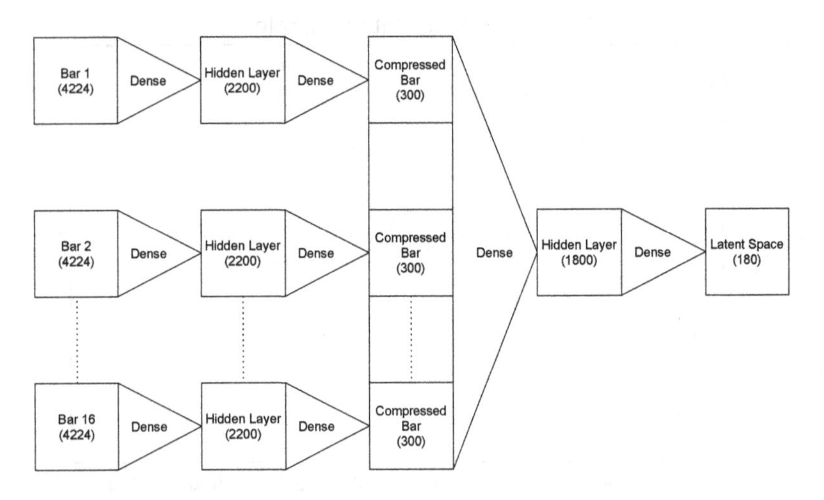

Fig. 3. The encoder model. The numbers in the brackets indicate the size of each part. The decoder is a mirrored version of the encoder. Bars 3–15 are represented with a dotted line.

We applied a process of trial-and-error to construct an autoencoder that would be useful for evolutionary music composition. This included attempts to use variational autoencoders [6], dense autoencoders, bidirectional LSTM autoencoders [16] and transformer models [10,11].

Ultimately, inspired by similar work by HackerPoet or CodeParade [7], we applied a standard (i.e., not variational) autoencoder. Figure 3 shows the encoder model layout. Each bar is compressed from a size of 4224 (pixels in each bar using piano roll notation) to a size of 300 with the assistance of a hidden layer.

The compressed bars are then concatenated (to a size of 4800) before being compressed again to the latent space representing the song with a size of 180. The decoder is an exact mirror of the encoder. Further, each transition between layers is done with a dense layer and the ReLU activation function. This is with the exception of the last layer which uses a sigmoid function, and the latent space layer which uses no activation function. In addition to this, dropout and batch normalisation layers were applied to negate overfitting.

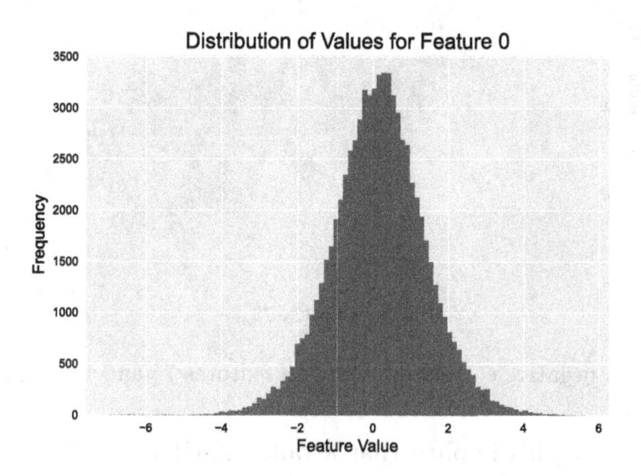

Fig. 4. Each feature in the latent space was found to follow a normal distribution.

Preliminary experiments in randomising the latent variables to generate new music led to "noisy" and hence unpleasant results. As such, the training set was fed into the encoder to investigate the nature of the latent space. The analysis showed that although each feature followed a normal distribution (an example of which is shown in Fig. 4), they, for the most part, did not follow a standard normal distribution (i.e., with $\mu = 0$, $\sigma^2 = 1$) as Fig. 5 reveals.

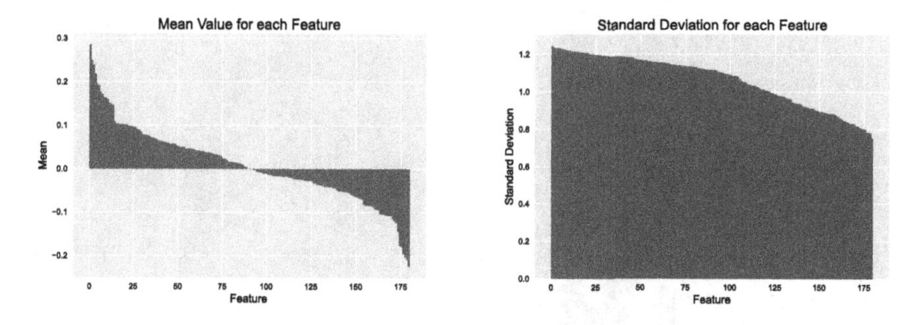

Fig. 5. The mean value and standard deviation of features in the latent space (ordered) from the training data.

Further analysis revealed that some features of the latent space were correlated with one another. This is illustrated in Fig. 6.

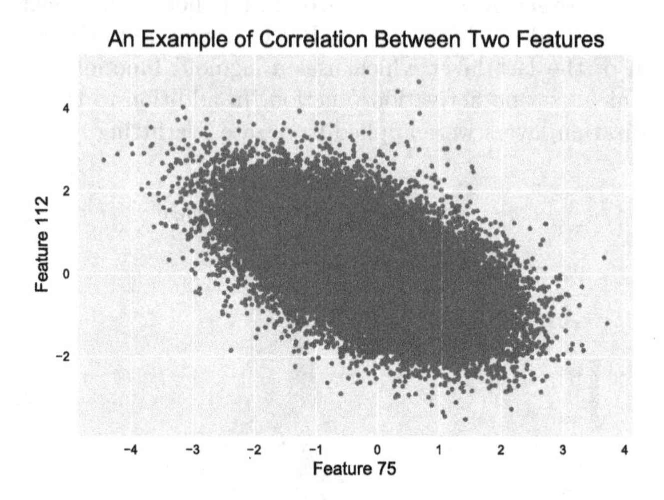

Fig. 6. There is a negative correlation between features 75 and 122 of the latent space.

In combination, this implied that a randomised vector for the latent space was likely to lie outside the space the decoder had been trained on. To rectify this issue, Principal component analysis [22] was applied to map a standard normal distribution to the latent space. The importance of each principal component can be visualised by looking at its associated eigenvalue in Fig. 7.

Diminishing returns are observed for the principal components with the first 15 or so being the most important before holding steady and dipping off again soon after 125. Applying principal component analysis allows us to map a standard normal distribution to the latent space of the autoencoder, forming a basis for evolution.

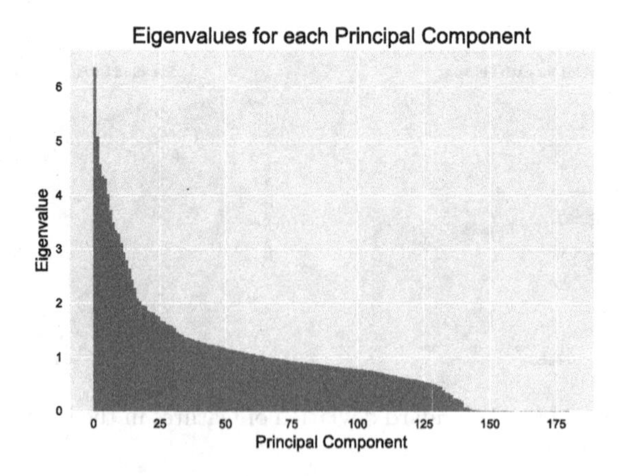

Fig. 7. The ordered eigenvalues associated with each principal component.

3.3 Evolutionary Mechanism

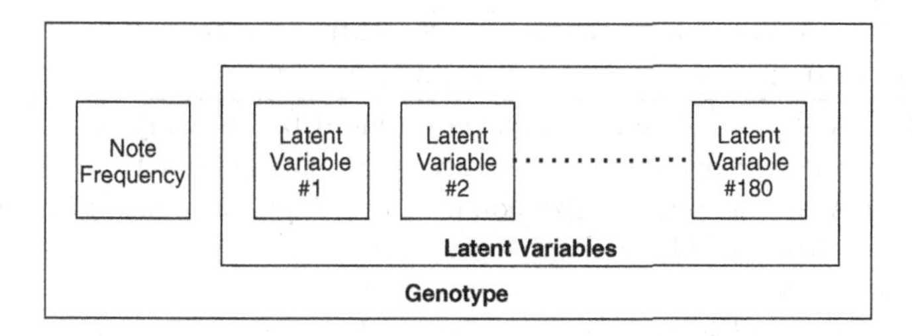

Fig. 8. The latent variables together with the note frequency threshold form the genotype for evolution.

To generate new music, a random vector of dimension 180 (the dimension of the latent space) can be sampled from a standard normal distribution. This is mapped to the latent space using principal component analysis before being sent through the decoder component of the autoencoder. Unlike the input, however, which is discrete for the elements of the piano roll (0 and 1 corresponding to the absence and presence of a note respectively), the decoder produces continuous values in the range $[0, 1]$. This can be thought of as the likelihood of a note being played—how confident the autoencoder is in thinking a note of a given pitch should be played at a given time. With this in mind, another value can be used for generating music which can be thought of as a threshold representing note frequency. This value, in $[0, 1]$, represents how confident the autoencoder must be about a note for it to be played; otherwise, it is ignored. Through this method, the decoder can now generate music. The threshold (referred to later as note frequency) in combination with the 180 latent variables form the genotype for evolution depicted in Fig. 8.

The evolutionary process can be defined as follows:

1. A population is initialised by creating individuals with random genotypes
2. The genotype of each individual is fed through the decoder to generate music
3. The user listens to individual songs and removes any that they do not like
4. New members of the population are created by applying Algorithm 1 to the remaining songs
5. Repeat the evolutionary process from step number 2

This process completes when the user is satisfied with the evolved population.

Algorithm 1: Evolutionary Process
Default Parameters: populationSize ← 8, nfMutChance ← 0.2, nfMutFac ← 0.02, mutChance ← 0.1, mutFac ← 0.08

```
   Input : A list, S, of songs (each represented by an array of 181 values)
           selected by the user to evolve with size greater than 1
   Output: A list of songs of size equal to the population size formed by
           adding newly evolved songs to S
 1 newSongs ← S;
 2 while newSongs.length < populationSize do
       // Get unique pair of songs to evolve, pairs not yet
          chosen will be selected where possible
 3     song1, song2 ← getUniquePairOfSongs(S);
 4     crossover ← 181-dimensional array, each position filled with
        Bernoulli(0.5);
 5     newSong ← song1 * crossover + song2 * (1 - crossover);
 6     mutations ← 181-dimensional array, each position filled with
        Bernoulli(mutChance) * N (0, mutFac);
 7     mutations[0] ← Bernoulli(nfMutChance) * N(0, nfMutFac);
 8     newSong ← newSong + mutations;
 9     newSongs.append(newSong);
10 end
11 return newSongs
```

3.4 The Application

A significant constraint with initial prototypes of our method was how time-consuming it is for humans to judge songs manually. As such, it was decided to create an application to help streamline this process and increase usability.

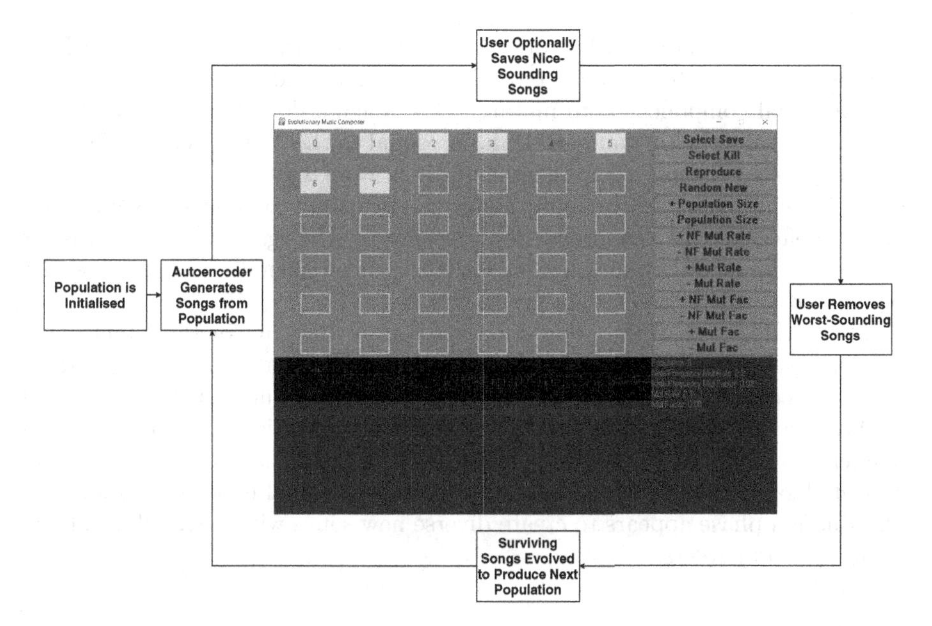

Fig. 9. The evolutionary music composition application shown inside a flowchart of its core operations.

The application, seen in Fig. 9, is divided into four main sections. The upper left area is used to display the population of songs generated by the algorithm (each number represents the ID of a song). A user can select a honeydew green box to hear the respective song played. At the same time, it is displayed in piano roll notation inside the black rectangle on the screen. The upper right purple area displays the controls of the application. The first four of which are the most important. Once the user evolves a song they enjoy, they can use the "Select Save" option to save the song. For each generation, the user can remove songs that they do not enjoy with the "Select Kill" button. After this is done, they can click the "Reproduce" control to generate new songs from those that survived. In the event that the user enjoys none of the songs, or just wishes for something new, the "Random New" button can be pressed. This introduces a new random song to the population.

The remaining ten controls either increment or decrement the parameters used for evolution. The parameters are displayed in the bottom right zone of the screen. "± Population Size" modifies the size of the population—the number of songs in each generation. The next four boxes modify two different types of mutation rates for evolution. "NF Mut Rate", short for note frequency mutation rate, modifies how likely the note confidence threshold is to change. The mutation is performed by adding Gaussian noise to the value with standard deviation equal to the "Note Frequency Mut Factor" (which is changed with the "± NF Mut Fac" buttons). The "Mut Rate" button works similarly, instead represented how likely each principal component is to mutate. This is again done by adding Gaussian noise to the values with standard deviation equal to the "Mut Factor". The note frequency and principle component mutations were kept independent because the equivalent change in the mutation of the principal components would have a drastic effect on the note frequency, creating poor songs. The intention is for the user to modify the mutation values if they feel generated songs are either too similar or too different from previous songs.

The initialisation of the population of songs is done by, for the latent space genotype part, drawing randomly from a standard normal distribution, and for the note frequency part, drawing from a uniform distribution between 0.2 and 0.4. From here, the user simply interacts with the application in the above-stated manner to create music they find appealing. The application, along with pre-generated samples is available online[1]. Through personal experimentation, the reproduction phase appears to create diverse new songs with each still sounding inspired by its parents.

3.5 Experiments

An experiment was conducted to verify that the application can create a variety of music with respect to tonality. The application was run several times, with selections made by the first author, to compose a dataset of music containing 49 songs. For each of these songs, a pitch class histogram was created [24]. Pitch-class histograms measure the frequency of types of notes. In standard music, there are 12 unique notes, and these are repeated through octaves. These 12 unique notes form our pitch-classes.

The distribution of the pitch-classes for four songs in the dataset is shown in Fig. 10. These were selected to demonstrate that a variety of pitch-class distributions is possible between separate runs. As can be seen, the distribution of pitches is quite different between the songs. Consider songs 1 and 2. Even though pitch-class 4 is the most common, the second most common for each seldom appears in the other (pitch-classes 9 and 12). This is among other differences, such as the presence of pitch-class 5. Additionally, 2 is the most common pitch-class in song 4 but is rare in all of the others. This essentially means that the four excerpts from the dataset have quantitatively different tonalities and keys. This implies that the application can create a variety of music tonality-wise.

[1] Source code for the music composition application: http://doi.org/10.5281/zenodo.4497829.

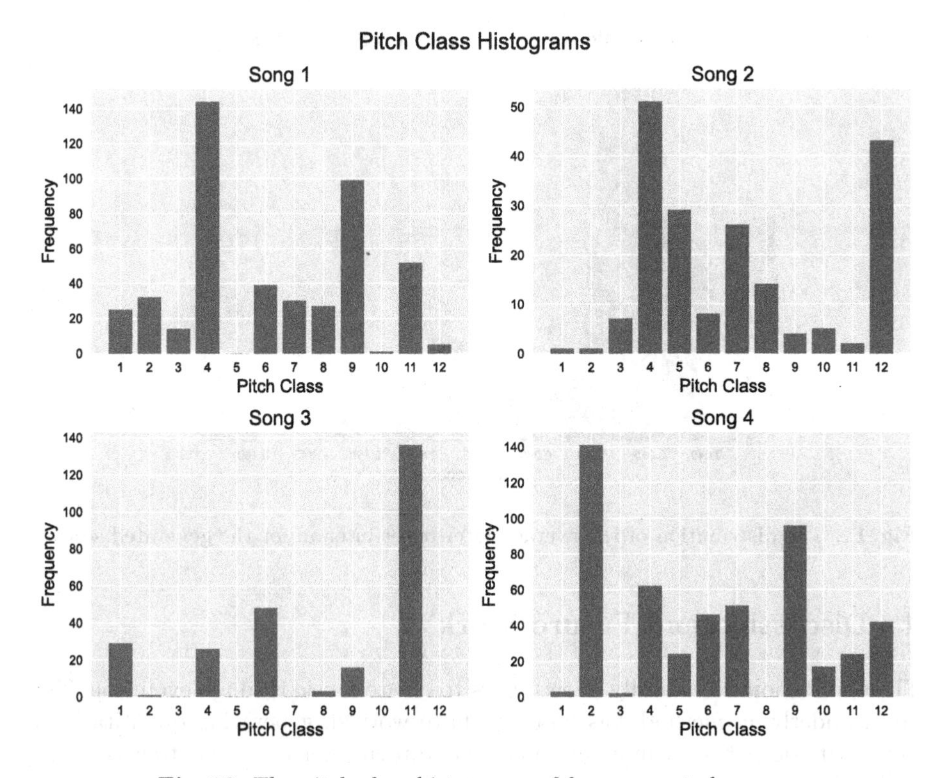

Fig. 10. The pitch-class histograms of four generated songs.

Another experiment was performed to establish that the application can also produce a range of rhythms. To quantify this, the average inter-onset-interval (IOI) [24] was calculated over the same dataset. The average IOI measures the mean time in seconds between two consecutive notes. All generated music is played at the default MIDI tempo of 120 bpm. The IOI can be used as a measure of how fast the rhythms are moving inside each song.

Figure 11 shows a histogram created to measure the distribution of the average IOI between the songs. From the numeric results, the IOI varies from less than 0.04 to more than 0.16 s in our dataset. This effectively means that the average time between notes of the slowest moving piece is more than four times that of the fastest. It is noted that there is a skewing in the histogram towards faster moving pieces. This was likely caused by the first author preferring faster-moving compositions to the slower moving ones when evolving the music. Regardless, the histogram shows that the application can indeed generate a range of rhythmic compositions.

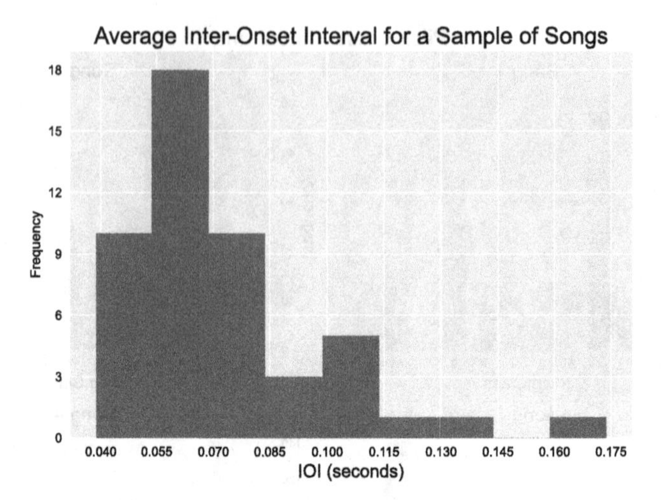

Fig. 11. The distribution of the average inter-onset-interval for the generated songs.

4 Discussion and Future Work

The application successfully allows users to create music through evolution. The novel underlying method was found to have worked, generating quantitatively diverse music between different runs. The autoencoder component successfully finds a lower-dimensional structure of music for the evolution part to run, albeit with simplified data input.

One clear limitation of the system is the simplification of the data to ignore duration so that notes are instantaneously played and released. In practice, this leads to situations where nothing is played for a long time as long notes in the training set translate into long rests. This can sometimes lead to occurrences of empty bars in the output (though usually only early on in the evolutionary process). Applying an autoencoder that can represent duration in our system would likely improve the quality, continuity and diversity of the composed music.

While our evaluation has examined the autoencoder's latent space and characterised the musical output of our system quantitatively, we have not undertaken a human-centred evaluation. Future user studies could examine both the quality of the music generated as well as the affordances of our evolution interface.

Although our system streamlines the process of human evaluation, a potential extension could be to include a discriminator neural network to supplement these decisions. This could allow the system to automatically evolve towards a target feature such as mood or genre. Multiple autoencoders could also allow a user to change the style. These features could be used for dynamic music generation as is used in video game soundtracks.

5 Conclusion

This paper presented an interactive application which generates music to a user's liking inspired by the process of biological evolution. The application uses a novel combination of autoencoder neural networks and evolutionary algorithms driven by explicit human feedback. The autoencoder was successfully used to capture musical structure from pre-existing music and represent this at a much lower dimension. Principal component analysis was then used to normalise the latent space and find a meaningful encoding of features. This provided a genotype which, for the purposes of evolution, could easily be edited to gradually change the music which is generated based off of other songs. The results show that the application is capable of creating a diverse range of music with regards to tonality and rhythm.

References

1. Biles, J.A., Miranda, E.R.: Evolutionary Computation for Musical Tasks, pp. 28–51. Springer, London, London (2007). https://doi.org/10.1007/978-1-84628-600-1_2
2. Bontrager, P., Roy, A., Togelius, J., Memon, N., Ross, A.: Deepmasterprints: generating masterprints for dictionary attacks via latent variable evolution*. In: 2018 IEEE 9th International Conference on Biometrics Theory, Applications and Systems (BTAS), pp. 1–9 (2018). https://doi.org/10.1109/BTAS.2018.8698539
3. Bontrager, P., Lin, W., Togelius, J., Risi, S.: Deep interactive evolution. In: Liapis, A., Romero Cardalda, J.J., Ekárt, A. (eds.) EvoMUSART 2018. LNCS, vol. 10783, pp. 267–282. Springer, Cham (2018). https://doi.org/10.1007/978-3-319-77583-8_18
4. Choi, K., Hawthorne, C., Simon, I., Dinculescu, M., Engel, J.H.: Encoding musical style with transformer autoencoders. In: Proceedings of the 37th International Conference on Machine Learning, ICML. Proceedings of Machine Learning Research, vol. 119, pp. 1899–1908. PMLR (2020). http://proceedings.mlr.press/v119/choi20b.html
5. Farzaneh, M., Mahdian Toroghi, R.: Music generation using an interactive evolutionary algorithm. In: Djeddi, C., Jamil, A., Siddiqi, I. (eds.) Pattern Recogn. Artif. Intell., pp. 207–217. Springer International Publishing, Cham (2020)
6. Foster, D.: Generative Deep Learning: Teaching Machines to Paint. Compose, Write and Play. O'Reilly Media, Newton (2019)
7. HackerPoet: Composer (2018). https://github.com/HackerPoet/Composer
8. Hawthorne, C., et al.: Enabling factorized piano music modeling and generation with the MAESTRO dataset. In: International Conference on Learning Representations (2019). https://openreview.net/forum?id=r1lYRjC9F7
9. Hoover, A.K., Szerlip, P.A., Norton, M.E., Brindle, T.A., Merritt, Z., Stanley, K.O.: Generating a complete multipart musical composition from a single monophonic melody with functional scaffolding. In: Proceedings of the Third International Conference on Computational Creativity (ICCC-2012), pp. 111–118 (2012)
10. Huang, C.Z.A., et al.: Music transformer: generating music with long-term structure. In: International Conference on Learning Representations (2019). https://openreview.net/forum?id=rJe4ShAcF7

11. Huang, Y.S., Yang, Y.H.: Pop music transformer: beat-based modeling and generation of expressive pop piano compositions. In: Proceedings of the 28th ACM International Conference on Multimedia, pp. 1180–1188. Association for Computing Machinery, New York, NY, USA (2020). https://doi.org/10.1145/3394171.3413671
12. Husbands, P., Copley, P., Eldridge, A., Mandelis, J., Miranda, E.R., Biles, J.A.: An Introduction to Evolutionary Computing for Musicians, pp. 1–21. Springer, London (2007). https://doi.org/10.1007/978-1-84628-600-1_1
13. Kaliakatsos–Papakostas, M.A., Floros, A., Vrahatis, M.N.: evoDrummer: deriving rhythmic patterns through interactive genetic algorithms. In: Machado, P., McDermott, J., Carballal, A. (eds.) EvoMUSART 2013. LNCS, vol. 7834, pp. 25–36. Springer, Heidelberg (2013). https://doi.org/10.1007/978-3-642-36955-1_3
14. Kaliakatsos-Papakostas, M.A., Floros, A., Vrahatis, M.N.: Interactive music composition driven by feature evolution. SpringerPlus **5**(1), 1–38 (2016). https://doi.org/10.1186/s40064-016-2398-8
15. Newman, M., Evans, S., Carroll, M., Hill, J., Camarena, D.: Vgmusic (2020). https://vgmusic.com/
16. Roberts, A., Engel, J., Raffel, C., Hawthorne, C., Eck, D.: A hierarchical latent vector model for learning long-term structure in music. In: Dy, J., Krause, A. (eds.) Proceedings of the 35th International Conference on Machine Learning Proceedings of Machine Learning Research, pp. 4364–4373. PMLR Stockholmsmässan, Stockholm, Sweden (2018). http://proceedings.mlr.press/v80/roberts18a.html
17. Rogozinsky, G., Shchekochikhin, A.: On VAE latent space vectors distributed evolution driven music generation. In: Proceedings of the 11th Majorov International Conference on Software Engineering and Computer Systems. MICSECS (2019). http://ceur-ws.org/Vol-2590/paper22.pdf
18. Romero, J.J.: The Art of Artificial Evolution: A Handbook on Evolutionary Art and Music. Springer Science & Business Media, Berlin (2008)
19. Scirea, M., Togelius, J., Eklund, P., Risi, S.: MetaCompose: a compositional evolutionary music composer. In: Johnson, C., Ciesielski, V., Correia, J., Machado, P. (eds.) Evolutionary and Biologically Inspired Music, Sound, Art and Design, pp. 202–217. Springer International Publishing, Cham (2016)
20. Secretan, J., Beato, N., D Ambrosio, D.B., Rodriguez, A., Campbell, A., Stanley, K.O.: Picbreeder: evolving pictures collaboratively online. In: Proceedings of the SIGCHI Conference on Human Factors in Computing Systems, pp. 1759–1768. CHI 2008, Association for Computing Machinery, New York, NY, USA (2008). https://doi.org/10.1145/1357054.1357328
21. Waschka II, R., Miranda, E.R., Biles, J.A.: Composing with Genetic Algorithms: GenDash. Springer, London (2007). https://doi.org/10.1007/978-1-84628-600-1_6
22. Wold, S., Esbensen, K., Geladi, P.: Principal component analysis. Chemom. Intell. Lab. Syst. **2**(1), 37–52 (1987). https://doi.org/10.1016/0169-7439(87)80084-9
23. Yamshchikov, I.P., Tikhonov, A.: Music generation with variational recurrent autoencoder supported by history. SN Appl. Sci. **2**(12), 1–7 (2020). https://doi.org/10.1007/s42452-020-03715-w
24. Yang, L.-C., Lerch, A.: On the evaluation of generative models in music. Neural Comput. Appl. **32**(9), 4773–4784 (2018). https://doi.org/10.1007/s00521-018-3849-7

A Swarm Grammar-Based Approach to Virtual World Generation

Yasin Raies[(✉)] and Sebastian von Mammen

Julius-Maximilians University, Würzburg, Germany
yasin.raies@stud-mail.uni-wuerzburg.de,
sebastian.von.mammen@uni-wuerzburg.de

Abstract. In this work we formulate and propose an extended version of the multi-agent *Swarm Grammar* (SG) model for the generation of virtual worlds. It unfolds a comparatively small database into a complex world featuring terrain, vegetation and bodies of water. This approach allows for adaptivity of generated assets to their environment, unbounded worlds and interactivity in their generation. In order to evaluate the model, we conducted sensitivity analyses at a local interaction scale. In addition, at a global scale, we investigated two virtual environments, discussing notable interactions, recurring configuration patterns, and obstacles in working with SGs. These analyses showed that SGs can create visually interesting virtual worlds, but require further work in ease of use. Lastly we identified which future extensions might shrink required database sizes.

Keywords: Procedural content generation · Multi-agent systems · Terrain generation · Swarm intelligence · Swarm art

1 Introduction

In computer games, procedural content generation (PCG) techniques are used, e.g., to algorithmically create small assets like weapons [4], large scale terrains [9] or entire planets [1,6]. The latter are examples of *world generation*, yielding vast and diversely populated terrains. Carefully balancing the PCG mechanics can afford playable, diverse and detailed contents at comparatively small workload. Combining the use of sub-generators and pre-generated assets gives the designer a lot of control but also requires a high amount of input data to achieve desirable results. As an alternative, we propose to define a small database of generative agents whose interactions, in turn, unfold into a complex world. This approach bears similarities to the complex interaction networks of enzymes and proteins through interpretation of cellular RNA and DNA that result in complex multicellular development [2]. This approach allows for adaptivity of generated artifacts to their environment through simple rules. For instance, vegetation can sprawl naturally in the presence of water. It also allows for interactivity—the generation can be stopped and resumed, or iteratively influenced. I.e., the user or

J. Romero et al. (Eds.): EvoMUSART 2021, LNCS 12693, pp. 459–474, 2021.
https://doi.org/10.1007/978-3-030-72914-1_30

player immersed into this generated world can remove, place, and modify actors anywhere at any time [19]. Lastly, the generation can be perpetual, yielding an open-ended evolution of growth and change of a virtual world.

We chose SGs [20], an agent-based extension of L-Systems [8], as the basic model to drive the envisioned agent-based world generation. In SGs, each agent senses its local surroundings and has the means of production of static artifacts as well as differentiation and proliferation into other agents. We adjust and extend the previously introduced SG models [20] into *virtual environment generating Swarm Grammars* (vSGs), focusing on the creation, population, and manipulation of terrains, considering topography, vegetation and bodies of water. We will provide a short introduction of related works in Sect. 2. Section 3 will formally introduce vSGs. Section 4 will investigate the capabilities of minimal vSG instances based on sensitivity analyses, while Sect. 5 will focus on the analysis of larger scale structures and link them to the relationships of agents involved in their generation. Section 6 will summarize our findings and conclude with a look on potential future work.

2 Background

In [7], Hendrikx et al. have created a hierarchy of game content, that can be procedurally generated. As can be seen in Fig. 1, they distinguish between six classes of content. In this work, we focus on generation of basic building blocks of virtual worlds (*Game Bits*) and their arrangement in a larger scale space (*Game Spaces*). Following the taxonomies in [7] or [17], works that are seminal to our approach are referenced. In particular, these are noise functions for the generation of terrains, L-systems for branched structures, and Boids to model behaviours.

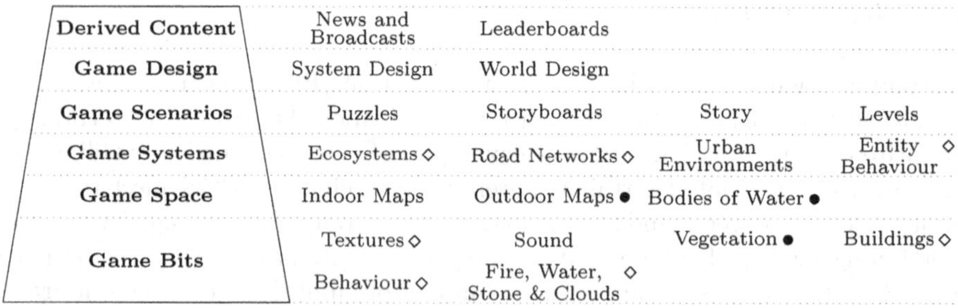

Derived Content	News and Broadcasts	Leaderboards		
Game Design	System Design	World Design		
Game Scenarios	Puzzles	Storyboards	Story	Levels
Game Systems	Ecosystems ◇	Road Networks ◇	Urban Environments	Entity Behaviour ◇
Game Space	Indoor Maps	Outdoor Maps ●	Bodies of Water ●	
Game Bits	Textures ◇	Sound	Vegetation ●	Buildings ◇
	Behaviour ◇	Fire, Water, Stone & Clouds ◇		

Fig. 1. Hierarchy of game content, as defined by [7]. •-markers denote classes that we generated through vSGs in this work, while ◇-markers denote classes where we see viable applications of SGs. Figure adapted from [7, p. 4, Fig. 1]

Sequences of random numbers produced by Pseudo-Random Number Generators (PRNGs) introduce variety in procedurally generated content. As PRNGs are driven by deterministic functions, feeding them with one and the same seed value ensures reproducibility of the sequences. While random numbers can, e.g., be used to choose, place or scale objects [5], there are methods to structure randomness in multi-dimensional space [10,21]. Such "procedural coherent noise" assigns random values to points in space, for instance to define height-maps of terrains [16]. Creating more cohesive structures, L-Systems, introduced by [8], are parallel string rewrite-systems, often used and initially conceived to describe the growth of plants. They consist of sets of grammatical rules to generate a string of symbols which in turn can be interpreted. There are numerous extensions to basic, context-free L-Systems that solely replace a single character by a string, e.g., context-sensitivity, stochasticity or parameterisation [12]. The established graphical interpretation of L-Systems has been introduced by [11] and parses a string one symbol at a time, informing a drawing agent (a "turtle") to move forward, to rotate or to leap to a previous location. As a result, branching structures as seen in plant morphologies can be retraced.

Unlike the approaches mentioned above, *Birdoids*, *Birdoid Objects* or *Boids* are used to model the behaviour of actors. Introduced in [13], Boids are a method to model the movement behaviour of swarms like flocks of birds or schools of fish. This is achieved through an agent-based approach, where each individual sees their immediate surroundings and moves according to simple urges, like wanting to be close its neighbours or gaining distance to avoid collisions. Due to its efficacy and ease of extendability, the Boid model has received much attention in the artificial life community, but also has been featured in media, like the 1992 Tim Burton film *Batman Returns* [14].

Finally, Swarm Chemistries (SCs) and SGs build on the foundations outlined so far and represent approaches that are rather close to our proposed model. Swarm Chemistries are a Boid-like model where agents are considered chemical reactants, with "spatio-temporal patterns considered as the outcome of chemical reaction[s]" [15]. SCs yield more complex patterns by increasing the number and heterogeneity of deployed agents. SGs, on the other hand, extend L-Systems by interpreting each symbol of the generated string as a Boid agent. As a result, the L-System rules determine the generative behaviour of an SG, whereas the Boid behaviour determines the spatial interaction and interpretation of the generative processes. SGs were created by von Mammen to model dynamic growth, governed by an ecology of interacting agents. An extended form of SGs have been studied in the context of architecture and interactive evolution [20].

3 World-Building Swarm Grammars

To generate virtual worlds, we extended Swarm Grammars with respect to five different aspects: (1) Generalisation of the model, (2) introduction of context-sensitivity in the generative rule sets, (3) terrain-formation capabilities, (4) extension of sensing to create terrain-awareness, (5) introduction of a more expressive energy system.

We subsumed agents and artifacts by the generic concept of an *actor*. Actors are represented by a location, a type-identifier, an energy level, the iteration they have been generated (or age) and their *predecessor*. In order to facilitate adaptivity to the environment not only in movement but also in reproduction, we introduced context-sensitivity in the generative rule sets, thereby defining dependencies between actors within a given neighbourhood range. While von Mammen introduced *extended Swarm Grammars* with similar operational capabilities [18], we incorporated them into the formal model, building on context-sensitive L-Systems. By means of context-sensitivity, markers, so-called stigmergic cues, can be placed in the environment to inform the next steps in a construction process. In order to generate a terrain, following the idea of a height-map, we defined an implicit surface through all current actor positions. We further enabled vSG agents to probe the terrain below them, informing them about its slope and their height. This allowed us to implement new urges: to move along the terrain's downhill-slope, as water droplets would, a self-propulsion tendency as seen in SCs, to prevent stalling, and an axis-constraint, targeting the effective shaping of artifacts. For finer grained control in a world with many different actors, we also added the means to weigh actor types differently on urge calculations. Finally, we diversified the energy system. Previously, SG agents would always inherit the full energy value from their parent and be removed when depleted. In an effort towards finer-grained control of generated structures, vSGs feature new modes of energy distribution/inheritance and an optional final rewrite step on zero energy.

3.1 The vSG Model

We define the starting configuration of an vSG model to be the tuple $SG = (CL, \Delta, \Gamma)$. CL denotes a context-sensitive L-System as described below, $\Delta = \{\Delta_1, \Delta_2, \ldots\}$ a finite set of agent type specifications and $\Gamma = \{\Gamma_1, \Gamma_2, \ldots\}$ a finite set of artifact type specifications. For each agent type Δ_i, parameters of form c_{name,Δ_i} will act as coefficients to movement urges. Parameters e_{name,Δ_i} correspond to energy calculations. a_{max,Δ_i} and v_{name,Δ_i} represent acceleration and velocity parameters, d_{name,Δ_i} and β_{Δ_i} influence the field of view of agents. The artifact types Γ_i only feature a single parameter $\eta_{\Gamma_i,terrain}$, their influence on the terrain. As this parameter is also featured in and between agent types, we have to set all influences of a on b $\eta_{a,b}$, where $a \in \Delta \cup \Gamma$ and $b \in \Delta \cup \{terrain\}$. As vSGs describe an iterative process, we will call $SG(n) = (CL, \Delta, \Gamma)_n = (SG, \mathbb{D}_n, \mathbb{G}_n)$ the n-th iteration of SG, wherein $\mathbb{D}_n = \{d_{1,n}, d_{2,n}, \ldots, d_{k,n}\}$ and $\mathbb{G}_n = \{g_{k+1,n}, g_{k+2,n}, \ldots, g_{l,n}\}$ describe all agents and artifacts present at iteration n. Every element $c_{i,n}$ of the set of all actors in iteration n $\mathbb{D}_n \cup \mathbb{G}_n = \{c_{1,n}, c_{2,n}, \ldots, c_{l,n}\}$ features five properties: Their position $p_{i,n}$, energy $e_{i,n}$, unique identifier $\iota_{i,n}$, back-reference $\rho_{i,n}$ and their type $c^*_{i,n} \in \Delta \cup \Gamma$. Agents additionally feature their velocity $v_{i,n}$ and an individual world center $s_{i,n}$. Finally, the function $h_n(p)$ shall be a representation of the height-map, describing the terrain height below or above p after iteration n.

With these definitions, we will describe the integration steps of the vSG model, next: (1) rewrite, (2) movement, and (3) terrain calculations. Note, that iteration n in indices may be omitted for clarity in writing.

Rewriting. The L-System $CL = (\alpha, P)$ is at the core of the rewrite step. At iteration $n = 0$, the *axiom* α is converted into $\mathbb{D}_0$ and $\mathbb{G}_0$, while at every second iteration with $n > 0$ the productions in P are applied. α may be a word over the alphabet $\Delta \cup \Gamma$. If α were the empty word λ, there would be no actors to simulate. Potential starting positions of actors are not encoded in α. For ease of use, however, implementations might feature a means to set $\mathbb{D}_0$ and $\mathbb{G}_0$ directly. As an implementation default, we place all instantiations of symbols featured in α at $p_{i,0} = \mathbf{0}$, with $e_{i,0} = 10$ energy, a unique id $\iota_{i,0}$ and a velocity vector $v_{i,0} = \mathbf{0}$, if c_i is an agent.

The reproduction rules $(p, c, d, \theta, s) \in P$ consist of the *strict predecessor* $p \in \Delta$ and an optional context $c \in (\Delta \cup \Gamma \cup \{\lambda\})^*$ which has to exist within distance d for the rule to be applicable. If c equals the empty word λ, a rule is always applicable. The successor word s may be a word over the alphabet $\Delta \cup \Gamma \cup \{\lambda, \psi\}$, where the optional *persist* symbol ψ may only occur once. ψ will be interpreted as the same agent type as p, but gives the predecessor an identity, allowing it to "live" through a rewrite step and to retain its ι_i. Further, it receives special treatment during energy distribution. In full, productions can be written as $p <_d c \xrightarrow{\theta} s$; if $c = \lambda$ this can be shortened to $p \xrightarrow{\theta} s$. In contrast to SGs, the rule selection probability θ does not represent an absolute probability of rule application. Rather, it describes the frequency of rule selection relative to other currently applicable rules for an agent d_i. This stochastic process may be overridden, if d_i has $e_i \leq 0$ and $e_{zero,d_i^*} \neq ()$. A replacement word akin to the right side of a rule may be provided as s in $e_{zero,i} = (s, v)$, with an additional energy value v assigned to each actor instantiated from s.

For each production (p, c, d, θ, s) applied at iteration n, actors corresponding to symbols in s are instantiated. We will refer to the sets of actors as $\mathbb{D}_{n'}$ and $\mathbb{G}_{n'}$. Calling their predecessor $d_{i,n-1}$. All actors $c_{j,n}$ inherit position and velocity, gain a new unique id and set $\rho_{j,n} = \rho_{i,n-1}$. If $c_{j,n}$ is an agent and its predecessors $c_{seed,d_{i,n-1}^*} \neq 0$, we set the individual world center of $c_{j,n}$ to $s_{j,n} = p_{j,n}$. This is useful to allow for unbounded worlds. If $c_{j,n}$ originated from a ψ symbol and any artifact $g_{h,n}$ has been placed by the same rule, we set $\rho_{j,n} = \iota_{h,n}$. This facilitates the tree-like structure of back-references. Concerning energy, all generated actors c_i are assigned a new energy value. Depending on the parameter $e_{suc,d_{j,n-1}^*}$ of their predecessor $d_{j,n-1}$, we provide four modes, how a successor's energy might be calculated: Constant energy, energy proportional to $e_{j,n-1}$, equal distribution of $e_{j,n-1}$ to successors, with optional losses, and constant energy, if $e_{j,n-1}$ is not consumed by this, or else equal distribution. Notably, the latter two modes do not create but only redistribute or consume energy from the system. For the persisted predecessor, there are four analogue modes for $e_{persist,d_{j,n-1}^*}$: Constant energy loss, energy loss proportional to successor count, as well as the counterparts to the latter tow modes for e_{suc}.

Movement. The second step of each iteration relies on Boid behaviour. All the following calculations are computed for every $d_{i,n'} \in \mathbb{D}_{n'}$. Should $\mathbb{D}_{n'}$ be empty, because the rewriting step has been skipped, we will operate on $\mathbb{D}_{n'} = \mathbb{D}_{n-1}$. Before calculating steering urges, we require the set of neighbours N_i of d_i. All actors within d_i's view cone, defined by d_{view,d_i^*} and $\beta_{d_i^*}$, similar to the one described in [18, Sect. 2.6.2], are considered neighbours. Through N_i, we can deduce the set S_i of neighbours deciding the separation distance d_{sep,d_i^*}. We implemented the three basic boid urges as follows: Alignment aims to match d_i's neighbours' velocities in magnitude and direction; cohesion aims to reach the average position of actors in N_i; separation aims to move away from actors in S_i, weighting those closer more heavily. These urges all consider the influence $\eta_{c_j^*,d_i^*}$ of $c_j \in N_i$ in their averages. In contrast to the basic urges the following urges do not depend on the neighbourhood of d_i. The center urge $V_{cen,i} = s_i - p_i$ drives agents to their world center, keeping them interacting. Variety is introduced by $V_{rnd,i}$, a random but deterministic steering vector within the unit sphere. The constant bias vector $V_{bias,i} = \mathbf{1}$, together with its coefficient vector, describes an urge towards a certain direction. It may be used, e.g., to model gravity and thus gravitropism [3]—the phenomenon which results in many plant shoots growing upwards, overcoming gravity. Lastly, we introduced urges reacting to the local terrain. $V_{floor,i} = (0 \; (p_{i,y} - h(p_i))^2 \; 0)^\mathsf{T}$ describes a downward pressure, that increases with distance to the underlying terrain. Together with $V_{bias,i}$, this could be used to set a desired height, where both cancel each other out. Concerning the shape of the terrain, the following urges build primarily on the gradient. We compute them by use of an utility offset vector $o = (0.5 \cdot t_s \; 0 \; 0.5 \cdot t_s)^\mathsf{T}$, which depends t_s, describing how far agents look, to determine a gradient. $V_{grad,i}$ describes the gradient vector within the xz-plane, while $V_{slope,i}$ adds the downward hill component $V_{dws,i}$. In detail: $V_{grad,i} = (h(p_i + o_x) - h(p_i - o_x) \; 0 \; h(p_i + o_z) - h(p_i - o_z))^\mathsf{T}$, $V_{slope,i} = (V_{grad,i} + V_{dws,i}) \cdot c$, where $c = -1$ if $|V_{dws,i}| < 0$ else $c = 1$ and $V_{dws,i} = (0 \; h(p_i - \frac{V_{grad,i} \cdot t_s}{|V_{grad,i}| \cdot 2}) - h(p_i + \frac{V_{grad,i} \cdot t_s}{|V_{grad,i}| \cdot 2}) \; 0)^\mathsf{T}$. They allow agents to move towards valleys and mountain tops, either on the xz-plane or as if they walked or rolled. Lastly, $V_{norm,i} = (0 \; h(p_i + o_z) - h(p_i - o_z) \; 1)^\mathsf{T} \times (1 \; h(p_i + o_x) - h(p_i - o_x) \; 0)^\mathsf{T}$ represents the normal vector of the terrain and is, therefore, perpendicular to it. Through summation of all $V_{name,i}$, weighted by corresponding coefficients c_{name,d_i^*}, an acceleration vector is calculated. This vector is then multiplied per element with the constraint vector $\mathbf{c}_{const,d_i^*}$. Finally, the acceleration is truncated to a maximum amount of a_{max,d_i^*} and added to the agent's velocity, which, in turn, is clipped to V_{max,d_i^*}. Pace-keeping, a desire to move at v_{pace,d_i^*} controlled by c_{pace,d_i^*} and further described in [15], is applied to d_i. As the penultimate step in actual movement, the resulting velocity is stored in $v_{new,i,n'}$ and added to the current position, resulting in $p_{i,new,n'}$. If this position is below the terrain and $c_{noclip,d_i^*} = 0$, it is moved onto the terrain surface. To finalize the movement step we reduce each d_i's energy either by a constant amount or proportionally to the distance traveled, depending on the mode of e_{move,d_i^*}. After these calculations, we copy each $d_{i,n'} \in \mathbb{D}_{n'}$ as $d_{i,n}$ into our final result set $\mathbb{D}_n$, setting $p_{i,n} = p_{i,new,n'}$, $v_{i,n} = v_{i,new,n'}$ and $e_{i,n} = e_{i,new,n'}$.

Terrain Calculations. This height of $h_n(p)$ is calculated by averaging the y-position of all actors c_i, weighted by their influence factor $\mu(c_i, p_j)$ on the terrain. This factor depends on the distance of c_i and p_j on the xz-plane and the influence $\eta_{c_i^*,terrain}$. We calculate it as $\mu(c_i, p_j) = (1 + \|p_j - p_i\|_{xz})^{-\eta_{c_i^*,terrain}}$. To disable influence of actor types, we set $\mu(c_i, p_j) = 0$, if $\eta_{c_i^*,terrain} = 0$. Notably, we employed a regularly spaced grid of sample points with spacing t_s for a more performant, cached, implementation. Therefore implementations might replace the height function in the movement step by the bilinear interpolation of the nearest sample-points. This representation also lends itself to visualization and export as a 3D-asset, due to easy triangulation.

4 Analysis of Local Model Behaviour and Dynamics

We investigated the space of local behaviour and dynamics of vSGs by sensitivity analyses of three minimal model configurations SG_1, SG_2, and SG_3 (Table 1).

4.1 Minimal Terrain Generation

SG_1 represents a minimal configuration capable of generating a basic terrain featuring plateaux, cliffs or mountains. We ran analyses on on a 300×300 area with $t_s = 2$. For SG_1, four agents of type Δ_1 occasionally placed Γ_1 artifacts at their position until iteration $n = 100$. Figure 2 shows the influence of $\eta_{\Gamma_1,terrain}$: Uniformly low values produce pointy features, values between 2 to 4 resulted in smooth features, and high values resulted in very distinct plateaus with cliffs between them, reminiscent of Voronoi diagrams.

The second property with large impact was the density of terrain-influencing actors. This has been discovered while examining sensitivity of the center urge coefficient. c_{cen,Δ_1} values above just 0.02 result in agents circling around their world center tightly, resulting in mostly flat terrains with high Γ_1-density hotspots, as seen in Fig. 2. Such *detail clusters* could be used intentionally to mimic rocks. The observed behaviour with higher c_{cen,Δ_1} is due to the dominance of V_{cen} over other urges. Management of dominant urges was a recurring difficulty, especially in situations, where one urge was dominant but then is cancelled out by another.

4.2 Minimal Tree Generation

SG_2 entertains Δ_1 agents that place "wood" artifacts Γ_1 wherever they move, resulting in a 3D trace. Seen in Fig. 3, the separation urge and distance d_{sep} control how new branches turn away from predecessors. The visualization shows that $|V_{sep}|$ is proportional to d_{sep} allowing for quick and slow separation behaviours with fixed c_{sep,Δ_1}. A normalized $\hat{V}_{sep}$ could simplify the interactions with the alignment urge, which counteracts the undirected growth by V_{sep}.

Table 1. Parameter and production definitions for vSGs SG_1, SG_2, and SG_3. All parameters not mentioned are set to zero. $\mathbb{D}_0$ was given directly, for each.

Parameter	Value	Parameter	Value	Parameter	Value	Productions
SG_1	Terrain generation by agents Δ_1 through terrain forming artifact Γ_1					
c_{const,Δ_1}	$(1\ 0.5\ 1)^\intercal$	a_{max,Δ_1}	0.5	d_{sep,Δ_1}	5	$\Delta_1 \xrightarrow{11} \psi$
c_{cen,Δ_1}	0.001	v_{max,Δ_1}	2	$e_{persist,\Delta_1}$	$(const, 0)$	$\Delta_1 \xrightarrow{1} \psi\Gamma_1$
c_{rnd,Δ_1}	0.2	v_{norm,Δ_1}	1	e_{suc,Δ_1}	$(const, 10)$	
c_{pace,Δ_1}	0.1	d_{view,Δ_1}	100	e_{move,Δ_1}	$(const, 0.1)$	
c_{noclip,Δ_1}	1	β_{Δ_1}	180	e_{zero,Δ_1}	$(\Delta_1 \rightarrow \lambda)$	
$\eta_{\Gamma_1,terrain}$	8	t_s	2			
SG_2	Tree generation with tree agents Δ_1 and wood artifacts Γ_1					
c_{bias,Δ_1}	$(0\ 0.02\ 0)^\intercal$	c_{pace,Δ_1}	0.1	β_{Δ_1}	170	$\Delta_1 \xrightarrow{6} \psi\Gamma_1$
c_{const,Δ_1}	$(0.9\ 1\ 0.9)^\intercal$	η_{Δ_1,Δ_1}	1	d_{sep,Δ_1}	10	$\Delta_1 \xrightarrow{3} \psi\Gamma_1\Delta_1$
c_{sep,Δ_1}	0.9	η_{Γ_1,Δ_1}	0.1	$e_{persist,\Delta_1}$	$(const, 0)$	
c_{ali,Δ_1}	1	a_{max,Δ_1}	0.2	e_{suc,Δ_1}	$(inherit, 0.85)$	
c_{cen,Δ_1}	0.0001	v_{max,Δ_1}	0.6	e_{move,Δ_1}	$(dist, 0.2)$	
c_{rnd,Δ_1}	−0.002	v_{norm,Δ_1}	0.4	e_{zero,Δ_1}	$(10, \Delta_1 \rightarrow \Gamma_1)$	
c_{norm,Δ_1}	0.1	d_{view,Δ_1}	20	t_s	5	
SG_3	Water generation with droplets Δ_1, clouds Δ_2, traces Γ_1 and water Γ_2					
c_{coh,Δ_1}	0.04	v_{max,Δ_1}	2.5	e_{suc,Δ_2}	$(inherit, 1)$	$\Delta_1 \xrightarrow{3} \psi$
c_{sep,Δ_1}	1.3	a_{max,Δ_1}	0.3	e_{zero,Δ_1}	$(0, \rightarrow \lambda)$	$\Delta_1 \xrightarrow{1} \psi\Gamma_1$
c_{ali,Δ_1}	0.5	β_{Δ_1}	120	e_{move,Δ_2}	$(const, 0)$	$\Delta_1 <_7 \Delta_1\Delta_1\Gamma_1 \xrightarrow{2} \psi\Gamma_2$
c_{rnd,Δ_1}	1.5	d_{view,Δ_1}	30	$e_{persist,\Delta_2}$	$(const, 0)$	
c_{floor,Δ_1}	0.1	d_{sep,Δ_1}	8	e_{suc,Δ_2}	$(const, 10)$	$\Delta_2 \xrightarrow{6} \psi$
c_{pace,Δ_1}	0.2	η_{Δ_1,Δ_1}	1	e_{zero,Δ_2}	$()$	$\Delta_2 \xrightarrow{1} \psi\Delta_1$
c_{slope,Δ_1}	0.3	e_{move,Δ_1}	$(const, 0.02)$	$\eta_{\Gamma_3,terrain}$	2	
v_{norm,Δ_1}	1.5	$e_{persist,\Delta_2}$	$(const, 0)$	t_s	10	

Low values of η_{Γ_1,Δ_1} result in an appearance of bushes, where higher values straightened branches by driving agents away. Cohesion particularly impacts the width of generated trees. Variations of both parameters can be seen in Fig. 4. η_{Δ_1,Δ_1} and c_{rnd} were both required to be non-zero for interesting results, as some perturbation is required and $\eta_{\Delta_1,\Delta_1} = 0$ would mean Δ_1 agents do not see and therefore do not interact with each other. Investigating the impact of energy on tree growth, we varied energy consumption on movement and energy distribution on replication. The former controls overall tree size, while the latter only influences the size of offshoots (Fig. 4). In order to evaluate the impact of terrain urges, we added two artifacts into our test scene, which generated a slope for the tree to grow on. Setting $c_{floor} = 0.003, c_{norm} = 0.3, c_{bias,y} = 1.6$ and $d_{sep} = 6$, we generated sweeps of V_{floor}, V_{norm} and V_{bias} for this new $SG_{2'}$. The changes were required to establish a new reference tree, as the initial SG_2 did not yield insightful results in the revised environment. As shown in Fig. 4, V_{floor} had a strong impact, especially as V_{norm} pushed agents towards the valley, indirectly increasing agents height. Besides this and an insensitive upwards push, we did not notice significant contribution of V_{bias} and V_{norm}.

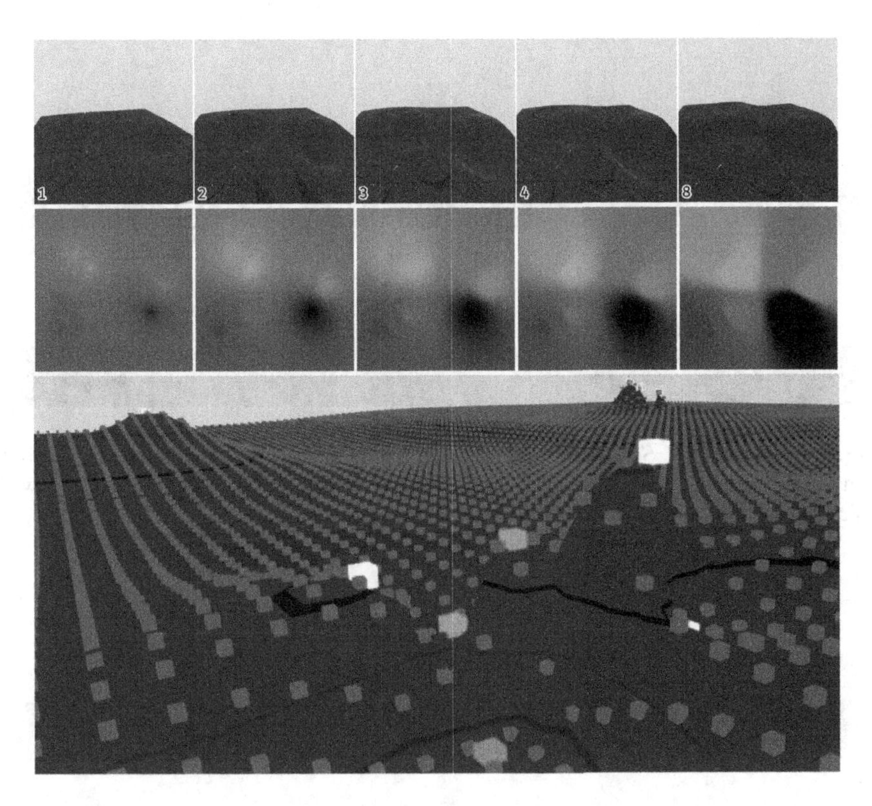

Fig. 2. Series of terrains and heightmaps of $SG_1(100)$, with varying $\eta_{\Gamma_1,terrain}$ (top and middle). Low values resulted in small peaks, while high values resulted in a clearly separated plateaus. The Γ_1 artifacts are shown in magenta. Screenshot of terrain details, generated by a high density of Γ_1 artifacts (magenta), placed by Δ_1 agents of SG_1 with $c_{center,\Delta_1} = 0.025$ (bottom). (Color figure online)

4.3 Minimal Water Simulation

SG_3 features water droplet-like agents Δ_1 that leave wet spots Γ_1 and are stochastically dropped by cloud agents Δ_2. To visualize where water accumulated, we employed the production $\Delta_1 <_7 \Delta_1\Delta_1\Gamma_1 \xrightarrow{2} \psi\Gamma_2$. As the context distance is smaller than d_{sep,Δ_1}, Γ_2 are only placed, if agents collide forcefully. Particularly, this can happen by accumulating agents with $c_{slope} > 0$ in a basin. The sensitivity analysis showed that the physically inspired approach, i.e. following V_{slope}, works across a broad value spectrum as long as some perturbation is introduced. Two intuitively insightful parameters were c_{coh} and c_{sep}, seen in Fig. 5. Low cohesion and high separation resulted in agents not deceeding the context-distance, resulting in few Γ_2 artifacts. High cohesion resulted in groups of agents overcoming V_{slope}, wandering freely, and low separation resulted in very dense puddles.

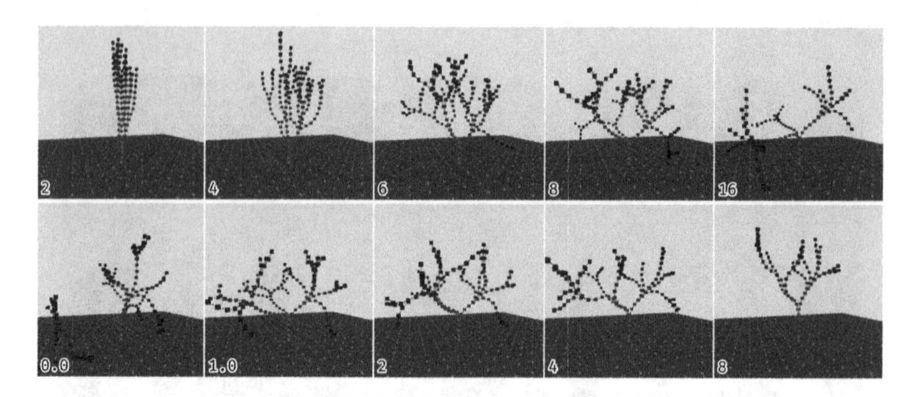

Fig. 3. Depictions of $SG_2(100)$ with varying values of d_{sep,Δ_1} (top) and c_{ali,Δ_1} (bottom). They showed a spectrum of orderly growth and uncontrolled sprawling with sharp turns, where V_{sep} dominated, highlighting that $|V_{sep}| \propto d_{sep}$.

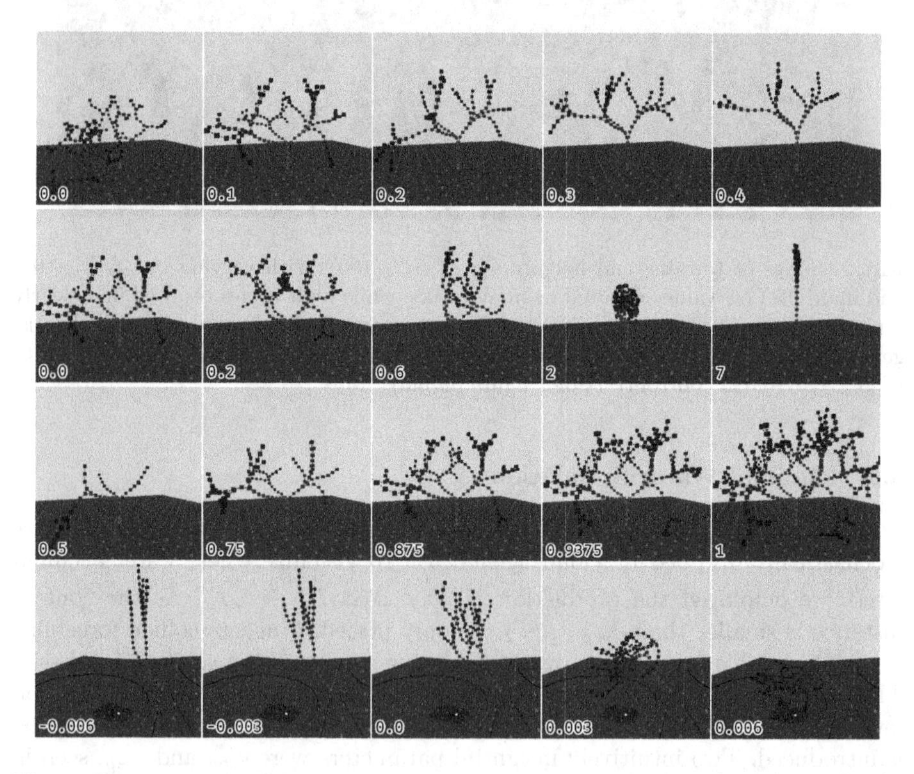

Fig. 4. Depictions of varying values of η_{Γ_1,Δ_1}, c_{coh,Δ_1}, and e_{suc,Δ_1} in $SG_2(100)$ and c_{floor,Δ_1} in $SG_{2'}(100)$ (top to bottom). Notably, c_{coh,Δ_1} controlled tree width, η_{Γ_1,Δ_1} straightened branches, and e_{suc,Δ_1} controlled offshoot length. Interactions between V_{floor}, V_{norm}, and V_{bias} can be observed in the last row as an oscillating motion, seen when $c_{floor,\Delta_1} > 0$.

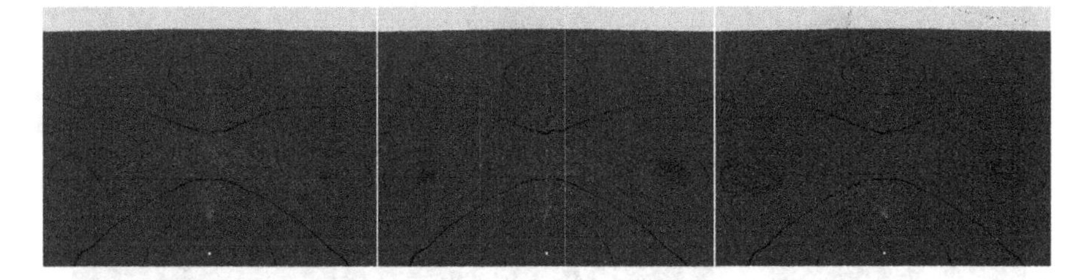

Fig. 5. $SG_3(1000)$ variations with $c_{sep,\Delta_1} = 0$ (left), $c_{sep,\Delta_1} = 0.5$ (middle), and $c_{coh,\Delta_1} = 0.4$ (right). Γ_1 are shown in brown, Δ_1 in magenta and Γ_2 in blue. Notably sizes of puddles change and agents moved out of the basin on the right. (Color figure online)

5 Global System Analysis

The analysis of minimal models highlights the inner workings of vSGs. However, an analysis with broader scope was needed to gauge the model's potential to generate worlds. Hence, as a first step, we combined models from Sect. 4 into a coherent albeit basic virtual environment, discussing required modifications for successful integration. Next, we created a large space from the ground up, with the goal of higher visual interest and more interactions between species.

5.1 Combining Minimal Models

The goal of merging the previous vSGs is to design a working, self-organising eco-system. The challenges involved are to link their respective agent types, to adjust influences and to distribute pre-existing agent types. We conceived two ways to distribute tree agents and link them to water droplets: (1) Design sapling agents, with a rule context-sensitive to water or (2) presume presence of seeds in the terrain and let water spawn trees occasionally. Here, we present the latter approach, as the movement of saplings increases the simulation time. Either method generated trees laying flat, as seen in Fig. 6. We ascribed this to inherited velocity and resolved the issue partially by introduction of an intermediary agent with $V_{max} = 0$, to immediately be replaced by tree agents. We called this intermediary agent a *biaser*, as it biases the velocity of a succeeding agent. Proximity to existing trees posed another problem as the separation urge of developing trees interfered a proper the growth processes (also seen in Fig. 6). Hence, we introduced exclusivity, only allowing for tree growth with sufficient space. As there are no model mechanics for this yet, we encoded this in spawning frequencies of rules, effectively disabling those with low probability. Finally, as terrain can be reshaped any time, we encountered trees buried below terrain. Future revisions might implement *relative positioning* to fix such structures, e.g., by maintaining height or the relative position to predecessors when the topography changes. For now, we coordinated the proper placement order through timed reproduction and energy depletion of the involved agents (Fig. 6).

Fig. 6. Multiple trees (red) generated generated by infrequent tree placement through water droplets (magenta), without intermediary biasers (left), and without minimal tree generation distance (right). The final configuration (bottom) had a file size 2 times that of SG_3. (Color figure online)

5.2 Creating a World from Ground Up

Aiming to generate a more detailed terrain, we created a regularly spaced grid of artifacts with $\eta_{terrain} \neq 0$, layered with other terrain shaping artifacts, placed by agents. Yielding results similar to those seen in Fig. 7, instead of the aspired smoother and more natural topography, we removed the regular artifacts. To keep the layering approach, two more stages of agents were added, which are placed together with artifacts of the previous stage and in turn place other terrain shaping artifacts themselves. The results of stage 2 and 3 can be seen in Fig. 8. As another layer of detail, we aimed to introduce grooves similar to hydraulic erosion. To disperse agents which might place such groves, we introduced hill-climbing agents with $c_{slope} < 0$. Reaching zero energy, they are replaced by an agent with $c_{slope} > 0$. Since terrain shaping artifacts placed by this second stage, would build walls instead of carving groves, we introduced a third stage with $c_{noclip} = 1$ and downwards bias, which finally placed artifacts. Results of both stage 2 and stage 3 agents are seen in Fig. 8. Proceeding with water generation, we again created a three stage process of cloud spawners, clouds and water droplets, each generating one after the other. To have the simulation terminate, we limited the agents' lifespans through energy consumption. To enhance simulation performance, we limited water artifact density by making use of the aforementioned indirect implementation of rule disabling. The same technique has been used to limit water droplet generation rate, as they tended

Fig. 7. Unexpected result of an attempt to create a smooth terrain through a grid of artifacts (white) and artifacts which were irregularly placed by agents (blue). (Color figure online)

to stack below clouds, generating water artifacts mid-air. Concerning the flora, we took a hybrid approach of those mentioned in Sect. 5.1. By spawning saplings through droplet agents, we increased the probability of them being near water, while not requiring them to be attracted by droplets. If we were to choose a pure sapling approach, we would have had required an unsatisfactory tradeoff between saplings not reaching water and clusters of saplings, as separation and cohesion urges apply equally to both actors types. This could be addressed by introducing per-actor type urge coefficients, view and separation distances.

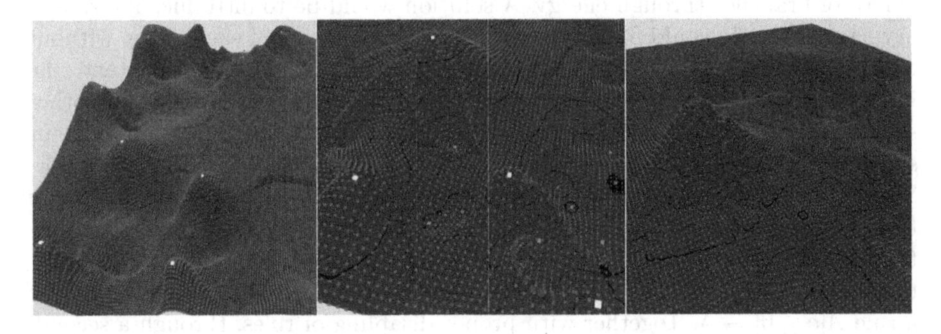

Fig. 8. Result of the layering terrain generation process after rough shaping (left), with unintended ridges (middle) and desired grooves (right).

The result so far is shown in Fig. 9. We increased its appeal by introducing a separated environment to generate a "tree farm" (also seen in Fig. 9). Here we could inspect the phenotypic plasticity of 100 identical genotypes exposed to variations in initial energy and the surrounding terrain. Generating firs through an upwards biased agent, which spawns other agents with heavily constrained vertical movement, was straight forward. A second pollard willow like tree, however, proved difficult. As seen in Fig. 9, they tended to collapse when introduced into the scene, which emphasized the strong impact of the creation environment.

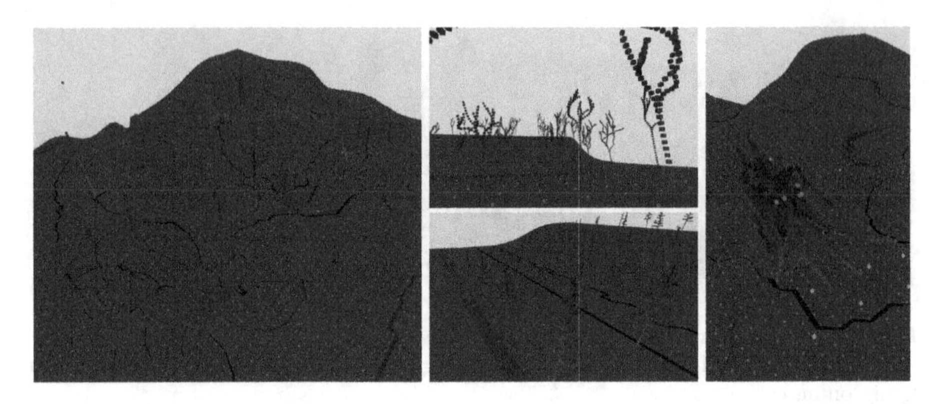

Fig. 9. A view into a more detailed vSG generated environment (left), views into a willow (top) and fir (bottom) farm, with changing terrain on one and varying starting energy on the other axis, and unexpected results of willow introduction into the more detailed environment (right).

As the farm environment ensured wide spaces between trees, cohesion had little effect, resulting in straight trees. A more severe problem arose, as we aimed to rework the willow into a weeping willow by means of a three stage agent production process (upwards, outwards, downwards). Consuming the initial energy to control the height of the stem, we could not scale the width of the tree or length of branches through energy. A solution would be to introduce intermediary agents, which would follow the artifact placing agents of each stage, without expending energy. With them dying and a resulting unsatisfied context, the intermediaries would be replaced by the agents of the next stage. This and other solutions did not work robustly and cluttered up the database our environment stems from. A remedy could stem from further expansion of the underlying L-System towards a parametric L-System, where rules could read from and write to actor parameters. This extension would generalize energy and allow for custom depletion functions. Agent type A could, for example, implement a timer through $A \rightarrow A[t = t - 1]$, while death on zero energy could be implemented by a rule $A[e \leq 0] \rightarrow \lambda$. Together with proper disabling of rules, through a second, *exclusive* context, such an extension would increase implementation complexity, but would be in line with our core goal of world generation from a small initial database.

6 Conclusion

In this work we proposed a novel agent-based model for generation of virtual environments, featuring adaptive, unbounded world generation from a small database. We extended the existing SG model by increasing interaction with the environment, allowing agents to form and react to the terrain, by abstracting generated data, and by diversifying the energy system.

We analysed the proposed vSG model, through observation of changes in parameters in isolated agent types, gaining a basic understanding of possible behavioural impacts of changes. On this basis, we documented two of our attempts of creating a database for world generation, both featuring terrain, tree generation and water simulation. They showed promising first results, as seen in Fig. 10, warranting further research. Throughout these very early analyses, we highlighted obstacles, e.g. flat trees and opaque configuration of water droplets, and resulting configuration patterns, e.g. timer and biaser intermediaries. As a result we identified exclusive context, relative positioning and parametric extension of the underlying L-System as promising additions.

Concerning future work, further manual exploration of the creation spaces of this model would prove beneficial. This could be done in the form of user studies or questionnaires, which could also verify our findings more throughly. Besides, an evolutionary approach to exploring and optimising vSG configurations might circumvent usability issues. Finally, more elaborate and automated interpretation of generated point data would allow this model to advance from its experimental state, to be used in digital media.

Fig. 10. A successful attempt to introduce firs into an existing environment. Configuration file size was 6 times that of SG_3, the most complex minimal model we used.

References

1. Bay 12 Games: Dwarf Fortress (2006). http://www.bay12games.com/dwarves/. Accessed on 19 Nov 2020. Current version 0.47
2. Costa, L.D.F., Rodrigues, F.A., Cristino, A.S.: Complex networks: the key to systems biology. Genet. Mol. Biol. **31**, 591–601 (2008). https://doi.org/10.1590/S1415-47572008000400001
3. Darwin, C.: The power of movement in plants. Appleton (1881)
4. Gearbox Software: Borderlands (2009)
5. Greuter, S., Parker, J., Stewart, N., Leach, G.: Undiscovered worlds-towards a framework for real-time procedural world generation. Fifth Int. Digital Arts Cult. Conf. **5**, 5 (2003)

6. Hello Games: No Mans Sky (2016)
7. Hendrikx, M., Meijer, S., Van Der Velden, J., Iosup, A.: Procedural content generation for games: a survey. ACM Trans. Multimedia Comput. Commun. Appl. **9**(1) (2013). https://doi.org/10.1145/2422956.2422957
8. Lindenmayer, A.: Mathematical models for cellular interactions in development i. filaments with one-sided inputs. J. Theor. Biol. **18**(3), 280–299 (1968). https://doi.org/10.1016/0022-5193(68)90079-9
9. Mojang Studios: Minecraft (2011)
10. Perlin, K.: An image synthesizer. In: Proceedings of the 12th Annual Conference on Computer Graphics and Interactive Techniques, pp. 287–296. SIGGRAPH '85, Association for Computing Machinery, New York, NY, USA (1985). https://doi.org/10.1145/325334.325247
11. Prusinkiewicz, P.: Graphical applications of l-systems. In: Proceedings on Graphics Interface '86/Vision Interface '86, pp. 247–253. Canadian Information Processing Society (1986)
12. Prusinkiewicz, P., Lindenmayer, A.: The Algorithmic Beauty of Plants. Springer, Berlin, Heidelberg (1990)
13. Reynolds, C.: Flocks, herds, and schools: a distributed behavioral model. ACM SIGGRAPH Comput. Graph. **21**, 25–34 (1987). https://doi.org/10.1145/280811.281008
14. Reynolds, C.: Boids – background and update (2001). http://www.red3d.com/cwr/boids/. Accessed on 19 Nov 2020
15. Sayama, H.: Swarm chemistry. Artif. Life **15**(1), 105–114 (2009). https://doi.org/10.1162/artl.2009.15.1.15107
16. Schaal, J.: Procedural terrain generation. A case study from the game industry. In: Game Dynamics, pp. 133–150. Springer (2017)
17. Smelik, R.M., Tutenel, T., Bidarra, R., Benes, B.: A survey on procedural modelling for virtual worlds. Comput. Graph. Forum **33**(6), 31–50 (2014). https://doi.org/10.1111/cgf.12276
18. von Mammen, S.: Swarm grammars: modeling computational development through highly dynamic complex processes. Ph.D. thesis (2009)
19. von Mammen, S., Jacob, C.: Genetic swarm grammar programming: ecological breeding like a gardener. In: Srinivasan, D., Wang, L. (eds.) CEC 2007, IEEE Congress on Evolutionary Computation, pp. 851–858. IEEE Press, Singapore (2007)
20. von Mammen, S., Jacob, C.: The evolution of swarm grammars–growing trees, crafting art, and bottom-up design. Comput. Intell. Mag. IEEE **4**, 10–19 (2009). https://doi.org/10.1109/MCI.2009.933096
21. Worley, S.: A cellular texture basis function. SIGGRAPH '96, ACM, New York, NY, USA (1996). https://doi.org/10.1145/237170.237267

Co-creative Drawing with One-Shot Generative Models

Sabine Wieluch(✉) and Friedhelm Schwenker

Institute for Neural Information Processing, Ulm University,
89081 Ulm, Germany
sabine.wieluch@uni-ulm.de

Abstract. This paper presents and evaluates co-creative drawing scenarios in which a user is asked to provide a small hand-drawn pattern which then is interactively extended with the support of a trained neural model. We show that it is possible to use one-shot trained Transformer Neural Networks to generate stroke-based images and that these trained models can successfully be used for design assisting tasks.

Keywords: Transformer · Co-creative · One-shot

1 Introduction

To quickly capture a thought or explain something in a visual way, we often tend to draw a quick doodle. Sketching by hand is still one of the most common ways to support thought processes [6] or to create new design ideas [18]. Therefore it is an attractive idea to support sketching processes with technology.

A well known research project for collaborative sketching is Sketch-RNN [7], where a Recurrent Neural Network is trained on a large dataset (Quickdraw) of small hand-drawn sketch samples. Given the image category ("dog", "car", etc.) the model is able to complete a sketch that was started by a user. Collabdraw [5] builds upon Sketch-RNN and explores a turn-taking way to interact with the Quickdraw dataset.

Creativity Support Tools [17] are technologies that aim to enhance a user's creativity or support ideation processes. Sketch-based Creativity Support Tools have been explored by Davis et al. [3]. Here, a Q-Learning agent is steadily trained on user pen strokes to learn their preferred drawing style. The user is also able to give feedback to suggestions via the drawing interface to improve upcoming suggestions.

"Cobbie" [10], a drawing robot, is another interesting example of a sketch-based Creativity Support Tool: robot and user draw together on a sheet of paper. Participants in a user study were asked to design new products. Results show that Cobbie's drawings are often utilized as inspiration for shapes in the final product design.

J. Romero et al. (Eds.): EvoMUSART 2021, LNCS 12693, pp. 475–489, 2021.
https://doi.org/10.1007/978-3-030-72914-1_31

This research aims to explore sketch-based Creativity Support Tools for abstract pattern drawing. A pattern is defined as a (repeated) decorative element used in design.

However, for designers it is often difficult to gather a large dataset of their drawings to train a generative model, like the Quickdraw dataset for Collabdraw. Also the step-by-step training of a learning agent is a large amount of additional work for a designer. Therefore we focus on a One-Shot approach where the designer only needs to provide one or few examples to train a generative model.

Also we aim to create support tools that produce path-based and not pixel-based data. Path-based data is important in domains like digital fabrication [21] or robotics like the above mentioned drawing robot Cobbie: machines like laser cutters or pen plotters need the path information to move along these lines.

2 Model Architecture: Transformer

To train a generative model on pen strokes, we first need a neural representation for stroke paths. A sketch consists of a sequence of pen strokes and each stroke can be described by small pen movements along straight lines. This sequence of sequences can be easily flattened to one large sequence. So, learning a neural representation for sketches can be formulated as a sequence generation task. Transformer Neural Networks [19] are a novel approach to processing sequential data in machine learning. They are mainly used in Natural Language Processing for translation tasks [4] or for text generation [12]. But Transformers also have been used for sketch image recognition [23,24] and recently Carlier et al. showed that Transformers outperform Recurrent Neural Networks for Vector Graphics [1].

When Transformers are used for sequence-to-sequence tasks, a encoder-decoder architecture is used. As we only want to generate a sequence in our setting, we only need the encoder module. At first, this seems like an unusual decision, as typically the decoder would be used. But in case of a Transformer, encoder and decoder are very similarly structured. The main difference is that the decoder receives additional input from the encoder in a translation setting. Since we can not provide this input (we do not translate and only generate), we are left with the encoder architecture: the decoder without the additional input capabilities equals the encoder architecture.

A sequence of straight pen moves will be used as input to our Transformer. Though, these pen moves can not be used as input directly and need to be converted to a vector representation first. In Natural Language Processing, word sequences are used as neural net input, which poses a very similar problem. Here, word embeddings [9] are used to encode words into vectors. Pen moves and the embedding process which will be explained in detail in the next section.

The Transformer input also receives a mask that prevents the neural net from seeing "future" sequence elements. In addition to the embedded pen move sequence, a positional encoding is needed to provide relative positional information to the Transformer network, as it does not contain any recurrence or convolution. In our research, we use the standard positional encoding defined in [19] where the *sine* function is used for even input indices and *cosine* function is used for odd indices:

$$PE_{(pos,2i)} = \sin(\frac{pos}{10000^{\frac{2i}{N}}}) \tag{1}$$

$$PE_{(pos,2i+1)} = \cos(\frac{pos}{10000^{\frac{2i}{N}}}) \tag{2}$$

where pos is the position in the input sequence, i the dimension (the position in the positional encoding vector) and N equals the embedding dimension, so embedding and positional encoding can be summed. 10000 was chosen by [19], so the sinusoid wavelengths form a geometric progression from 2π to $10000 \cdot 2\pi$.

Transformer encoder layers can be stacked on top of each other any number of times. They consist of a masked Multi-Head Attention layer with normalization and a feed-forward network with normalization.

The Multi-Head Attention layer is the most important part: it consists of multiple parallel Self-Attention layers, whose output is concatenated and finalized with a linear layer. The Self-Attention layers give the neural net the ability to focus or ignore certain elements in the sequence. Running multiple Self-Attention layers in parallel prevents bad random initialization. Recent research shows, that a majority of the heads can be pruned without significant drop in performance [20]. The Attention mechanism for each head is defined as:

$$X \cdot W_Q = Q \tag{3}$$

$$X \cdot W_K = K \tag{4}$$

$$X \cdot W_V = V \tag{5}$$

$$Attention(Q, K, V) = softmax\Big(\frac{QK^T}{\sqrt{d_k}}\Big)V \in \mathbb{R}^{S \times d_v} \tag{6}$$

Q (Query), K (Key) and V (Value) with $Q, K \in \mathbb{R}^{S \times d_k}$ and $V \in \mathbb{R}^{S \times d_v}$ are calculated by multiplying the embedding matrix $X \in \mathbb{R}^{S \times N}$, which contains all embedding vectors for all embedded pen moves with the trained weight matrices $W_Q, W_K \in \mathbb{R}^{N \times d_k}$ and $W_V \in \mathbb{R}^{N \times d_v}$. Where S is the input sequence length and d_k and d_v are attention projection dimensions.

With the trained weight matrices Q and K, a score for each sequence element is calculated and normalized by the square root of the dimension of K. With $softmax$, these scores are further normalized so they sum up to 1. Finally these scores are multiplied with the values V, which results in weighted values.

After the Attention layer, the data is processed in a feed-forward neural net and after passing all stacked encoder layers, it reaches a linear and softmax layer. These scale the output to the embedding size and give a vector of probabilities for each pen move. From these probabilities, a single pen-move is sampled via top-k sampling. Figure 1 gives a visual overview of the Transformer encoder we used and input and output vectors are visualized in Fig. 3b.

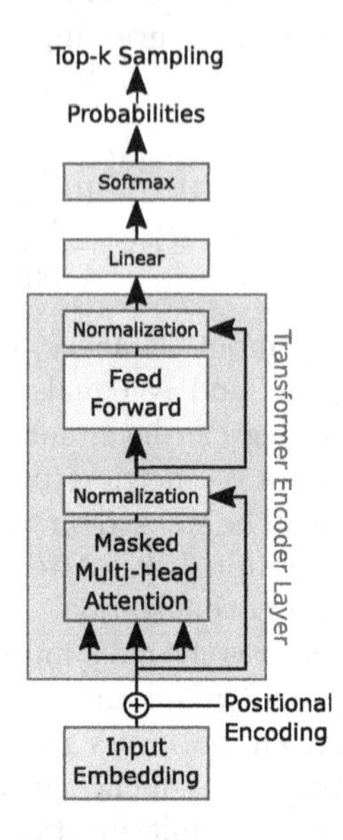

Fig. 1. Transformer Encoder: input pen moves are embedded and so transformed into a vector representation. They also receive a relative positional encoding and are then processed through multiple parallel Self-Attention layers and a feed-forward net. Encoder layers can be stringed together multiple times. The output is then transformed back to the embedding size with a linear layer. Finally a softmax layer is used to receive the probabilities from a pen move is sampled with top-k.

In this research, we use the Adam optimizer [8] with the described changes in [19], where the learning rate is first linearly increased for the first warm-up steps and thereafter decreased again.

3 Data Set Generation

In One- or Few-Shot learning with natural images [16], the original pixel image is transformed into multiple altered, smaller images called "patches" which then form the training dataset. These patches are often produced by scaling, rotating, sheering or other manipulations.

To train our sketch-based generative model, we use the method proposed in our latest work [22]: The whole stroke-based image is manipulated by mirroring, rotating, scaling and translating. Because each path consists of a list of points, a path has an implicit direction. In our setting, the path direction is not important, so it can be reversed to generate new patches. After all paths have been manipulated, they are rearranged in a new order. We sort them in a greedy way

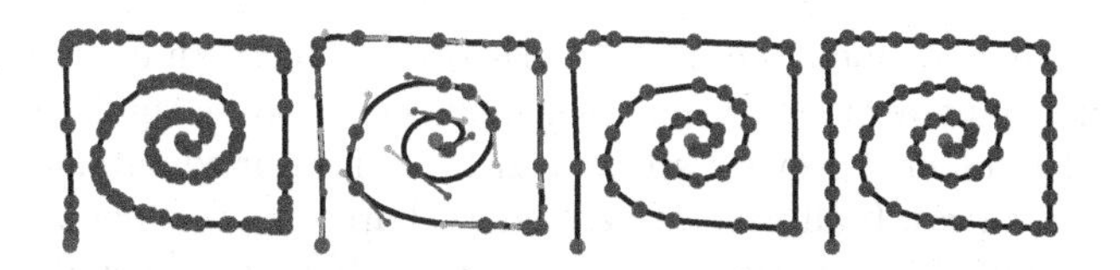

Fig. 2. Path processing from stroke to pen moves: first a hand-drawn stroke is recorded by adding a new point to the path on every mouse move event. This point cluttered lines is simplified substituting points with curves. These curves are again flattened with a certain error and too long sections are split into multiple smaller path segments.

by distance, so that the pen travel is as short as possible. This step is important to ensure that the next stroke will be generated close to the previous stroke. If the sorting step is not performed, strokes in the generated images appear very scattered and incoherent.

In our experiments, we use hand-drawn images in a 180×180 unit boundary box. The images are drawn directly on a computer screen using a digitizer pen. In the drawing process, the current pen location is recorded on every frame update. This results in a very point cluttered path, which can be seen in the example in Fig. 2. The path is then simplified by fitting Cubic Bézier Curves [13] through the points with an allowed maximum error. The algorithm used is described in [14] and the result can be seen in the second image in Fig. 2. In the next step, the simplified curves are flattened: the curves are approximated by straight lines with a given maximum error. If a line is too long, it can be divided into multiple shorter lines. These resulting short straight lines form the actions our Transformer is trained on and will further be called "pen moves". A visual representation of the stroke curve flattening and pen move generation can be seen in the last two images in Fig. 2.

A pen move is defined with the following attributes:

- Position: relative x and y coordinates to the last point.
- Pen State: binary state if the virtual pen is drawing or only moving.

The pen state attribute enables the possibility to move the virtual pen without drawing. These later invisible moves are needed to end a stroke and move the pen to the beginning of the next following stroke. With the current pen move definition, it is not possible for the Transformer to indicate if the image is finished. Using an embedding as seen in Fig. 3 allows us to add special event moves like an "imageEnd" token. A path-end token is not needed, as it can be read from the pen state. Embeddings work like a lookup table where a list of predefined pen moves is mapped to vector representations, which are trained along with the Transformer. Figure 3a shows a short example of the embedding process. Because only valid pen moves are embedded, the Transformer can also only predict valid pen moves:

If the pen move attributes were used directly as Transformer input, for example in the form of $(x, y, penState, imageEnd)$, it would allow for invalid pen moves where the pen state is 1 (line will be drawn) but the *imageEnd* flag is also activated.

To make the embedded pen moves cross compatible with every input image, the maximum pen travel length for each move needs to be small and should be set to a fixed number. So every stroke can be well approximated by one set of predefined pen moves. If all generative models share the same pen move set, they have the possibility to interact with each other. If models would use different pen move sets, certain pen moves might not be contained in the other model's embedding. In this research, we chose a maximum pen move length of 15 units.

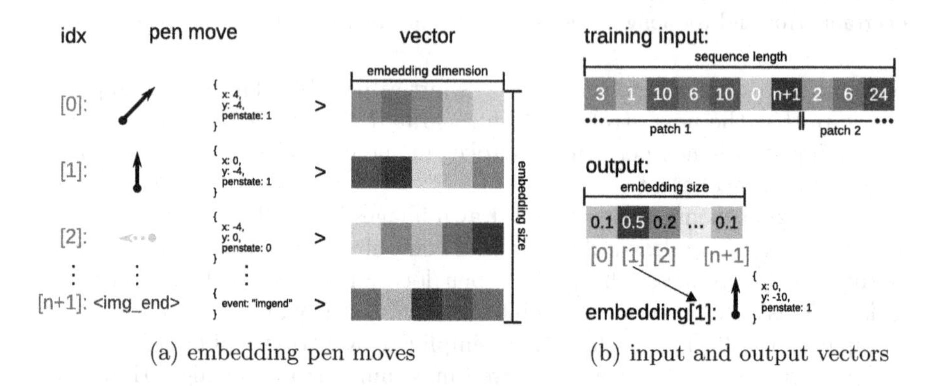

(a) embedding pen moves (b) input and output vectors

Fig. 3. a) Pen moves (their object representations) are embedded. The embedding acts as a lookup table, where a list of pen move is mapped to a list of trained vector representations. So with a known index, either the pen move object or the vector representation can be retrieved. b) The input vector to the Transformer contains a sequence of pen moves represented by their embedding index. The vector representation will then be looked up by the Transformer. The output vector has the length of the embedding size. It contains prediction probabilities for each embedded pen move. For a greedy sampling, one needs to search the highest probability value and use the value's position in the output vector to look up the embedded pen move.

Fig. 4. "Boxes" and "Spirals" dataset recorded for our experiments. Each path was assigned a random Color for better distinction. (Color figure online)

4 Generative Sketches

For our experiments in co-creative drawing, we recorded two template images as can be seen in Fig. 4: The first image "Boxes" consists of rectangles that are drawn into each other. They all have the same orientation and lines do not cross. The second image is called "Spirals" and consists of multiple spiral-shapes that turn into different directions. It also contains one little circle that fills some open space. The images differ in the number of paths (11 and 6) and also path lengths.

From each image, we created a new patch set of size 500 for each epoch to prevent overfitting (the benefits of rotating patch sets has been evaluated by Wieluch and Schwenker [22]). We used the following settings for our Transformer:

- Epochs: 200
- Batch Size: 200
- Sequence Length: max. template image sequence length
- Embedding Dimension: 52
- Encoder Layers: 3
- Attention Heads: 4
- Feed Forward Size: 2048

The sequence length was chosen to give the neural net the possibility to learn the whole image as context. If the sequence length was chosen shorter, the neural net often did not produce image-end tokens in our experiments. Also we do not train the Transformer image by image, but instead use all 500 patch images as one long sequence and move a sliding window along this sequence. With this technique, the Transformer learns to generate a new image after another image is finished. So the input vector may contain the end of patch one and the beginning of patch 2 as can be seen in Fig. 3b.

Figure 5 shows two generated images from our trained models.

It is clearly visible that the generated paths represent the intention (spirals, boxes, circles) of the template image. Most shapes appear close together, the box paths are also stacked into each other and the boxes also show the same direction. This represents the template image very well.

However, it can be seen that some paths cross each other, which should not happen as no path crosses another in the template image.

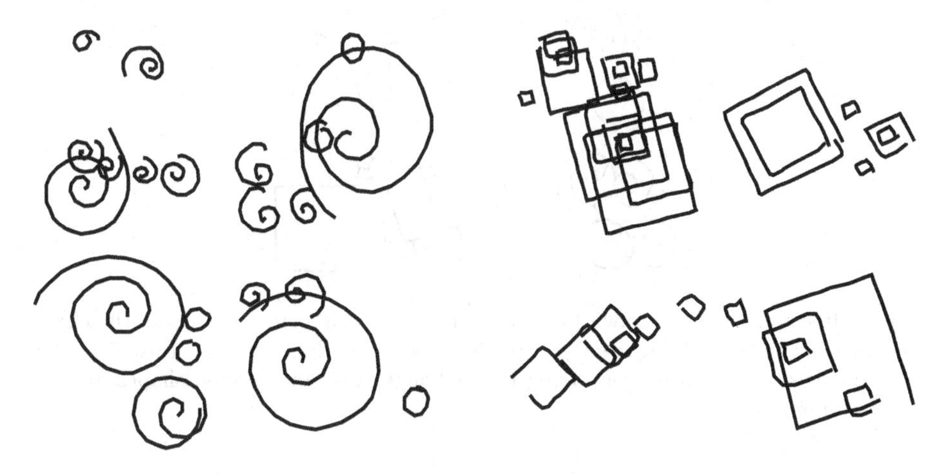

Fig. 5. Generated images from the "Spirals" (left) and "Boxes" (right) dataset.

5 Co-creative Drawing

In this section, we present three co-creative drawing tools to interact with our generative models. Recordings of each interaction technique can be seen in the additional material.

5.1 Autocompletion

In this scenario, the user starts by drawing a line and the model completes the given path. The results of four such iterations can be seen in Fig. 6: the red stroke is drawn by the user and the black part is the completed path drawn by the model. On the left side the "Spirals" model is used and on the right side the "Boxes" model.

Both models work very well with completing the stroke. But it is also visible that the unknown user input "confuses" the models and shapes emerge that are not intended by the template image. This can be best observed in the last "Spirals" image where the user starts to draw a small circle, but the model draws a rather odd loop. Also the boxes on the right side tend to have unusual interlinked ends.

We expect that these unusual generated shapes will be used in creative drawing processes, as unexpected outcomes support creative thinking [11].

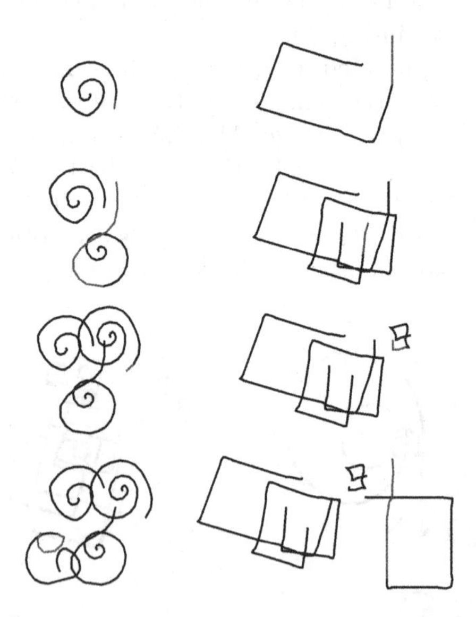

Fig. 6. Iterations of user-drawn lines (red), finished by a generative model (black) in one image. The left side utilizes the "Spirals" model, the right side the "Boxes" model. Both models work well on continuing lines, though some unexpected shapes appear. (Color figure online)

5.2 Generative Stamps

In this experiment, the user suggests a point where the model should place a new stroke (in our setting by clicking via mouse). If the user does not like the created shape, the stamping process can be undone and a new shape can be generated at the same position.

This co-creative interaction loop can be used to quickly produce large pattern-filled areas by a generative type of "stamping" new shapes. This way the user is in control of where the next stroke should be placed but not of what exactly is drawn. The drawing process is very quick, because the user does not need to draw and specify the beginnings of lines. Accordingly the resulting images will look more coherent to the template image, because the user does not draw lines and the model is also not "confused" by user-drawn lines. But this also results in less novelty in the generated strokes.

On the technical side, this interaction loop can be created by calculating pen moves from the last drawn stroke point to the suggested new stroke position and adding these to the neural net input sequence.

An example of this procedure is shown in Fig. 7: The red line is already placed in the image. The red circles indicate possible positions and the corresponding black lines show the strokes that would be drawn at these suggested positions.

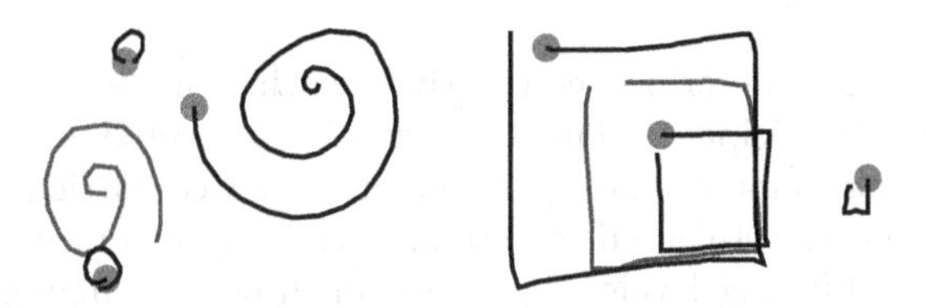

Fig. 7. Generative Stamping: The user can choose a position (red circle) on where the model should draw a new stroke (corresponding black line), considering the already drawn lines (red line). (Color figure online)

5.3 Suggestions

In our last experiment, we let the user draw a line and the model will suggest multiple possibilities on how to continue this line. Though inferring from a neural net is a deterministic process: the model will only give one result on one certain input. Though sampling mechanics like top-k can be used to receive different outcomes: only the top k predictions are used and their probabilities normalized. According to these new probabilities, a prediction is sampled.

Figure 8 shows examples of such a generative processes: In red the currently drawn line is shown. The black lines are the possible continuations suggested by the model and can be accepted by clicking on the suggestion of choice.

Depending on the stroke, the number of distinguishable predictions changes. Also the length between suggestions might change as the model is able to predict a path end before the maximum suggestion length is reached.

This co-creative interaction loop gives more freedom to the user as the generative stamping experiment but also more control over the actual outcome. In this scenario, novelty or unexpected shapes that are not directly part of the template input image, will also occur.

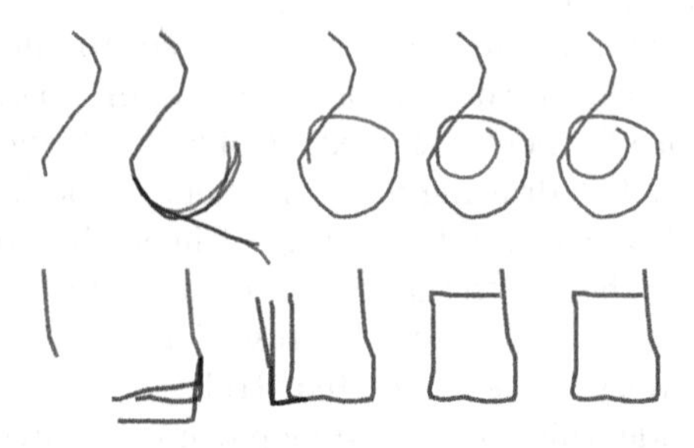

Fig. 8. Model-suggestion supported drawing with the "Spirals" and "Boxes" dataset: the user (red) starts to draw a line and receives suggestions by the model on how the stroke could be continued (black). (Color figure online)

6 User Study

To evaluate the three described co-creative drawing methods, we conducted a user study with 8 participants. The study was implemented as a website, so the majority of five participants took part remotely. Three participants conducted the study on site and were asked to verbalize their thoughts while drawing. The participants had differing levels of practice in drawing, ranging from "daily", "once a week", "once a month" to "never". Also, different input devices were used: one half used a mouse, whereas the other half used a digitizer pen.

The study consisted of two drawing tasks and surveys in between. In the first task, the participants were asked to draw any pattern of their liking. They were allowed to use any of the three drawing tools and could also choose to draw without any AI support. Each drawing tool could be used with two template patterns: the "Spirals" and "Boxes" template images introduced in Fig. 4. No time limit was given and the drawing canvas could be cleared and drawing steps could also be undone to create an experimentation friendly environment. When the participant was satisfied, the result was saved and the survey started.

The second task was similar, with the exception that the participant was asked to choose one of the two template images to draw a pattern in the same style. This task was introduced to evaluate if certain co-creative drawing tools are especially useful to create novel but style preserving images from a template. The surveys consisted of five questions from the Creativity Support Index (CSI) [2] to evaluate Exploration, Collaboration, Engagement, Effort and Expressiveness of all three tools in both drawing tasks. Additionally, the participants were asked to describe situations in which they did and did not like to use a certain tool.

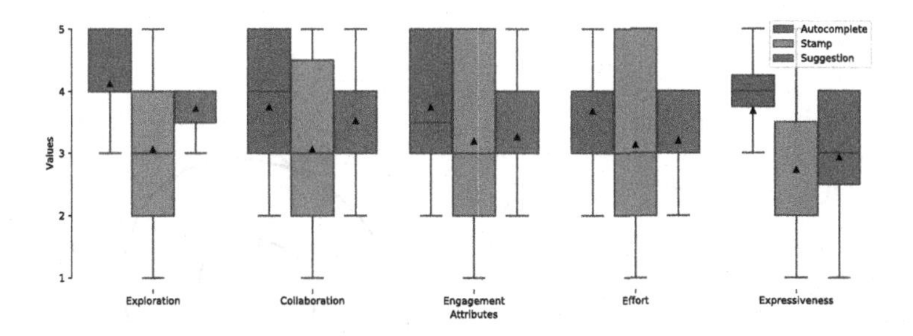

Fig. 9. Creative Support Index survey results for all three generative drawing tools.

The co-creative drawing tools were configured as follows:

- Autocomplete: the tool completes a user-drawn line in context of the next closest line, if available (the pen moves from the next closest line and the new pen moves from the user stroke are used as input sequence). The moves in the path are sampled in a greedy manner ($k = 1$ for top-k sampling).
- Stamp: the tool creates a new line at a given point in context of the next closest line, if available. The pen moves are sampled with $k = 2$ top-k sampling to give the user slightly different results for the same drawing spot.
- Suggestion: the tool creates suggestions utilizing top-k sampling with $k = 50$ for a broad variety of suggestions. The suggested sequence has a maximum length of 10 pen moves.

6.1 Quantitative Results

The survey results are visualized as box-plots in Fig. 9: "Autocomplete" shows the highest mean score in all five CSI attributes, though we could not find a significant difference between tools in a paired test.

Pen users showed a significantly better assessment than mouse users in the collaboration aspect of the "autocomplete" drawing tool. This might be due to the fact that it is hard to draw precise lines with a mouse, especially curves. Those mouse-drawn lines might be more likely to "confuse" the trained model and so the model might more likely suggest to end the user-drawn line than continue it. Of course this behaviour will be seen as un-collaborative.

We also found a significant difference between frequent drawers (once a week or more often) and non-frequent drawers in the "stamp" tool: frequent drawers rated the "stamp" tool significantly better than non-frequent drawers.

We could not find any significant differences in the tool usage perception between the free and the style preserving drawing task.

6.2 Qualitative Results

In the following, we will summarize results from observations or written assessments from the questionnaires:

(a) free (b) style preserving

Fig. 10. Example images drawn by participants for the free drawing and the style preserving drawing task: Autocompletion (blue), Stamp (orange), Suggestion (green). (Color figure online)

- Autocomplete:
  - positive: quick, better suggestion than suggestion tool. Used for: exploring shapes, idea generation, filling in details, being lazy and letting the algorithm finish, fine and detailed lines.
  - negative: shapes overlap, it added unnecessary edges (Boxes template), tool has problems with unspecific shapes.
- Stamp:
  - positive: fun to use. Used for: filling in blank areas, loosen up the image, quick repetition of small shapes.
  - negative: only made rather small shapes, hard to control.
- Suggestions:
  - positive: easier to use with mouse. Used for: style imitation while maintaining control, experimentation and idea generation.
  - negative: suggestions sometimes feel useless or too similar, suggested path segments too short, similar suggestions are hard to distinguish.

From observing three participants, we also found that unexpected results from the model often were used as inspiration and so were included in the drawing rather than deleted. Users also recognized quickly that the model is trained on shapes of a certain size and that starting a shape larger than the training data will not adequately be completed. It was also recognized that tools depend on previous lines. This was especially clear to users when using the "Boxes" template: newly creates boxes aligned in their orientation with previous drawn boxes. Example drawings for the free and style preserving drawing task can be seen in Fig. 10.

In summary, "Autocomplete" and "Suggestions" are used in similar situations: for experimentation and idea generation. In style preserving tasks, also both tools were rated similar useful. Negative aspects from both drawing tools could be decreased my combining both tools into one: the combined tool could auto complete a user-drawn line multiple times to give more than one solution. So the user still has control over the look of the shape but does not need to click multiple times to complete the line.

The "Stamping" tool was rated very controversially ranging from "did not like to use it at all" to "it was fun and useful for filling blank spots". The rating differed significantly between frequent (positive rating) and non-frequent drawers (negative rating). The cause of these drastically differing perceptions is not clear and needs further investigation in future research.

7 Conclusion

In this research, we evaluated three co-creative drawing tools where a user collaboratively creates sketches with a generative model. The model is trained on only one template image (One-Shot).

The three tools (line autocompletion, stamping and line continuation suggestions) where evaluated in a user study utilizing the Creative Support Index for evaluating exploration, collaboration, engagement, effort and expressiveness.

Results showed that "autocompletion" and "suggestions" were well received and supported creative sketching and ideation. Both tools could be combined into one for future usage where multiple suggestions for line autocompletions are given. This would erase most negative aspects in both tools.

The "stamping" tool was rated controversially and we found significant differences between frequent (low ratings) and non-frequent (high ratings) drawers. The tool was most often used for filling in blank spaces or loosening up a image.

As the model's context view is limited by the sequence length, it would be very interesting to use hierarchical approaches in our future work, as they have been successfully used in other domains like dialogue generation [15] to provide a broader structure to the generative process. We also would like to experiment with a forced widening of the drawing area, so that the model will fill a large canvas with shapes of the input dataset and essentially create a large image from a small sample. This could be a very useful application in several design domains.

References

1. Carlier, A., Danelljan, M., Alahi, A., Timofte, R.: Deepsvg: a hierarchical generative network for vector graphics animation (2020)
2. Carroll, E.A., Latulipe, C.: The creativity support index. In: CHI '09 Extended Abstracts on Human Factors in Computing Systems, pp. 4009–4014. CHI EA '09, Association for Computing Machinery, New York, NY, USA (2009). https://doi.org/10.1145/1520340.1520609
3. Davis, N., Hsiao, C.P., Yashraj Singh, K., Li, L., Magerko, B.: Empirically studying participatory sense-making in abstract drawing with a co-creative cognitive agent. In: Proceedings of the 21st International Conference on Intelligent User Interfaces, pp. 196–207. IUI '16, Association for Computing Machinery, New York, NY, USA (2016). https://doi.org/10.1145/2856767.2856795
4. Devlin, J., Chang, M.W., Lee, K., Toutanova, K.: Bert: Pre-training of deep bidirectional transformers for language understanding. arXiv:1810.04805 (2018)
5. Fan, J.E., Dinculescu, M., Ha, D.: Collabdraw: An environment for collaborative sketching with an artificial agent. In: Proceedings of the 2019 on Creativity and Cognition, pp. 556–561. Association for Computing Machinery, New York, NY, USA (2019). https://doi.org/10.1145/3325480.3326578
6. Fish, J., Scrivener, S.: Amplifying the mind's eye: sketching and visual cognition. Leonardo **23**(1), 117–126 (1990)
7. Ha, D., Eck, D.: A neural representation of sketch drawings. arXiv:1704.03477 (2017)
8. Kingma, D.P., Ba, J.: Adam: a method for stochastic optimization. arXiv:1412.6980 (2014)
9. Levy, O., Goldberg, Y.: Neural word embedding as implicit matrix factorization. In: Advances in Neural Information Processing Systems, pp. 2177–2185 (2014)
10. Lin, Y., Guo, J., Chen, Y., Yao, C., Ying, F.: It is your turn: Collaborative ideation with a co-creative robot through sketch. In: Proceedings of the 2020 CHI Conference on Human Factors in Computing Systems, pp. 1–14. CHI '20, Association for Computing Machinery, New York, NY, USA (2020). https://doi.org/10.1145/3313831.3376258

11. Macedo, L., Cardoso, A.: Assessing creativity: the importance of unexpected novelty. Structure **1**(C2), C3 (2002)
12. Radford, A., Wu, J., Child, R., Luan, D., Amodei, D., Sutskever, I.: Language models are unsupervised multitask learners. OpenAI Blog **1**(8), 9 (2019)
13. Salomon, D.: Curves and surfaces for computer graphics. Springer Science Business Media (2007)
14. Schneider, P.J.: An algorithm for automatically fitting digitized curves. In: Graphics Gems, pp. 612–626. Academic Press Professional, Inc. (1990)
15. Serban, I.V., et al.: A hierarchical latent variable encoder-decoder model for generating dialogues. In: Thirty-First AAAI Conference on Artificial Intelligence (2017)
16. Shaham, T.R., Dekel, T., Michaeli, T.: Singan: learning a generative model from a single natural image. In: Proceedings of the IEEE International Conference on Computer Vision, pp. 4570–4580 (2019)
17. Shneiderman, B.: Creativity support tools: accelerating discovery and innovation. Commun. ACM **50**(12), 20–32 (2007)
18. Suwa, M., Gero, J.S., Purcell, T.: The roles of sketches in early conceptual design processes. In: Proceedings of Twentieth Annual Meeting of the Cognitive Science Society, pp. 1043–1048. Lawrence Erlbaum Hillsdale, New Jersey (1998)
19. Vaswani, A., et al.: Attention is all you need. In: Advances in Neural Information Processing Systems, pp. 5998–6008 (2017)
20. Voita, E., Talbot, D., Moiseev, F., Sennrich, R., Titov, I.: Analyzing multi-head self-attention: Specialized heads do the heavy lifting, the rest can be pruned. In: Proceedings of the 57th Annual Meeting of the Association for Computational Linguistics, pp. 5797–5808. Association for Computational Linguistics, Florence, Italy (2019). https://doi.org/10.18653/v1/P19-1580, https://www.aclweb.org/anthology/P19-1580
21. Wang, J., He, Y., Tian, H., Cai, H.: Retrieving 3d cad model by freehand sketches for design reuse. Adv. Eng. Inform. **22**(3), 385–392 (2008)
22. Wieluch, S., Schwenker, F.: Strokecoder: path-based image generation from single examples using transformers (2020)
23. Xu, P., Joshi, C.K., Bresson, X.: Multi-graph transformer for free-hand sketch recognition. arXiv:1912.11258 (2019)
24. Xu, P., Song, Z., Yin, Q., Song, Y.Z., Wang, L.: Deep self-supervised representation learning for free-hand sketch. arXiv:2002.00867 (2020)

Author Index

Printed in the United States
by Baker & Taylor Publisher Services